SECURITY AND PRIVACY IN THE AGE OF UBIQUITOUS COMPUTING

IFIP – The International Federation for Information Processing

IFIP was founded in 1960 under the auspices of UNESCO, following the First World Computer Congress held in Paris the previous year. An umbrella organization for societies working in information processing, IFIP's aim is two-fold: to support information processing within its member countries and to encourage technology transfer to developing nations. As its mission statement clearly states,

> IFIP's mission is to be the leading, truly international, apolitical organization which encourages and assists in the development, exploitation and application of information technology for the benefit of all people.

IFIP is a non-profitmaking organization, run almost solely by 2500 volunteers. It operates through a number of technical committees, which organize events and publications. IFIP's events range from an international congress to local seminars, but the most important are:

• The IFIP World Computer Congress, held every second year;
• Open conferences;
• Working conferences.

The flagship event is the IFIP World Computer Congress, at which both invited and contributed papers are presented. Contributed papers are rigorously refereed and the rejection rate is high.

As with the Congress, participation in the open conferences is open to all and papers may be invited or submitted. Again, submitted papers are stringently refereed.

The working conferences are structured differently. They are usually run by a working group and attendance is small and by invitation only. Their purpose is to create an atmosphere conducive to innovation and development. Refereeing is less rigorous and papers are subjected to extensive group discussion.

Publications arising from IFIP events vary. The papers presented at the IFIP World Computer Congress and at open conferences are published as conference proceedings, while the results of the working conferences are often published as collections of selected and edited papers.

Any national society whose primary activity is in information may apply to become a full member of IFIP, although full membership is restricted to one society per country. Full members are entitled to vote at the annual General Assembly, National societies preferring a less committed involvement may apply for associate or corresponding membership. Associate members enjoy the same benefits as full members, but without voting rights. Corresponding members are not represented in IFIP bodies. Affiliated membership is open to non-national societies, and individual and honorary membership schemes are also offered.

Contents

SECURITY AND PRIVACY IN THE AGE OF UBIQUITOUS COMPUTING

IFIP TC11 20th International Information Security Conference
May 30 – June 1, 2005, Chiba, Japan

Edited by

Ryoichi Sasaki
Tokyo Denki University
Japan

Sihan Qing
Chinese Academy of Sciences
China

Eiji Okamoto
University of Tsukuba
Japan

Hiroshi Yoshiura
University of Electro-Communications
Japan

 Springer

Library of Congress Cataloging-in-Publication Data

A C.I.P. Catalogue record for this book is available from the Library of Congress.

Security and Privacy in the Age of Ubiquitous Computing, Edited by Ryoichi Sasaki, Sihan Qing, Eiji Okamoto and Hiroshi Yoshiura

p.cm. (The International Federation for Information Processing)

ISBN 978-1-4419-3819-0   e-ISBN 978-0-387-25660-3

Printed on acid-free paper.

Printed in the United States of America.

9 8 7 6 5 4 3 2 1
springeronline.com

Preface

This book contains the proceedings of the 20th IFIP International Information Security Conference (IFIP/SEC2005) held from 30th May to 1st June, 2005 in Chiba, Japan. It was the 20th SEC conference in the history of IFIP TC-11. The first one was held in Stockholm, Sweden in May 1983. After that, IFIP/SEC conferences have been in various countries all over the world. The last IFIP/SEC conference held in Asia was IFIP/SEC2000 in Beijing, China.

In IFIP/SEC2005, we emphasize on "Security and Privacy in the Age of Ubiquitous Computing". Even in the age of ubiquitous computing, the importance of the Internet will not change and we still need to solve conventional security issues. Moreover, we need to deal with the new issues such as security in P2P environment, privacy issues in the use of smart cards and RFID systems. Therefore, in IFIP/SEC2005, we have included a workshop "Small Systems Security and Smart Cards" and a panel session "Security in Ubiquitous Computing".

This book includes the papers selected for presentation at IFIP/SEC2005 as well as the associated workshop. In response to the call for papers, 124 papers were submitted to the main conference track. These papers were evaluated by members of the Program Committee in terms of their relevance, originality, significance, technical quality and presentation. From the submitted papers, 34 were selected for presentation at the conference (an acceptance rate of 27%). We also include 6 short papers selected by the Workshop committee.

We would like to thank Mr. Leon Strous, chair of IFIP TC-11, Professor Norihisa Doi, Professor Hideki Imai, Professor Tsuneo Kurokawa and Professor Shigeo Tsujii, members of the SEC2005 Advisory Committee for their continuous advice. We are grateful to the members of the Program Committee for their voluntary efforts to review manuscripts. We are also grateful to the members of the Local Organizing Committee for their efforts in preparing this conference, especially Professor Yuko Murayama, chair of this committee.

Ryoichi Sasaki, Tokyo Denki University
(General Chair)
Sihan Qing, Chinese Academy of Science
Eiji Okamoto, University of Tsukuba
(Program Chairs)

IFIP/SEC2005 Conference Committees

Conference General Chair
Ryoichi Sasaki, Tokyo Denki University, Japan

Advisory Committee
Norihisa Doi, Chuo University, Japan
Hideki Imai, University of Tokyo, Japan
Tsuneo Kurokawa, Kokugakuin University, Japan
Shigeo Tsujii, Chuo University, Japan

Programme Committee co-Chairs
Sihan Qing, Chinese Academy of Sciences, China
Eiji Okamoto, University of Tsukuba, Japan

Programme Committee
H. Armstrong, Curtin University of Technology, Australia
W. Caelli, Queensland University of Technology, Australia
E. Chaparro, Fundacion Via Libre, Argentina
B. de Decker, K. U. Leuven, Belgium
Y. Deswarte, LAAS-CNRS, France
M. Dupuy, SGDN/DCSSI/CERTA, France
M. El-Hadidi, Cairo University, Egypt
J. Eloff, University of Pretoria, South Africa
S. Fischer-Huebner, Karlstad University, Sweden

D. Gollmann, Technische Universitat Hamburg-Harburg, Germany
D. Gritzalis, Athens University of Economics and Business, Greece
H. Inaba, Kyoto Institute of Technology, Japan
K. Iwamura, Canon Inc., Japan
S. Jajodia, George Mason University, USA
S. Katsikas, University of the Aegean, Greece
H. Kikuchi, Tokai University, Japan
K.-Y. Lam, PrivyLink, Singapore
C. Landwehr, CISE/CNS, USA
W. List, Partner, Independent, UK
J. Lopez, University of Malaga, Spain
K. Matsuura, University of Tokyo, Japan
Y. Murayama, Iwate Prefectural University, Japan
T. Nakanishi, Okayama University, Japan
M. Nishigaki, Shizuoka University, Japan
G. Papp, Vodafone Hungary, Hungary
M. Park, Mitsubishi Electronic Corporation, Japan
H. Pohl, ISIS - InStitute for Information Security, Germany
R. Posch, Graz Univ. of Technology, Austria
K. Rannenberg, Goethe University Frankfurt, Germany
K. Sakurai, Kyushu University, Japan
P. Samarati, University of Milan, Italy
I. Schaumuller-Bichl, Johann Wilhelm Kleinstrase 11, Austria
L. Strous, De Nederlandsche Bank, NL
K. Tanaka, Shinshu University, Japan
S. Teufel, Universite de Fribourg, Switzerland
D. Tygar, University of California, Berkeley, USA
V. Varadharajan, Macquire Univ., Australia
I. Verschuren, TNO ITSEF, NL
J. Vyskoc, VaF, Slovakia
M. Warren, Deakin University, Australia
T. Welzer, University of Maribor, Slovenia
H. Yoshiura, University of Electro-Communications, Japan
S. Furnell, University of Plymouth, UK
J. Knudsen, Copenhagen Hospital Corporation, Denmark
I. Ray, Colorado State University, USA
T. Virtanen, Helsinki University of Technology, Finland
R. Solms, Port Elizabeth Technikon, South Africa
Local Organizing Committee Chair
Yuko Murayama, Iwate Prefectural University, Japan

Local Organizing Committee

Steering Chairs
Yuko Murayama, Iwate Prefectural University, Japan
Yoshito Tobe, Tokyo Denki University, Japan

Program Chairs
Eiji Okamoto, Tsukuba University, Japan
Hiroshi Yoshiura, University of Electro-Communication, Japan
Mi Rang Park, Mitsubishi Electronic Corporation, Japan

Finance Chair
Masatoshi Terada, Hitachi Limited, Japan

Publicity Chairs
Masakatsu Nishigaki, Shizuoka University, Japan
Ryuya Uda, Tokyo University of Technology, Japan

Local Arrangement Chairs
Hiroaki Kikuchi, Tokai University, Japan
Moriaki Itazu, Tokyo Denki University, Japan

Publication Chair
Kanta Matsuura, University of Tokyo, Japan

Liaison Chairs
Seiichiro Hayashi, Japan Internet Payment Promotion Association, Japan
Kouji Nakao, KDDI Corporation, Japan
Satoru Torii, Fujitsu Limited, Japan

Workshop on Small Systems Security and Smart Cards

Working Conference Programme Committee

PART I REGULAR PAPERS

ACCOUNTABLE ANONYMOUS E-MAIL

Vincent Naessens*, Bart De Decker, Liesje Demuynck[1]
K.U.Leuven, Department of Computer Science, Celestijnenlaan 200A, 3001 Leuven, Belgium,
*K.U.Leuven, Campus Kortrijk (KULAK), E. Sabbelaan 53, 8500, Kortrijk, Belgium

Abstract: Current anonymous e-mail systems offer unconditional anonymity to their users which can provoke abusive behaviour. Dissatisfied users will drop out and liability issues may even force the system to suspend or cease its services. Therefore, controlling abuse is as important as protecting the anonymity of legitimate users when designing anonymous applications.

This paper describes the design and implementation AAEM, an accountable anonymous e-mail system. An existing anonymous e-mail system is enhanced with a control mechanism that allows for accountability. The system creates a trusted environment for senders, recipients and system operators. It provides a reasonable trade-off between anonymity, accountability, usability and flexibility.

Key words: privacy, anonymity, accountability, control

1.   INTRODUCTION

Anonymous e-mail systems serve many purposes ranging from making public political statements under oppressive governments to discussing topics that might otherwise lead to embarrassment, harassment or even loss of jobs in more tolerant political environments [1]. However, unconditional anonymous e-mail systems (such as remailers) can also be used for sending offensive mail, spam, black mail, copyrighted material, child pornography...

[1] Research assistant of the Research Foundation – Flanders (FWO – Vlaanderen)

Legal actions against this service may even force it to shut down. Therefor, controlling abuse is as important as protecting the anonymity of legitimate users. Both considerations play a central role in the design of the accountable anonymous e-mail system (AAEM). Anonymity should always be guaranteed to legitimate users but extra controls are necessary to prevent or at least discourage abuse.

The paper is organized as follows: section 2 describes a general anonymous credential system; these credentials will be used in the e-mail system that is designed in section 3. Section 4 describes and evaluates a prototype of the system. An overview of related work is given in section 5. We conclude in section 6 with a summary of the major achievements.

2.   ANONYMOUS CREDENTIALS

This section describes Idemix [11,12], a general anonymous credential system. Idemix helps to realize anonymous yet accountable transactions between users and service providers. The credential system is used to introduce anonymity control in the AAEM system. A simplified version of the Idemix protocols is presented here. Not all inputs of the protocols are described. The outputs of interactive protocols are not always visible to all parties involved in the protocol.

Nym registration. A nym is the pseudonym by which the user wants to be known by an organization. Idemix has two kinds of nyms: ordinary nyms and rootnyms. The user establishes a nym with an organization based on his master secret[2].

    Nym registration (for registering ordinary nyms). $Nym_{UO}=RegNym()$. Note that the rootNym (see below) is hidden in every nym of the user.

    RootNym registration (for registering rootnyms). $RootNym_{UR} = RegRootNym(Sig_{UR}, Cert_{UC})$. The user signs the established nym with his signature key, which is certified by an external certificate (which links the user's public key with his identity). Hence, the organization holds a provable link between the rootnym and the identity certified by the certificate.

Credential issuing. $Cred_{UI} = IssueCred(Nym_{UI}, CredInfo_{UI})$. After establishing a nym with an organization, the user may request a credential on

---

[2] Note that all the user's nyms and credentials are linked to the user's master secret. Hence, sharing one credential means sharing all other credentials as well.

that nym. Credentials can contain additional information: show options (e.g. one/limited/multi-show) and attributes (e.g. age, citizenship, expiration date, ...).

Credential showing. *(Transcript$_{UV}$,Msg$_{UV}$)* = *CredShow(Nym$_{UV}$, Cred$_{UI}$, CredShowFeatures$_{UV-I}$)*. The user proves to the verifying organization V that he owns a credential issued by organization I. A credential show can have several features:

- The credential is shown relative to another nym. The (anonymous) user proves that the nym on which the credential is issued and a nym by which he is known by the verifier, belong to the same user (they are based on the same user secret).
- Local and/or global deanonymization is allowed (cfr. below).

In addition, a user may choose to prove any attribute (or a property of these attributes). Showing a credential results in a transcript for V which can be used later in double spending checking and deanonymization protocols. During a credential show, a message can be signed, which provably links the message to the transcript of the credential show. The following anonymity properties are valid:

- Two or more credential shows of the same credential can not be linked together (unless the credential is a one-show credential).
- A credential show does not reveal the nym on which it was issued.

Deanonymization. Transcripts of anonymous credential shows can be deanonymized by including a verifiable encryption of the user's nym (local deanonymization) or rootNym (global deanonymization). Thus, only parties that have the deanonymization keys can deanonymize credential shows

Local deanonymization.
(Nym$_{UI}$, DeAnonTranscript$_{D-UV}$) = *DoLocalDeanon(Transcript$_{UV}$)*. If a credential show is locally deanonymizable, the nym for which the credential was issued can be revealed by a deanonymizer D. A deanonymization transcript contains a provable linking between the transcript and the nym.

Global deanonymization.
(RootNym$_{UR}$,DeAnonTranscript$_{D-UV}$)=DoGlobalDeanon(Transcript$_{UV}$).
If the credential show was globally deanonymizable, the user's rootnym can be revealed.

Credential Revocation. *RevokeCred(Nym$_{UI}$, Cred$_{UI}$).* The issuer can also revoke credentials issued on a nym.

3.   ACCOUNTABLE ANONYMOUS E-MAIL

First, the requirements of the players in the system are described. The requirements analysis results in the design of an anonymous mail system enhanced with a control mechanism that allows for accountability. The roles in the system are described in the second paragraph. Third, we describe the protocols used in the different phases. Finally, we evaluate the trust properties in the system.

3.1   Requirements

User requirements. These requirements depend on the role the users play in the the mail system. Senders want their privacy to be protected whereas recipients mainly have control requirements. Whereas current remailer systems mainly focus on the former, our design considers the concerns of both parties.

The control requirements of *recipients* are twofold:

Spam control measures. Recipients don't want spam in their mailbox. The mail system should take measures to discourage this kind of abuse.

Accountability for criminal mail. It should be possible to prosecute the sender of a criminal mail. The mail system should guarantee that the identity of the sender can be revealed (i.e. the user is accountable).

The anonymity requirements of mail *senders* will only be met as long as the sender abides by the rules (no spam or criminal mail).

Unlinkability of a mail to the initiator.

Unlinkability of different mails from the same initiator.

System requirements. The system requirements are twofold:

Offering good service to users. The mail system wants to offer a good service in order to attract users. Therefor, the mail system contracts its users to meet the requirements of their users.

Limited use of resources. To be able to meet the accountability requirements, evidence will be stored in the system. The amount of evidence stored by the AAEM-system itself should be minimal.

Law Enforcement requirements. The government may require accountability of misbehaving users. Such users have sent illegal mails (such as illegal music files, child porn, ...) or criminal mails (such as death threats, bribery, blackmail ...).

3.2   Roles

Registrar (for registration). The registrar knows a provable link between a user's true identity and his rootNym. To increase trust in the system, the registrar is independent of the AAEM-system. Its cooperation is required to identify the sender of an anonymous e-mail.

AAEM system infrastructure (core of the mail system).

 - **Activation Manager** (for activation). The Activation Manager handles the (anonymous) user registrations, and if the AAEM-system is not for free, also deals with payments. Payment can be anonymous; however, it is not a pre-requisite to fulfil the requirements.
 - **Mail guard** (for mailing). The Mail Guard imposes strict access control to the AAEM-system (only registered users are authorized to use the services of the system) and adds verifiable proofs to the message guaranteeing that the sender of the message can be identified under certain conditions.
 - **Complaint handler**. The Complaint Handler handles suspected (unacceptable/criminal) mail that is sent through the mail system. Complaints are sent by recipients of such mail. The Mail Guard can also pass suspected mail to the Complaint Handler.

Deanonymization granting/handling infrastructure.

 - **Law Enforcement entity (or Justice).** The role of the Law Enforcement Entity (LE) is to verify whether the identity of a user behind a nym may be revealed.
 - **Arbiter.** The arbiter's role is to verify whether an e-mail fulfills the local or global de-anonymization condition.
 - **Deanonymizer.** This authority can retrieve the pseudonym of the sender of an anonymous e-mail message.

Communication infrastructure.

- **Anonymous connection system (AC).** The connection between the sender and some of the other participants need to be anonymous.
- **Re mailers (REM).** The remailers constitute the existing anonymous e-mail system. The AAEM system forwards mail towards the recipient through a remailer system.

Users of the AAEM system.

- **(Anonymous) Sender.** The sender is entitled to send e-mail anonymously (when he is registered to the system). As long as he abides by the rules, his anonymity will be respected.
- **Recipient.** The owner of a mailbox and e-mail address. The recipient may refuse to accept anonymous e-mails unless they are "stamped" by a trusted Mail Guard. If the email is spam or contains illegitimate or criminal content, the recipient may file a complaint with the AAEM-system.

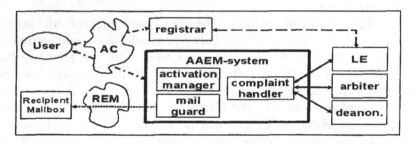

Figure 1. Overview of the AAEM system.

3.3　　　Protocols

This section describes the protocols used in different phases.

Registration. In this phase, the user contacts the registrar to obtain a root credential. The user first establishes a rootNym and signs that rootNym with an external certificate (issued by a trusted CA). The registrar stores the identity proof and issues a root credential on the rootNym. The root credential can be shown multiple times to the Activation Manager.

U: Cert$_{UC}$ from C (an external Certification Authority)

$U \leftrightarrow R$: $RootNym_{UR} = RootNym(Sig_{UR}, Cert_{UC})$
R: stores $[Sig_{UR}, Cert_{UC}, RootNym_{UR}]$
$U \leftrightarrow R$: $Cred_{UR} = IssueCred(RootNym_{UR}, ['ACT'])$

Activation of Anon-Email Service. Each user has to activate the anonymous mail service with the Activation Manager before he can send mail. The Activation Manager issues a *send credential*, required to send mail. Before the user receives a send credential, he must prove that he has previously registered (possesses a valid root credential) and solve an activation puzzle.

To achieve these goals, the user first establishes an anonymous communication channel and registers a nym with the Activation Manager. The user then shows his root credential relative to that nym. This prevents unregistered users to activate the mail service. The credential show is undeanonymizable. The Activation Manager verifies the credential show. The user then solves the activation puzzle. The puzzle discourages users to activate the service several times[3].

$U \leftrightarrow M$: $Nym_{UM} = RegNym()$
$U \leftrightarrow M$: $(Transcript_{UM-R}, Msg_{UM}) =$
    $CredShow(Nym_{UM}, Cred_{UR}, CredShowFeatures_{UM-R})$
        with $CredShowFeatures_{UM-R}=[LDeDanon=null, GDeAnon=null]$
        and $Msg_{UM} = $ contract between U and AAEM, may contain
        explanation of acceptable use policy.
$U \leftrightarrow M$: activation procedure (solving puzzle)
$U \leftrightarrow M$: $Cred_{UM} = IssueCred(Nym_{UM}, ['SEND'])$
M: stores $[Transcript_{UM-R}, Nym_{UM}, Cred_{UM}]$

Sending anonymous mail. Sending mail is conditionally anonymous. A user is anonymous as long as he abides by the rules. If the system is used to send spam mail, the user's send credential will be revoked; if criminal mail is sent through the system, the sender can be identified and prosecuted.

The sender is responsible for removing identifying headers before contacting the mail guard, who will verify the sender's send credential. During the credential show, the mail is signed, which provably links the mail to the transcript of the credential show. The transcript is locally and globally

---

[3] The activation puzzle can be omitted if the user has to go through a prior payment phase, in which he receives a one-show payment credential. In that case, the user must pay in order to activate the mail service.

deanonymizable. The Mail Guard verifies the credential show and attaches the transcript to the mail. The Mail Guard then forwards the mail to the recipient mailbox over a remailer network.

$U \leftrightarrow G$: (Transcript$_{UG\text{-}M}$, Msg$_U$)=
　　CredShow(null , Cred$_{UM}$, CredShowFeatures$_{UG\text{-}M}$)
　　　with CredShowFeatures$_{UG\text{-}M}$ =
　　　　　[LDeAnon=[DCond=Unacceptable|Criminal, Arbiter=A],
　　　　　　GDeAnon=[DCond=Criminal,Arbiter=A]]
　　and Msg$_U$ = message to be sent anonymously to the recipient
G: forwards [Msg$_U$, Sig$_G$(Transcript$_{UG\text{-}M}$)] to recipient r through REM

Receiving anonymous mail. The recipient checks the validity of the Transcript$_{UG\_M}$ with respect to the message Msg$_U$. If the verification fails, the message is discarded. If the verification is successful, the user reads the message. If the message is abusive (unacceptable or criminal), the recipient forwards the mail to the complaint handler.

Unacceptable behavior (Spam, ...). If a user has sent spam, the send credential of that user should be revoked. Revoking the send credential consists of three steps:

Decision of Arbiter. The recipient sends the suspected mail (mail contents and transcript) to AAEM system. The Complaint Handler signs the mail and forwards the request to the Arbiter. The Arbiter first verifies the validity of the mail w.r.t. the transcript. It then verifies whether the mail is really unacceptable. The Arbiter returns his signed decision. If the mail is unacceptable, the Complaint Handler informs the Deanonymizer.

Disclosing Nym. The Deanonymizer receives a signed message from the Complaint Handler. The message contains the Arbiter's decision (i.e. "Unacceptable"), the mail and the transcript. The Deanonymizer verifies the advice, and if positive, locally deanonymizes the transcript. He then returns the nym and a deanonymization transcript to the Complaint Handler. The Deanonymizer also stores the Arbiter's signed decision.

Credential revocation. When the Complaint Handler receives the nym from the Deanonymizer, the mail system actually revokes the credential issued on the nym. The victim is also kept informed.

$r \rightarrow C$: *[ComplaintSpam, Mail$_U$]*
   with *Mail$_U$ = [Msg$_U$, Transcript$_{UG-M}$]*
   and *Transcript$_{UG-M}$ contains LDeAnon=[Unacceptable|Criminal, A]*

$C \rightarrow A$: *Sig$_C$(Mail$_U$, Unacceptable?)*
$C \leftarrow A$: *Sig$_A$(Mail$_U$, Unacceptable)*

$C \rightarrow D$: *Sig$_A$(Mail$_U$, Unacceptable)*

D: *(Nym$_{UM}$, DeAnonTranscript$_{D-UG}$)= DoDeAnonLocal(Transcript$_{UG-M}$)*
D: stores *Sig$_A$(Mail$_U$, Unacceptable)*

$C \leftarrow D$: *[Nym$_{UM}$, DeAnonTranscript$_{D-UG}$]*

C: *RevokeCred(Nym$_{UM}$, Cred$_{UM}$)*
C stores: *[Sig$_A$(Mail$_U$, Unacceptable), Nym$_{UM}$, DeAnonTranscript$_{D-UG}$]*

$r \leftarrow C$: *Sig$_C$(Sender=BANNED, Sig$_A$(Mail$_U$, Unacceptable))*

Criminal behavior. Criminal behavior can be detected by the recipient (e.g. blackmail, stalking, ...) or by a mail system component (e.g. illegal content...). In both cases, the identity of the mail sender should be revealed. In addition, the user's send credential can be revoked. Revocation of a send credential is described above. Revealing the identity of the sender requires the following steps:
   Decision of Arbiter(see above).

Disclosing RootNym. If the mail is criminal, the Arbiter convinces the deanonymizer to reveal the rootNym behind the transcript. The deanonymizer globally deanonymizes the transcript and returns the rootNym and the deanonymization transcript to the Complaint Handler.

Revealing identity. The Complaint Handler forwards the evidence to LE. LE then orders the Root Authority to reveal the identity of the user behind the rootNym. LE stores the evidence that proves the link between the sender and the criminal mail.

$r \rightarrow C$: *[ComplaintCriminal, Mail$_U$]*
   with *Mail$_U$ = [Msg$_U$, Transcript$_{UG-M}$]*
   and *Transcript$_{UG-M}$ contains GDeAnon=[Criminal, A]*

$C \rightarrow A$: *Sig$_C$(Mail$_U$, Criminal?)*

$C \leftarrow A: Sig_A(Mail_U, Criminal)$

$C \rightarrow D: Sig_A(Mail_U, Criminal)$

$D: (RootNym_{UR}, DeAnonTranscript_{D\text{-}UG}) =$
$$DoDeAnonGlobal(Transcript_{UG\text{-}M})$$
$D: stores \ Sig_A(Mail_U, Criminal)$

$C \leftarrow D: [RootNym_{UR}, DeAnonTranscript_{D\_UG}]$

$C \rightarrow LE: [Sig_A(Mail_U, Criminal), RootNym_{UR}, DeAnonTranscript_{D\text{-}UG}]$

$LE \rightarrow R: Sig_{LE}(RootNym_{UR}, GETIDENTITY)$
$LE \leftarrow R: [RootNym_{UR}, Sig_{UR}, Cert_{UC}, Msg_{UR}]$

$LE \ stores \ [RootNym_{UR}, Sig_{UR}, Cert_{UC}, Msg_{UR}]$

3.4   Properties

This section focuses on the trust properties in the system. The mail system creates a trusted environment for senders, recipients and administrators of an AAEM system.

Sender. First, the sender may trust that different mails cannot be linked by the AAEM system. Although send credentials are issued by the AAEM system, credential shows are unlinkable. Therefore, different mails from the same user can not be linked by AAEM. Note that an anonymous communication infrastructure is required. Second, the sender may trust that his send credential will not be revoked as long as he does not send abusive mail. Three parties are involved in revoking send credentials: AAEM, A and D. D will only locally deanonymize the transcript after permission of A. AAEM can only revoke the send credential related to the transcript after local deanonymization. Moreover, AAEM needs A's permission to revoke the credential. Nevertheless, trust is required in a righteous Arbiter. AAEM and user can possibly negotiate which Arbiter to involve before a credential show. Third, revealing the user's identity is only possible with cooperation of external entities: A, LE, D and R. D only globally deanonymizes a transcript after A's approval. R only reveals the link between the rootNym and the identity of the user after approval of LE.

Recipient. A valid transcript guarantees that a mail is locally and globally deanonymizable. Recipients also know the verifier of valid transcripts (i.e. AAEM) and the deanonymization conditions. Moreover, the recipient can block mail containing no or invalid transcripts.

AAEM, D and R. AAEM can not be liable as long as it observes the rules (i.e. respects the decision of the Arbiter). AAEM stores evidence about unacceptable/criminal mail and the judgement of A about the mail. D also stores such evidence. In case of criminal mail, R stores LE's judgement.

4. PROTOTYPE

4.1 Description

Client side infrastructure.
The *application layer* consist of three components. The registration/activation module receives registration/ activation requests from the user and passes them to the credential layer. The mail client is configured to forward mail to a local SMTP server (running at the sender's machine). The mail server filters any identifying headers and passes the message to the credential layer.

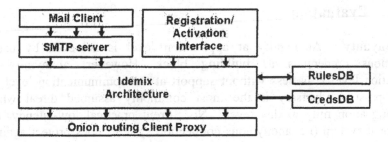

Figure 2. Client side infrastructure

The *credential layer* implements Idemix credential protocols (credential showing, credential issuing, ...). The credential layer requires two databases in order to execute a credential protocol. First, the CredsDB stores credentials (root credential, activation credentials, mail credentials). This database is updated as (new) user credentials are retrieved/revoked. Second, the RulesDB specifies the rules for credential showing/credential issuing during each phase (registration, activation, sending mail). The RulesDB is configured at set up time. Idemix requests are passed to the communication layer.

The *communication layer* deals with anonymous connections to the AAEM system.   An onion routing proxy[2,3] is inserted at communication level. In the current implementation, the client composes an anonymous path to the AAEM system.   In an alternative implementation, the communication layer can contact an external communication proxy that sets up anonymous connections. However, the latter implementation has different anonymity properties.   Access configurations to anonymous connection systems are discussed in [2,3]. The client side also consists of a module that verifies transcripts and sends complaints to the AAEM system.

Core of AAEM system.

The *activation manager* and the *mail guard* are implemented as two different Idemix organizations. The Mail Guard verifies mail, stores the transcripts as attachment and passes them to a Mixmaster remailer proxy (running at the same machine). The remailer proxy chooses a chain of remailers, recursively encrypts the message and forwards the message through the remailer system.

The *Complaint Handler* receives complaints from recipients. The Complaint Handler forwards them to an arbiter and/or a law enforcement entity. This depends on the type of complaint.

4.2   Evaluation

Anonymity.   Anonymity at application level is achieved by using anonymous credentials as building block. However, anonymity at application level is useless without support at the communication level. A global passive adversary is the most commonly assumed threat when analyzing anonymity at this level.   No current practical low-latency, bi-directional system (i.e. anonymous connection system) does protect against such a strong adversary.

The prototype implements anonymous connections between the sender and the AAEM system.   The credential protocols require a real-time, bidirectional communication channel. However, sender and recipient are the real endpoints of communication.   The AAEM system forwards mails to recipients over a remailer system that resists global attacks.   Thus, global attackers cannot link the endpoints of communication.

Usability/Deployability. To be deployed and used in the real world, the system is not expensive to run:

   The design does not place a heavy liability burden on AAEM
   operators (as discussed in section 3.4).

- Decentralized storage of mail transcripts reduces the number of disk space required by AAEM. The number of stored activation transcripts is linear to the number of activations. The amount of evidence stored by AAEM is linear to the number of accepted complaints. However, the system discourages multiple activations and abusive behaviour.
- The system extends an existing infrastructure. AAEM uses a pre-existing network of anonymous remailers and anonymous connections.
- Once a user has installed the client software, he only has to change the location of the outgoing SMTP server in his mail client.

Flexibility/transparency. The components are loosely coupled by a layered design. Transparency is achieved between the mail component and the communication component. The communication component can easily be replaced by another implementation. Second, the system foresees a loose coupling between different entities: the Arbiter and the Law Enforcement entity do not require any knowledge about Idemix and the structure of the mail system to judge complaints. Even the deanonymizer doesn't require knowledge about the structure of the mail system. To simplify the complaint handling procedure, the deanonymizer itself can be the Arbiter. This requires additional trust in the deanonymizer.

5.   RELATED WORK

Our work on AAEM was largely motivated by the problems of current anonymous mail systems and tries to be a reliable extension of current remailer systems. The Mail Guard functions as front end to a remailer system. Our implementation uses Mixmaster[5, 10] remailers. However, only the communication proxy at the Mail Guard has to be re-implemented to work well with other types of remailers. If replies should be supported, the SMTP server at the client side also has to be re-implemented. This server must know the available remailers in order to build a reply structure. The current implementation does not support replies: the SMTP server just removes the "return-path" header.

The first anonymous mail system open to the public was *anon.penet.fi* [9]. Unfortunately, penet did not use encryption. Moreover, only one machine needs to be compromised in order to break the anonymity. *Type-1 remailers*, also called Cypherpunk remailers, were developed to address many shortcomings of the penet system. Type-1 remailers have public keys with which incoming messages are encrypted. A message can be sent through a chain of type-1 remailers, having been successively encrypted for each of them. Each remailer in a chain knows only the identity of the previous remailer and the next remailer in the chain. The system also supports reply functionality.

Type-2 remailers[5,10] offer several improvements in security over type-1 remailers. These improvements make hop-by-hop analysis considerably harder. They include fixed size messages, replay detection and better reordering of messages at remailers. Type-2 remailers do not support replies to unknown destinations.

Type-3 remailers[8], also called Mixminion remailers, support secure single-use reply blocks. Mix nodes cannot distinguish Mixminion forward messages from reply messages. Directory servers allow users to learn public keys and performance statistics of participating remailers. Mixminion provides a mechanism for each node to specify an exit policy (open exit nodes versus middleman exit nodes) and describes a protocol which allows recipients to opt-out of receiving mail from remailers. However, this requires recipients to send an opt-out request to each open exit node. This is very difficult to realize in practice as new remailers become available. Moreover, if receiving mail is opt-out, non-abusive mail is also retained. Our approach is to discard only anonymous messages without a valid transcript. Senders of abusive mail can be held accountable.

Nymserv[1] is an e-mail pseudonym server: the server keeps a public key and a reply block for every nym. Nymserv also functions as front end and back end to a remailer system. Mail sent from the server to a user leaves through a chain of Cypherpunk remailers; requests to create nyms and to send mail from them arrives through a chain of Cypherpunk remailers. Nyms of abusive users can be revoked. Nymserve also uses a high-latency anonymous communication system. However, different mails from the same user can be linked. Moreover, only a limited amount of control is possible: users can not be accountable for sending abusive mail.

6.   CONCLUSION

The presented mail system considers both anonymity requirements of senders and accountability requirements of recipients. A reasonable trade-off between these interests is achieved.

An acceptable level of anonymity at communication level is achieved by reusing existing solutions: anonymous connections and remailers. An anonymous credential system is used as building block for accountability of application specific data/actions. Moreover, the credential system is loosely coupled to the application.

Trust is achieved by splitting responsibilities over different entities and accurate complaint handling procedures. However, a trusted external party is still required in applications where conditional anonymity is a design issue.

REFERENCES

[1] David Mazieres and M. Frans Kaashoek. The design, Implementation and Operation of an Email Pseudonym Server. In Proceedings of the 5th ACM conference on Computer and communications security, p.27-36, November 02-05, 1998, San Francisco, California, United States.

[2] P. Syverson, M.Reed and D. Goldschlag. Onion routing access configurations. In DARPA Information Survivability and Exposition (DISCEX 200), volume 1, p.34-40. IEEE CS Press, 2000.

[3] P. Syverson, G. Tsudik, M. Reed and C. Landwehr. Towards an Analysis of Onion Routing Security. In H. Federrath, editor, Designing Privacy Enhancing Technologies: Workshop on Design Issue in Anonymity and Unobservability, p.96-114. Springer-Verlag, LNCS 2009, July 2000.

[4] M. Reed, P. Syverson and D. Goldschlag. Anonymous connections and onion routing. IEEE Journal on Selected Areas in Communications, 16(4): 482-494, May 1998.

[5] U. Moller, L. Cottrel, P. Palfrader and L. Sassaman. Mixmaster Protocol - Version 2. Draft, July 2003, *http://www.abditum.com/mixmaster-spec.txt.*

[6] B. Levine, M. Reiter, C. Wang and M. Wright. Timing analysis in low-latency mix-based systems. In A. Juels, editor, Financial Cryptography. Springer-Verlag, LNCS, 2004.

[7] C. Gulcu and G. Tsudik. Mixing E-mail with Babel. In Network and Distributed Security Symposium(NDSS 96), P.2-16. IEEE, February 1996.

[8] G.Danezis, R.Dingledine and N. Mathewson. Mixminion: Design of a type-3 anonymous remailer protocol. In 2003 IEEE Symposium on Security and Privacy, p.2-15. IEEE CS, May 2003.

[9] J. Helsingius. anon.penet.fi press release. *http://www.penet.fi/press-english.htm.l*

[10] Cottrel. Mixmaster and remailer attacks.
http://www.obscura.com/~loki/remailer/remailer-essay.html.

[11] Jan Camenisch, Els Van Herreweghen: Design and Implementation of the Idemix Anonymous Credential System. Research Report RZ 3419, IBM Research Division, June 2002. Also appeared in ACM Computer and Communication Security 2002

[12] Els van Herreweghen, Unidentifiability and Accountability in Electronic Transactions. PhD Thesis, KULeuven, 2004.

PROTECTING CONSUMER DATA IN COMPOSITE WEB SERVICES

Craig Pearce, Peter Bertok, Ron Van Schyndel
RMIT University, Melbourne, Australia
{crpearce,pbertok,ronvs}@cs.rmit.edu.au

Abstract: The increasing number of linkable vendor-operated databases present unique threats to customer privacy and security intrusions, as personal information communicated in online transactions can be misused by the vendor. Existing privacy enhancing technologies fail in the event of a vendor operating against their stated privacy policy, leading to loss of customer privacy and security. Anonymity may not be applicable when transactions require identification of participants. We propose a service-oriented technically enforceable system that preserves privacy and security for customers transacting with untrusted online vendors. The system extends to support protection of customer privacy when multiple vendors interact in composite web services. A *semi-trusted processor* is introduced for safe execution of sensitive customer information in a protected environment and provides accountability in the case of disputed transactions.

Key words: Electronic commerce; privacy; security; web services.

1.   INTRODUCTION

Many vendors have shown poor security of customer databases, leading to intrusions, loss of customer privacy and even identity theft [internetnews.com, 2003].

When back-end customer databases are copied, sold or linked with databases of other vendors, the wealth of available customer information rapidly increases. In some cases, customers trust a vendor with personal information, however the information is collected for processing by other (untrusted) parties along the chain, as seen in outsourcing and supply chain management [Medjahed et al., 2003].

Currently, private information that customers choose to release to vendors, such as medical information or credit card details, cannot be fully controlled by the customer once released. In addressing this issue, we have designed a generalised application-layer privacy platform, named: TEPS, the Technically Enforceable Privacy and Security system. TEPS protects from customer privacy violations at the vendor-side by preventing an untrusted vendor from ever holding customer personally identifiable information (PII) in plain view. The customer decides which of their personal attributes to protect and we introduce a semi-trusted processor (STP) that is trusted not to disclose customer PII within local execution of vendor-provided business logic. Full trust of the STP is not required as accountability and code watermarking [Collberg and Thomborson, 2002] can detect other forms of STP abuse. Mobile code is utilised as a method of communicating messages of varying protection levels amongst the entities of the service-oriented electronic commerce architecture.

TEPS is a generalised model, and is suitable within the Web Services architecture, where multiple vendors can interact to fulfill customer requests, typically seen with a front-end web service broker that outsources back-end activities to other web services.

Our results from a fully scaled implementation within wired and wireless networks, and the possibility of mobile clients, show that TEPS is suitable within service-oriented transactions, enforcing consumer privacy as a value-added service.

2.   BACKGROUND AND RELATED WORK

Traditionally, once a vendor has access to plain-text (non-encrypted) customer information, there are no technical methods available to restrict its use of that information.

Anonymising layers, such as [Chaum, 1981, Jakobsson and Juels, 2001, Dingledine et al., 2004], help protect the customer source identity, and sometimes vendor destination, but once personally identifiable information has been captured by the vendor it can no longer be controlled. Identity Management systems, such as [Waldman et al., 2000, Campbell et al., 2002, Jendricke et al., 2004], act as an intermediary between customer and vendor and provide a pseudonym of the customer instead of the customer's real identity. This establishes privacy as long as pseudonyms cannot be linked to the customer's real identity. However, pseudonyms cannot be used when a vendor is required to authenticate a customer in environments that provide services both in electronic and traditional environments, such as banking, voting and payment. Credential-carrying pseudonyms [EU FP6 PRIME Project, 2005] could be considered an alternative to strong authentication, but require globally present identity management mechanisms.

Non-traceable anonymous payment systems, such as [Chaum, 1982, Chaum et al., 1990] for transactions requiring authentication remain to be problems, such as medical subscriptions and large order requests.

The Secure Electronic Transactions (SET) protocol used hashing techniques to preserve privacy of payment and order information, although overheads of client-side certificates, implementation difficulties and lack of extensibility for multiple vendors within integrated transactions made it unsuitable for complex environments, such as Web Services [Medjahed et al., 2003].

The Secure Sockets Layer (SSL/TLS) [Dierks and Rescorla, 2004] provides communication channel authentication, message confidentiality and integrity but protects only the communication channel between customer and vendor. Customer privacy from untrusted vendors is not protected once data has reached the vendor.

Protection of a customer's personally identifiable information (PII) has been proposed [Kenny and Korba, 2002] but does not offer assurance of enforceability in global e-commerce. Furthermore, the proposed PII-protecting model [Kenny and Korba, 2002] requires full trust in the data controller, which is also responsible for accountability. Personnel are required to manually check data processing activity and the security of data controllers is simplified to a question of reputation. Extensible support for multiple vendors interacting within a transaction has not been addressed.

Encrypting digital identifiers and enforcing associated privacy policies through trusted computing technologies [Casassa et al, 2003] has been suggested, however all participants are required to operate within the confines of a globally unified trusted computing platform.

Recent developments in XML-based privacy between customer and vendor has seen the emergence of Platform for Privacy Preferences (P3P)

[W3C, 2002, Berthold and Köhntopp, 2001] for the Internet and Enterprise Privacy Authorization Language (EPAL) [Ashley et al., 2003] for organisations. P3P and EPAL provide a standardised way for the vendor to represent their privacy policy and allow the customer to specify their privacy needs but cannot provide technical assurance that the vendor will not digress from their stated privacy policy. EPAL provides logging and reporting capabilities and enforces privacy access within an organization [Goldberg, 2002] using network privacy monitors, however, is not appropriate for complex transactions as customers are required to unconditionally trust resources governed by vendor organisations. Furthermore, P3P and EPAL were designed for web-based applications, using the traditional client-server model, and are not suitable for Web Services [Medjahed et al., 2003].

Issues of vendors digressing from their stated privacy policy, lack of identification and non-repudiation in anonymous payment systems, overheads of client identity certificates and legal factors due to globalisation have encumbered electronic commerce with privacy concerns. In many jurisdictions, revelation of customer databases to third parties is legally punishable if detected, but is still prevalent due to limitations in tracking down the perpetrator. Globalisation increases this problem as privacy laws in some jurisdictions are weak or non-existent.

The "Technically Enforceable Privacy and Security" (TEPS) system helps solve these core issues by operating as a generalised service at the application-level protocol layer, and is suitable in a service-oriented architecture to prevent vendors from ever gaining access to customer privacy information.

3.   SCENARIOS: HOW ONLINE TRANSACTIONS AFFECT CUSTOMER PRIVACY

In this section we describe two realistic scenarios currently threatening customer privacy that TEPS aims to alleviate.

3.1   Scenario 1: Online brokers

A customer uses on online bookseller web service as the vendor to locate a textbook. After finding a suitable match, the customer decides to purchase the package from the vendor. Current practices require customers to log into the vendor's website with a previously established account that probes for customer identity information. SSL/TLS is used for encrypting credit card information, which is generally handled by a payment gateway, not the vendor. The vendor redirects customers to a payment gateway, and once payment is complete, the payment gateway returns an outcome to the vendor. Despite what may be stated within the vendor's privacy policy, SSL/TLS does not prevent the vendor from disclosing consumer spending habits to other parties.

3.2 Scenario 2: Composite web services

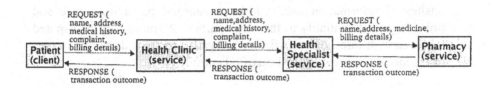

Figure 1. Composite web services

A customer seeks medication by lodging a request to an online health clinic and must log in for identification. As with Scenario 1, the previously established account may require a number of personally identifiable customer attributes deemed private in nature. The health clinic is a front-end only, outsourcing medical knowledge to a specialist back-end service, as shown in Figure 1. Furthermore, if medicine is required, the specialist outsources prescription services to a pharmacy. The customer may not be aware of multiple vendors operating to fulfil their transaction. Each of these back-end services will request customer details from the front-end service to perform their business activity, possibly without customer knowledge. Privacy policies of back-end services may be independent to the health clinic privacy policy agreed to by the customer.

4. SYSTEM DESIGN

TEPS is composed of the following entities:
- **Customer (CUS):** Operates a client (CL) machine through a web browser;
- **Client (CL):** Computer used by customer in transacting with a vendor;
- **Vendor (V):** Service-oriented online store (for example, travel agent, weather service);
- **Semi-Trusted Processor (STP):** Partially trusted intermediary between client and vendor in processing vendor business logic on

customer PII data. Example STPs include payment gateways, identity verifiers and marketing bureaus to name a few;

- **Certificate Authority (CA):** Trusted certificate server used for distribution and revocation of digital certificates to the entities communicating in an online transaction. The CA can be used throughout online transactions for verification of certificates with public key encryption and signing;

- **Accountability Authority (AA):** Used in disputed transactions to provide accountability of participants in case of abuse. The *AA* stores hashes of information used within a transaction, saving space and providing confidentiality to the other parties. A transaction is disputed when enough threshold certificates are gathered from disputing parties or if requested by an external certified entity.

The *AA* and *CA* are essential services for a technically-enforceable system that guarantees privacy and accountability. The current approach to online transactions (Section 3, Scenario 1) uses SSL/TLS encryption and X.509 Certificates signed by certificate authorities (CAs) to communicate vendor certificates to clients. An accountability service is not provided, limiting the types of transactions performed online due to lack of defined dispute resolution mechanisms.

4.1 Assumptions

In formulating our system, we considered the following assumptions:
- *STPs* will not knowingly reveal PII data to another entity (with the exception of an accountability authority in pre-defined legal circumstances);
- *STP, AA* and Certificate Authority (*CA*) services are who they claim to be; host security has not been breached;
- Vendors comply with the privacy system by programming their business logic in a way that is executable by the *STP*;

These assumptions show the proposed solution to be useful in providing customer privacy protection in scenarios where vendors are willing to program and communicate their business logic to STPs. This is not a major overhead, as vendor business logic should be a direct implementation of the action stated publicly in their privacy policy. In cases of rigid intellectual property agreements, non-disclosure agreements (NDAs) or outsourcing could be negotiated between vendor and semi-trusted processor.

Additional privacy requirements, such as data minimisation and purpose binding can be met by the customer proactively reading the vendor's privacy policy and discontinuing the transaction if the collection purpose or amount of requested information is not appropriate.

We plan for TEPS to utilise existing privacy and security services where possible. While TEPS is a generalised model, this paper explores TEPS in a service-oriented environment, with Figure 2 showing the communication stack layering TEPS on top of web services, as web services alone do not protect customers from misbehaving vendors. SSL/TLS can be used for underlying channel communication security.

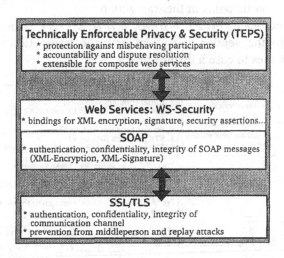

Figure 2. TEPS communication stack: privacy and security for Web Services

4.2 Processing of an online transaction

Figure 3 shows the functional steps taken in a transaction using TEPS. Each phase within Figure 3 is described here:

1. Whenever a vendor's form requests an input that has been marked PII, the client privacy reference monitor will transparently request a list of *STPs* from vendor. The vendor will compile a list of *STPs* (consulting a business registry *(BR)* if needed) and return this to the client with vendor's privacy policy *(VP)*. The client hashes the *VP* and stores it safely in case of a disputed transaction;

2. From a given a list of *STPs*, the client will choose one, and then contact it to download the PII protector mobile code, providing a name certificate for transfer of a temporal public key. The *STP* generates K_{STP-CL}, a shared secret key, encrypting it with the client's public key for confidentiality. The PII-protecting mobile code is signed by the *STP*;

3. The customer fills out the vendor's HTML form. The client executes *STP's* mobile code which protects the customer's PII by encrypting it with K_{STP-CL};

4. Upon receiving the PII-Protecting mobile code, the vendor executes the mobile code which prompts for a business process activity (*BPA*);

5. Once the mobile code cycles back to the STP, the *BPA* is processed with customer's PII data in a safe environment;

6. Threshold certificates are provided by *CA* after providing the name certificates of participants in the transaction

7.*STP* communicates *h(VP), h(BPA), h({PII}K_{STP-CL})* hashes to *AA*. STP then responds to the vendor and client with the transaction outcome and threshold certificates in case a dispute arises;

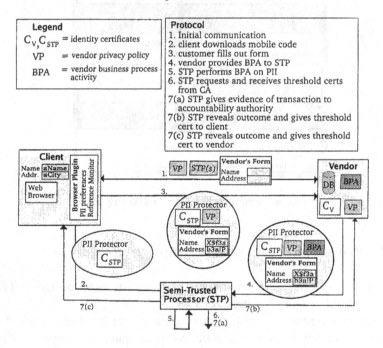

Figure 3. Privacy in transactions

The transaction will be aborted if the client is not satisfied with the list of *STPs* provided by the vendor in Figure 3 Step 1. If a party stops responding during the processing of a transaction, the transaction will time out and be aborted.

4.3   Composite web services

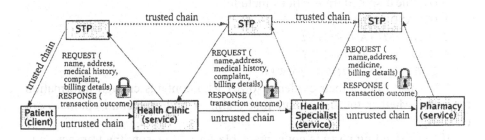

Figure 4. Technically enforced privacy and security in composite web services

Scenario 2 of Section 3 described a transaction involving a customer and multiple vendors. Web Services privacy is an open question when each vendor performs a separate business process activity, integrated to form a composite web service [Medjahed et al., 2003]. We address this issue by forcing the front-end web service to clearly state the need of back-end vendors in their privacy policy, and the client agreeing to transitivity of semi-trusted processing of personal information. The TEPS protocol is then performed recursively for each back-end vendor. For instance, the example composite web service in Figure 1 involves a separate invocation of TEPS for the Health Specialist and Pharmacy services, as shown here in Figure 4. Each subsequent vendor has an associated, possibly different, semi-trusted processor to perform its business process activity, preserving privacy for the previous vendor. A tree-based structure is formed and includes two chains of information flow: (1) untrusted vendor chain which has no access to client personally identifiable information or adjacent vendor privacy information and (2) trusted chain for semi-trusted processors to communicate customer personally identifiable information (PII) from top STP to bottom STP. While trust management of the STP chain is not addressed here, we assume clients to explicitly agree to adjacent STPs in a chain exchanging privacy information between themselves (transitivity).

4.4   Accountability and disputed transactions

Transactions may be disputed when two or more parties out of three submit a dispute request with their allocated threshold certificate.

Alternatively, external certified entities *(ECEs)* can initialise a disputed transaction by submitting a signed request with appropriate certification. An example scenario for *ECE* involvement would be law enforcement officers with reason to believe one of the parties committed fraud.

Possible disputed transactions include:

1. (*CL* AND *STP*) AGAINST *V*
2. (*V* AND *STP*) AGAINST *CL*
3. (*CL* AND *V*) AGAINST *STP*
4. *ECE* AGAINST (*STP* OR *CL* OR *V*)

Each party gives their evidence to *AA* who contains enough information to judge whether the defendant, first claimant and/or second claimant are cheating.

If the defending party is not contactable for any reason, the transaction is logged as 'in dispute' by *AA* and claimants.

The dispute resolution mechanism is a two-step protocol , with the AA firstly attempting to reach an outcome without knowledge of the PII-protecting key, K_{STP-CL}. If an outcome cannot be determined at this point, only then will the *AA* request submission of K_{STP-CL} as evidence; both client and STP are asked to provide the shared secret key as either party may be a suspect.

5.   SECURITY CONSIDERATIONS

We have relaxed trust on the STP to not reveal customer PII and properly execute PII within the vendor business process activity. This opens up hostile STP possibilities, such as:

- STP falsifying the transaction outcome: client and vendor could request a dispute, resulting in the AA detecting an anomaly in the transaction;
- STP leaking vendor's business process activity: vendor can mitigate risk by code watermarking [Collberg and Thomborson, 2002] the business process activity for detection of misuse, such as disclosure or reverse-engineering;
- External denial of service (DoS) attacks: it is expected that the STP provides a list of replicated services to alleviate bottleneck and single point of failure concerns.

Collusion between two parties (for example, vendor and STP) prevents the remaining party from issuing a disputed transaction request. The remaining party could still contact an external certified entity (ECE) for further investigation.

We have assumed the STP will not knowingly disclose customer PII, however, in the case of compromise, a noticeable amount of information may accumulate over time. Customers can mitigate potential risk by choosing an STP that operates within the same data privacy laws and we expect that finding a reputable STP is easier than finding a reputable vendor.

Although privacy principles of 'data minimisation' and 'purpose binding' are not technically enforced by TEPS, compliance has been placed in the customer's domain. Customers can check vendor PII requests against their stated privacy policy before opting to continue with the transaction. Customers and STPs can check vendor purpose binding and is considered a legal issue if not followed, pre-empting a transaction dispute.

6.   FORMAL ANALYSIS OF THE PROTOCOL

TEPS has been formally verified with the Casper protocol compiler and FDR2 model checker [Donovan et al., 1999] to prove confidentiality on customer PII data, vendor business process activities hold against all currently known communication channel attacks.

Due to combinatorial explosion of the search space, privacy assertions for composite web services could not be formally verified by FDR2. However, as simple web services privacy is formally verified, and composite web services are iterated simple web services, induction suggests TEPS provides technically-assured privacy of composite web services.

7.   IMPLEMENTATION AND RESULTS

TEPS was implemented in Java with Web Services support for SOAP messaging and WSDL documents. Our system offers flexibility of public key certificate representations, supporting X.509 and SPKI/SDSI formats. X.509 is the industry standard, providing identity certificates but it requires hierarchies of fully-trusted certificate authorities and cannot handle threshold certificates. SPKI/SDSI is a simplified and flexible certificate system allowing identity and authorisation certificates, fine-grained access control and, most importantly, supports threshold certificates. We implemented a secured SPKI/SDSI framework, that was reported in [Pearce et al., 2004a, Pearce et al., 2004b], which allows for naming, access control and thresholding.

TEPS services use thread-based concurrency to support multiple transactions simultaneously. Business process activities (BPAs) are compiled Java bytecodes packaged as '.jar' archives. Vendors could possibly provide BPAs to semi-trusted processors in an encrypted form for confidentiality.

Experiments were conducted on Intel(R) PIII 1GHz machines, with separate machines for each service, communicating over a wired 100Mbps switched network. We measured client connectivity on both the 100Mbps switched network and a wireless 802.11g network at speeds of 1Mbps and 11Mbps. The wireless access point used media access control (MAC) filtering and Wired Equivalency Privacy (WEP) based encryption for additional security.

Table 1. Total client-wait times using TEPS with and without TLS

CONFIGURATION	TIME (sec)
TEPS, Wired 100Mbps	7.67
TEPS, Wired 100Mbps, SSL/TLS	8.35
TEPS, Wireless 1Mbps	8.01
TEPS, Wireless 11Mbps	7.67

Table 1 shows protocol performance in the client perspective by measuring total client-wait time over the entire length of a transaction. Vendor privacy policies and business process activities were fixed at one kilobyte each. Timing of business process execution by STPs were not performed as they gave a constant time among each experiment and, pragmatically speaking, are highly dependent on the business purpose of the vendor. Results from Table 1 indicate that TEPS is efficient at servicing simple web services transactions for both wired and wireless clients, with overheads of around seven to eight seconds per web services transaction. In fact, transaction times did not significantly differ for either wireless or wired network speeds, never exceeding 5% of total transaction times. This suggests that transaction performance will remain satisfactory as network speeds scale down further. Tunnelling TEPS over SSL/TLS incurred a penalty of nearly one second for total client wait-times. Service start-up times took an

Table 2. Processing and communications costs for participant

Party	Number of Messages		Total Message Sizes (kb)	Processing + Communication Times (sec)			Cryptographic Operations	
	Send	Recv		Send	Recv	Total	Encrypts	Decrypts
CL	4	4	~92	3.01	3.64	6.65	1 (symm)	1 (asymm)
V	3	4	~56	0.01	6.58	6.59	-	-
STP	5	3	~136	0.05	3.53	3.58	-	1 (symm)
CA	1	1	~60	3.09	0.04	3.13	3 (asymm sig)	-
AA	0	1	~0.8	-	0.11	0.11	-	-

additional three to five seconds for SSL/TLS enabled sockets due to key randomisation and secure socket establishment.

For a deeper understanding of practicalities within TEPS-enabled web services transactions, we measured processing and communication costs incurred by each party for each communicated message. This was collated to give an overview on how much work is performed by each participant, as shown in Table 2.

Client and vendor have the highest costs in terms of time, due to encryption, communication and awaiting responses from other parties respectively. The STP, as is evident with the vendor, spends almost all of its time waiting to receive messages, whereas the certificate authority incurs most of its costs in generating and communicating threshold certificates.

Our results suggest a linear extension of composite web services yields linear growth in time complexity. For example, the Health Clinic service detailed in Scenario 2 of Section 3 would involve three iterations of TEPS, each iteration being interleaved within its adjacent iteration with a total client wait-time approximately three times longer than a single iteration.

8. DISCUSSION AND FURTHER WORK

Through the use of a semi-trusted processor, TEPS guarantees protection of customer personally identifiable information (PII) against untrusted vendors in the application layer. This also prevents vendors from linking up databases and identifying customers on seemingly unlinkable attributes (triangulation). Introducing an accountability authority allows for externally certified entities to follow up unlawful activities.

TEPS supports execution of business process activities for (1) once-off transactions (for example, customer using an online broker) and (2) transactions requiring multi-vendor integration, that being composite web services.

In the first scenario, described in Section 3, the business process activity may require access to the vendor database (for example, an inventory table). It is the responsibility of vendor and semi-trusted processor to agree on appropriate mobile code and dependent parameters to satisfy business logic for execution of business process activities. One solution can involve the vendor attaching required data from its own database to the business process. Alternatively, both vendor and STP can agree on a common link for respective scrambled PII and plain-text PII database entries. The second scenario is addressed by iterating the TEPS protocol for each additional back-end vendor web service, creating a trusted chain for semi-trusted processors and an untrusted vendor chain. Complexity is linear which suggests that the system is extensible for transactions of growing numbers of interacting services. However, for large business processes or a large number of co-operating vendors, long running transactions (LRTs) may be required to provide acceptable client wait-time.

We expect to alleviate vendor reluctance of outsourcing full business processes to STPs by the use of code watermarking: detecting STP misuse, such as disclosure or reverse-engineering. More comprehensive solutions may be more applicable, such as source code escrow agreements.

TEPS prevents vendors from profiling clients, which is another privacy issue. However, if customers choose to allow profiling of their activities, the STP can profile customers based on gathered information, anonymise (by removing identifiable elements) and pass it back to the vendor.

We have not investigated programming challenges of aggregation and separation of business processes into activities that can be processed by separate parties. Furthermore, aggregation and separation of privacy policies among co-operating vendors is an area of future work.

Investigation into the benefits and trade-offs of caching vendor business policies with identity and authentication details will help decide whether additional performance gains are worth the risk against obsolescence. Vendor policies negotiated on a client-by-client basis presents an open problem in this approach.

9.   CONCLUSION

In this paper we proposed the Technically-Enforceable Privacy and Security (TEPS) system that prevents vendors from ever obtaining customer personally identifiable information. Major components of the system were the following:

- semi-trusted processor to (1) protect customer personally identifiable information (PII) and (2) execute vendor-provided business processes with customer PII data in a protected environment;
- accountability service to provide recourse when one or more parties abuse the protocol;
- resolution mechanism for transaction disputes;

Furthermore, we showed how TEPS is extensible in supporting composite web services by iterating the protocol for multiple back-end vendors.

TEPS has been verified to ensure customer privacy is maintained against untrusted vendors or external attackers and that vendor business process activities are not accessible to parties other than the semi-trusted processor.

Our results indicated that the solution was suitable for web services as client wait-times for transactions were within an acceptable range. TEPS also performed well in slower wireless networks and transaction times grew in a linear fashion as complexity of interactions rose in composite web services scenarios.

TEPS gives privacy and security guarantees to prevent untrusted vendors from obtaining private customer information within traditional transactions and composite multi-vendor web services. In helping alleviate consumer concerns and address open issues of privacy within composite web services, service-oriented transactions can become a safer practice.

ACKNOWLEDGEMENTS

We would like to thank Formal Systems for providing a license to freely use FDR2. We would also like to thank the anonymous reviewers for their useful suggestions.

REFERENCES

[Ashley et al., 2003] Ashley, P., Hada, S., Karjoth, G., Powers, C., and Schunter, M. (2003). Enterprise Privacy Authorization Language (EPAL). Research Report 3485, IBM Research.

[Berthold and Köhntopp, 2001] Berthold, O. and Köhntopp, M. (2001). Identity management based on P3P. In *Lecture Notes in Computer Science*, volume 2009, pages 141–160.

[Campbell et al., 2002] Campbell, R., Al-Muhtadi, J., Naldurg, P., Sampemane, G., and Mickunas, M. Dennis (2002). Towards security and privacy for pervasive computing. In *Proceedings of the International Symposium on Software Security, Keio University*, Keio University, Tokyo, Japan.

[Casassa et al, 2003] M. Casassa Mont, S. Pearson, P. Bramhall. Towards Accountable Management of Privacy and Identity Information. ESORICS 2003: 146-161

[Chan et al., 2002] Chan, H., Lee, R., Dillon, T., and Chang, E. (2002). E-Commerce: Fundamentals and Applications. pages 287–298. ISBN: 0-471-49303-1.

[Chaum, 1981] Chaum, D. (1981). Untraceable Electronic Mail, Return Addresses and Digital Pseudonyums. *Communications of the ACM*, 24(2):84-90.

[Chaum, 1982] Chaum, D. (1982). Blind Signatures for Untraceable Payments. *Crypto*, pages 199–203.

[Chaum et al., 1990] Chaum, D., Fiat, A., and Naor, M. (1990). Untraceable electronic cash. *Proceedings on Advances in cryptology. California, United States*, pages 319–327.

[Collberg and Thomborson, 2002] Collberg, C. and Thomborson, C. (2002). Watermarking, Tamper-Proofing, and Obfuscation - Tools for Software Protection. In *IEEE Transactions on Software Engineering*, volume 28, pages 735–746.

[Dierks and Rescorla, 2004] Dierks, T. and Rescorla, E. (2004). The TLS Protocol Version 1.1. Internet Draft http://www.potaroo.net/ietf/ids-wg-tls.html.

[Dingledine et al., 2004] Dingledine, R., Mathewson, N., and Syverson, P. (2004). Tor: The Second-Generation Onion Router. In *In Proceedings of the 13th USENIX Security Symposium*.

[Donovan et al., 1999]Donovan, B., Norris, P., and Lowe, G. (1999). Analyzing a library of security protocols using Casper and FDR. In Workshop on Formal Methods and Security Protocols.

[Goldberg, 2002] Ian Goldberg. Privacy-enhancing Technologies for the Internet, II: Five Years Later. Workshop on Privacy Enhancing Technologies. April 2002

[internetnews.com, 2003] internetnews.com, Staff: (2003). Acxiom Hacked, Customer Information Exposed. Website:
www.internetnews.com/storage/article.php/2246461.

[Jakobsson and Juels, 2001] Jakobsson, M. and Juels, A. (2001). An Optimally Robust Hybrid Mix Network. In *Proceedings of the twentieth annual ACM symposium on Principles of distributed computing*, pages 284–292. ACM Press.

[Jendricke et al., 2004] Jendricke, U., Kreutzer, M., and Zugenmaier, A. (2004). Pervasive Privacy with Identity Management. In *Proceedings of ACM Symposium on Applied Computing*, pages 1593–1599. ACM Press.

[EU FP6 PRIME Project, 2005] PRIME: Privacy and Identity Management for Europe. Website: http://www.prime-project.eu.org/ Last accessed: 15-11-2004.

[Kenny and Korba, 2002] Kenny, S. and Korba, L. (2002). Applying digital rights management systems to privacy rights management. *Computers & Security*, 21(7):648–664.

[Medjahed et al., 2003] Medjahed, B., Benatallah, B., Bouguettaya, A., Ngu, A. H. H., and Elmagarmid, A. K. (2003). Business-to-business interactions: issues and enabling technologies. *The International Journal on Very Large Data Bases*, 12(1):59–85.

[Pearce et al., 2004a] Pearce, C., Bertok, P., and Thevathayan, C. (2004a). A Protocol for Secrecy and Authentication within Proxy-Based SPKI/SDSI Mobile Networks. AusCERT Asia Pacific Information Technology Security Conference ISBN: 1864997745.

[Pearce et al., 2004b] Pearce, C., Ma, Y., and Bertok, P. (2004b). A Secure Communication Protocol for Ad-Hoc Wireless Sensor Networks. IEEE International Conference on Intelligent Sensors, Sensor Networks & Information Processions, Melbourne, Australia.

[W3C, 2002]W3C (2002). Platform for Privacy Preferences (P3P). W3C Recommendation www.w3c.org/TR/2002/REC-P3P-20020416/.

[Waldman et al., 2000] Waldman, M., Rubin, A., and Cranor, L. (2000). Publius: A robust, tamper-evident, censorship-resistant, web publishing system . In *Proc. 9th USENIX Security Symposium*, pages 59–72.

A DECISION MATRIX APPROACH
to prioritize holistic security requirements in e-commerce

Albin Zuccato

Karlstad University, Department of Computer Science, Universitetsgatan 2, 65188 Karlstad, Sweden

Abstract: In security management, the concept of security requirements has replaced risk analysis when assessing appropriate measurements. However, it is not clear how elicited requirements can be prioritized? State of the art methods to prioritize the holistic nature of security requirements are applicable only after major revisions. This dilemma is the starting-point for proposing a qualitative decision matrix approach which is quick and where the results are reproducible and sufficiently accurate. This article describes how the parameters for a prioritization are derived and how the prioritization is carried through.

Key words: decision matrix, holistic security requirement, security requirement prioritization

1.   INTRODUCTION

In recent years the term security requirement has become more and more popular in the security management community. The purpose of a security requirement is to guide the implementation and ongoing administration in security management [ISO 13335-1, 1996]. In earlier years, a security requirement was mainly interpreted as a factor that had to be derived from a risk analysis process – see [ISO 13335-1, 1996], [ISO 17799, 2000]. The risk value then clearly indicated the importance of the requirement. The more severe the risk was, the higher was the incitement to realize the requirement. In that manner a priority order, dependent on the risk value, can be established and the resources can be dedicated to the most important requirements. This is necessary as we assume that only limited resources are available which are insufficient for realizing all security requirements.

However, for to e-commerce applications [Zuccato, 2004] suggests that also the stakeholder and the environment can in addition to impacts of risks on assets also provide valuable inputs to the holistic security requirements. This broadening of a security requirement implies that the conventional mechanism for prioritization is no longer suitable. Therefore we propose the decision matrix approach, which relies on a strategic management method in order to prioritize business activities, called the Boston Consulting Group (BCG) Matrix, and adapt it to the security area. The proposed approach is described later in the article, where also an application example is provided.

Apart from the functional demands we proposed in [Zuccato, 2002a] that an approach that works in an e-commerce environment should also fulfill additional demands. One demand for each decision method should be that the results can be reproduced later[4]. Another demand that is specifically important in e-commerce is short time-to-market cycles – therefore a ranking method must be fast.

To justify the proposal of a new approach we will start by discussing related work on requirement prioritization approaches from the security field as well as the software engineering community. Shortcomings that make those approaches unsuitable for the discussed problem will be pointed out.

2.   REQUIREMENT PRIORITIZATION TODAY

The concept of requirements is a recent trend and currently heavily influenced by the previous approach of risk analysis. [ISO 17799, 2000] mentions for example security requirements, but has risk analysis as the only source. This implies that risk management concepts can be applied for prioritization. [ISO 17799, 2000] and, based on that, [CCTA CRAMM, 1996], argue that the asset value and the savings indicate the risks that should be mitigated. The problem with this assumption is that risks are taken as the only source for security requirements. [Zuccato, 2004] states that security in e-commerce cannot solely rely on risk analysis. Additional input from business and stakeholder have to be included in order to cover a broader picture. Such requirements are then no longer expressed in terms of risks for an asset. Therefore the old prioritization (higher risks first) is inappropriate.

---

[4]One argument in favor of that is liability claim to a court. With a reproducible process it is easier to prove an honest and negligent behavior.

As an alternative to risk concept, sometimes business metric systems are used – see [Gordon et al., 2004]. Prominent examples used in the security field are Return of Security Invest (RoSI) [Wei et al., 2001] or Net Present Value (NPV).

RoSI conducts a cost-benefit analysis almost in the same way that we are going to propose it. However, the fundamental difference is that that RoSI was designed to evaluate the effectiveness of security safeguards. The approach chooses a risk and then evaluates in how far a given safeguard prevents it. RoSI implicitly assumes that all risks (or mainly the most prominent risks) are considered. A similar approach is presented in [Pfleeger and Pfleeger, 2003], where risks are processed in order of their magnitude. In [Zuccato, 2002b] we argue that security (requirements) can "earn" money as a business enabler (i.e. generate a positive cash flow) and it would be wise to consider that in the cost-benefit analysis.

The NPV approach in security anticipates the occurrence of future cash flows when a risk is mitigated by a safeguard. Such cash flows would represent the annual spending and the annual savings for the anticipated risks – it would be possible to replace a risk with a requirement. However, apart from the risk related problem mentioned above, we have another problem with NPV which is that future cash flows and future interest-rates (for discounting) must be known in advance. In a highly volatile area that information security constitutes such a long term prognoses seems to be almost impossible[5].

A third alternative is to rely on the requirement prioritization schemes from the software engineering community. Three of these approaches should be discussed as representatives.

We start with the eXtreme Programming (XP) [Beck, 2000] as a representative for the agile methods. With XP the customer is requested to define a priority for each requirement (called story). When it comes to security this implies a specific problem, namely that many customers do not realize the importance of security [Hitchings, 1995] and therefore rank it very low – as current experiences with security problems indicate. A second problem is that such decisions are hardly reproducible.

The second approach is to ask the stakeholder how (a) satisfied with the availability of a security feature or (b) unhappy with its absence they would be [Robertson and Robertson, 1999]. This approach is better than just simply asking the customer, as it probably mitigates the "dislike -factor" when

---

[5]NPV is also a quantitative method and as [Moses, 1992] argued, quantitative methods imply problems of data generation in the security field.

distributed to various stakeholders. Regarding the reproducibility, however, it is only slightly better.

Finally, we look at the requirement prioritization carried through in the Unified Process (UP) [Jacobson et al., 1999] as a representative for the monumental processes. [Leffingwell and Widrog, 1992] indicates that two different prioritizations are required in the UP. The first one lies on the customer's side, where he/she has to define the features required. The assumption is, in conformity with approaches presented earlier, that the decision maker possesses some kind of oracle that supports the decision making. However, it can be questioned whether this is true for security, as we assume that the decision maker seldom has enough knowledge to conduct such decisions. The second prioritization in the UP is carried through by the software architect, who decides, based on the first prioritization, which requirements should be implemented first and which ones will be postponed to later iterations or versions. It is therefore necessary to assume that they are initially ranked highly enough when considering security requirements, so that they will be implemented also after the second prioritization. It is obvious that this assumption is doubtful as the same decision restrictions as above can be applied.

These problems with each of the above mentioned methods indicate that they are not entirely suitable and could only be applied after major adoptions. We therefore propose a different approach used in strategic management when deciding which products (features) are required on the market which also is suitable for the security field and security requirements.

3.   PORTFOLIO ANALYSIS

In strategic management, one of several important tasks in order to survive in the competitive market and to maximize the profits is to find the optimal product portfolio. As a result of that, the portfolio analysis was proposed in the 1970ies to find out the actual product's position on the market. Based on that information the further steps were planned.

The first approach came from the [[Boston-Consulting-Group, 1972] (BCG) and today, thanks to its simplicity, it is still the most frequently used one, and it will be investigated further on in this article.

3.1 Boston Consulting Matrix

The BCG Matrix is based on two criteria: the reference market's growth rate (acting as an indicator for the attractiveness) and the market share in relation to the firm's largest competitor (measuring competitiveness). A large market growth means that the product is mostly at the begin of its life cycle and has the potential to get large parts of the market although not having it yet. In the matrix – Fig. 1 – these two criteria form the axes. Additionally the matrix is divided into 4 zones, where each of them intuitively represents the products position on the market.

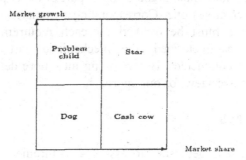

Figure 1. Product portfolio matrix after BCG

After defining the duple for each product, the value pair is going to be drawn in the matrix. Based on the position, different strategies are proposed (see for example [Lambin, 1997]).

Cash cow   (a well situated, profitable product) The priority strategy is to earn money.

Dog   (an old product for divestment) The priority strategy is to divest.

Star   (a young product with market potential) Investment is recommended to make the product a cash cow.

Problem child   (product in start-up phase, which needs placement) Depending on the relative position in the quadrant, two strategies are possible: an investment strategy to make the product a cash cow, or a divestment strategy to make the product a dog.)

[Lambin, 1997] argues that although the initial assumptions may be restrictive – but assumably correct – an accurate and valuable recommendation can be generated. An advantage worth to emphasize is that the matrix is straight forward and intuitive and therefore easy to understand and apply.

4.   DECISION MATRIX FOR SECURITY REQUIREMENTS

In the previous section, the BCG matrix was introduced as a tool when deciding how a portfolio should be developed further. The problem is similar when it comes to security requirements: how do we decide which ones should be developed first and which ones can wait? To conduct this decision we first need to position the requirements in the matrix. Then we can derive a priority list. Additionally, the position in the matrix can suggest a course of action for the treatment.

The positioning mentioned is the difficult part of the approach as it is the non-mechanistic (creative) one. Corresponding parameters to attractiveness and competitiveness must be derived for each requirement. When the requirements are parameterized, the mechanistic part of the priority generation must be conducted. Before going into more detail for each step we will provide an overview for our approach.

4.1   Approach

To begin with, it is necessary to assess a requirement according to its potential, i.e. to generate something similar to the tuple of attractiveness and competitiveness used in the BCG Matrix. Each requirement should be represented as a tuple containing the perceived security benefit and cost-complexity of the realization.

Security benefit   To reflect competitiveness of a requirement we propose to use the perceived security benefit. Security benefit should mean either (a) that the requirement provides high protection of own resources and/or (b) that the requirement will increase the security benefit as it enables business. This is based on the underlying assumption for holistic security requirements, where they not only insure company resources but also enable the selling of the product because of a competitive advantage gained from the satisfaction of security needs from customers – for a more elaborate discussion of these security drivers see [Zuccato, 2002b]. Then we can say that the higher the security benefit of a requirement is, the more competitive it is in respect to other requirements.

Cost/Complexity   We think that attractiveness of the requirement is represented best by its costs of realization and the associated complexity. These factors represent in how far the requirement is likely to fulfill its perceived function. The more it costs and the harder it is to realize, the higher the stake is. However, the cost-complexity measure makes only sense in relation to the intended security level. It is important to mention that a requirement that is easier and cheaper to enforce than a second one with he same benefit should be prioritized, and it most definitely does not mean that the cheap and easy way is always the best solution.

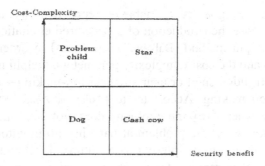

Figure 2. Requirement prioritization matrix

Before applying the matrix concept, the meaning of the quadrants needs to be set into relation to the input values. This is necessary as the complexity and costs are indirect proportional to the benefit. More benefits and limited costs are preferable. Additionally the quadrants must be redefined to reflect the scope of security requirements.

Dog means that not only the complexity and the costs are low, but also are the benefits. The requirement has an indifferent potential.

Problem child means that the complexity and the costs are high and the expected benefit is low. The potential of the requirement is low.

Cash Cow means that the complexity and the costs are low but the expected benefit is high. Such a requirement is very promising to realize as it has high potential.

Star means that both the cost-complexity and the benefits are high. Although such a requirement is interesting its realization is also highly risky. Therefore, as for the dog, the potential is indifferent.

4.2   Input data elicitation

The approach for every security requirement is to elicit the perceived security benefit and the cost-complexity level. Due to several reasons of impracticability of quantitative methods we will use a qualitative approach. Firstly we think that the required empirical data for a quantitative estimation is hard to provide due to the high dynamics in the security field – see [Moses, 1992]. Secondly, we think that most quantitative estimates require an "oracle" – most likely statistical prediction or a simulation – which would not necessarily provide more accuracy than a qualitative estimate (i.e. an expert guess). And thirdly, we think that an empirical method is more prone to violate our quickness requirement for a prioritization method.

However, as the goal is to achieve reproducibility and acceptable accuracy, we propose the conduction of a structured elicitation. We suggest the use of the Delphi method [Dalkey and Helmer, 1963] in order to predict the security benefits and the cost-complexity parameters. Delphi is a method that is used to support judgmental or heuristic decision-making – i.e. creative or informed decision making. According to [Adler and Ziglio, 1995], Delphi is a suitable method when "(a) the problem does not lend itself to precise analytical techniques; (b) the problem at hand has no monitored history nor adequate information on its present and future development [and]; (c) addressing the problem requires the exploration and assessment of numerous issues". We think that all these factors are accurate in concern of our elicitation problem.   Alternative approaches to Delphi could be brainstorming or questionnaires.   However, both alternatives can create problems in the reliability and are eventually subjects to the "dislike" problem mentioned above.

"The Delphi method is based on a structured process for collecting and distilling knowledge from a group of experts by means of a series of questionnaires interspersed with controlled opinion feedback" [Adler and Ziglio, 1995]. In the beginning a questionnaire is sent to selected experts. The filled-in questionnaires are collected and aggregated as a second step. Different ways to derive the aggregates are possible, but here a mean value approach has been used. The mean-value should then be rounded to the next integer to avoid positioning problems in the evaluation. The aggregates constitute feedback to the experts, and in case of to big variation – decided by the method performer – the experts are requested to further state or revise their opinions. This process is conducted until the intended accuracy is achieved. Note that the higher the accuracy demand is, the higher the cost will be – which holds true for all decision methods.

The design of the questionnaire mentioned above is important in order to achieve satisfactory inputs for the result generation – i.e. the requirement prioritization.   To perform the subsequent prioritization process efficiently we need to have sufficient parameter information without adding much complexity to the prediction – which would require additional time. We therefore propose the use of an ordinal scale for the parameter. To derive the scale, according to [Fowler, 1995], one must design the granularity to (a) achieve validity, and (b) make the elements of the distribution distinguishable. This would indicate that the higher the granularity is, the better. However, [Fowler, 1995] says that 5 to 7 categories are probably as many categories as most respondents can use meaningfully. This means that we will aim for a six value scale as our scale must be a multiple of two to correspond with the quadrant structure of the matrix. The quadrants should be made explicit to the respondents by introducing a neutral point in between, as [Fowler, 1995] says that neutral points help to reduce ambiguity. Therefore we introduce a neutral point between 3 and 4 where 1 - 3 represent low and 4 - 6 represent high. [Fowler, 1995] suggests to use numbers for reliability reasons, but to provide adjectives for clarification of the categories' meaning. Based on that we propose the following scale for each parameter:

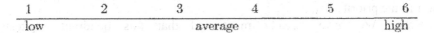

Figure 3. Ordinal scale for the survey

To derive the categories for each requirement, two questions should be asked for each requirement.
- • How much security benefit do you associate with the requirement?
- • How much complexity and cost do you associate with the realization of the requirement?

The result is then represented as a tuple:

Requirement(Benefit, Cost-Complexity)

4.3 The prioritization activity

The aim of the decision matrix is to derive, which requirements should be implemented at first. The position in the matrix suggests the priority of the requirements.

To derive the priority, we suggest two different methods which should be used dependent on the accuracy demand, the quality of the inputs and the application place. For the first method we suggest the use of the quadrants. Based on the result a priority can be derived. The second method will rely on a more formal prioritization that eventually could be automatized.

4.3.1 Informal prioritization

For the start, we assume that the requirements are placed in the matrix. The quadrants can then be used to derive a requirement priority list. This list suggests which requirements should be considered first.

In general we can say that the closer a requirement is to the right lower corner, the more preferable it is. Given the quadrants we therefore suggest the following prioritization:

Requirement list = (Cash Cows, lower Stars, lower Dogs, higher Stars, higher Dogs, Problem child)

Problem child   These requirements are likely to be problematic in the implementation. The expected benefit will not justify that and they will end up low in the priority list.

Cash Cow   These requirements are of great priority as much benefit is expected for the associated costs and complexity. They will all end up high in the priority list.

Star   We have already mentioned that this quadrant suggests indifference. However, we can derive a priority in the way that we imagine a diagonal from the source to the upper right corner. All requirements that are below will have higher priority than the requirements above it. Therefore the "lower Stars" will follow directly after the Cash cows and the "higher Stars", and end up in the middle of the priority list.

Dog   The "Dog requirements" are similar to the stars when it comes to indifference. The same diagonal as mentioned above can be used to derive priorities.   "lower Dogs" will come after the "lower Stars" and "higher Dogs" after the "higher Stars" just before the "problem child" requirements.

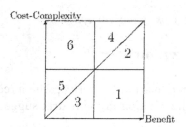

Figure 4. Informal prioritization matrix

Although the informal method can be less accurate we propose it because:

1. it is a good way to visualize the requirement prioritization for the decision maker;
2. in cases where the input variables do not provide high accuracy – because the Delphi method was abandon in favor of a faster or more suitable method in specific situations – the informal method do not introduce a fictive accuracy and;
3. in some situations – e.g. a requirement engineering workshop with stakeholders [Zuccato, 2004] – a visual and less technology dependent method is preferable

4.3.2 Formal prioritization

The formal approach starts by deriving a value for each requirement, which defines the position in the matrix. This value is calculated by dividing the Benefit with the Cost-Complexity – Equ. 1. To position that in the matrix ($a_{i,j}$) we assume that i=*Benefit* and j=*CostComplexity*.

$$a_{i,j} = \frac{Benefit}{CostComplexity} \tag{1}$$

For a 6×6 matrix we can construct generic values as shown in table 1.

Table 1. Priorities for a 6x6 Matrix

Cost \ Benefit	1	2	3	4	5	6
6	0.16	0.33	0.5	0.66	0.83	1
5	0.2	0.4	0.6	0.8	1	1.2
4	0.25	0.5	0.75	1	1.25	1.5
3	0.33	0.66	1	1.33	1.66	2
2	0.5	1	1.5	2	2.5	3
1	1	2	3	4	5	6

For prioritization we construct – as in the informal approach – a preference set. We compare two requirements with each other until we have processed all requirements[6] . This comparison leads to a preference set where *a≻b* means that a is preferred to b, and *a∼b* means that they are indifferent.

When the prioritization value of one requirement ($a_{i,j}$) is different to the other requirement ($a_{k,l}$), we construct a preference order by following the Equ. 2.

$$a_{i,j} > a_{k,l} \Rightarrow a_{i,j} \succ a_{k,l} \tag{2}$$

The prioritization value can be equal under two circumstances. In these cases a preference order should be achieved dependent on the requirement

---

[6]Note that this is a classical sorting problem. Therefore sort algorithms should be used to process all requirements.

parameter. If the parameters are equal, the requirements are indifferent and receive the same priority[7] – see Equ. 3.

$$a_{i,j} = a_{k,l} \wedge i = k \wedge j = l \Rightarrow a_{i,j} \sim a_{k,l} \tag{3}$$

If the parameters are different from each other, we define that more security benefit ($i>k$) is preferable, as our overall goal is to improve the system security – see Equ. 4. However, when having a limited budget this interpretation must not correspond with the truth and could be reconsidered ($j<l$).

$$a_{i,j} = a_{k,l} \wedge i > k \Rightarrow a_{i,j} \succ a_{k,l} \tag{4}$$

5.   E-COMMERCE SCENARIO

We start by looking at some security requirements proposed in [Zuccato, 2004]. Those requirements have an Internet-banking scenario as a background, where the customers access their accounts and make money transfers.
1. Sensitive user data (passwords, keys ...) in a database needs to be stored bi-directionally (not hashed) encrypted due to requirements of the voice recognition system.
2. A demand of internal audit means that audit logs for the intrusion detection system must be stored for three months.
3. An activity log for each transaction should be kept for six months.
4. When saving personal information for statistical purposes, user pseudonyms should be used whenever possible to comply with the data protection legislation.
5. User authentication for accessing bank accounts and services via the internet is necessary.
6. The privacy policy must define customer profiling as one purpose for the activity logs.

We start by preparing the questionnaire for the Delphi method. Then we select some experts representing different areas to cover all aspects of the

---

[7]Note that this is an intended behavior as those requirements then form a priority group, where the requirements are of the same importance and the selection can be conducted based on project planning considerations.

holistic requirements. A few examples could be: a security officer, a product owner, a bank manager, a security implementer...

In this example we assume that we will receive the following parameter values after a number of Delphi iterations.

Requ.	Benefit	CostComp.		Requ.	Benefit	CostComp
1	4	2		4	2	6
2	2	2		5	6	2
3	4	4		6	2	1

Informal method

When we conduct the informal approach we must transfer the requirements to the matrix – see Fig. 5.

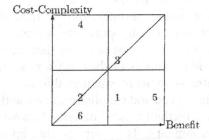

Figure 5. Qualitative security requirement decision matrix

From there we can follow our algorithm and derive a priority list. We get the following priority set: ({5,1},3,{6,2},4). Note that according to this priority list there is no priority between 5 and 1 and 6 and 2.

Formal method

To conduct the formal approach we need to calculate the values for each requirement.

Requ.	Value		Req.	Value
1	2		4	0.33
2	1		5	3
3	1		6	2

When applying the algorithm we will end up with a priority set as follows: (5,1,6,3,2,4). The difference to the informal method is that due to the higher granularity we achieved a greater accuracy in the result. Requirement 6 would have gained less attention in the informal approach than in the formal one as it is in the wrong quadrant.

6.   CONCLUSIONS

It is a difficult task to make requirement prioritization easily understandable and reconstructible. In a market environment, where time-to-market cycles are measured in weeks instead of months, the speed of such a method is of considerable importance. The method presented in this article is supposed to solve these problems by enabling a prioritization based on a matrix approach common in strategic management.

By choosing this matrix approach, large parts of the prioritization work become mechanistic and therefore easy to reproduce. The non-mechanistic part uses an established prediction method to derive parameter values and can therefore be more easily reproduced. Concerning the speed, a final judgment can only be made after extensive testing. However, from a theoretical perspective, properties as simplicity and the mechanistic prioritization imply acceptable speed behavior.

In future research it would be interesting to transform this qualitative method into a quantitative one by providing means to derive the input parameters by calculatory means – as we hope that the progress in security management will provide a sufficient database for statistical prediction. It would be of great interest to learn whether this could enhance accuracy further by not decreasing speed and simplicity significantly.

A predecessor of this method was, as described above, applied once in an Internet banking environment. However, as this approach has changed partially, we plan further application in order to verify the presented ideas in this article.

REFERENCES

[Adler and Ziglio, 1995]   Adler, M. and Ziglio, E. (1995). Gazing into the oracle: the Delphi method and its application to social policy and public health. London - Jessica Kingsley.

[Beck, 2000] Beck, K. (2000). extreme programming explained. Addison Wesley.

[Boston-Consulting-Group, 1972]   Boston-Consulting-Group (1972). Perspectives and Experience. Boston, Mass.

[CCTA CRAMM, 1996]   CCTA CRAMM (1996). CCTA Risk Analysis and Management Method. Central Computer and Telecommunication Agency, United Kingdom, user manual edition.

[Dalkey and Helmer, 1963]   Dalkey, N. and Helmer, O. (1963). An experimental application of the delphi method to the use of experts. Management Science, (9): 458–467.

[Fowler, 1995]   Fowler, F. (1995). Improving Survey Questions, volume 38 of Applied Social Research Methods Series. Sage Publications.

[Gordon et al., 2004]   Gordon, L. A., Loeb, M. P., Lucyshyn, W., and Richardson, R. (2004). 2004 csi/fbi computer crime and security survey. Technical report, CSI/FBI.

[Hitchings, 1995]   Hitchings, J. (1995). Achieving an integrated design: the way forward for information security. In Eloff and von Solms, editors, Information security – the next decade, pages 369 – 383. IFIP, Chapman & Hall.

[ISO 13335-1, 1996]   ISO 13335-1 (1996). ISO/IEC TR 13335-1: 1996 Information technology – Guidelines for the management of IT Security – Part 1: Concepts and models for IT Security. International Standard Organization.

[ISO 17799, 2000]   ISO 17799 (2000). ISO/IEC 17799: 2000, Information technology – Code of practice for information security management. International Standard Organization.

[Jacobson et al., 1999]   Jacobson, I., Booch, G., and Rumbaugh, J. (1999). The Unified Software Development Process. Object technology series. Addison Wesley Longman.

[Lambin, 1997]   Lambin, J.-J. (1997). Strategic marketing management. McGraw-Hill.

[Leffingwell and Widrog, 1992]Leffingwell, D. and Widrog, D. (2000). Managing software requirements: a unified approach. Addison Wesley.

[Moses, 1992]   Moses, R. H. (1992). Risk Analysis and Management. In K.M.Jackson and J.Hruska, editors, Computer Security Reference Book, pages 227 – 263. Butterworth Heinemann.

[Pfleeger and Pfleeger, 2003] Pfleeger, C. and Pfleeger, S. L. (2003). Security in Computing. Addison & Wesley, 3ed edition.

[Robertson and Robertson, 1999]Robertson, S. and Robertson, J. (1999). Mastering the requirements process. Addison-Weseley.

[Wei et al., 2001]   Wei, H., Frinke, D., Carter, O., and Ritter, C. (2001). Cost-benefit analysis for network intrusion detection systems. In 28th Annual Computer Security Conference.

[Zuccato, 2002a]   Zuccato, A. (2002a). A modified mean value approach to assess security risks. In Labuschagne, L. and Eloff, M., editors, 2nd annual conference of Information Security for South Africa – ISSA-2 Proceedings. South African Computer society.

[Zuccato, 2002b]   Zuccato, A. (2002b). Towards a systemic-holistic security management. Licenciate thesis, Karlstad Unviersity Studies.

[Zuccato, 2004] Zuccato, A. (2004). Holistic security requirement engineering for electronic commerce. Computers & Security, 23/1: 63 – 76.

ASSIGNMENT OF SECURITY CLEARANCES IN AN ORGANIZATION

Lech J. Janczewski, Victor Portougal
Department of Management Science and Information Systems, The University of Auckland, Private Bag 92019, Auckland, New Zealand

Abstract: The paper discusses the assignment of security clearances to employees in a security conscious organization. New approaches are suggested for solving two major problems. First, full implementation of the 'need-to-know' principle is provided by the introduction of Data Access Statements (DAS) as part of employee's job description. Second, for the problem of setting up border points between different security clearances, the paper introduces a fuzzy set model. This model helps to solve this problem, effectively connecting it with the cost of security. Finally, a method is presented for calculating security values of objects security clearances for employees when the information objects are connected to each other in a network structure.

Key words: Information security, data security, security models, security clearances, fuzzy sets, Data Access Statement.

1. INTRODUCTION

Managing Information Security depends on business environment, people, information technology, management styles, and time – to list the most important. An analysis of the chain of security arrangements shows a significant weak point. It is the issue of assigning security clearances to an individual. This paper presents an attempt to solve this problem by the optimisation of an information security system subject to cost constraints. As a result, an optimisation procedure that assigns formally the security clearances to all employees of an organisation has been developed. In a typical business environment this procedure is based on the position of a given person within the hierarchy of an organisation. The general principle is that "the higher you are within the company hierarchy the highest security clearance you must have". Such an approach clearly incurs significant problems. In the one extreme a person might have a security clearance that is too high for his/her job, which increases the total cost of the security system. The higher the security clearance, the higher the cost (for instance, of security training). On the opposite side a person with a security clearance too low for his/her job must obtain temporary authority for accessing specific documents. Such a procedure could be costly, time consuming and decrease the efficiency of operations. Portougal & Janczewski (1998) demonstrated the consequences of the described approach in complex hierarchical structures.

A competing and more logical idea is to apply the "need to know" principle. Unfortunately, this principle does not give adequate guidance to the management as to how to set-up security clearances for each member of the staff. Amoroso, (1994) describes the "principle of least privilege". The recommended application is based on subdividing the information system into certain data domains. Data domains in the main contain secret or confidential information. Users have privileges (or rights to access) to perform operations for which they have a legitimate need. "Legitimate need" for a privilege is generally based on a job function (or a role). If a privilege includes access to a domain with confidential data, then the user is assigned a corresponding security clearance. It is easy to see the main flaw of this approach is that a user has access to the whole domain even if he/she might not need a major part of it. Thus the assigned security clearance may be excessive. A similar problem arises regarding the security category of an object. A particular document (domain) could be labelled "confidential" or "top secret" even if it contains a single element of confidential (top secret) information. In this paper we suggest another realisation of the "need to know" principle. Our method is based on the Data Access Statements (DAS), defined for every employee as part of their job description. DAS lists all data elements needed by an employee to perform her/his duties effectively. Thus we shift the assignment of security clearance from the domain level to the element level.

Our approach allows not only the solving of the difficult problem of defining individual security clearances. It also connects this problem to more general problems of the security of the organisation as a whole, to the problem of security cost and cost optimisation.

2. DATA ACCESS STATEMENT

There is a lot of attention in literature to employee specifications and job analysis. It is strange though, that the one of the most important aspects of the job analysis, which is information use, is completely out of specification. We suggest that in addition to the main content of a job description a Data Access Statement (DAS) for every employee is added.

Schuler (1992) defined the following components of a job description:
- Job or Payroll title,
- Job number and job group to which the job belongs,
- Department and/or division where the job is located,
- Name of incumbent and name of job analyst,
- Primary function or summary of the job,
- Description of the major duties and responsibilities of the job,
- Description of the skills, knowledge and abilities,
- Relationship to other jobs.

The job description is the best place to define the security clearance of employee through DAS. It could be, for instance, an additional "bullet point' in the above list.

DAS was introduced earlier by (Portougal & Janczewski, 1998), and was defined as follows:

1. *Data Access Statements* (DAS) of a staff member is a vector, containing *Data Access Statements Elements* (DASE) as its components.
2. Each DASE defines what type of access to information/data is allowed (read, write, delete, etc)
3. Each DASE is defined as a result of the analysis of the job description document related to the given position
4. Each DASE has a confidentiality parameter CP assigned (being an element of the organization's database it should have the same value CP, e.g. 1, if we think they are all of equal value).

An example of DAS statements is presented in Table 1. At the bottom of the column the total value of information accessible is shown. We shall call it SCV – *Security Clearance Value*, thus tying the assignment of a security clearance to the volume of accessible information.

Any production facility has an information system. Table 1 lists all the data elements used within an organization. Every data element has an assigned confidentiality parameter (CP), which characterizes its importance from the point of view of security. For more about assigning CP's refer (Portougal & Janczewski, 1998).

Table 1. Database elements listing and DAS for all employees of the production facility

	CP	A	B	C	D	E	F	G	H	I
Volume of products	1	√	√	√		√				√
Sales	1	√		√		√				√
Labour cost	1	√	√	√						
Material cost	1	√	√	√	√				√	
Cost	1	√	√	√		√				
Labour cost (of N)	1	√	√	√			√	√		
Materials cost (of N)	1	√	√	√	√		√	√	√	
Sales (by products)	1	√		√		√	√	√		√
Volume of products (by product)	1	√	√	√		√	√	√		√
Costs (by products)	1	√	√	√			√	√		
Costs (by materials)	1	√	√	√	√		√	√	√	
TOTAL (SCV)		11	9	11	3	5	6	6	3	4

Positions Codes:

A: General Manager   F: Production Unit N
B: Operations   G: Account N Manager
C: Accountant General   H: Raw material Store Manager
D: Purchasing   I: Finished Goods Store
E: Sales and Marketing

In this example we assume that each data element is independent, so knowledge of a particular element does not allow one to find the value of the other. In order not to overcomplicate the example we assume all CP equal to 1.

3.   MODELLING SECURITY CLEARANCES

The security clearance allows a person to access a certain part of a database. We can assume that the optimum security clearance is assigned strictly in accordance with the "need to know" principle. Unfortunately, the "need to know" principle assigns to every employee a specific area of the database, and generally there will be as many different areas as the number of employees. At the same time, there are always a limited (2-4) number of security clearances. Thus the assigned clearance will practically always be different from optimum, below or above that optimal point.

Clearly, the probability of an information leak goes up, when the difference between the actually assigned clearance and the optimum clearance is increasing. At the same time assigning extra security clearance involves extra cost. Let us analyse the cost of assigning security clearances to particular persons in a more detailed way.

There is a correlation between security of the system, numbers of security measures, and their costs, i.e.

more security measures ⇒ more secure system ⇒ more costs

Many sources, e.g.: (Frank, 1992) indicate the above correlation is not linear but has a tendency to grow exponentially. Similar situations exist in the case of assigning security clearances. The higher security clearance of an employee means a higher expenditure to the employer. The structure of costs would be somehow different from the security measures listed above. The costs like those listed below would be of significance:
- Examination of candidate credentials,
- Security training,
- Security equipment (especially for accessing protected zones, either physical or system),
- Management of the system controlling the security clearances.

Again one might expect that there is a correlation of security clearances with costs:

higher security clearance granted ⇒ higher costs for the organization

The security clearances should be directly related to the jobs and should follow the "need to know" principle. The security clearances are designed to subdivide the employees of the organization into classes according to data access privileges, e.g. secret, confidential and general. Following the usual approach, borderlines should be drawn, defining the minimum amounts and importance of data in use for each category. It was analyzed in the Introduction that, before our development of quantitative measures of confidentiality (CP), this subdivision was performed either by employees' position or by assigning security categories to data domains, and then using these categories for defining clearances. With the CP and SCV defined the problem becomes much easier and more logical to solve.

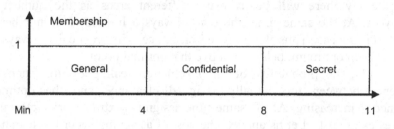

Figure 1. Security categories (crisp representations)

In our example (Figure 1), let us have three security categories: *general*, *confidential* and *secret*. We shall define the borderline SCV between *general* and *confidential* as 4, and the borderline SCV between *confidential* and *secret* as 8. If the total SCV of information in use by an employee is less or equal to four, then this person is not required to follow special security procedures at all, and he/she would be assigned a general clearance. If the total SCV is between five and eight, then the *confidential* clearance should be applied, meaning that this employee is under an obligation to use and follow all the security procedures defined for this clearance. Similarly, if the SCV of data in use is more than eight, then this employee should be assigned the *secret* clearance.

Though this procedure is simple and easy to understand, nevertheless it has two weak points:

1. This procedure implies that the security experts will be able to define the borderlines. In reality it is not so easy, and sometimes the decision about the borderlines is provided by reasons well outside the model, for example by position.

2. Under this procedure it is hard to explain why employees with SCV close to the borderline from different sides have different clearances. What is the crucial reason for an employee with SCV equal to 0.79 has a clearance *confidential*, but his colleague with SCV = 0.81 has *secret* clearance?

Both points indicate an inadequacy in our security clearance modeling. Basically, the inadequacy comes from using a classical *crisp set* for modeling, like this used by (Pfleeger, 1997). The crisp set is defined in such a way as to dichotomise the individuals into two groups: members (those that certainly belong to the set) and not members (those that certainly do not). A sharp distinction should exist between members and non-members of the class. This is definitely not so in our case. The classes of security clearances do not exhibit this characteristic. Instead, the transition from member to non-member of one class appears gradually rather than abruptly. This is the basic concept of fuzzy sets.

In the first fuzzy model we shall assume only two security clearance classes: *general* (set G) with no security cost and *secret* (set S) with a security cost A for each member of the class. The membership functions of class S are given in Figure 2. The vertical lines on Figure 2 represent the employees of the example company and the value of their membership function in the set S. General Manager and Accountant General have the value equal to 11 (A,B), Operations Manager has it equal to 9/11 (C), Purchasing Manager has it equal to 3/11 (I), etc.

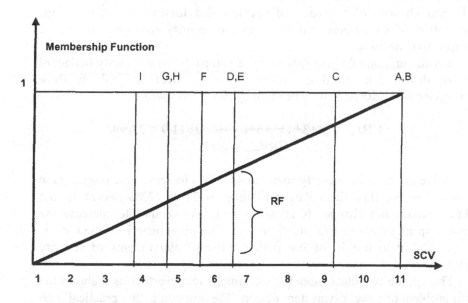

Figure 2. Membership function for the fuzzy set "secret"

If we assign to every manager the security clearance *secret*, then the cost of the security system will be equal to 9A (As there are 9 managerial positions in the company). If this is not affordable, then some of the managers will be put into G class. This involves a risk of information leak.

Let us assume that this risk is proportional to SCV (the more a person knows the higher is the risk). We shall introduce the *risk factor* (RF) for an employee *i* as:

$$RF_i = SCV_i / SCV_{max}$$

A good estimate for the *company risk factor* (CRF) would be either:

$$CRF_{max} = \max_i RF_i, \text{ or}$$
$$CRF_{av} = \Sigma_i RF_i / N,$$

where N is the total number of employees.

CRF_{max} characterises the risk of information leak from the most informed employee. It is better for evaluation than CRF_{av}, when the SCV_i of the employees are diverse. Sometimes both are useful.

The risk factor can not be used directly for the evaluation of real security threats. It is only a coefficient in a more complex equation with unknown chances of a breach of security and losses from it. But the assumption of its proportional value to the security risk gives it a good comparative meaning.

Let in our example postulate that the company has a security budget of 3A, or that it can afford to assign the secret clearance only to three employees: GM, AG and OP. The security risk factors will be:

$$CRF_{av} = (3+3+4+5+6+6+0+0+0)/11/9 = 27/99.$$
$$CRF_{max} = 6/11$$

If we increase the security spend to 4A (33% increase, one more person classified as S), then the CRF_{av} will drop to 21/99 (22% decrease), but CRF_{max} would not change. It is worth to think whether to increase the security spend or not in this situation. Thus, the main benefit of the CRF is the possibility to use it for comparing different assignments of security clearances.

Though the two class model is too simplistic, nevertheless it shows the main problem of a security system design. The problem is that practically no organisation can afford a security system with a zero risk factor, and it is forced to look for a suitable trade-off between the cost and the risk factor

We shall show that introduction of intermediate classes helps in security improvement without cost increases.

Let we introduce an intermediate clearance *confidential* (set C). We shall assume that the security procedures designed for this clearance eliminate the risk of data leakage for all employees with SCV_i no more than 4. Let the cost of these procedures be B = A/3, and the security budget as before is 3A. The possible variant of assigning clearances to employees is shown in Figure 3.

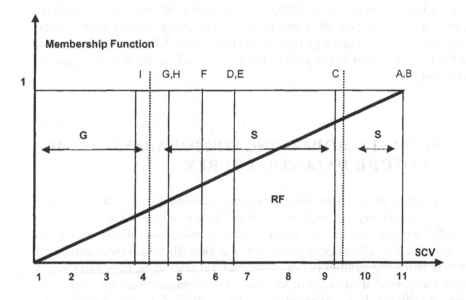

Figure 3. Risk function and its cover by 3 classes G, C and S

In this variant we sacrifice the clearance S for the Operations Manager (OP), changing it to C, which incurs a security risk factor of (9 – 4)/11. It allows the provision, within the limits of our budget, two more employees with higher risk factors, PUN and ANM, with the same clearance C. This will decrease their risk factors by 4/11 each. The security risk factors will be:

$$\mathbf{CRF}_{av} = (3+3+4+5+2+2+5+0+0)/11/9 = 24/99.$$
$$\mathbf{CRF}_{max} = 5/11.$$

This shows a significant improvement in security estimates.

The next step will be to sacrifice the S clearance of either GM or AG and to provide C clearances for him/her and additional two employees with higher risk factors. This will leave only 2 persons in the G class; 6 persons will be in the C class, 2 of them having a non-zero risk factor. The security risk factors for this distribution show the following improvement:

$$CRF_{av} = (3+3+0+1+2+2+5+6+0)/11/9 = 22/99.$$
$$CRF_{max} = 6/11.$$

The first criterion has decreased, but the second shows an increase. We can choose either the previous variant of clearances distribution if we prefer the second criterion, or to go further on if we prefer the first one. Then the final logical step is to use the budget for assigning all employees a uniform clearance C. In our case this does not show further improvement. Generally, an analysis of both company risk factor functions CRF_{av} and CRF_{max} show the best way for their optimisation, but this analysis is outside the scope of this paper.

4. SECURITY MODELLING FOR DATABASES WITH NETWORK DATA STRUCTURES

In any of these cases some important problems were not addressed, which is commonly referred to as the "jigsaw puzzle" intelligence. A watchful analysis of the officially collected materials allows to obtain sometimes highly classified information. For example, close monitoring and analysis of the registration plates of military vehicles at barrack gates could detect movement of military units. The same principle may be applied to business facilities. It is estimated that 80% to 90% of the data collected by intelligence gathering organisations (both civilian and military) originate from entirely legal sources and are obtained without any security breakage.

Reconstruction of classified information (without adequate security clearance) through an analysis of accessible data is based on the fact that the most of data used in business, production, services and military are logically connected. Knowledge of that connection algorithm allows one to reconstruct a relatively accurate model of the reality with the use of only few components.

The main argument presented in this paper is that individual clearances should not be based on the position of an individual within the organizational hierarchy. Rather the individual clearances should be defined by the confidentiality of the documents the employees use in their everyday managerial activity. The methods for defining security clearances depend on the information structure in the organization. For a hierarchical model Portougal & Janczewski, (1998), suggested an algorithm that assigns crisp security categories. Later on they expanded (Portougal & Janczewski, 2000) the algorithm for the case when the security categories are treated as fuzzy intervals.

The models with hierarchical data structures showed some unexpected results. Naturally, the security clearances depended on the form of the data tree. A perfect data tree, having at least two branches on each organizational level and no overlapping of data (i.e. any data unit is known to only one person on a given organizational level) would produce security clearances directly adequate to the position of the person with respect to this level. Otherwise, with real-life trees, the differences were dramatic. The   specific features of the models are as follows.

- The set of key indicators (which need to be protected).
- Data structures of all key indicators. Basically, all performance indicators have a hierarchical structure and thus every key indicator can be modelled as a tree. However, because all indicators are formed from elementary data feedback, the general structure of the data flow in the company will be a network.
- An algorithm that assigns Confidentiality Parameter (CP) to every element works for every DAS sequentially, matching it against data network also sequentially.

An experimental implementation in a company showed the following results.

Before implementation of the model described above, the company had three security categories: "Secret", "Confidential" and "Internal use" defined in Table 2.

Key indicators showing total production volume, sales and costs were considered "secret". The security clearances were related to the position of given employee within the organizational framework.

Table 2. Structure of the security clearances

Security level	Associated security clearance
Secret	General manager
Confidential	Level 2 management
Internal use	Level 3 management

After the introduction of the algorithm calculating the real security clearance values, and the proper definition of security clearances, the structure of security clearances changed (Table 3, last column). Analysis of the information flow in the company shows that in a number of cases the security clearances are not matching the amount and confidentiality of information the employees are having an access to. "Finished good store" manager is perhaps the best example. By the virtue of that person's activities he/she could have a total knowledge of production and sales volumes, while this person security clearance was set-up only on the "internal use" level.

Table 3. List of individual security clearances

Level	Position name	Initial clearance	Calculated clearance
1	General Manager	S	S
2	Operations manager	C	S
3	Production Unit 1 mgr	IU	IU
3	Production Unit 2 mgr	IU	IU
3	Production Unit 3 mgr	IU	IU
2	Accounting mgr	C	S
3	Product 1 accountant	IU	IU
3	Product 2 accountant	IU	IU
3	Product 3 accountant	IU	IU
2	Purchasing mgr	C	C
3	Raw material store mgr	IU	C
2	Marketing mgr	C	C
3	Finished goods store mgr	IU	C

5.   CONCLUSION

The main results of this paper may be summarized as follows:

1. For the full and complete implementation of the 'need-to-know' principle we introduced data access statements (DAS) as part of employee's work description. Thus the access to the information is granted to every employee on the data element level as opposed to the existing practice of granting access on a domain level.

2. We suggest changing the existing practice of assigning security categories to data base domains, and to assign instead a confidentiality parameter (CP) to every element of the data base. The data base will be characterized from the confidentiality point of view in more detail.

3. We showed that current crisp models of assigning security clearances do not include cost and efficiency optimization. Instead we developed optimization models, based on fuzzy sets theory.
4. As a measure of efficiency of the security system we introduced the company risk factor (CRF), which makes possible to compare different ways of security organization under a limited budget.
5. Most of information processed and stored in a database is related to each other. These relationships may allow calculation or estimation of, sometimes quite confidential, information. Knowledge of these relationships therefore might influence significantly content of security labels attached to objects and subjects. We addressed this problem under assumption that the database has a network structure.

Further research in this direction might include the development of optimization models, based on analysis of both company risk factor functions CRF_{av} and CRF_{max} and the structure of the set of feasible solutions. Another direction of research includes the development of models optimizing costs of the security system under risk constraints.

REFERENCES

Amoroso, E., (1994), Fundamentals of Computer Security Technology, Prentice Hall, USA.

Frank, L., (1992), EDP-Security, Elsevier Science Publishers, The Netherlands.

Pfleeger, C., (1997), Security in Computing, Prentice Hall, USA.

Portougal, V., Janczewski, L., (1998), Industrial Information-weight Security Models, Information Management & Computer Security, Vol. 6. No 5, Great Britain.

Portougal, V. & Janczewski, L., (2000), "Need-to-know" principle and fuzzy security clearances modeling, Information Management & Computer Security, Vol. 8. No 5, Great Britain

Schuler R. et all, (1992), Human Resource Management in Australia, Harper Educational, Australia.

TOOL SUPPORTED MANAGEMENT OF INFORMATION SECURITY CULTURE
Application in a Private Bank

Thomas Schlienger and Stephanie Teufel
international institute of management in telecommunications (iimt), University of Fribourg, Switzerland

Abstract: In this paper, we present a management process we have developed for an Information Security Culture. It is based theoretically on action research and practically on expert interviews and group discussions. A Decision Support System, which supports the process, allows quick survey of the existing Information Security Culture in an organization and analysis of the results, thus discovering strong and weak points. This tool recommends, based on stored measures and rules, actions to improve the weak points. It helps security officers to do their work and to improve the Information Security Culture in their organizations. The application of the process and the Decision Support System in a Private Bank is presented here and major findings are discussed.

Key words: Information Security; Information Security Culture; Awareness; Assessment; Decision Support System.

1. INTRODUCTION

The intensified dependence on information processing in recent years has increased the organizational risk of becoming a victim of computer abuse. This risk will continue to rise within the coming years. Existing technical and procedural countermeasures can be enhanced by socio-cultural measures to increase the security awareness and the security knowledge of staff within an organisation, thus improving the security level of the whole organization (Martins, Eloff 2002; Schlienger, Teufel 2002). Potential losses by cyber attacks, computer abuse and industrial espionage can be prevented. Security culture should support all activities in such a way that information security becomes a natural aspect of the daily activities of every employee. It can help to build the necessary trust between the different actors and should become part of the organizational culture, which defines how an employee sees the organization (Ulich 2001: 503). It is a collective phenomenon that grows and changes over time and can, to some extent, be influenced or even designed by the management.

This paper discusses first our management process for analyzing, maintaining and changing Information Security Culture. We then present a Decision Support System that supports this management process. This tool is designed to quickly analyze the existing culture and to automatically propose measures to improve weaknesses. It also allows comparison of the Information Security Cultures between different organizations (benchmarking) or that of a Culture within the same organization over different points in time. In this instance, the management process and tool were applied in a project at a Private Bank. We discuss the settings and findings of this project and the lessons learned.

2.   MANAGEMENT OF INFORMATION SECURITY CULTURE

Information Security Culture, like organizational culture, cannot be created once and then used indefinitely without further action or modification. To ensure that it corresponds with the targets of an organization, culture must be maintained or modified continuously. It is a never ending process, a cycle of analysis and change. The first step is to analyze the actual Information Security Culture (diagnosis). If the culture does not fit with the organization's targets, the culture must be changed. If it fits, it should be reinforced. The necessary actions must be chosen (planning) and realized (implementation). The success of the actions taken must then be checked and learning specified (evaluation). The process is illustrated in Figure 1.

2.1   Process Description

In the following section, we give a short overview of these four management steps.

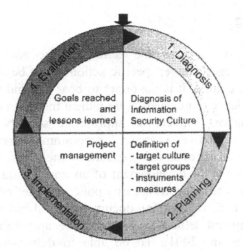

Figure 1. Information Security Culture management process

2.1.1   Diagnosis

In order for security culture to make a substantial contribution to the field of information security, it is necessary to have a set of methods for its study. Bearing in mind the difficulties in comprehending culture at all, the use of a combination of measurement tools and methods as proposed among others by (Rühli 1991; Schreyögg 1999) seems evident. This allows verification of the results with other methods and the use of different viewpoints in interpreting them. The researcher is thus able to pick the appropriate methods, which help him/her assess the security culture in his/her organization. In our research we use:

- Analysis of security specific documents, e.g. security policy
- Questionnaires with employees
- Interviews or questionnaires with security officers
- Observation, e.g. clean desk policy verification

A more detailed discussion of the evaluation items and methods can be found in (Schlienger, Teufel 2003). In this paper, we concentrate only on questionnaires, as they are the instruments best suited for a tool supported assessment. We have developed a standardized questionnaire on the basis of the organizational behavior model of (Robbins 2001), see also (Martins, Eloff 2002). This divides organizational behavior into three layers: organization, group and individual, with in all twenty areas (e.g. work and technology design, communication, attitude etc.). The questionnaire has 42 questions, which are answered on a five point Likert scale from 1 (I strongly agree) to 5 (I strongly disagree).

2.1.2   Planning

The diagnosis step reveals the actual culture and its weaknesses. Depending on the target culture, specific actions must be taken to maintain or even change the culture. It is important to bear in mind that changing an existing, inappropriate culture needs more radical measures than maintaining an appropriate culture. Whereas an appropriate security culture can be maintained by an effective awareness programme, changing a culture involves the reengineering of all existing cultural measures.

Clear objectives for the development of an appropriate security culture must be set. We propose using the security policy as a definition of the target security culture. It is an overarching document for all measures concerning information security and defines the basics for security behaviour, see also (von Solms, von Solms 2004). To be able to define the right cultural measures, it is also essential to know which people one wishes to influence. A widely used approach is to define three groups: IT-staff, managers and lower-level employees/support staff, and to implement special measures for each group. In our research, segmentation by function (IT vs. business) or hierarchical position (managers vs. lower-level employees/support staff) revealed statistically significant differences that suggest the need to define special cultural measures for specific departments or management levels.

Comparing the actual with the target security culture, one can choose the right instruments to implement the target culture. Culture cannot be decreed by regulations; more subtle actions are possible and necessary. A number of possible instruments exist to influence Information Security Culture, the most important ones are: responsibilities, internal communication (awareness campaigns), training, education and exemplary action of managers.

2.1.3   Implementation

The planned actions must now be implemented. This phase can be organized as for every other project: it is essential to define detailed activities, responsibilities and resources, the schedule and the budget. We will not go into details concerning this phase.

2.1.4   Evaluation

Evaluation is the last step in our Information Security Management process. It provides valuable information about the efficiency and effectiveness of the actions implemented. It helps to improve the actions taken, to define necessary follow-up and also to legitimate investment in

Information Security Culture. This is especially important in applying for the following year's budget.

To highlight the changes achieved in a culture, the same instruments, in our case the same questionnaire, should be used. This questionnaire can be complemented by specific questions on the actions taken to reveal its effectiveness. Evaluation also reinforces organizational learning (Argyris, Schon 1978):

1. single loop learning ("adaptation"): the actions taken are evaluated to be improved in the future, e.g. the educational programme can be improved, knowing the strengths and weaknesses.
2. double loop learning ("change"): the evaluation also has an impact on the Information Security Culture itself. Undertaking an evaluation affirms the importance of information security. Employees pay attention to this topic once again.
3. deutero learning ("learn how to learn"): evaluation also helps to improve the evaluation process itself. Experiences from carrying out an evaluation will change and improve further evaluations of Information Security Culture.

2.2   Scientific and Practical Foundation

The proposed management cycle has its roots in a scientific research method and in practical exchange of ideas and experience.

The scientific root lies in action research. It is an established research method, used in social sciences since the mid-twentieth century, and it gained much interest in information systems research toward the end of the 1990s (Baskerville 1999; Björck 2001; O'Brien 2001). Action researchers assume that complex social systems, like an organization and its information systems, cannot be reduced to components for meaningful study. They can be best studied by introducing changes in social processes and then observing the effects of these changes. This involves five steps: diagnosis, action planning, action taking, evaluating and specifying learning. In our management process we use the same steps, but have integrated the steps evaluation and learning, since learning normally accompanies all steps but is most important in the evaluation.

The process has also been checked concerning practicability during discussions within the Working Group "Information Security Culture" of the FGSec (information security society Switzerland). The group consists of nine researchers, security officers and security consultants with experience in socio-cultural measures in information security. The process has been proved practical in this expert round and is now the recommended procedure of managing Information Security Culture.

3.   A DECISION SUPPORT SYSTEM FOR THE MANAGEMENT OF INFORMATION SECURITY CULTURE

The complexity and the interdependence of information systems and of information security management are steadily growing. Providing tool support to security officers helps them to cope with complex decision making under time pressure. Computer based tools impart knowledge, which can provide the necessary foundation for decisions. Information systems that help to analyze the existing culture and to propose possible actions for improving weaknesses can be a major asset for the information security officer. The problem field of Information Security Culture management is either not structured, or, at the least, badly structured, and therefore not suited to automated decision taking. It is therefore not possible to build complete decision trees with all actions and consequences. Although a tool for Information Security Culture management is therefore not a Decision Support System in its narrow sense, it is one in a broad sense.

Decision Support Systems are not decision automatons, but they can help the user to prepare for decision making by surveying, filtering, completing and aggregating information. Decision Support Systems help to (Hättenschwiler, Gachet 2003):
- make decisions faster,
- improve the quality of decisions,
- reach the goals with fewer resources and
- make more rationale, robust and replicable decisions.

The tool supports in its first stage, see also (Krieger 2004), the management of Information Security Culture in the steps of diagnosis, planning and evaluation. The architecture is illustrated in Figure 2. It surveys the Information Security Culture with two questionnaires, one for all staff (survey component A) and one for the security officer (survey component B). It automatically analyzes statistically the survey results, discovers weaknesses in the culture and proposes actions to improve weak points (reporting component). Thus the security officer quickly obtains status information and knowledge about the Information Security Culture of his/her organization. He/she can then choose actions from the proposition list and implement them. The survey component can also be used to carry out the evaluation. An administration component allows administrators and researchers to manage surveys, questionnaires, best practices and users.

In a second stage further functions are planned. It is planned to support benchmarking survey results relating to one company with those of other companies and to improve the planning stage.

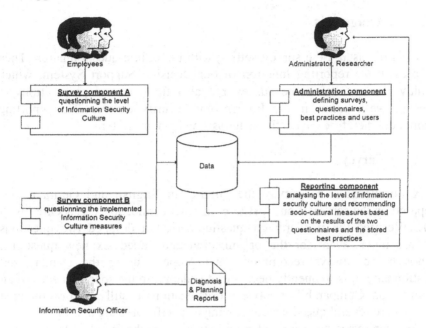

Figure 2. Architecture of the Information Security Culture Decision Support System

The tool has been developed on the web technology html and ASP.NET; results are stored in the free runtime version of Microsoft SQL server, and the analysis and reporting is undertaken with Crystal Reports. The web based client/server technology allows easy distribution of the application.

4. DEMONSTRATION AND CASE STUDY: APPLICATION TO A PRIVATE BANK

The tool was applied in a Private Bank in November 2004. Whereas the bank has employees worldwide, most of the staff works in Switzerland. The company has already carried out an information security awareness programme this year and is now starting to analyze its Information Security Culture in a systematic way. The project was supported by the top-level management and was signed by the CEO, CFO, COO, Head of Enterprise Risk and Head of Information Security. The project was thus backed by the highest hierarchical levels.

4.1   Diagnosis

We first discuss the survey setting with the online questionnaires. Then we present the reporting function of our Decision Support System, which displays the results of the survey but also findings and propositions to improve weak points in the Information Security Culture. The reporting component therefore covers the diagnosis and planning steps.

4.1.1   Survey

A questionnaire to survey the culture, in English and German, was prepared on the server of our Institute. Internet connection was secured with SSL. We used the standardized questionnaire, but dropped two questions that are not relevant for the organization and added six new questions. Although we always recommend and propose using the standardized questionnaire, it is frequently necessary to adapt it to the specific needs of an organization. Comparability between organizations is still given on the area level, where several questions concerning a specific area are aggregated.

The employees were invited by an email from the Head of Information Security to fill out the questionnaire. They also received the URL to the questionnaire with an anonymous company login and password. On the questionnaire, and prior to answering the questions, each employee first has to authenticate him-/herself and also to indicate his/her position (3 levels), his/her function (7 functions) and his/her region (4 regions). The questionnaire then consists of 46 mandatory questions and a section for optional comments. Cookies are set to anticipate multiple answers from the same account.

The Head of Information Security answered the security officer's questionnaire, which surveys the measures already taken to create and support an appropriate Information Security Culture. The database currently stores 87 answers from other security officers of Swiss organizations. Comparing an organisation's results with those of other Swiss organizations gives valuable information about the maturity of the Information Security Culture from the security officer's viewpoint.

Approximately 19% of all staff in Switzerland and Liechtenstein responded to the survey. The confidence interval of 7.36 at a confidence level of 95% provides enough accuracy for a statistical analysis of this group. However the feedback of only 0% to 5.5% from the other three regions gives us not enough data for a statistical analysis of these branches. One main problem of the two branches in Asia was the poor Internet access. In spite of that, the general problem of all three branches apparently is the lack of interest in information security.

4.1.2 Reporting

The reporting section is designed for the security officer and the senior management. It shows the answers on different aggregation layers:

- Overview: all questions aggregated, to give an overall picture (see Figure 3).
- Level: the questions concerning Organization, Group and Individual are aggregated to give level information.
- Area: the questions concerning an area are aggregated to give a more detailed picture of the areas. This analysis is called the Information Security Culture Radar and gives a wide range of information at a glance. It is the favourite aggregation level for benchmarking (see Figure 4).
- Single question: the results of a single question give the most detailed information.

The results can be filtered according to position, department and region to receive more details and to be able to define specific actions for target group. The report can be exported to different formats (PDF, Word and Html). Figure 3 shows the navigation of the reporting component and the entry screen with the overview. On the left side is a navigation tree, where the user can jump directly to the different levels, areas or questions. On the top are the filters and also the export function. The second top line offers

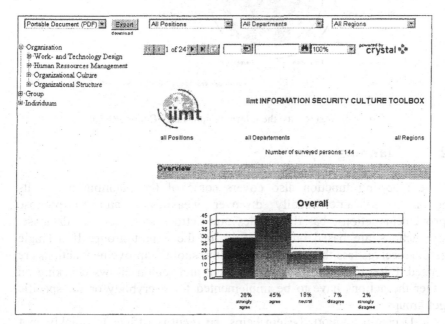

Figure 3. Overall results and Navigation

possibilities for searching and navigating through the pages.

The overall result (Figure 3) of 73% agreement is good. In discussion with the working group, we set the threshold for a satisfactory Information Security Culture at 60% agreement for each question. This number can be adjusted by the organisation for each question if wanted. In our survey the CSO agreed to the recommended threshold. Although the overall result is good, the analysis of areas and individual questions reveals improvement points.

The Information Security Culture Radar (Figure 4) shows the results on the area aggregation layer. It shows at once where the strengths and weaknesses are. Weaknesses are on the areas Human Resources Management, Organisational Culture and Problem Management. Actions should focus on improving the worst areas and maintaining the good areas at the same level.

Information Security Culture Radar

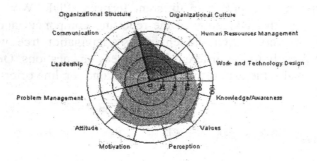

Figure 4. Area results: the Information Security Culture Radar

4.2   Planning

The reporting function also covers some of the planning phase. Its function is to automatically discover weaknesses and to propose improvement actions, based on rules and actions stored in the database. These actions are based on the results of the expert group. If a single question receives less than the agreement threshold, improvement actions are proposed. Filtering on position, department and region allows checking on whether the actions have to be implemented for everybody or for specific target groups.

The Decision Support System helps the security officer to quickly spot weaknesses and to retrieve possible measures. Depending on the specific situation and specific needs, he or she can then choose preferred actions and implement them in his or her organization.

Figure 5 shows the result of the question "I receive training (courses, presentations, self-study etc.) in security applications and procedures I need for my work." The threshold is not reached, so the system reveals a problem and proposes improvement actions. In this case, employees do not receive the necessary information security training. The proposed measures focus on general information security education and specialised training in security procedures and tools.

4.3   Future steps: Implementation and Evaluation

The bank is going to implement the most promising improvement measures during 2005. It is also planned to evaluate the actions taken in an evaluation survey at the end of 2005 or beginning of 2006. The evaluation step is necessary to a systematic management of Information Security Culture. It gives valuable information about the effectiveness and efficiency of the implemented measures and supports organizational learning.

We expect valuable improvement of information security in this bank and hope that systematically managing Information Security Culture will become a part of its organizational culture.

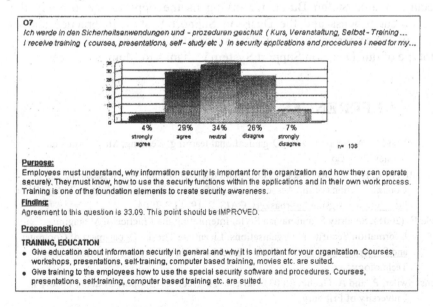

Figure 5. Question results with problem description and improvement propositions

5. CONCLUSIONS

In our survey on Information Security Culture in Swiss Organizations (Schlienger, Rues Rizza 2004) we discovered that most of the organizations rate the socio-cultural dimension of information security as very important. However, they encounter problems in proving its value in terms of improvement in information security and return on investment. The proposed method and tool helps to bridge this gap by allowing organizations to systematically analyze their information security culture, to quickly identify weaknesses and improvement actions and to prove progress in Information Security Culture. The application in a real world project shows its usefulness and the experience shows that we have reached our goals. In future development we will extend the functions of the Decision Support System to provide benchmarking and better support in the planning phase.

ACKNOWLEDGMENTS

We thank all members of the Working Group "Informationssicherheits-kultur" (Information Security Culture) of the FGSec (information security society Switzerland) for their valuable discussions and cooperation. We especially thank Stefan Burau for giving us the opportunity to apply the management process and the Decision Support System in practice and to gain valuable experience with it. The successful implementation of the first prototype of our Decision Support System is thanks to Manuel Krieger.

REFERENCES

Argyris, C. and D. A. Schon (1978). Organizational learning. Reading, Mass., Addison-Wesley Pub. Co.

Baskerville, R. L. (1999). Investigating Information Systems with Action research, Communications of the Association for Information Systems. Volume 2, Article 19. http://www.cis.gsu.edu/~rbaskerv/CAIS_2_19, 11.3.2004.

Björck, F. (2001). Security Scandinavian Style: Interpreting the Practice of Managing Information Security in Organisations. Licentiate Thesis. Department of Computer and Systems Sciences. Stockholm, Stockholm University & Royal Institute of Technology.

Hättenschwiler, P. and A. Gachet (2003). Skriptum in Decision Support Systems Theory I. University of Fribourg.

Krieger, M. (2004). Ein Decision Support System für das Management der Informationssicherheitskultur. Masterarbeit. Lehrstuhl für Management der Informations- und Kommunikationstechnologie, Universität Freiburg i.Ue.

Martins, A. and J. H. P. Eloff (2002). Information Security Culture. In: M. A. Ghonaimy, M. T. El-Hadidi and H. K. Aslan, Eds. Security in the information society: visions and perspectives. IFIP TC11 International Conference on Information Security (Sec2002), Cairo, Egypt, Kluwer Academic Publishers: 203-214.

O'Brien, R. (2001). Um exame da abordagem metodológica da pesquisa ação [An Overview of the Methodological Approach of Action Research]. In: R. Richardson, Ed. Teoria e Prática da Pesquisa Ação [Theory and Practice of Action Research]. João Pessoa, Brazil, Universidade Federal da Paraíba. English version: http://www.web.ca/~robrien/papers/arfinal.html (11.3.2004).

Robbins, S. P. (2001). Organizational Behavior. New Jersey, Prentice Hall.

Rühli, E. (1991). Unternehmungskultur - Konzepte und Methoden. In: E. Rühli and A. Keller, Eds. Kulturmanagement in schweizerischen Industrieunternehmungen. Bern und Stuttgart, Paul Haupt Verlag: 11-49.

Schlienger, T. and R. Rues Rizza (2004). Befragung zur Informationssicherheitskultur in CH Organisationen, Arbeitsgruppe "Informationssicherheitskultur" der FGSec (information security society switzerland). www.fgsec.ch/ag/isk/Marktbefragung2p.pdf, 9.11.2004

Schlienger, T. and S. Teufel (2002). Information Security Culture - The Socio-Cultural Dimension in Information Security Management. In: M. A. Ghonaimy, M. T. El-Hadidi and H. K. Aslan, Eds. Security in the information society: visions and perspectives. IFIP TC11 International Conference on Information Security (Sec2002), Cairo, Egypt, Kluwer Academic Publishers: 191-201.

Schlienger, T. and S. Teufel (2003). Analyzing Information Security Culture: Increasing Trust by an Appropriate Information Security Culture. Proceedings of the International Workshop on Trust and Privacy in Digital Business (TrustBus'03) in conjunction with 14th International Conference on Database and Expert Systems Applications (DEXA 2003), September 1-5 2003, Prague, Czech Republic, IEEE Computer Society.

Schreyögg, G. (1999). Organisation: Grundlagen moderner Organisationsgestaltung. Wiesbaden, Gabler Verlag.

Ulich, E. (2001). Arbeitspsychologie. Zürich, vdf, Hochschulverlag an der ETH Zürich.

von Solms, R. and B. von Solms (2004). "From policies to culture." Computers & Security 23(2004): 275-279.

ERPSEC - A REFERENCE FRAMEWORK TO ENHANCE SECURITY IN ERP SYSTEMS

Prof. S.H. von Solms, M.P. Hertenberger
Rand Afrikaans University

Abstract: This paper proposes a method of integrating the concept of information ownership in an Enterprise Resource Planning (ERP) system for enhanced security. In addition to providing enhanced security, the reference framework ERPSEC developed for this study provides better manageability and eases implementation of security within ERP software packages. The results of this study indicate that central administration, control and management of security within the ERP systems under investigation for this study weaken security. It was concluded that central administration of security should be replaced by a model that distributes the responsibility for security to so-called information owners. Such individuals hold the responsibility for processes and profitability within an organization. Thus, they are best suited to decide who has access to their data and how their data may be used. Information ownership, coupled with tight controls can significantly enhance information security within an ERP system.

Key words: Database security, security policy, misuse detection, authentication, information flow

1.   INTRODUCTION

The concept of information ownership has been around for some time. However, its full benefit has never been harnessed in the ERP software space[5]. In ERP software systems, security is critically important. ERP systems are fairly complex and integrate functions and data across an entire enterprise. The fact that human resources data and financial information is integrated with production planning and sales data should illustrate the requirement for stringent security subsystems. Additionally, ERP systems in use by an organization contain critical business data. Hence, it is essential that such information be protected from unauthorized access. Unauthorized access to the data within the ERP system's database must be prevented, especially since a large percentage of fraud takes place within the organization[4].

In the study completed by the authors[1], various ERP software products where evaluated to determine how security is implemented. In all cases, the security subsystem forces centralized control by one or more central security administrators. It is the view of the authors that this approach, though practical and widely used, weakens security. In the study, a framework for implementing the information ownership approach to strengthen and enhance security within ERP software packages is proposed. This paper briefly summarizes some of the findings.

2.   THE TRADITIONAL APPROACH AND ITS PROBLEMS

To provide the reader with sufficient information on the traditional way of implementing security within an ERP environment, the following brief discussion is provided.

ERP implementation projects require many skilled resources from various disciplines. To ensure adequate knowledge transfer, staff members from the organization for which the ERP system is being configured are included in the project team. The technical skills required to implement and configure the software are quite different to the business and process knowledge that is required to change the workings of the software components to support the business processes and add value to the organization. Technical skills are generally required to assist in the implementation and realization of a security policy. Briefly stated, the reason for this is due the fact that:
−   security is generally considered an administrative, and therefore a
    technical role
−   the implementation of an adequate security infrastructure requires
    specific knowledge relating to the ERP system's technical architecture. In
    all the ERP systems reviewed, detailed knowledge of system objects and
    their use and function is a prerequisite.

– the ERP systems available provide only a centralized way of implementing and maintaining security objects and settings

2.1   Specialization by discipline and resources

Hence, the traditional approach to the implementation of security within ERP systems is based on a centralized approach. There is nothing physically wrong with this approach. However, the centralized approach provided by ERP software packages does not allow the organization to expand its security infrastructure to comply with information ownership principles. To illustrate this in a different way, consider that ERP software systems contain a huge variety of functions and configuration possibilities. To understand all facets of a single system in detail is virtually impossible. Hence, specialization of skills takes place almost naturally. Business-oriented users are more concerned with the real-world application of the ERP software and how the configuration can be changed to mirror the processes within the organization. In contrast, technical experts and administrators delve deeply into the architecture and structure of the system; they are more concerned with how the system has been built. The knowledge divide becomes apparent when a business process owner requests the configuration of a security object from a technical security administrator. As the focus of both parties is different, understanding from both sides may be lacking.

2.2   Translation of business requirements into technical terms

The requirements of the business process owner for increased security in order to protect the organization from fraud, for example, must be translated into a technical specification by the security administrator. Though this process may be fairly simple in some cases, more complex requirements may not be easy to implement technically. An example of a simple security requirement may be the restriction of permitting only certain users print to a certain printer in the organization. The requirement can be fairly easily understood and translated into the technical format required by the system. Similarly, testing such an access restriction is fairly simple and does not provide too many possibilities for failing. A far more complex requirement may include access restrictions to data for certain material types, cost centers and locations. In a large organization many material types and locations may be present. Ensuring that all users have been allocated the correct security objects becomes far less trivial to implement and configure than the preceding example involving only a single printer.

2.3   Possible introduction of errors and hence weakened security

To restate the above concept, the assumption that a central system or security administrator has the ability to understand all nuances and specifics of each functional area is often incorrect. Instead, the security administrator must gather information from each area of the business. Once all these details have been gathered, the security administrator is able to translate the requirements of each business area into the appropriate roles and profiles within the ERP system. In many cases, the security administrator has to select objects manually to create the appropriate access authorization for the user. It should be clear that such a process is often completed with a number of errors and omissions.

2.4   The security administrators as a bottleneck

The security administrator in an ERP environment must contend with numerous business areas and functional areas. These include sales, finance, human resources and so on. Adding the various organizational layers on top of this, together with various stakeholders the business may have to support, creates an environment in which the centralized security administrator becomes a central bottleneck. Figure 1 illustrates this more vividly:

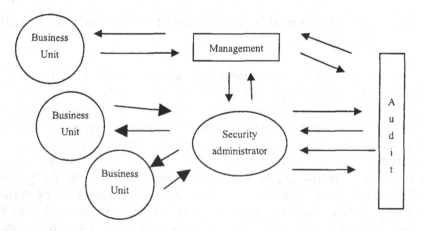

Figure 1. Centralized security within an ERP environment

Figure 1 indicates one of the main problems with the implementation of security within an ERP environment, namely that of creating a bottleneck by having to route all requests for change through one or more administrators who possess the technical knowledge on how to deal with the request.

3. THE APPROACH USING INFORMATION OWNERSHIP

The authors are of the opinion that the support of information ownership can assist in enhancing the configuration of security objects in ERP software environments[8]. In addition, the approach using information ownership provides various additional benefits that are useful to organizations implementing ERP software packages to support their business processes. To this end, the ERPSEC framework has been developed and will be briefly discussed during the remainder of this paper. Prior to the discussion relating to the ERPSEC framework, it should be clear why the authors consider the approach using information ownership to be beneficial.

3.1 Reduction of complexity

Within traditional ERP environments, the centralized approach to implementing access control and access restrictions enables one or more security administrators to create and maintain profiles, roles and user master records. As has been mentioned above, this approach suffers from a number of problems, most notably that the security administrator cannot and usually does not understand the complexities of the actual business processes within the organization and how these have been mapped to the functionality of the selected ERP software package. To combat this problem and to promote more rigid and adequate security within an ERP environment, it is necessary to deal with complexity within the system as a whole. The provision of an integration layer to reduce complexity is a definite requirement. Such an integration and simplification layer is not available in any of the currently available ERP software packages. The ERPSEC framework proposes an integration layer to simplify the creation of ERP security objects. Figure 2 illustrates the integration layer and should be compared to Figure 1 above.

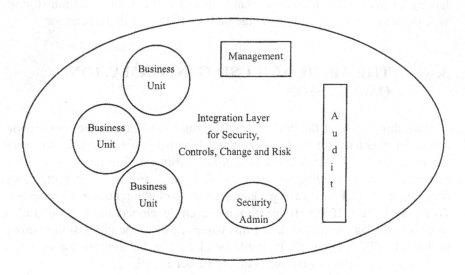

Figure 2. Providing an integration layer to support information ownership

As an aside, the addition of a suitable integration layer that has the ability to translate and present technical security objects to non-technical users makes the decentralization of security configuration a distinct possibility. This may not always be recommended, but goes a long way in supporting segregation of duties issues and need-to-know principles in certain organizations. Further comment regarding this will be made once the ERPSEC framework has been discussed in more detail.

3.2   Simplification

The sudden popularity of ERP software packages stemmed primarily from their ability to integrate all data within the organization, to deliver real-time results and reporting and to make specific functionality available to the user at the desktop level. In stark contrast to the adaptability and flexibility of being able to configure the ERP software package to the needs of the business, the configuration of security related objects is completed by technical staff. Mention has already been made of the complexity of ERP software packages. This complexity is necessary for the software package to be adaptable to different industries and legal requirements. The ERP software packages investigated during the course of this study provide full support for the configuration of the software to adapt it to support various business processes. Similar functionality for the configuration of security objects is missing. In fact, all security-related objects are generally grouped together and are not easily distinguished from one another. The ability to document the necessary access restrictions and security objects is hampered by very technical naming conventions. In ERP systems targeted at organizations with a smaller user population, the configuration of the security subsystem is often fairly trivial, offering the security administrator very little flexibility. The ERPSEC framework attempts to simplify the creation and maintenance of security related objects within an ERP environment.

4. WHY INFORMATION OWNERSHIP?

To enhance the concept of information ownership, the concept of an information owner must first be explained. The concept of permitting individuals within the business to manage and maintain their own information security is termed information ownership. Information owners are individuals in charge of a certain business area within the organization. Generally, these individuals are already in charge of a division, such as finance or sales, for example. In other words, these individuals are stakeholders within the business and carry some form of responsibility. It is the goal of the information ownership approach within the ERPSEC framework to provide the tools to individuals who are held liable or responsible for certain actions taking place within the business. If these individuals are not provided with the tools to support their decision-making process and the ability to ensure that their data is safe, they cannot be held responsible for anything that occurs within their sphere of responsibility. This concept aligns closely with that of segregation of duties. One definition[2] states that segregation of duties is a method of working whereby tasks are apportioned between different members of staff in order to reduce the scope for error and fraud.

Prior to the ERPSEC framework being discussed, some advantages of the information ownership approach are listed here:
− Technical security administrators are experts at maintaining technical security objects, but often lack the necessary knowledge relating to the impact these objects have when allocated to the wrong user. Information owners are aware of their business area and know the impact of incorrect allocation of one or more security objects
− Technical security administrators are generally not aware of the staff members in various organizational units. Therefore, the creation of roles and allocation of security objects is done based purely on feedback and information received from the relevant organizational unit or division. In

contrast, information owners are focused on their business area, know
their staff and can make informed decisions based on their capabilities
and possible weaknesses
- The ability to compartmentalize security objects based on their
applicability to various sections of the business can radically reduce the
time and effort required to implement and configure the ERP security
subsystem. As each information owner can take care of his own section
of the business, this permits the security administrators to take on a role
that examines security in more detail across the enterprise. The
integration layer provided by the ERPSEC framework supports this
compartmentalization.
- Information ownership goes a long way to promote and control
segregation of duties issues and improve corporate governance. Due to a
large number of legislative requirements, this is a very important topic
for organizations at present. At present, very little support is provided by
existing ERP software products to assist organizations in dealing with
these complexities.

5.   THE ERPSEC REFERENCE FRAMEWORK

The ERPSEC reference framework has one primary aim, namely to
enhance and increased the security and access control within an ERP system.
Existing ERP system already contain a centralized security subsystem.
Hence, retrofitting the ERPSEC framework to an existing product may not
be an easy task. The definition of ERPSEC in the study provides an object-
oriented definition that should ease a possible physical implementation
sometime in the future.

In addition to enhancing security within an ERP system, ERPSEC will
attempt to cater for the following:
- a reduction in complexity of the security configuration;
- the ability to increase responsibility and accountability within the
organization;
- a faster implementation time by providing decentralized access to
security objects;
- to improve the quality of the security configuration as a whole;

These goals can be realized by considering the current state of security
subsystems in existing ERP software packages. A brief review is provided
below.

5.1   Traditional ERP security subsystems

The discussion presented in this paper provides the most basic details regarding the ERPSEC framework. The complete study by [1] contains a detailed description including object and table definitions for the creation of the framework in "real-life".

A mention of the centralized nature of the security subsystems of existing ERP software products has already been made. In the model employed by these products, a single administrator modifies and maintains the security objects for all users, regardless of their place within the organization.

This is depicted in Figure 3.

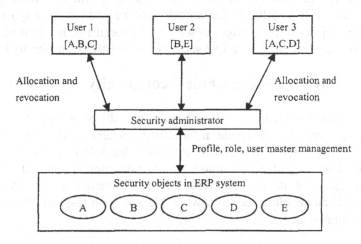

Figure 3. Centralized security within an ERP environment

A simple solution to include the concept of information ownership for purposes of the ERPSEC framework is to define individual information owners. From the preceding discussion of information ownership it should be clear that information ownership implies a form of decentralization. The decentralization is such that individual stakeholders become responsible for groups of users within the organization.

5.2   Information owners

The information owner is an important part in the ERPSEC framework. Not only does the information owner know what access is required in order for the department or organizational unit to function, but also has an in-depth knowledge of the jobs and tasks performed within that department. Hence, the translation of a particular task to its security and access restriction requirements within the ERP system is far simpler to determine. A further advantage is that the requirements do not have to be communicated to a third party, such as the central security administrator.

The information owner thus plays the role of a decentralized security administrator, albeit only for the area of the business the information owner belongs to. For this to be possible, the ERPSEC framework must cater for some additional requirements. From the preceding discussion, it was made clear that the translation of access restrictions and security requirements was a major factor that inhibited and decreased security within a traditional ERP system. The requirement of an information owner within the ERPSEC framework cancels this complication, but does not fully solve all problems. To be useful, the ERPSEC framework must provide some way of allowing security objects to be configured without the need for the detailed technical understanding of the system that security administrators generally have.

5.3   Dealing with technical complexity

The previous section has dealt briefly with the requirement the ERPSEC framework has for dedicated information owners in the organization. In order to permit these information owners to be able to create and maintain their own security objects, a high level of abstraction is required. Abstraction of the technical details regarding the configuration and maintenance of security is an absolute necessity when placing such responsibilities with the information owners.

Abstraction can be achieved in the proposed ERPSEC framework. Instead of relying on a central security administrator who must know all technical details to create and maintain security objects, ERPSEC introduces an additional layer in the security subsystem that allows security objects to be configured and maintained in a very simple yet powerful fashion. It should be clear that retrofitting existing ERP software packages with such an additional software layer may not be practical. However, future versions of current ERP software packages could easily incorporate such an abstraction layer to promote and support the concept of information ownership. Figure 4 below depicts the additional abstraction layer. The abstraction layer can also be considered a simple interface layer. The layer has the responsibility of translating the input of the non-technical information owner into technical object names and function codes. In effect, the interface translates technically detailed security objects and presents them to the information owner in a very simplistic fashion. Ideally, the interface for the information owner should be a point-and-click environment in which allocation of security objects and settings can be made quickly and easily. A technical implementation of such an interface is beyond the scope of this paper. A description may be found in the full study relating to ERPSEC.

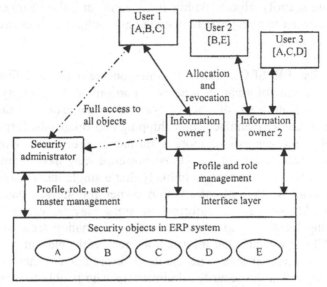

Figure 4. Decentralized security within an ERP environment

Though the inclusion of an interface in the ERPSEC framework allows the concept of information ownership to be supported more fully, additional requirements still exist. Most importantly, ERPSEC must validate and handle information owners in a slightly different way to ordinary users of the system. In addition, the interface for the information owner should be restrictive enough to ensure that only appropriate security objects for that information area can be configured and maintained.

5.4 Validating information owners

As mentioned in the section above, validation of information owners is important to ensure that access restrictions can be defined. The ERPSEC framework does not require too many special mechanisms to validate information owners. An information owner within the ERPSEC framework is simply another user of the system. The crucial difference is that an information owner has some additional access rights that an ordinary user would not have.

ERPSEC requires information owners to have special security settings added to their user master record that identifies them as information owners. In addition, the information they are responsible for is also identified, as well as the users they should be permitted to administer. This solves two problems, namely the ability of information owners to be able to configure and maintain security objects within the system, and the restriction of the information owner to being able to operate only within a set information area of the organization.

Technically, ERPSEC requires information areas to be defined. In the simplest sense, an information area is a portion of the ERP system that corresponds to an area of the real-world organization. Examples of information areas are manufacturing, shipping and financials. Depending on the size of the organization, more information areas may exist; as an example, an organization with a global presence may have manufacturing capacity in various countries. It is unlikely that a single individual would be able to perform the task of information owner for the all manufacturing divisions worldwide. Hence, information areas may be created for each manufacturing location. Regardless what the information areas are deemed to be, ERPSEC associates the information areas with relevant users in that location or information area. The assigned information owner is the only individual other than the security administrator who is able to maintain and configure access restrictions for those users. The assignment of users to information areas and the creation of information areas themselves are tasks that can be completed by the security administrator. This task is not technically complex and does not involve detailed knowledge of business

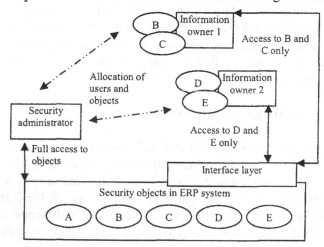

Figure 5. ERPSEC compartmentalizes users and information areas

processes. This information can be gleaned from the organization's structure.

Figure 5 attempts to represent the restricted access of the information owners. Note that the interface layer is responsible for ensure that the access to all security objects takes place in a compartmentalized fashion. In technical terms, the security subsystem may contain a number of tables that list all information owners with their associated information areas. A second table contains the information owner and allocated system users. The ERPSEC framework performs numerous checks whenever access to a security object is required by any user. In the case of an information user, ERPSEC verifies that the user in question is identified as an information owner. Once this check has successfully been completed, ERPSEC queries the table containing the information areas for that information owner. Access is not permitted to any object that is not within the list permitted for that user. At a higher level, ERPSEC ensures that no information area is accessible by more than one information owner. This ensures concurrency control and integrity.

5.5   Maintaining integrity

The concept of integrity has been briefly mentioned above. Integrity is a very important concept that has to be adhered to. As has been discussed, ERPSEC changes the traditional centralized security model to a decentralized one. In this decentralized approach, more than one user is able to modify security settings for ordinary system users. In effect, ERPSEC proposes multiple security administrators with one significant difference: each information owner is restricted to a very narrow set of users and security objects that may be configured and maintained.

Integrity is maintained within the ERPSEC framework by controlling the allocation of information areas to information owners. As discussed previously, it is very important that each information area be allocated to a single information owner. The structure of ERPSEC ensures this by providing a list of available information areas that have been configured, and by permitting the allocation of each one to only a single information area. If required, a single physical user could be the information owner for more than one information area. This is a matter of preference and does not compromise the operation of ERPSEC. However, the combination of information areas for a single information owner reverts back to the traditional centralized approach with most of its inherent problems. For this reason, a single information owner for each information area is recommended.

The allocation of information areas to individual information areas is performed by a security administrator. Once this has been completed, security objects within the ERP system that belong to that information area have to be allocated to that information area. This is not a trivial exercise and must be performed by the information owners.

5.6 Extending information areas to include security objects

Figure 3 briefly introduced the notion of allocating security objects to individual information owners. To facilitate the transfer of functionality to ordinary users within the system, the information owner must be able to allocate physical security objects to users. The concept of allocating such security objects is similar to the allocation of information areas to information owners. In the case of security objects, the organization determines which functional blocks within the ERP system belong to which information area. As has been determined in the study of existing ERP systems, a large number of function or menu codes are present in an ERP system. The ERPSEC framework assumes that all functions within an ERP system can be represented by function codes. In technical terms, a function code can be a menu item, shortcut or text entry that is linked to a particular program or functionality within the software system. Once the user selects a menu item or enters the function code, that program or functionality is executed. In this way, the user is able to perform tasks such as the entry of an order or the creation of an invoice.

To provide information owners with the ability to distribute the required functionality to the users within their information area, the information owner must be able to allocate function codes to relevant users in the system. It is logical to assume that function codes can be grouped to form segments of basic functionality within an ERP system. This fact is supported by the study of existing ERP systems: groups of function codes represent functionality within a particular module of the ERP system. In this way, all material management functions are associated with a particular group of function codes, for example. Hence, the allocation of function codes to individual information areas and hence to information owners is not an impossible task.

In the ERPSEC framework, the information owners determine which function codes belong to their information area. There may be cases where the ownership of a particular function code cannot be determined precisely. In this case, the organization should allocate that function code to the most likely information area. The result of this exercise within the ERPSEC framework is an additional data structure within the security subsystem that contains all function codes that have been allocated to a particular information owner. By employing the integrity and concurrency rules discussed earlier, only one information owner is able to allocate any particular function code. This has important consequences for the tightening and enhancing of security within the ERP system.

5.7 Enhancing security through ERPSEC

The operation of the ERPSEC framework has been briefly discussed in the above sections. The allocation of information areas to information owners, and the allocation of particular function codes to these information areas enhances security automatically. The reason for this is the fact that only the stakeholders or owners of the information are permitted to allocate access to it. This has important consequences for the overall security of the ERP system: as the information owner is responsible for the performance of his business unit, restricting access to only those users in the organization that require access is sensible. In contrast to the traditional centralized approach, the approach of allocating access at the information area level adds an extra layer of trust and security. As the information owner is aware of the users to whom certain function codes are being allocated, the allocation of the function codes takes place in a more secure environment. In contrast, a centralized security administrator is not always fully aware of all users within the organization and cannot determine why a certain function code may have to be allocated.

5.8 Support for segregation of duties

The fact that the information ownership approach increases support for segregation of duties has been mentioned briefly. Segregation of duties is seen to be one of the most important aspects to prevent fraud and heighten security[7].It is important to note that this requirement is becoming a legislative requirement for many organizations to adhere to. Legislation at present mandates accountability.

The ERPSEC framework assists by providing a non-technical platform for stakeholders to configure and define what security is required within their area. This in turn requires the stakeholder or information owner to accept the responsibility for the configuration and content of the resulting security object. Though it has to be stated that the ERPSEC framework cannot guarantee increased security, the adherence to the information ownership principles provides a platform from which stricter security measures can be put in place.

6.   CONCLUSION

The paper has briefly introduced the reader to the problems traditionally faced when implementing security within an ERP software environment. The most pressing problems were identified and discussed. To provide a solution to these problems, the paper introduced the concept of information ownership as a means to increasing and enhancing the security subsystem of an ERP software package. The proposed ERPSEC framework was introduced. The requirements of the ERPSEC framework to support the concept of information ownership were highlighted. It was briefly shown how ERPSEC assists in supporting the following:
– a reduction in complexity of the security configuration;
– the ability to increase responsibility and accountability within the organization;
– a faster implementation time by providing decentralized access to security objects;
– to improve the quality of the security configuration as a whole;

REFERENCES

1. M. Hertenberger, Prof. S.H. von Solms, *A framework for ERP security*, PhD study in progress, 2005
2. The Information Security Glossary, http://www.yourwindow.to/information-security
3. Internal controls, Committee of Sponsoring Organizations of the Treadway Commission (COSO), http://audit.ucr.edu/internal_controls.htm
4. Joseph R. Dervaes, *Internal Fraud and Controls*, Washington Finance Officer's Association, 48th Annual Conference, 19 September 2004
5. K. Vuppula. *BW security approaches*, http://www.intelligenterp.com/feature/2002/12/0212feat1_1.shtml, 2002
6. P. Manchester, Financial Times, 12 November 2003
7. Elizabeth M. Ready, Emerging Fraud Trends, State of Vermont, 2003
8. M. Hertenberger, Prof. S.H. von Solms, *A case for information ownership in ERP systems*, Kluwer Publishing, 2004

A NEW ARCHITECTURE FOR USER AUTHENTICATION AND KEY EXCHANGE USING PASSWORD FOR FEDERATED ENTERPRISES

Yanjiang Yang[1,2], Feng Bao[1] and Robert H. Deng[2]

[1]Institute for Infocomm Research, 21 Heng Mui Keng Terrace, Singapore 119613; [2]School of information Systems, Singapore Management University, Singapore 259756

Abstract: The rapid rise of federated enterprises entails a new way of trust management by the fact that an enterprise can account for partial trust of its affiliating organizations. On the other hand, password has historically been used as a main means for user authentication because of operational simplicity. We are thus motivated to explore the use of short password for user authentication and key exchange in the context of federated enterprises. Exploiting the special structure of a federated enterprise, our proposed new architecture comprises an *external server* managed by each affiliating organization and a *central server* managed by the enterprise headquarter. We are concerned with the development of an efficient authentication and key exchange protocol using password, built over the new architecture. The architecture together with the protocol well addresses off-line dictionary attacks initiated at the server side, a problem rarely considered in prior effort.

Key words: federated enterprise; password authentication; dictionary attack; key exchange; public key cryptosystem.

1.   INTRODUCTION

Driven by the promise of cost saving, expansion of market share and quality improvement of service provision through consolidation and cooperation, industry has seen a rapid rise of federated enterprises. Specifically, a federated enterprise consolidates under one corporate umbrella multiple divisions, branches and affiliations serving different aspects of business continuum and service coverage. For example, in the banking sector, a central bank has numerous branches distributed across a city, a region. Another example is in the healthcare area, where a federated hospital integrates many inside and outside units, e.g., clinical laboratories, departments, outpatient clinics, managed care organizations, pharmacies and so on. In a federated enterprise, each affiliating organization has its own business interest, providing service to a distinct group of users.

Following conventional ways, each affiliating organization has to work independently on trust management, maintaining by itself account information of its users. However, this may not be optimal in practice. First, affiliating organizations may lack sufficient expertise and funds for a secure maintenance of user account. This situation deteriorates with the trend that organizations are becoming increasingly fond of outsourcing IT management to some specialized service providers. In such circumstances, system administrators may present themselves as a big threat to system security [1]. Second, from the users' perspective, a user apparently prefers assuming the higher credit of the entire enterprise rather than that of an individual affiliating organization. For these reasons, new paradigm of trust management that well takes advantage of the special structure of federated enterprises is of interest and urgency.

On the other hand, human memorable password has historically been used as a main means for user authentication, due to operational simplicity. In particular, no dedicated device is required for storing password, which is deemed of particular importance as users are becoming increasingly roaming nowadays. We are thus interested in exploring the use of short passwords for user authentication and key exchange in the context of federated enterprises. Towards this, we are faced to first address the weaknesses inherent in password-enabled systems: because of a limited *dictionary* space, password is susceptible to brute-force *dictionary attack,* and more precisely *off-line dictionary attack*[8] [2]. Specifically, in off-line dictionary attack, an attacker records the transcript of a successful login between a user and the server, and then enumerates and checks every possible password against the transcript, until eventually determine a correct password. Tremendous effort has been dedicated to resisting off-line dictionary attack in password-enabled protocols (e.g., [3, 4, 5, 6, 7, 8, 9, 10]). An assumption common to these methods is that the server is completely reliable, so users share with the server clear passwords or some easily-derived values of passwords for subsequent user authentication. As such, while these protocols are sufficiently robust to off-line dictionary attacks by the outside attackers, they

---

[8] In contrast to off-line attack is on-line dictionary attack, which turns out to be easily thwarted by restricting the number of unsuccessful login attempts made by a user.

are not intended to defend against the server, e.g., in the event of penetration by outside attackers.

However, for the reasons discussed earlier (limited expertise and funds, outsourcing, etc.), threats posed by the server become clearer in the setting of federated enterprises. As a consequence, servers maintained by the affiliating organizations of an enterprise are no longer deemed fully trusted. Adapting password-enabled systems to federated enterprises has to additionally mitigate the concern of *unreliable* servers, in addition to outside attackers. To this end, we propose a new architecture for user authentication and key exchange using password, geared to the needs of federated enterprises. In its simplest configuration, the architecture consists of an *external server* managed by each affiliating organization and a *central server* administrated by the enterprise; each server only keeps partial information on a user password, such that no single server can recover the password by means of off-line dictionary attacks user authentication is accomplished together by the two servers. Our attention is given to the development of an efficient authentication protocol for the new architecture, rather than formal provable security. The proposed architecture together with the protocol enjoys several attractive features

The rest of the paper is organized as follows. In Section 2, we review related work. We then present our new architecture and discuss extension in Section 3. In Section 4, we propose an efficient user authentication and key exchange protocol well attuned to the new architecture, and security of the protocol is examined. Finally, Section sec5 concludes the paper.

2. RELATED WORK

Resistance to off-line dictionary attack has long been in the core of research on password-enable systems. It is a proven fact that *public key techniques* are absolutely necessary so as to resist off-line dictionary attacks, whereas the involvement of *public key cryptosystems* is not essential [7]. Accordingly, two separate lines of research have been seen in the literature: combined use of password and public key cryptosystem, and password only approach. For the former, asymmetry of capacity between the users and the server is considered, so that a user only uses a password while the server has a public/private key pair at its disposal. Examples of such public key-assisted password authentication include [7, 8, 11]. With no surprise, the use of public keys entails the deployment of PKI for certification, adding to the users the burden of checking key validity. To eliminate this drawback, password-only authenticated key exchange (PAKE) protocols have been extensively explored (e.g., [3, 4, 5, 6, 9, 10]). The PAKE protocols do not involve any public key cryptosystem whatsoever.

What common to the above methods is the assumption that the server is totally trustful, so a user shares a password or some password-derivative value with the server. In other words, these methods are by no means resilient to the off-line dictionary attack initiated by the server, e.g., in the event of break-ins by outside attackers. To address this problem, [12] first proposed the so-called *password hardening* technique by transforming a weak password into a strong one through several servers, thereby eliminating a single point of vulnerability. Afterwards, [13] improved this multi-server model. Further and more rigorous extensions were due to [14] and [15], where the former built a t-out-of-n threshold PAKE protocol and gave formal security proof under the Random Oracle Model [16], and the latter presented another two provably secure threshold PAKE protocols under the standard model. A limitation of the protocols in [14] and [15] is their low efficiency, so they may not be practical to resource-constrained users, e.g., mobile phones. Moreover, notice that in the above multi-server setting, password is still susceptible to a *weaker* form of a single point of vulnerability, in the sense that passwords are eventually reconstructed by a *dealer* at the time of user authentication.

By contrast, in our architecture no trust exists between the central server and the external server, thus a single point of vulnerability is completely eliminated. This however adds substantial challenges to the design of the underlying authentication protocol. Our basic architecture (one central server is involved) is similar to a recent novel two-server model in [17], which was to overcome the deficiency of complex and computation extensive protocols in [14, 15]. The protocol in [17] assumes that servers use SSL to establish secure communication channel with users. Distinctions between our work and [17] include: (a) we achieve mutual (bilateral) authentication as well as key exchange, whereas [17] considered merely unilateral authentication of the user to the servers, and no key exchange; (b) we develop a different protocol for testing the equality of two numbers under our presumed adversary model. Without a similar explicitly specified adversary model, the protocol in [17] may cause trouble in case that one of the two servers deliberately disrupts the protocol, attempting to gain advantages over the other; (c) our architecture is tailored to federated enterprises, whereas the model in [17] was suggested for an organization outsoucing a part of its trust management to a security service provider; (d) we also suggest extending the basic architecture to an architecture including a group of central servers, solving the possible bottleneck caused by one single central server.

3.   A NEW AUTHENTICATION ARCHITECTURE FOR FEDERATED ENTERPRISES

Our basic architecture is shown in Figure 1: at the server side, an *external server* and a *central server* coexist; the external server is the actual one providing service to the users, thereby standing front-end; the central server is to assist the external server for user authentication, staying transparent to the users. A main objective of this architecture is to "harden" a user's short password into two long secrets and each is hosted by a server, so that neither of the two servers can launch off-line dictionary attacks. As discussed earlier, a uniqueness of this architecture is represented by the fact that no trust exists between the two servers. This on the one hand totally eliminates a single point of vulnerability, while on the other hand makes the design of the underlying password-enabled protocol particularly challenging. In particular, the two servers together validate user passwords, whereas no extra information should be leaked to each other in facilitating off-line dictionary attack.

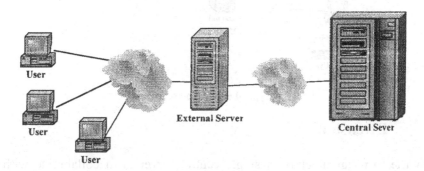

Figure 1. Basic two-server architecture for federated enterprises

Figure 2 shows the scenario when applying this basic structure to a federated enterprise: the headquarter of the enterprise manages the central server, and each affiliating organization operates an external server, providing service to a group of users of its own. This architecture offers several benefits:

1. First of all, of particular advantage is that neither the central server nor the external servers can compromise user passwords by means of off-line dictionary attack.
2. Affiliating organizations are relieved from strict trust management to some extent, so they can dedicate their limited expertise and resources to enhancing service provision to the users.

3. Users are afforded to assume the higher credit of the enterprise, while engaging business with individual affiliating organizations.
4. The enterprise is provided a way to monitor the affiliating organizations, deterring them from cheating.
5. As the central sever is hidden from the public, the chance for it under attacks is substantially minimized, thereby increasing the overall security of the architecture.

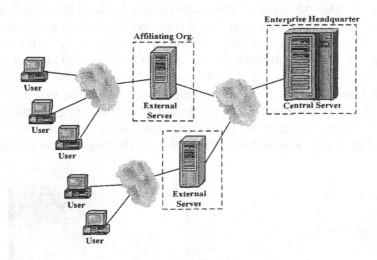

Figure 2. Typical application scenarios

3.1   Extension

In the basic architecture, a single central server is to collaborate with numerous external servers in a federated enterprise. It is thus possible that the central server becomes a system bottleneck. To mitigate this concern, we suggest extending to include several central servers as a group as shown in Figure 3. The group of central servers work under a t-out-of-n threshold secret sharing scheme (e.g., [18]), so that each holds a share of the secret that would be otherwise kept by a single central server (n is the total number of the central servers). When an external server requests for user authentication, one of the servers volunteers to manage the reconstruction of the secret among the group. This voluntary central server is the only one interacting with the requesting external server, thus the external server feels nothing different as with a single central server. This extension not only solves the potential bottleneck problem, but also addresses the issue of failures or break downs of a single central server.

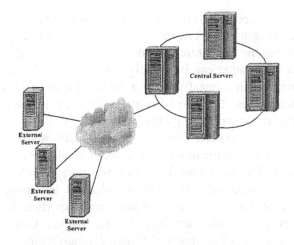

Figure 3. Extended architecture including a group of central servers

4. AN AUTHENTICATION AND KEY EXCHANGE PROTOCOL USING PASSWORD

In this section, we shall detail the authentication and key exchange protocol using password, which is geared to the setting of federated enterprises. The protocol is built over the basic architecture in Figure 1. Extension to the extended architecture in Figure 3 is also discussed. A central building block of our protocol is a sub-protocol for testing the equality of two numbers. For ease of reference, we start with listing the notations that are used in the sequel.

Table 1. Notations

p, q	two large primes such that $p = 2q+1$.
g	a generator of a subgroup of Z_p^* of order q.
π	a user's password.
$h(.), H(.)$	cryptographic hash functions which are modelled as random oracle [16].
U, ES, CS	identity of a user, the external sever and the central server, respectively.
$E_X(.), D_X(.)$	encryption and decryption functions (of a semantic secure public key cryptosystem) by entity X's public key and private key, respectively.

4.1 The setting

Three types of entities are involved in our system, i.e., the users, the external servers and the central server. For the purpose of authentication, each user U has a short password π, and π is transformed into two long secrets, each of which is held by the external server ES that U belongs to and the central server CS, respectively. CS stays transparent to the public, and ES acts as the relaying party between U and CS. In such a scenario, it is reasonable to assume authentic communications between ES and CS. Definitely, this assumption can be readily accomplished by the two parties sharing a secret, which is used to key a MAC. Considering the close tie between the two parties, it is also convenient for them to periodically (e.g., once a week) update this secret, e.g., the headquarter (CS) sends a new secret to the affiliating organization (ES) by normal mail weekly. CS has an authentic key pair that corresponds to a semantically secure public key cryptosystem, with the encryption function (resp. decryption function) $E_{CS}(.)$ (resp. $D_{CS}(.)$). As discussed earlier, in a federated enterprise CS clearly assumes more trust than ES because of sufficient expertise, funds, and the fact that CS is not directly exposed to the public. Considering such asymmetry in terms of trust upon CS and ES, adversary model in our protocol is that CS is *semi-honest* and ES is *malicious*, with respect to their desire for off-line dictionary attack, and they do not collude. More specifically, CS is honest-but-curious [19], i.e., it follows the protocol, with the exception that it may try to derive extra information by analyzing the protocol transcript; on the contrary, ES may act arbitrarily for uncovering user passwords.

4.2   High level description

Central to our protocol is to fight against off-line dictionary attack by the servers (Note that outside attackers are clearly no more powerful than the external server in this regard). The intuition behind our authentication and key exchange protocol is as follows: in an out-of-band *registration* phase, a user U "hardens" his password π into two random long secrets s and $\pi + s$, and registers them to the external server ES and the central server CS, respectively, where s is a random number. In *authentication* phase, U picks another long random number r and sends r and $\pi + r$ to the two servers, respectively. Upon receiving the messages, ES computes $a = r - s$, and CS computes $b = (\pi + r) - (\pi + s) = r - s$. Afterwards, the two servers engage into an interactive protocol to test $a \overset{?}{=} b$. Note that $a = b$ holds if and only if user U knows π. Upon the servers validating the user, ES and U negotiate a common session key for subsequent data exchanges. Clearly, from s and r (resp. $\pi + s$ and $\pi + r$), ES (resp. CS) is unable to gain anything useful on π. It is thus of crucial importance to ensure the protocol for testing $a \overset{?}{=} b$ could not facilitate the servers for off-line dictionary attack. In what follows, we first propose a protocol allowing two parties to test $a \overset{?}{=} b$, which will be invoked by our final authentication and key exchange protocol that follows.

4.3   **A protocol testing *a* ?= *b***

A protocol for simply testing *a* ?= *b* by two parties *A* (possessing *a*) and *B* (possessing *b*), while without disclosing *a and b* may be quite straightforward. See a simple example: *A* sends *h*(*a*) to *B* and *B* sends *h*(*b*) to *A*, where *h*(.) is a hash function as defined in Table 1; each party checks *h*(*a*) ?= *h*(*b*). A variant is that *A* sends $G_a = g^a$ mod *p* to *B* and *B* sends $G_b = g^b$ mod *p* to *A*, where *p* and *g* are defined as in Table 1; each party then checks G_a ?= G_b. Both methods however cannot avert off-line dictionary attack by the two parties in our case. Take the first example and *A* for instance, *A* chooses a random number *r* and computes *a* = *r* - *s* for himself, and simply sends *r* to *B*. *B* will return *h*(*r* - *s* - π) to *A*. It is easy to see that *A* can enumerate every possible password until find π', such that *h*(*a* - π') = *h*(*r* - *s* - π). In a same way, the attack applies to the variant example, although in normal cases, it is hard to get *a* (resp. *b*) given G_a (resp. G_b) according to *discrete logarithm assumption*. These examples convey to some extent the subtleties in designing a protocol in our case of withstanding off-line dictionary attack.

Let QR_p denote the group of quadratic residues modulo *p*, and a hash function be defined as $h: \{0, 1\}^{|p|} \rightarrow QR_p$, where *p*, *q*, *h*(.) are public parameters and as defined in Table 1. Note that in practice *h* can be achieved by squaring a one-way hash function, e.g., SHA1. We outline our proposed protocol for testing *a* ?= *b* in Figure 4 (all arithmetic operations are calculated modulo *p*).

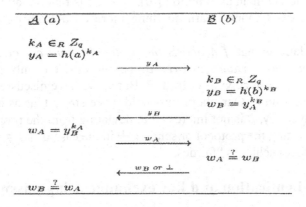

Figure 4. A protocol testing *a* ?= *b*

Specifically, *A* picks $k_A \in_R Z_q$ on the fly and computes $y_A = h(a)^{k_A}$ mod *p*. *A* initiates the protocol by sending y_A to *B*. Upon receiving the message, *B* chooses $k_B \in_R Z_q$, in turn computes $y_B = h(b)^{k_B}$ mod *p* and

$w_B = y_A{}^{k_B} \bmod p$, respectively. B then sends y_B to A. After receiving y_B, A computes $w_A = y_B{}^{k_A} \bmod p$ and sends w_A to B. With w_A, B tests $w_A \;?= w_B$. B then sends w_B to A if $w_A = w_B$, and a special label $\perp$ otherwise. A then tests $w_B \;? = w_A$ if w_B is received.

4.3.1   Security analysis

In line with the adversary model defined in our final protocol, we assume A a malicious adversary while B an honest-but-curious adversary. In addition, for the moment we assume A does not replay messages, and the communication between A and B is authentic. These assumptions will be clear when we come to our final protocol.

We start by by claiming that *upon completion of the protocol, either (1) A and B learn that a = b or (2) A and B learn a ≠ b but nothing more on the opposite side's secret.*

Clearly in the first case, if $w_A = w_B$ holds, A and B learn that $a = b$. We next show if $a \neq b$ (this may be due to that A cheats by intentionally using a, which is different from his original input), both parties learn nothing more. Consider A first: intuitively, A gets $y_B = h(b)^{k_B} \bmod p$ at the end of the protocol. It is easy to see that without knowing k_B, A is unable to obtain anything on b in an information theoretic sense. Next, consider the case of B: when the protocol terminates, of relevance to B is ($y_A = h(a)^{k_A} \bmod p$, $w_A = y_B{}^{k_A} \bmod p$). Notice intuitively that $(w_A, y_B = w_A{}^{k_A} \bmod p, y_A,$ $h(a) = y_A{}^{k_A^{-1}} \bmod p)$ is indistinguishable from (w_A, y_B, y_A, z) under the decisional Diffie-Hellman assumption [20], where z is random and $k_A k_A^{-1} = 1$ $\bmod q$. Therefore B cannot learn anything more on a from executing the protocol.

Our next claim is that *if A aborts the protocol before completion, he is unable to gain more advantages over B.* To see this, the only place A is likely to abort is after receiving y_B from B. But as we have discussed, y_B does not leak anything on b. Our claim thus holds. We stress that as an honest-but-curious adversary, B is not interested in deviating from the protocol, e.g., deliberately aborting the protocol or sending $\perp$ in the case of $w_A = w_B$. In this sense, our protocol achieves "fairness".

4.4   Authentication and key exchange using password

We now present an efficient authentication and key exchange protocol using password, built over the basic architecture in Figure 1. The earlier protocol for testing $a \;?= b$ is invoked in this protocol as a building block, where ES plays the role of A and CS takes the role of B. In the sequel, we occasionally omit "modulo p" in stating arithmetic operations as long as the context is clear.

System parameters are defined as follows: p and q are as defined in Table 1, and QR_p is the group of quadratic residues modulo p. A hash function $H(.)$ (e.g., SHA1) is employed. Moreover, g is picked from QR_p as $g \in_R QR_p\backslash\{1\}$. Clearly, g is of order of q. $H(.)$, p, q and g are public parameters.

To enrol as a legitimate user in a service, it is natural that at the beginning, a user must authenticate to the service provider and in turn establish a password with the organization for subsequent service access. In our case, U needs to register to not only the actual service provider ES but also the enterprise CS that ES is affiliated to.

4.4.1 Registration

Suppose U has already successfully authenticated to ES, e.g., by showing his identity card, U picks his password π and selects a random number $s \in_R QR_p$. U then registers in a secure way s and $\pi + s \bmod p$ to ES and CS, respectively. Here for purely simplicity reasons, we assume $(\pi + s \bmod p) \in QR_p^9$. Consequently, ES stores the account information (U, s) to its secret database, and CS stores $(U, \pi + s \bmod p)$ to its secret database. Someone may wonder how U registers $\pi + s$ to CS, as CS is supposed hidden from the public. This is in fact not a problem in practice: U can contact CS by normal mail, etc. Indeed, imagine that a user enrols in a branching bank, it is not strange at all that the user still needs to submit a secret to a higher authority of the bank so as to activate his account.

Upon completion of the registration, U can request service from ES, by exploiting the protocol in Figure 5 for authentication and establishment of a common session key.

4.4.2 The protocol

Let us follow the protocol (in Figure 5) step by step. To initiate the protocol, U picks x as $x \in_R Z_q$ and computes $e_x = g^x \bmod p$, which will be used for (Diffie-Hellman) key exchange. U also selects r as $r \in_R Z_q$, and encrypts e_x, $\pi + r \bmod p$ and T using CS's authentic public key as $e_0 = E_{CS}(e_x, \pi + r, T)$, where T is the current timestamp. U then sends in $M1$ the message of (U, e_x, r, e_0, T) to ES. Upon receipt of the message, ES first checks whether T is within a pre-defined time window: if T expires, ES simply

---

[9] In our protocol, we require $(\pi + s \bmod p) \in QR_p$. Indeed, if $(\pi + s \bmod p) \notin QR_p$, then it must hold that $(p - \pi - s \bmod p) \in QR_p$.

returns *reject* to U and aborts the protocol; otherwise, *ES* proceeds ahead. *ES* searches his secret database for *U*'s account. If no such an account is found, *ES* returns *reject* to U and stops the protocol; otherwise, *ES* fetches the secret s and computes $a = r - s \bmod p$; *ES* also keeps e_x in his *live* buffer. Afterwards, *ES* relays (U, e_0, T) to *CS* in *M2*.

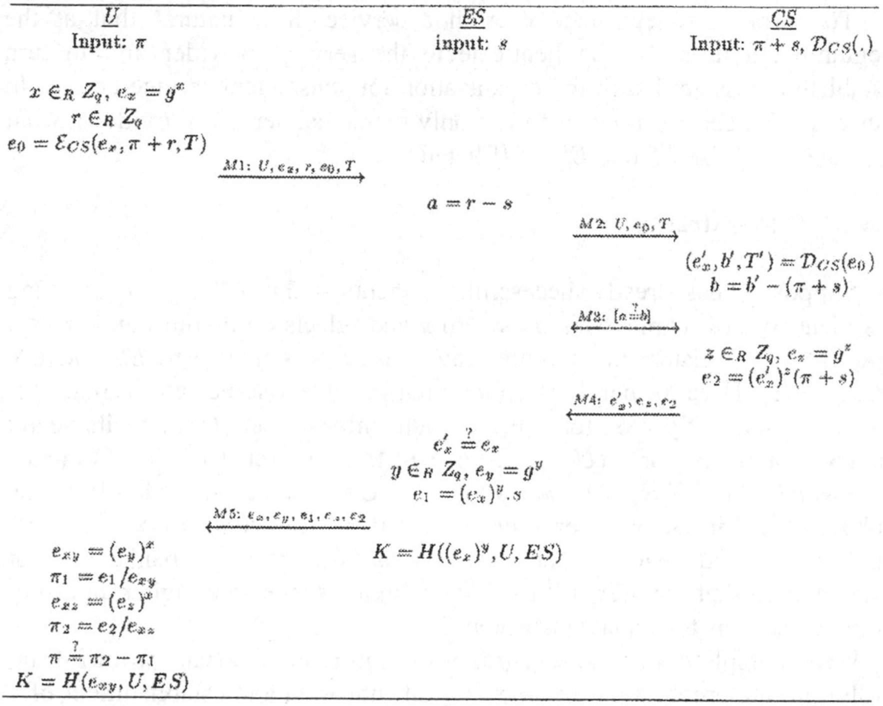

$\underline{U}$	$\underline{ES}$	$\underline{CS}$
Input: π	input: s	Input: $\pi + s, \mathcal{D}_{CS}(.)$

$$x \in_R Z_q, e_x = g^x$$
$$r \in_R Z_q$$
$$e_0 = \mathcal{E}_{CS}(e_x, \pi + r, T)$$

$\xrightarrow{\;M1:\; U, e_x, r, e_0, T\;}$

$$a = r - s$$

$\xrightarrow{\;M2:\; U, e_0, T\;}$

$$(e_x', b', T') = \mathcal{D}_{CS}(e_0)$$
$$b = b' - (\pi + s)$$

$\xleftarrow{\;M3:\; [a \overset{?}{=} b]\;}$

$$z \in_R Z_q, e_z = g^z$$
$$e_2 = (e_x')^z(\pi + s)$$

$\xleftarrow{\;M4:\; e_x', e_z, e_2\;}$

$$e_x' \overset{?}{=} e_x$$
$$y \in_R Z_q, e_y = g^y$$
$$e_1 = (e_x)^y \cdot s$$

$\xleftarrow{\;M5:\; e_x, e_y, e_1, e_z, e_2\;}$

$$K = H((e_x)^y, U, ES)$$

$$e_{xy} = (e_y)^x$$
$$\pi_1 = e_1/e_{xy}$$
$$e_{xz} = (e_z)^x$$
$$\pi_2 = e_2/e_{xz}$$
$$\pi \overset{?}{=} \pi_2 - \pi_1$$
$$K = H(e_{xy}, U, ES)$$

Figure 5. An authentication and key exchange protocol using password

In a similar way, *CS* checks the freshness of T and the account of U in his secret database. If both are correct, *CS* decrypts e_0 to get $(e_0', b', T) = \mathcal{D}_{CS}(e_0)$. *CS* then checks whether $T = T$: if not, *CS* rejects; otherwise, *CS* continues. *CS* takes out $\pi + s$ and computes $b = b' - (\pi + s) \bmod p$. Next, in *M3*, *CS* and *ES* engage in the protocol of Figure 4 to test $a \; ?= \; b$. If $a \neq b$, *CS* rejects and *ES* in turn replies *reject* to U. Otherwise, *CS* chooses $z \in_R Z_q$ and computes $e_z = g^z \bmod p$, which is in turn used to "encrypt" $\pi + s$ as $e_2 = e_x'^z(\pi + s) \bmod p$. *CS* then sends in *M4* the message of (e_x', e_z, e_2) to *ES*. Upon receiving the message, *ES* checks whether $e_x' = e_x$ (e_x is being kept alive in the buffer) to ensure that e_x received in *M1* has not been replaced by outside attackers. Interestingly, here e_x and e_x' are serving an extra purpose of "freshness nonce". If $e_x' \neq e_x$, *ES* notifies *CS* and sends *reject* to U. Otherwise, *ES* picks

$y \in _R Z_q$ and computes $e_y = g^y$ mod p. e_y is then used to "encrypt" s as $e_1 = e_x^y.s$ mod p. Afterwards, *ES* sends $(e_x, e_y, e_1, e_z, e_2)$ in *M5* to *U*. Here, e_x acts as a freshness nonce. *ES* also computes a common session key K as $K = H((e_x)^y, U, ES)$. Upon receiving the message, *U* does the following calculations: computes $e_{xy} = (e_y)^x$ mod p and obtains $\pi_1 = e_1/e_{xy}$ mod p; computes $e_{xz} = (e_z)^x$ mod p and gets $\pi_2 = e_2/e_{xz}$ mod p; tests π ?= $\pi_2 - \pi_1$ mod p: if the equality holds, *U* is assured of the authenticity of *ES* and computes the common session K as $K = H(e_{xy}, U, ES)$.

4.4.3 Security analysis

Our architecture is different from either the standard client-server model (e.g., [7, 10]) or the multiple-server model (e.g., [14, 15]), so formal provable security may be quite involved. We thus give an informal security analysis for the moment, yet we believe our analysis still suffices to guarantee the security of the protocol. Recall that the primary goal of this protocol is to resist off-line dictionary attacks by the two servers, where *ES* is a malicious adversary and *CS* is an honest-but-curious adversary under the adversary model that represents different levels of trust upon *ES* and *CS*. It is easy to see that outside attackers are no more powerful than *ES* in terms of the capability to uncover *U*'s password. Admittedly, outside attackers can act as man-in-the-middle between *U* and ES, resulting in a legitimate user being deemed illegitimate by the servers. Note that such attacks are inevitable in any protocol, and discussion of them is beyond the scope of this work as they are not relevant to the password attacks.

Resistance to CS:
In the protocol, what relevant to *CS* for off-line dictionary attack is ($\pi + r$ mod p, $\pi + s$ mod p), as well as the interactive protocol for testing a ?= b. Clearly, from $\pi + r$ mod p and $\pi + s$ mod p, *CS* is unable to learn anything on π; as discussed earlier, the protocol for testing a ?= b leaks nothing more on π. Consequently, as a passive semi-trusted adversary, *CS* cannot launch effective off-line dictionary attacks.

Resistance to ES:
Intuitively, if following the protocol, of help to *ES* regarding off-line dictionary attack are (r, e_0) and (s, e_2). However, $E_{CS}(.)$ is a semantic secure encryption, so the first pair does not help in dictionary attack; notice then that (e_z, e_2) is a standard ElGamal encryption. As widely known, it is also semantic secure when $g \in QR_p$ and ($\pi + s$ mod p) $\in QR_p$ as in our protocol. Therefore, *ES* is not effective in off-line dictionary attack as long as he follows the protocol (behaving as a passive adversary).

As an active adversary, *ES* can modify or forge protocol transcript. To see this, *ES* may pick x of its choice and computes e_x, and in turn makes a dubious e_0, in an attempt to deceive *CS* into replying with e_2 under his e_x. This cheating however cannot succeed, due to the fact that without knowing π, *ES* is not able to convince *CS* of $a = b$. We do notice that in such a way, *ES* can launch *on-line* dictionary attack by repeatedly guessing passwords, and engaging in the protocol for testing $a\ ?=\ b$: each time *CS* returns $\perp$ (reject), *ES* is assured to exclude a password from the dictionary. However, it is clear that such attacks are *unavoidable* in any password-enabled system, but can be readily thwarted by limiting the number of unsuccessful authentication attempts (regarding a same user) made by *ES*.

Security to outside adversaries:

While no more effective than *ES* in terms of dictionary attack, an outside adversary could attempt to acquire the common key K established between U and *ES*. This is another common attack to authentication and key exchange protocols. In our protocol, as the adversary does not know π, he has no way to negotiate a dubious common session key with *ES* in the name of U. What remains to consider is the scenario that the adversary derives the session key K by watching the protocol transcript between U and *ES*. This in our case is clearly equivalent to breaking the Diffie-Hellman assumption: by given g^x mod p and g^y mod p, to compute g^{xy} mod p without knowing x and y.

4.4.4   Extension

We introduce briefly how to adapt the protocol to the extended architecture in Figure 3 that includes a group of central servers. The extension turns out to be straightforward. The central servers work under a t-out-of-n threshold secret sharing scheme [18], each keeping a share of $\pi + s$ that would be otherwise preserved by a single central server. At the time an external server requests for user authentication, one of the servers volunteers to be the *dealer*, managing the reconstruction of $\pi + s$. The voluntary dealer is the only one interacting with the requesting external server. While the dealer could become a single point of vulnerability, compromise of it actually affects solely those $\pi + s$ that had ever been reconstructed on it. Furthermore, as already discussed, the chance of compromising a central server is practically minor. After all, this extension at the central server side is actually aimed at solving possible bottleneck problems and break-downs due to a single central server.

4.5   Discussions

The proposed protocol enjoys several advantages. Among others, first, the protocol is particularly efficient to the users in terms of both communication and computation. As to computation, a user only needs to compute 2 one-line exponentiations, and 1 off-line exponentiation and 1 off-line public key encryption. This is important when consider to support resource-constrained users, e.g., mobile phones. The communication to the users is optimal: only one round of interaction is involved. Second, a user can use the same password to register to different enterprises or to different affiliating organizations in a same enterprise (by varying s). This avoids a big inconvenience in traditional password-enabled systems (e.g., those reviewed in Section 2), where a user has to memorize different passwords for different applications.

We next clarify a possible argument why we do not simply rely on the central server(s) for full trust management of the affiliating organizations, a paradigm similar to Kerberos [21]. The reasons are as follows: first, each affiliating organization has its own business interest, so it has a stake to involve into the trust management of its own; second and more important, a main objective of our architecture is to avoid a single point of vulnerability.

Finally, while the assumption of CS being an honest-but-curious adversary well represents the different levels of trust upon an enterprise and its affiliating organization, it is a strong one. Design of an authentication and key exchange protocol in the case of CS being also a malicious adversary (e.g., allowed to wiretap the communication between U and ES) is an open problem.

5.   CONCLUSIONS

We explored applying authentication and key exchange using password to federated enterprises. Taking advantage of the special structure of a federated enterprise, a new architecture comprising an external server and a central server was proposed. A user authentication and key exchange protocol using password that is geared to the architecture was presented. Attention was focused on resisting off-line dictionary attacks by the servers, a topic rarely considered in previous effort. Our proposed architecture and protocol enjoyed several attractive features.

ACKNOWLEDGEMENTS

The authors would like to thank the anonymous reviewers for their valuable comments.

REFERENCES

[1] L. Bouganim, P. Pucheral, Chip-Secured Data Access: Confidential Data on Untrusted Servers, in: *Very Large Data Bases (VLDB)*, pp. 131-142, 2002.

[2] D. V. Klein, Foiling the Cracker - A Survey of, and Improvements to, Password Security, in: *2nd USENIX Security*, pp. 5-14, 1990

[3] E. Bresson, O. Chevassut, and D. Pointcheval, Security Proofs for an Efficient Password-Based Key Exchange, in: *ACM. Computer and Communication Security*, pp. 241-250, 2003.

[4] S. Bellovin, and M. Merritt, Encrypted Key Exchange: Password- Based Protocols Secure Against Dictionary Attacks, in: *IEEE Symposium on Research in Security and Privacy*, pp. 72-84, 1992.

[5] S. Bellovin and M. Merritt, Augmented Encrypted Key Exchange: A Password-Based Protocol Secure Against Dictionary Attacks and Password File Compromise, in: *ACM. Computer and Communication Security*, pp. 244-250, 1993.

[6] J. Katz, R. Ostrovsky, and M. Yung, Efficient Password-Authenticated Key Exchange Using Human-Memorable Passwords, in: *Advances in Cryptology, Eurocrypt'01*, LNCS 2045, pp. 475-494, 2001.

[7] S. Halevi, and H. Krawczyk, Public-key Cryptography and Password Protocols, in: *ACM. Computer and Communication Security*, pp. 122-131, 1998.

[8] M. K. Boyarsky, Public-key Cryptography and Password Protocols: The Multi-User Case, in: *ACM Conference on Computer and Communication Security*, pp. 63-72, 1999.

[9] J. Katz, R. Ostrovsky, and M. Yung, Forward Secrecy in Password-Only Key Exchange Protocols, in: *Security in Communication Networks*, 2002

[10] M. Bellare, D. Pointcheval, and P. Rogaway, Authenticated Key Exchange Secure against Dictionary Attacks, in: *Advance in cryptology, Eurocrypt'00*, pp. 139-155, 2000.

[11] L. Gong, M. Lomas, R. Needham, and J. Saltzer, Protecting Poorly Chosen Secrets from Guessing Attacks, *IEEE Journal on Seclected Areas in Communications*, 11(5), pp. 648-656, 1993.

[12] W. Ford, and B. S. Kaliski Jr, Sever-assisted Generation of a Strong Secret from a Password, in: *IEEE. 9th International Workshop on Enabling Technologies*, 2000.

[13] D. P. Jablon, Password Authentication Using Multiple Servers, in: *RSA Security Conference*, LNCS 2020, pp. 344-360, 2001.

[14] P. Mackenzie, T. Shrimpton, and M. Jakobsson, Threshold Password-Authenticated Key Exchange}, in: *Advances in Cryptology, Crypto'02*, LNCS 2442, pp. 385-400, 2002.

[15] M. D. Raimondo, and R. Gennaro, Provably Secure Threshold Password-Authenticated Key Exchange, in: *Advances in Cryptology, Eurocrypt'03*, LNCS 2656, pp. 507-523, 2003.

[16] M. Bellare, P. Rogaway, Random Oracles are Practical: A Paradigm for Designing Efficient Protocols, in: *ACM. Computer and Communication Security*, pp. 62-73, 1993.

[17] J. Brainard, A. Juels, and B. Kaliski, M. Szydlo, A New Two-Server Approach for Authentication with Short Secret, in: *USENIX Security*, 2003.

[18] A. Shamir, How To Share A Secret, *Communications of the ACM*, Volume 22, pp. 612-613, 1979.

[19] O. Goldreich, Secure Multi-party Computation, Working Draft, Version 1.3, June 2001.

[20] D. Boneh, The Decision Diffie-Hellman Problem, in: *3rd International Algorithmic Number Theory Symposium*, LNCS 1423, pp. 48-63, 1998.

[21] J. Kohl, and C. Neuman, RFC 1510: The Kerberos Network Authentication Service, 1993.

A SECURE QUANTUM COMMUNICATION PROTOCOL USING INSECURE PUBLIC CHANNELS

I-Ming Tsai[1], Chia-Mu Yu[2], Wei-Ting Tu[1], and Sy-Yen Kuo[1]
[1]Department of Electrical Engineering; [2]Graduate Institute of Computer Science and Information Engineering; National Taiwan University, No.1, Sec. 4, Roosevelt Road, Taipei, 106, Taiwan

Abstract: Due to the discovery of Shor's algorithm[1], many classical crypto-systems based on the hardness of solving discrete log and factoring problem are theoretically broken in the presence of quantum computers. This means that some of the classical secret communication protocols are no longer secure and hence motivate us to find other secure crypto-systems. In this paper, we present a new quantum communication protocol which allows two parties, Alice and Bob, to exchange classical messages securely. Eavesdroppers are not able to decrypt the secret messages and will be detected if they do exist. Unlike classical crypto-systems, the security of this protocol is not based on the hardness of any unproven mathematic or algorithmic problem. Instead, it is based on the laws of nature.

Key words: Quantum Cryptography; Encrypted Communication; Quantum Entanglement

1. INTRODUCTION

Quantum information science is a highly interdisciplinary field of research and hence has applications in nearly every field of computer science and electrical engineering. Cryptography, most notably key distribution, is one example. Classical cryptography enables two parties, Alice and Bob, to exchange confidential messages such that the messages are illegible to any unauthorized third party. The problem is that it is difficult to distribute the secret key securely through a classical channel. This is known as the key distribution problem. Classical key distribution protocols based on the hardness of mathematical or algorithmic problems[2,3] are conditionally secure *i.e.* theoretically insecure. However, quantum cryptography allows a number of applications that are not possible classically. An example is the Quantum Key Distribution (QKD) protocol -- a protocol dealing with secure key distribution using quantum mechanics.

Theoretical study and physical implementations of QKD have been developed rapidly after Bennett and Brassard proposed the standard BB84 protocol[4]. Basically, QKD schemes can be categorized into two classes -- non-deterministic QKD and deterministic QKD. For non-deterministic QKD, the sender and the receiver have no control over what bit string is used as the key. Typical non-deterministic QKD schemes include BB84, E91[5] and B92[6] protocols. In contrast, in a deterministic scheme, the sender and receiver have a total control of what bit string is used. This is actually, in classical cryptography terms, a secure communication, or an encryption/decryption process[7-10].

A secure communication protocol allows the sender (Alice) and the receiver (Bob) to exchange messages securely without running the risk of being decrypted by an eavesdropper (Eve). As a secure communication protocol, two requirements must be satisfied. First, upon a successful transmission process, the secret messages shall be able to be read out as its original form by the legitimate receiver. Second, in the presence of an eavesdropper, the encrypted message shall give her absolutely no information even if she may have total control of the channel. In the following sections, we present a protocol which not only fulfills these two requirements, but also can detect the eavesdroppers, if they do exist.

2.   BACKGROUND

The state of a single quantum bit can be written as a linear combination of two states in a two-dimensional complex vector space as

$$|\psi\rangle = \alpha|0\rangle + \beta|1\rangle,$$ (1)

where α and β are complex numbers and $|\alpha|^2 + |\beta|^2 = 1$. The two orthonormal states $|0\rangle$ and $|1\rangle$ forms a computational basis of the system and the contribution of each basis state to the overall state (α and β in this case) is called the probability amplitude. According to quantum mechanics, when the system is measured, the state *collapses* to one of the basis states

($|0\rangle$ or $|1\rangle$). The probability of collapsing to a particular basis state is directly proportional to the square of the probability amplitude associated with it. More specifically, when a measurement is performed on a quantum state, the probability of getting a result of $|0\rangle$ is $|\alpha|^2$ and the probability of getting a result of $|1\rangle$ is $|\beta|^2$. Obviously, due to the rule of probability, $|\alpha|^2 + |\beta|^2 = 1$. The symbol for a quantum measurement is shown in Fig.1(a).

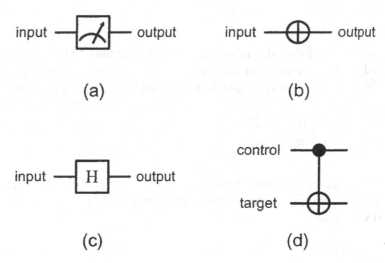

Figure 1. The symbols for quantum measurement and various quantum gates

The state described above exhibits a unique phenomenon called quantum *superposition*. When a particle is in such a state, it has a part corresponding to $|0\rangle$ and a part corresponding to $|1\rangle$, at the same time. However, when a measurement is performed, it collapses to one of the states in the basis (eigenstates). To distinguish the above system from a classical binary digit, such a unit is called a quantum binary digit, or *qubit*. An easy way to describe a qubit is to use column matrices. For example, Eq. (1) is equivalent to the notation

$$|\psi\rangle = \begin{pmatrix} \alpha \\ \beta \end{pmatrix}. \tag{2}$$

Similar to classical bits manipulated by classical logic gates, a qubit can be manipulated using *quantum gates*. Like a qubit, a quantum gate can also be written in matrix form. In its matrix form, a quantum gate G must be unitary, *i.e.* satisfying $GG^+ = G^+G = I$, where G^+ stands for the transpose conjugate of G. This is because any such gates can be pictorially described as a rotation on the Bloch sphere. When a qubit goes through some quantum gates, the state vector is rotated to another direction. An example of quantum gate is the quantum **NOT** gate, which has the matrix representation of

$$N = \begin{pmatrix} 0 & 1 \\ 1 & 0 \end{pmatrix}. \tag{3}$$

Using the matrix form, the new state after a quantum **NOT** gate can be calculated using matrix multiplication. For example, when a qubit $|\alpha|^2 + |\beta|^2 = 1$ goes through a quantum **NOT** gate, the state changes to

$$|\psi'\rangle = \begin{pmatrix} 0 & 1 \\ 1 & 0 \end{pmatrix} \begin{pmatrix} \alpha \\ \beta \end{pmatrix} = \begin{pmatrix} \beta \\ \alpha \end{pmatrix}. \tag{4}$$

The symbol for a quantum **NOT** gate is shown in Fig.1(b). Another important quantum gate is the **HADAMARD (H)** gate. The matrix form of a **HADAMARD** gate is

$$H = \frac{1}{\sqrt{2}} \begin{pmatrix} 1 & 1 \\ 1 & -1 \end{pmatrix}, \tag{5}$$

and is able to make the following state changes:

$$H(|0\rangle) = \frac{1}{\sqrt{2}} (|0\rangle + |1\rangle), \tag{6}$$

$$H(|1\rangle) = \frac{1}{\sqrt{2}} (|0\rangle - |1\rangle). \tag{7}$$

The symbol for a **HADAMARD** gate is shown in Fig.1(c).

The space of a multi-qubit system can be modeled by the tensor product of each individual space. For example, a two-qubit state is a linear combination of four basis states:

$$|\psi\rangle = \alpha|00\rangle + \beta|01\rangle + \gamma|10\rangle + \delta|11\rangle, \tag{8}$$

with α, β, γ, δ complex numbers and $|\alpha|^2 + |\beta|^2 + |\gamma|^2 + |\delta|^2 = 1$. Similar to the single qubit case, a two-qubit system can be represented using a 4×1 column matrix and a two-qubit gate can be represented using a 4×4 matrix. An example of two-qubit gate is the **CONTROL-NOT (CN)** gate, as shown in Fig. 1(d). A **CN** gate consists of one *control* bit x and one *target* bit y. The target qubit will be inverted only when the control qubit is $|1\rangle$. Assuming x is the control bit, the gate can be written as **CN** $(|x,y\rangle) = |x, x \oplus y\rangle$, where $\oplus$ denotes exclusive-or. This actually performs a permutation on the basis as follows: $|00\rangle \to |00\rangle$, $|01\rangle \to |01\rangle$, $|10\rangle \to |11\rangle$, and $|11\rangle \to |10\rangle$. In column matrix, this is equivalent to

$$|\psi'\rangle = \begin{pmatrix} 1 & 0 & 0 & 0 \\ 0 & 1 & 0 & 0 \\ 0 & 0 & 0 & 1 \\ 0 & 0 & 1 & 0 \end{pmatrix} \begin{pmatrix} \alpha \\ \beta \\ \gamma \\ \delta \end{pmatrix} = \begin{pmatrix} \alpha \\ \beta \\ \delta \\ \gamma \end{pmatrix}. \tag{9}$$

An interesting phenomenon in quantum mechanics is *entanglement*. Imagine that Alice and Bob share a two-qubit system in the state

$$|\psi\rangle = \frac{1}{\sqrt{2}} (|00\rangle + |11\rangle)_{ab}, \tag{10}$$

where a and b denote Alice and Bob respectively. According to quantum mechanics, if Alice takes a measurement on qubit a, the state of the qubit will collapse to $|0\rangle$ with a probability of $\frac{1}{2}$. Moreover, Alice immediately knows that the state of the other qubit (qubit b) must be $|0\rangle$. In other words, once the measurement result of one qubit is decided, the state of the other one is perfectly correlated and can be instantaneously decided, no matter how far away Alice and Bob are separated. A similar result happens if the result of Alice's measurement is $|1\rangle$. This non-classical correlation among multiple quantum systems is called quantum entanglement, because they can not be written as separable states and are considered to be entangled. Studies of different types of entanglement and their applications are an important issue in quantum information science.

3.   ENCRYPTED COMMUNICATION PROTOCOL

The proposed protocol uses one entangled qubit pair to transmit one encrypted classical bit, then an n-bit classical message can be transmitted bit-by-bit via this protocol. At the end of the transmission, an error checking process is employed to check the integrity of the whole message.

3.1   Resource requirement

In this paper we assume Eve has unlimited technological and computational power. She can perform any operation on the transmitting qubit as long as it is allowed by the laws of nature. Under these circumstances, the propose protocol can protect both the privacy and integrity of the message using a classical public channel and a quantum channel. The natures of these channels are described in the following paragraphs.

A classical channel is a communication path that can be used to transmit classical information from a sender to a receiver. For example, an optical fiber which allows Alice to send her voice to Bob is a typical classical channel. Depending on whether the channel is readable or writable by an unauthorized third party, classical channels can be further categorized into classical private channels and classical public channels.

A classical private channel is a channel, together with some appropriate mechanisms, which are capable of maintaining the privacy and integrity of the messages transmitted via that channel. The term privacy refers to the fact that the data carried in the channel cannot be read or revealed by anyone without authorization. It involves mainly data encryption algorithms and secret keys. An encryption mechanism, together with a secret key, can be used to translate the message into a form that is unreadable without the secret key. The term integrity means the message from the source can not be either accidentally or maliciously modified, altered, or destroyed. In other words, the messages exchanged between Alice and Bob are identically maintained during the transmission process.

As a contrast, a classical public channel is a classical channel that maintains only the data integrity, regardless of the privacy. In other words, a classical public channel can be used to transmit classical information from Alice to Bob without being modified by eavesdroppers. However, anyone, including eavesdroppers, can read the original message. Radio broadcasting in a non-jamming environment is an example of a classical public channel. In general, a classical public channel is a weaker assumption compared to a classical private channel.

A quantum channel is a communication channel which can be used to transmit quantum information from a sender to a receiver, as opposed to a classical channel transmitting only classical information. In other words, a quantum channel can be used to transmit a quantum state as described in Eq. (1), from the sender to the receiver. An example of quantum channels is an optical fiber that can be used to transmit and maintain the polarization of photons.

3.2   Bit encryption protocol

In the following paragraphs, we give the specific steps and associated examples of the encrypted quantum communication protocol. All the steps are illustrated in Fig. 2.

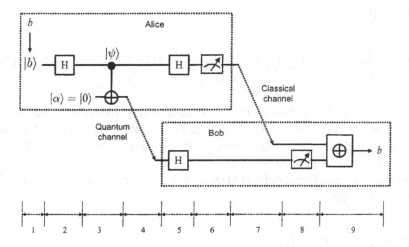

Figure 2. The encrypted quantum communication protocol with each step indicated

1. Assuming Alice has a classical secret bit $b \in \{0,1\}$ which she wants to send to Bob. To do this, Alice encodes her classical secret bit b into a quantum state $|0\rangle$ in case $b = 0$, or $|1\rangle$ in case $b = 1$.
2. Then Alice applies a **HADAMARD** gate on $|b\rangle$ to get a quantum state $|\psi\rangle$. Depending on the classical secret bit, the state will be

$$|\psi\rangle = H(|0\rangle) = \frac{1}{\sqrt{2}}(|0\rangle + |1\rangle) \qquad (11)$$

if $b = 0$, or

$$|\psi\rangle = H(|1\rangle) = \frac{1}{\sqrt{2}}(|0\rangle - |1\rangle) \qquad (12)$$

in case $b = 1$.

3. Alice then prepares an ancillary qubit $|\alpha\rangle = |0\rangle$ and applies a **CONTROL-NOT** gate $\mathbf{CN}\big(|\psi\rangle, |\alpha\rangle\big)$. The notation $\mathbf{CN}\big(|\psi\rangle, |\alpha\rangle\big)$ stands for a **CONTROL-NOT** gate with $|\psi\rangle$ as the control bit and $|\alpha\rangle$ as the target bit. This creates an entanglement between $|\psi\rangle$ and $|\alpha\rangle$, since

$$CN\left(\frac{1}{\sqrt{2}}\big(|0\rangle+|1\rangle\big)_{\psi}, |0\rangle_{\alpha}\right) = \frac{1}{\sqrt{2}}\big(|00\rangle+|11\rangle\big)_{\psi\alpha} \tag{13}$$

if $b = 0$, or

$$CN\left(\frac{1}{\sqrt{2}}\big(|0\rangle-|1\rangle\big)_{\psi}, |0\rangle_{\alpha}\right) = \frac{1}{\sqrt{2}}\big(|00\rangle-|11\rangle\big)_{\psi\alpha} \tag{14}$$

in case $b = 1$. The subscript ψ and α denote the order of the qubits.

4. Alice sends qubit $|\alpha\rangle$ to Bob through the quantum channel. After Bob gets qubit $|\alpha\rangle$, he tells Alice through the classical public channel that he has received the qubit.

5. Both Alice and Bob apply **HADAMARD** gates to their own qubits. If $b = 0$, this gives

$$H \otimes H\left(\frac{1}{\sqrt{2}}\big(|00\rangle+|11\rangle\big)_{\psi\alpha}\right)$$
$$= \frac{1}{\sqrt{2}}\left(\frac{|0\rangle+|1\rangle}{\sqrt{2}}\frac{|0\rangle+|1\rangle}{\sqrt{2}} + \frac{|0\rangle-|1\rangle}{\sqrt{2}}\frac{|0\rangle-|1\rangle}{\sqrt{2}}\right) \tag{15}$$
$$= \frac{1}{\sqrt{2}}\big(|00\rangle+|11\rangle\big)_{\psi\alpha}$$

However, in case $b = 1$, it gives

$$H \otimes H\left(\frac{1}{\sqrt{2}}\big(|00\rangle-|11\rangle\big)_{\psi\alpha}\right)$$
$$= \frac{1}{\sqrt{2}}\left(\frac{|0\rangle+|1\rangle}{\sqrt{2}}\frac{|0\rangle+|1\rangle}{\sqrt{2}} - \frac{|0\rangle-|1\rangle}{\sqrt{2}}\frac{|0\rangle-|1\rangle}{\sqrt{2}}\right) \tag{16}$$
$$= \frac{1}{\sqrt{2}}\big(|01\rangle+|10\rangle\big)_{\psi\alpha}$$

6. Alice takes a measurement of her qubit with respect to $|0\rangle$ and $|1\rangle$. According to Eq.(15) and Eq.(16), she will get a result of either $|p\rangle = |0\rangle$ or $|p\rangle = |1\rangle$ with a probability of $\frac{1}{2}$. Alice then translates the result $|p\rangle$ into the corresponding classical bit: $p = 0$ if $|p\rangle = |0\rangle$ or $p = 1$ if $|p\rangle = |1\rangle$.

7. Alice sends the result p to Bob through the classical public channel.

8. Similarly, Bob takes a measurement of his qubit with respect to $|0\rangle$ and $|1\rangle$. According to Eq.(15) and Eq.(16), he will get a result of either $|q\rangle = |0\rangle$ or $|q\rangle = |1\rangle$ with a probability of $\frac{1}{2}$. Bob then translates the result $|q\rangle$ into the corresponding classical bit: $q = 0$ if $|q\rangle = |0\rangle$ or $q = 1$ if $|q\rangle = |1\rangle$.

9. Unlike Alice, who sends her result through the classical public channel, Bob keeps the result secret and performs

$$b = p \oplus q \tag{14}$$

to recover the classical message b.

3.3   Protocol description

In the protocol described above, the information of the secret bit b is encoded as the phase of the entanglement state after Alice applies the **CONTROL-NOT** gate. This can be seen from the phase (plus vs. minus sign) in Eq.(13) and Eq.(14). If Alice sends only one qubit to Bob, the information is shared between them and can not be retrieved via any local operation. In other words, the only qubit sent by Alice via the quantum channel does not contain enough information to recover the secret bit b.

To recover the original secret bit, either a joint operation (for example, a **CONTROL-NOT** gate) or classical message exchange between the two parties is necessary. In this protocol, Alice does not send both qubits to Bob, she keeps one qubit in her hand to avoid a joint operation performed by the eavesdropper. Instead, two **HADAMARD** gates are performed by Alice and Bob separately. Since after these operations, the measurement results of these qubits become perfectly correlated (as in Eq.(15) and Eq.(16)) and the secret bit can be deduced by a simple calculation over the two classical bits according to Eq.(17). However, one of the two classical bits is now in Bob's hand. All Alice has to do is to reveal her classical bit p to Bob. To do this, Alice can send her classical bit p to Bob via the classical public channel. Note that the result announced by Alice is completely random, so it does not contain enough information for Eve to deduce the secret bit. At the end of the protocol, Bob can count the number of '1's and decrypt the secret bit according to Eq.(17). If the number of '1's is even, the message b is 0. On the other hand, if the number of '1's is odd, the message b is 1.

4.   ANALYSIS OF THE PROTOCOL

In this section, we assume the existence of an eavesdropper and show that the protocol is secure as long as the qubit sent by Alice reaches Bob.

4.1   Analysis on eavesdropping

As described previously, step (1)-(3) are performed by Alice locally. Basically these steps prepare an entanglement state depending on the secret bit b. The only chance Eve can get information from the channel is step (4) and (7). A typical attack is shown in Fig. 3.

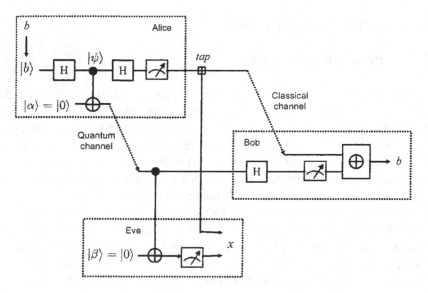

Figure 3. The encrypted quantum communication protocol with eavesdroppers

Since step (4) is the only chance for Eve to attack the quantum channel, we discuss this first. As Eve has the capability of performing quantum gates to that qubit, without lose of generality, we assume that Eve prepares an ancillary qubit $\beta = |0\rangle$ and performs a CN(α, β) to get some information from the flying qubit.

The state is now

$$CN_{\alpha\beta}\left(\frac{1}{\sqrt{2}}\left(|00\rangle+|11\rangle\right)_{\psi\alpha},|0\rangle_{\beta}\right)=\frac{1}{\sqrt{2}}\left(|000\rangle+|111\rangle\right)_{\psi\alpha\beta} \qquad (18)$$

for $b = 0$, and

$$CN_{\alpha\beta}\left(\frac{1}{\sqrt{2}}\left(|00\rangle-|11\rangle\right)_{\psi\alpha},|0\rangle_{\beta}\right)=\frac{1}{\sqrt{2}}\left(|000\rangle-|111\rangle\right)_{\psi\alpha\beta} \qquad (19)$$

in case $b = 1$. The notation $CN_{\alpha\beta}$ stands for a **CN** gate with α as the control and β as the target. In the following steps, if Eve performs a **HADAMARD** gate as Alice and Bob do in step (5), the state will evolve as follows.

1. If $b = 0$, it gives

$$H \otimes H \otimes H\left(\frac{1}{\sqrt{2}}\left(|000\rangle+|111\rangle\right)_{\psi\alpha\beta}\right)$$

$$=\frac{1}{\sqrt{2}}\left(\frac{|0\rangle+|1\rangle}{\sqrt{2}}\frac{|0\rangle+|1\rangle}{\sqrt{2}}\frac{|0\rangle+|1\rangle}{\sqrt{2}}+\frac{|0\rangle-|1\rangle}{\sqrt{2}}\frac{|0\rangle-|1\rangle}{\sqrt{2}}\frac{|0\rangle-|1\rangle}{\sqrt{2}}\right) \qquad (20)$$

$$=\frac{1}{2}\left(|000\rangle+|011\rangle+|101\rangle+|110\rangle\right)_{\psi\alpha\beta}$$

2. if $b = 1$, it becomes

$$H \otimes H \otimes H\left(\frac{1}{\sqrt{2}}\left(|000\rangle-|111\rangle\right)_{\psi\alpha\beta}\right)$$

$$=\frac{1}{\sqrt{2}}\left(\frac{|0\rangle+|1\rangle}{\sqrt{2}}\frac{|0\rangle+|1\rangle}{\sqrt{2}}\frac{|0\rangle+|1\rangle}{\sqrt{2}}-\frac{|0\rangle-|1\rangle}{\sqrt{2}}\frac{|0\rangle-|1\rangle}{\sqrt{2}}\frac{|0\rangle-|1\rangle}{\sqrt{2}}\right) \qquad (21)$$

$$=\frac{1}{2}\left(|001\rangle+|010\rangle+|100\rangle+|111\rangle\right)_{\psi\alpha\beta}$$

From Eq.(20) and Eq.(21), we see this still makes the total number of '1's even in case $b = 0$ and odd in case $b = 1$. After Alice announces her measurement result in step (7), if Bob knew the result of all three qubits he could still count the total number of '1's to deduce the secret bit.

Assuming the secret bit $b = 0$, the total number of '1's is even ($|000\rangle, |011\rangle, |101\rangle, |110\rangle$). However, there is a probability of $\frac{1}{2}$ ($|011\rangle$ and $|101\rangle$) that Eve has a '1' in her hand. The silent eavesdropper has no way to get rid of this bit and push this information back to Alice or Bob. This makes the total number of '1's belonging to Alice and Bob odd and will hence flip the secret bit. As for Eve's qubit, it carries no information because it can be either $|0\rangle$ or $|1\rangle$, each with a probability of $\frac{1}{2}$. In summary, the intrusion introduces an error but gives Eve no information. Similar analysis holds for other unitary operations performed by Eve.

Since the existence of eavesdropping will inevitably introduce errors, Alice and Bob can detect the intrusion by appending an error checking code in the message. A simple error checking algorithm that allows two parties to perform message encryption is shown in the following section.

4.2 Message encryption protocol

The bit encryption protocol allows two parties to transmit one classical bit each time. The result is either a successful transmission or a bit-flip induced by eavesdropping. With the bit encryption protocol described above, an n-bit message can be sent using the following procedure to protect its integrity.
1. Alice sends the message bit-by-bit using the bit encryption protocol.
2. They negotiate publicly to decide a hash function.
3. Alice sends the hash result, bit-by-bit, using the bit encryption protocol.
4. Bob gets both the message and hash result. He can check the integrity of the message using the hash. If they don't match, the message is corrupted. Otherwise, the message is valid.

4.3 Channel analysis

In this protocol, two communication channels are used. One is a classical public channel; the other is a quantum channel. As described previously, the classical channel is a public channel, so the data is public readable. However, we did not discuss whether the channel can be publicly writable. Actually, if the classical public channel is contaminated, the result decrypted by Bob will be flipped and hence cause an error. From this point of view, the classical public channel is publicly writable, but any incorrect value inevitably causes an error. This is because an attack in the classical channel is protected by the quantum channel and will be detected. Moreover, this implies that the protocol still works even if a man-in-the-middle exists only in the classical channel. Similarly, the quantum channel is publicly writable as long as the classical channel is not contaminated. This is because even the flying qubit is replaced by an uncorrelated new qubit, the eavesdropping will still be detected by the integrity checking process. However, if both classical and quantum channels are controlled by Eve, then she will be able to do whatever she likes as a man-in-the-middle. This becomes an authentication problem, which is outside the scope of this paper.

5. CONCLUSION

In this article, we propose a new cryptographic protocol based on a phase-encoding scheme. Local operation and classical communication can be used to achieve private communications between the sender and the receiver. In case eavesdroppers exist and have total control of the channel, the protocol not only gives absolutely no information but also can detect the existence of eavesdroppers. Unlike its classical counterpart, the security of the protocol does not depend on any unproven hard problems. It is based on the laws of physics.

REFERENCE

1. P. Shor, Algorithms for quantum computation: discrete logarithms and factoring, *Proceeding of the 35th Annual IEEE Symposium on the Foundations of Computer Science*, 124-134(1994).
2. R. Rivest, A. Shamir, and L. Adleman, A method for obtaining digital signatures and public-key cryptosystems, *Communications of the ACM*, vol. (2) 21, pp. 120-126(1978).
3. W. Diffie, and M. E. Hellman, Multiuser Cryptogrphic Techniques, *Proceeding of AFIPS National Computer Conference*, 644-654(1976).
4. C. Bennett, and G. Brassard, Quantum Cryptography: Public Key Distribution and Coin Tossing, *Proceedings of IEEE International Conference on Computers Systems and Signal Processing*, 175-179 (1984).
5. A. K. Ekert., Quantum Cryptography based on Bell's theorem, *Phys. Rev. Lett* 67(6), 661-663(1991).
6. C. Bennett, Quantum Cryptography: Uncertainty in the Service of Privacy, *Science* 257, 752-3 (1992).
7. K. Bostrom, and T. Felbinger, Deterministic Secure Direct Communication using Entanglement, *Phys Rev Lett.* 2002 Oct 28;89(18):187902.

8. Qing-Yu Cai, Deterministic Secure Direct Communication using Ping-Pong Protocol without Public Channel, http://xxx.lanl.gov/abs/quant-ph/0301048.
9. Qing-Yu Cai, Deterministic secure communication protocol without using entanglement, *Chin. Phys. Lett*, 21(4),601 (2004).
10. Z. Zhao, T. Yang, Z.-B. Chen, J.-F. Du, and J.-W. Pan, Deterministic and highly efficient quantum cryptography with entangled photon pairs, *Phys. Rev. Lett.*, http://xxx.lanl.gov/abs/quant-ph/0211098.

TRUSTED COMPONENT SHARING BY RUNTIME TEST AND IMMUNIZATION FOR SURVIVABLE DISTRIBUTED SYSTEMS

Joon S. Park[1], Pratheep Chandramohan[2], Ganesh Devarajan[3], and Joseph Giordano[4]

[1,2,3]*Laboratory for Applied Information Security Technology (LAIST), School of Information Studies, Syracuse University; [4]Information Directorate, Air Force Research Laboratory*

Abstract: As information systems became ever more complex and the interdependence of these systems increase, the survivability picture became more and more complicated. The need for survivability is most pressing for mission-critical systems, especially when they are integrated with other COTS products or services. When components are exported from a remote system to a local system under different administration and deployed in different environments, we cannot guarantee the proper execution of those remote components in the currently working environment. Therefore, in the runtime, we should consider the component failures (in particular, remote components) that may either occur genuinely due to poor implementation or the failures that occurred during the integration with other components in the system. In this paper, we introduce a generic architecture and mechanisms for dynamic component-failure detection and immunization for survivable distributed systems. We have also developed a prototype system based on our approaches as a proof of our ideas.

Keywords: Component Immunization; Recovery; Survivability.

1. INTRODUCTION

Although advanced technologies and system architectures improve the capability of today's systems, we cannot completely avoid threats to them. This becomes more serious when the systems are integrated with

Commercial Off-the-Shelf (COTS) products and services, which usually have both known and unknown flaws that may cause unexpected problems and that can be exploited by attackers to disrupt mission-critical services. Usually, organizations including the Department of Defense (DoD) use COTS systems and services to provide office productivity, Internet services, and database services, and they tailor these systems and services to satisfy their specific requirements. Using COTS systems and services as much as possible is a cost-effective strategy, but such systems—even when tailored to the specific needs of the implementing organization—also inherit the flaws and weaknesses from the specific COTS products and services used. Traditional approaches for ensuring survivability do not meet the challenges of providing assured survivability in systems that must rely on commercial services and products in a distributed computing environment[31, 29, 30].

The damage caused by cyber attacks, system failures, or accidents, and whether a system can recover from this damage, will determine the survivability characteristics of a system. A survivability strategy can be set up in three steps: protection, detection and response, and recovery[21, 16, 18]. To make a system survivable, it is the mission of the system, rather than the components of the system, to survive. This implies that the designer or assessor should define a set of critical services of the system to fulfill the mission. In other words, they must understand what services should be survivable for the mission and what functions of which components in the system should continue to support the system's mission[25].

2.   DEFINITION OF SURVIVABILITY

The definitions of survivability have been introduced by previous researchers[20, 23]. In this paper, we define survivability as the capability of an entity to continue its mission even in the presence of damage to the entity. An entity ranges from a single component (object), with its mission in a distributed computing environment, to an information system that consists of many components to support the overall mission. An entity may support multiple missions. Damage can be caused by internal or external factors such as attacks, failures, or accidents. If the damage suspends the entity's mission, we call it *critical damage* (CD), and if it affects overall capability, but the mission can still continue, we call it *partial damage* (PD). Since we believe survivability is a mission-oriented capability, there are basically three abstract states of the system: normal, degraded, and suspended. A system is in the normal state (S_0) when it is running with full capability. It is in the degraded state (S_1) when it is running with limited capability because of PD, which does not suspend the overall mission. Finally, the system is in the suspended state (S_2) when it cannot continue its mission because of CD. When partial recovery (PR) occurs to an infected component, only the mission-related service is recovered, so the service is still in a degraded mode with limited capacity. When there is a total recovery (TR) such as that resulting from component substitution, service is provided at full capacity. As understood intuitively, PR and TR on S_0, PD and PR on S_1, and PD and CD on S_2 do not change their current states. From the survivability point of view, we may put up with partial damages (PD) on the system as long as the mission carries on. We may simply insulate the damaged components from others instead of recovering them, although the performance of the overall system may degrade. However, if the damage is so critical that the system cannot continue its mission, we must recover the damaged components as soon as possible in order to continue the mission. We describe the concept of survivability using a finite state machine. Abstractly, we can consider the damages and recovery actions as inputs to a survivable entity. Furthermore, we can classify the outputs of the entity into two abstract cases, one for the outputs when the mission performed (M) successfully, and one for the outputs when the mission failed (F). This generates Table 1, which shows the transitions and outputs for each pair consisting of a state and an input. Based on this table, we generate a finite state machine in Figure 1.

Table 1. State Table for Survivable Systems

State (S)	Transition Function (f)				Output Function (g)			
	Next State				Output (O)			
	Input (I)				Input (I)			
	PD	CD	PR	TR	PD	CD	PR	TR
S_0	S_1	S_2	S_0	S_0	M	F	M	M
S_1	S_1	S_2	S_1	S_0	M	F	M	M
S_2	S_2	S_2	S_1	S_0	F	F	M	M

The finite-state machine $M = (S, I., O, f. g, s_0)$ consists of a finite set S of states (where S_0 is an initial state), a finite input alphabet I, a finite output alphabet O, a transition function f that assigns each state and input pair to a new state, and an output function g that assigns each state and input pair to an output. In this state diagram, we have three states (normal state (S_0), degraded state (S_1), and suspended state (S_2)), four types of inputs (partial damage (PD), critical damage (CD), partial recovery (PR), and total recovery (TR)), and two outputs (when mission performed (M), and when mission failed (F)).

To continue the mission, the system must stay in either S_0 or S_1. Some strict missions do not allow the critical components to stay even one moment in the suspended state (S₂) until the mission is completed. However, in reality, we believe most missions may allow critical components to stay in the suspended state (S₂) for a moment until they are recovered and the state is changed to the degraded state (S₁) or normal state (S₀).

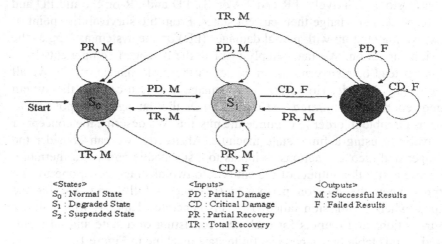

Figure 1. Abstract State Diagram for Survivable Systems

We could decompose S₁ and S₂ into sub-states to represent detailed transitions based on the actual missions and applications described in[20]. In this paper, however, we introduce a generic approach to describe the concept of survivability with the abstract inputs, states, and outputs. We believe the three states (S₀, S₁, and S₂), the four kinds of inputs (PD, CD, PR, TR), and the two kinds of outputs (M, F) can represent the state transitions of a survivable entity based on our mission-oriented survivability definition.

3.   RELATED WORK

3.1   Black-box and white-box testing

Currently, existing technologies for identifying faulty components are more or less static in nature. One of those approaches employs black box testing for the components. In this technique, the behavioral specification[2] is provided for the component to be tested in the target system. The main disadvantage of this technique is the specifications should cover all the detailed visible behavior of the components, which is impractical in many situations. Another approach employs a source-code analysis, which depends on the availability of source code of the components. Software testability analysis[35] employs a white-box testing technique, which determines the locations in the component where a failure is likely to occur. Unlike black box testing, white box testing allows the tester to see the inner details of the component and later help him to create appropriate test data. Yet another approach is software component dependability assessment[36], a modification to testability analysis where each component is tested thoroughly. These techniques are possible only when the source code of the components is available.

3.2   Fault injection

In the past[19] we have employed a simple behavioral specification utilizing execution-based evaluation. Their approach combines software fault injection[1, 24, 33, 34] at component interfaces and machine learning techniques to: (1) identify problematic COTS components; and (2) understand these components' anomalous behavior. In their approach of isolating problematic COTS components, they created wrappers and introduced them into the system under different analysis stages to uniquely identify the failed components and to gather information on the circumstances that surround the anomalous component behavior. Finally, they preprocess the collected data and apply selective machine learning algorithms to generate a finite state machine for better understanding and increasing the robustness of the faulty components. In other research[7] the authors have developed a dynamic problem determination framework for a large J2EE platform, employing a fault detection approach based on data clustering mechanisms to identify faulty components. This research also employed a fault injection technique to analyze how the system behaves under injected faults.

3.3   Bytecode instrumentation

Performing fault injection analysis and providing immunization to the components either by rewriting the existing code or by creating additional wrappers is a non-trivial task when the source code for the component is not readily available. Source code may not be available at all when we are dealing with COTS components and externally administered components downloaded dynamically in runtime at local machine. This is an issue that needs to be addressed before proceeding further. Providing immunization and performing fault injection at the component interfaces require modification of the component code; however, we assume that the source code is *not* available in a large disturbed application. Instead, we provide the immunization to the runtime code (e.g., JAVA Bytecode) by extending the code instrumentation technique[5, 6, 8, 10, 15, 17]. Instrumentation techniques have previously been used for debugging purposes; to evaluate and compare the performance of different software or hardware implementations such as branch prediction, cache replacement, and instruction scheduling; and in support of profile-driven optimizations[3, 9, 11, 22].

4.   RUNTIME COMPONENT TEST AND IMMUNIZATION

4.1   Generic system architecture

Figure 2 shows the generic architecture of our component failure detection and immunization system. It consists of a Monitoring Agent, an Immunization Agent, and a Knowledge Base. The monitoring agent is further divided into two subsystems: the fault injection subsystem and fault detection subsystem. Before a component is run on a host (especially a mobile component downloaded from another machine under different administration), the fault injection subsystem injects faults into the component, while the fault detection system analyzes component behavior in response to the injected faults. The component's internal structure information, such as method interface, arguments, local variables, etc., is accessible in runtime; thus, this information can be used in the dynamic component analysis.

If there is no abnormal behavior, the monitoring agent allows the component to run in the local machine. For the performance reason, we can finish this analysis with the local components and fix the failures in the source codes before the operation starts (if the source codes are available). However, this is not possible for the remote components because their source codes are usually not available to the local machines. When the monitoring agent detects abnormal behaviors in the mobile component through the fault injection analysis, the fault detection subsystem identifies the reason for failure and informs the immunization agent to immunize the faulty component accordingly.

The immunization agent builds and deploys immunized components to the target system. The immunization agent possesses a knowledge base that consists of a list of procedures for how to provide immunization for component failures. The immunization agent provides immunization and increases the survivability of the faulty components. Basically, there are two options to increase the survivability of the vulnerable components and to make it more robust[12]: (1) inform the vendor of the software problems and wait for a patch; or (2) immunize the components with wrappers or instrument the faulty methods with updated and modified methods for more robust behavior[4, 26]. The first technique is not feasible for dynamic runtime recovery from errors; consequently, we have adopted the second approach to provide immunization and increase the survivability of vulnerable components.

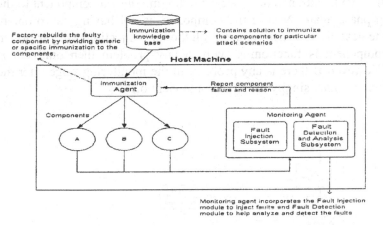

Figure 2. Component Failure Detection and Immunization

4.2   The strategy

Figure 3 summarizes the steps involved in the entire process of detecting and immunizing faulty components. When we download a component from the remote location we perform the first test to determine if there are any dependent components. If so, we also download the dependent component. The component that is downloaded is an executable file for which we don't have the source code. By using an additional tool in runtime (e.g. Jikes BT[15] for JAVA bytecode), however, we can determine most of the intricate structure details of the component that we have downloaded. The test as well determines the structure of the code (including the data flow and the interdependencies of the functions inside the component) that is required to do a runtime test in the local environment. Then, we go into the next phase of monitoring the component performance.

In the next phase we inject the faults and observe the performance of the component. The fault injection module injects test inputs (faults) and analyses the behavior of the component. Different machines (or applications) may have different fault injection modules based on their test criteria. For instance, one module may test internal failures, while another may test the robustness against cyber attacks. After the test inputs are injected we collect evidences and reasons for the failures, specific methods, classes that are affected. If there are any failures detected we check if we can provide some immunization to that failure from the knowledge base that we have built and updated regularly. If we have a specific solution for the failure we provide it from the knowledgebase, otherwise we provide it a generic immunization[27, 28]. After the immunization is done we send the immunized component to the monitoring phase again. Now if the component is not having any problem we go to the next phase where we see if all the fault injections are performed and the component is functioning without any problem then our goal is achieved. However, if there is any problem in the monitoring stage after the immunization we may simply drop the component off.

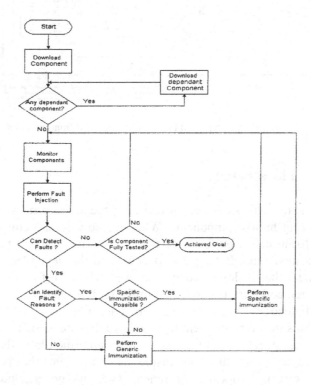

Figure 3. Strategy for Identifying Component Failures and Immunization

We can provide component immunization in runtime by either class-level modifications or method-level modifications. By class-level modification the references to the original class definitions are replaced by another subclass of the original class. By method-level modification, we modify some of the runtime (executable) code in the original method by adding new runtime code (i.e. Java bytecode in our implementation) or deleting some runtime code or both at the same time. The latter provides more flexibility to build more powerful immunized class. At the same time method level modification is more difficult to implement than class level modifications because it involves direct modification of already existing Java bytecode whereas the class-level modification just involves rearranging references in the class file. The main advantage of using method level instrumentation techniques is that all the modifications are transparent to other components, which make calls to the modified components because the semantics and syntax are still maintained after modifications.

5. PROTOTYPE DEVELOPMENT

Although the detailed techniques for component-failure detection and immunization are slightly different based on the programming languages, applications, and local policies, we believe our approach is applicable to most of distributed systems, which require survivability. We focus on the component failure scenarios here, but we believe our approach can be extended with cyber attacks. In our experiment, we detect component failures such as naming collisions, infinite loops, multi-threading problems, and array out-of-bound problems, and successfully immunized them in runtime so that the component's service can continue in a reliable manner. In the following description, we mainly concentrate on the problems of naming collisions because they cannot be rectified in the programming time and this particular paper has a space constraint. The other problems might be avoided when the programmer takes extra care during programming. However, we still need to check those problems in a remote component during runtime under a strict component-sharing policy.

5.1 Detection and immunization of naming collisions

When we perform tests for a local component, naming collision across other spaces cannot be detected. However, when we perform the test after the component is downloaded from a remote machine and integrated with local components there can be naming collisions occurring. There are possibilities that two or more components, which are being integrated together, might have the same variable name or even within the same component the same variable name can be used in different contexts. When the client program tries to access these variables there are possibilities that it might get the wrong value.

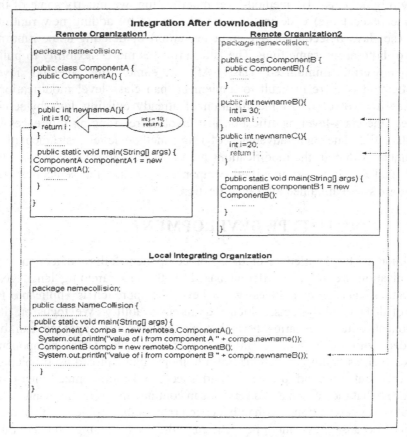

Figure 4. Detection and Immunization of Naming Collision

The downloaded component's internal structure information such as method interface, arguments, local variables, etc. is collected in runtime after analyzing the Java bytecode. Using the structure information and fault injection module, the local machine performs a fault injection test to determine all the return values in the component. This enables the local machine to figure out the architecture of the component and then to decide, which are all the function values required. Once the functions are selected the component is passed into the naming collision test stub where we test if there is any other component with the same variable name returning the value. If the testing says there are no variables with the same variable name then integration is taken to the next level.

Now if there are variables with the same variable name from different component then our immunization code for this scenario will be creating an instance for the remote class. Using this instance we access the method name and through that we access the variable value(e.g. compa.newnameA()). The renaming process to avoid naming collision is to be performed mainly when we convert the private function to public function. The original source code writer's intension would have been that the function was a private function its scope is well defined and hence he can reuse the variable name. If there is a private function then this will not affect our processing as that variable it limited to the scope of that class. However, if there are two variables from the same component with the same name then we can go about changing the name as per the naming convention so that it becomes easier for the programmer who is working with the source code generated from the bytecode to differentiate from the other common variable named item.

The main advantage of this system is that we can get access to the variables which where initially not possible to access and then by renaming them we are able to distinguish between the two similarly name variable. This as well helps in the optimal code re-usage. In reality, performing instrumentation is a non-trivial task because it involves precision handling of instructions. In most of the cases the instrumentation requires dealing with intermediate-level code (e.g., Java bytecode) or low-level code (e.g., Assembly), which requires ultra care when modifying these kinds of code. Basically our principle can be applied to more complex problems but the complexity of the immunization code increases quite considerably when dealing with complex problems. An important point should be noted here that it is not always possible to apply immunization by changing the code (Java bytecode in this case). In some cases the reason for the failure is not known even after performing thorough fault injection analysis. In other cases code segments can be inherently complex to be discerned for further modifications (immunization). In such scenarios specific immunization is not possible, so we need to provide generic immunization by rebuilding the faulty component or deploying it in a new conducive environment.

As depicted in Figure 4, we download two components from remote location A and remote location B. After the download we first modify the package name so that the downloaded component can also become a part of the new component being developed. Supposing the programmer is interested in the method newnameB() after looking into the component's architecture. He simply modifies the private method to a public method and then finds out that there exists a naming collision within the same component. In order to access the variable value the method has to be made public. Now that the fault injector has made the method public with a return value, he can access that variable value by simply creating an instance of the remote object in the local component and hence being able to access that newly converted public methods' return variable value.

5.2 Evaluation results

We implemented the prototype for the component evaluation phase of our fault detection and runtime immunization approach to determine the existence of naming collisions. After we generate the source codes we perform three stages of tests to: (1) identify the variables in use; (2) ascertain the scope of each variable; and (3) determine if naming collisions will occur when their respective intermediate values are accessed.

There are two scenarios of accessing the variables in other components. Suppose component A tries to access a variable "i" in component B, and they both are in the same package, where class1 is in component A and class2 is in component B. The procedure followed to access that variable is by classname.methodname.variablename—in our example, class2.func2.i. Through this method component A will be able to access the variable "i" in component B. Still, there is a possibility that the variable "i" may not be accessible as it could be in the private member function of the component B. For this reason, we need to extend the scope of that method to public. When we extend the variable's scope there is a chance that there is another variable "i" in the same component, which is globally defined or within the same method with another initialization to the same variable. Consequently, the accessing component might be getting the last assigned value to that variable. In order to access the initial value, we will have to assign different names to those variables that cause naming collisions.

The second scenario occurs when a component is trying to access the variables from different components. Suppose component A is accessing the variable "i" from component B, as well as variable "i" from component C. The first step for the component to access the variables from different components will be to put them all into the same package. After this, we have to check the scope of the variable to determine if it is possible for another component to access this variable; if not, then we will have to extend the scope of the variable and then verify it doesn't cause any naming collisions, and then provide access to the component attempting to access that variable. Suppose class 2 is in component B and class 3 is in component C, and methods func2 is in class2 and func3 is in class3, to access the value of the "i" in component B, the code will be class2.func2.i. Similarly, the variable "i" in component C can be accessed using the code class3.func3.i. To avoid further confusion, we can assign these variables to different names after abstracting them in component A so that naming collisions do not occur in the root component.

Table 2. Naming Collision Results

Number of Components Tested	Number of Components with Naming Collision	Total Number of variables reused	Naming collisions without Scope Extension	Naming Collisions with Scope Extension	Detected and Immunized
81	37	104	30	36	66

Table 2 shows the test results for the components that were tested in our experiment. Most of the components that where tested were downloaded from IBM's Alpha works website, while the rest were from various other sources. Each component has a minimum of 100 lines of code or more.

The total number of components tested was 81. Out of the 81 components, 37 components experienced naming collision problems, both before and after their respective scopes were extended. A total of 104 variable names were reused in different scopes in the various components. Out of these 104 variables, 30 variables had scopes that were not well defined, causing naming collisions even without an extension in scope. There were a total of 36 variables that caused naming collisions after their scope was extended. We were able to detect all 66 instances where variables caused naming collisions.

6.   CONCLUSION AND FUTURE WORK

Although many current systems are being developed using Java, there are also many other distributed software components developed using other technologies such as Windows COM[14] (e.g., DLLs), Unix Shared Libraries (e.g., SO files), and even .Net libraries. The .Net platform is relatively new and is a major competitor for Sun's Java. The .Net uses Intermediate Language (IL), which is very similar to the Java Intermediate Bytecode and uses a Common Language Runtime (CLR) also very similar to Java Virtual Machine (JVM) to load the code in to memory. Since .Net and Java share common object oriented model, memory models, semantics and architecture. Our instrumentation and immunization techniques can be directly applied with little modifications. In contrast, DLLs and Shared Libraries are quite different from the bytecode (intermediate code) because these are libraries in assembly code (low level). In the past there has been some research conducted in this area, and in[13] they have formulated a technique to intercept and instrument COM applications. We can apply our methodologies theoretically to these platforms but in reality we may face some technical challenges. Instrumenting Windows COM applications is more difficult than instrumenting Java bytecode because of the inherent complexity of the COM technology. Our future goal is to apply our current immunization techniques to other platforms by overcoming these complexities.

Furthermore, we consider that cyber attacks may involve tampering of existing source code to include undesired functionality, replacing or modifying a genuine component with a malicious one. Software components can be subject to two major kinds of attacks, (1) An attack involving the modification of existing source code to introduce additional malicious functionality, and, (2) An attack involving the introduction of a malicious piece of code independent of the original program that can be started when the original component is used and run independent of it (e.g. a Trojan Horse). Our goal is to detect this unauthorized integrity change in code by extending our previous work[32] and extract the malicious parts out of the component while retaining its originally expected functionality.

ACKNOWLEDGMENTS

This work was supported by the US Air Force Research Lab's extended research program, based on the SFFP (Summer Faculty Fellowship Program) award in 2004, sponsored by the Nation Research Council (NRC) and the Air Force Office of Scientific Research (AFOSR).

REFERENCE

1. Dimiter R. Avresky, Jean Arlat, Jean-Claude Laprie, Yves Crouzet. *Fault Injection for the Formal Testing of Fault Tolerance.* The Twenty-Second Annual International Symposium on Fault-Tolerant Computing, July 8-10, 1992: 345-354.
2. Abadi and L. Lamport. *Composing Specications.* ACM Transactions on Programming Languages, 15(1): 73-132, January 1993.
3. Anant Agarwal, Richard Sites and Mark Horwitz. *ATUM: A New Technique for Capturing Address Traces Using Microcode.* In Proceedings of the 13th International Symposium on Computer Architecture, 119-127, June 1986.
4. Amitabh Srivastava and Alan Eustace. "*ATOM A System for Building Customized Program Analysis Tools.*" In Proceedings of the SIGPLAN '94 Conference on Programming Language Design and Implementation (PLDI), pages 196-205, June 1994.
5. BCEL - Bytecode Engineering Library http://bcel.sourceforge.net/

6. BIT: Bytecode Instrumenting Tool http://www.cs.colorado.edu/~hanlee/BIT/index.html
7. M. Chen, E. Kiciman, E. Brewer, and A. Fox. Pinpoint: *Problem Determination in Large, Dynamic Internet Services*. In Proceedings of the IEEE International Conference on Dependable Systems and Networks, DSN, 2002.
8. Ajay Chander, John C. Mitchell, Insik Shin. *Mobile Code Security by Java Bytecode Instrumentation*. In Proceedings of the 2001 DARPA Information Survivability Conference & Exposition (DISCEX II), pages 1027-1040, Anaheim, CA, June 2001.
9. Brian Bershad et al. *Etch Overview*. http://etch.cs.washington.edu/
10. James R. Larus and Eric Schnarr. *"EEL: Machine-Independent Executable Editing."* In proceedings of the SIGPLAN '95 Conference on Programming Language Design and Implementation (PLDI), pages 291-300, June 1995.
11. Susan J. Eggers, David R. Keppel, Eric J. Koldinger, and Henry M. Levy. *Techiques for efficient Inline Tracing on   a Shared-Memory Multiprocessor*. In Pro-ceedings of the 1990 ACM Sigmetrics Conference on Measurement and Modelings of Computer Systems, 8(1), May 1990.
12. A. Ghosh, J. Voas. *Inoculating Software for Survivability*. Communications of the ACM, July 1999.
13. Galen Hunt and Doug Brubacher. *Detours: Binary Interception of Win32 Functions*. Proceedings of the 3rd USENIX Windows NT Symposium, pp. 135-143. Seattle, WA, July 1999.USENIX.
14. Galen Hunt and Michael Scott. *Intercepting and Instrumenting COM Applications*. Proceedings of the Fifth Conference on Object-Oriented Technologies and Systems (COOTS'99), pp. 45-56. San Diego, CA, May 1999. USENIX.
15. Jikes Bytecode Toolkit - IBM Alpha Works http://www.alphaworks.ibm.com/tech/jikesbt.
16. S. Jajodia, C. McCollum, and P. Ammann. *Trusted Recovery*. Communications of the ACM, 42(7), pp. 71-75, July 1999.
17. JOIE - The Java Object Instrumentation Environment http://www.cs.duke.edu/ari/joie/
18. J. Knight, M. Elder, and X. Du. *Error Recovery in Critical Infrastructure Systems*. Proceedings of the 1998 Computer Security, Dependability, and Assurance (CSDA'98) Workshop, Williamsburg, VA, November 1998.
19. G. Kapfhammer, C. Michael, J. Haddox, R. Coyler. *An Approach to Identifying and Understanding Problematic COTS Components*. The Software Risk Management Conference, ISACC 2000.
20. J. Knight and K. Sullivan. *Towards a Definition of Survivability*. Proceedings of the 3rd Information Survivability Workshop (ISW), Boston, MA, October 2000.
21. P. Liu, P. Ammann, and S. Jajodia. *Rewring Histories: Recovering from Malicious Transactions*. Distributed and Parallel Databases, 8(1), pp. 7-40, January 2000.
22. James R. Larus and Thomas Ball. *Rewriting Executable Files to Measure Program Behavior*. Software, Practice and Experience, 24(2), February 1994.
23. H. Lipson and D. Fisher, *Survivability -- A New Technical and Business Perspective on Security*. Proceedings of the New Security Paradigms Workshop (NSPW'99), Caledon Hills, Ontario, Canada, September 21-24, 1999.
24. Henrique Madeira, Diamantino Costa, Marco Vieira. *On the Emulation of Software Faults by Software Fault Injection*. International Conference on Dependable Systems and Networks (DSN 2000). New York, New York, June 25 - 28, 2000.
25. N. Mead, R. Ellison, R. Linger, et al. *Survivability Network Analysis Method*, SEI Technical Report: CMU/SEI-00-TR-013, September 2000.
26. Amitabh Srivastava and David Wall. *"A Practical System for Intermodule Code Optimization at Link-Time."* Journal of Programming Languages, vol 1, no 1, pages 1-18, March 1993.
27. Joon S. Park. *Component Survivability for Mission Critical Distributed Systems*. Technical Report, NRC/Air Force SFFP (Summer Faculty Fellowship Program), 2004.

28. Joon S. Park and Pratheep Chandramohan. *Component Recovery Approaches for Survivable Distributed Systems.* 37th Hawaii International Conference on Systems Sciences (HICSS-37), Big Island, Hawaii, January 5-8, 2004.
29. Joon S. Park, Pratheep Chandramohan, and Joseph Giordano. *Survivability Models and Implementations in Large Distributed Environments.* 16th IASTED (International Association of Science and Technology for Development) Conference on Parallel and Distributed Computing and Systems (PDCS), MIT, Cambridge, MA, November 8-10, 2004.
30. Joon S. Park, Pratheep Chandramohan, and Joseph Giordano. *Component-Abnormality Detection and Immunization for Survivable Systems in Large Distributed Environments.* 8th IASTED (International Association of Science and Technology for Development) Conference on Software Engineering and Application (SEA), MIT, Cambridge, MA, November 8-10, 2004.
31. Joon S. Park and Judith N. Froscher. *A Strategy for Information Survivability.* 4th Information Survivability Workshop (ISW), Vancouver, Canada, March 18-20, 2002.
32. Joon S. Park and Ravi Sandhu. *Binding Identities and Attributes Using Digitally Signed Certificates.* 16th IEEE Annual Computer Security Applications Conference (ACSAC), New Orleans, Louisiana, December 11-15,   2000.
33. Ted Romer, Geoff Voelker, Dennis Lee, Alec Wol-man, Wayne Wong, Hank Levy, Brian Bershad, and Brad Chen. *Instrumentation and Optimization of Win32/Intel Executables Using Etch. In Proceedings of the 1997USENIX Windows NT Workshop.* August 1-7, 1997.
34. Jeffrey Voas. *Software Fault Injection.* IEEE Spectrum, appeared in 2000.
35. Jeffrey Voas, Keith W. Miller, and Jeffrey E. Payne. *Pisces: A tool for predicting software testability.* In the Proceedings of the Symposium on Assessment of Quality Software Development Tools, pages 297-309, New Orleans, LA, May 1992.
36. Jeffrey Voas and Jeffrey Payne. *Dependability certification of software components.* Journal of Systems and Software, 2000.

DESIGN AND IMPLEMENTATION OF TPM SUP320

Jiangchun Ren, Kui Dai, Zhiying Wang, Xuemi Zhao and Yuanman Tong
School of Computer, National University of Defense Technology, Changsha, Hunan, P.R.China, 410073

Abstract: Security of computer in network is becoming more and more challengeable. The traditional way of applying a common smart card to application can not meet the requirement of high degree of security in critical systems. Trust Computing Group (TCG) drafts out specifications on trust computing platform, which have been acknowledged by specialists in this field. Following these specifications, we designed and implemented a chip named SUP320 with SOC technology. This paper gives the chip's hardware architecture, firmware modules and method for low power. Performance of SUP320 is tested in the end. We find that SUP320 is better than traditional smart cards in both security and efficiency.

Key words: TCPA(TCG); TPM; SUP320; SOC; low power; smart card; keys management.

1. INTRODUCTION

Computer systems in network are often attacked by viruses and trojan horses, the basal platform can not build up a trust and secure environment for applications on it. Most systems resist hacker's attack by technologies such as digital certificates and public key infrastructure to authenticate participants and provide cryptographic keys. But the arithmetic of cryptography reduces system's efficiency heavily. A smart card then is used to accelerate the arithmetic by hardware component.

The smart card surely increases the system security to some extent, but it can not meet the requirement of high degree of security. For it has some obvious disadvantages: firstly, the communication protocol between the card and the host is too easy that users' privacy may be hijacked in the middle; Secondly, the card's processor and memory are isolated and just connected by wires, the data in memory is possibly decrypted by hacker who can steal the card; and thirdly, the mode of smart card is single, administrator can not apply different security policies to different applications. Just for these reasons, Compaq, IBM, Intel, HP and Microsoft launched and formed Trust Computing Platform Alliance (TCPA, renamed TCG later), TCG drafts out a specification on the subsystem for security in universal platform[1]. The specification suggests a Trust Computer Base (TCB) should be used in the platform and the whole system's security infrastructure should be built up on it. The TCB combines a highly secure chip with its outside circuit. The secure chip is often named Trust Platform Module (TPM)[1][2][3].

In this year, we designed and implemented a TPM chip named SUP320, and designed an architecture of subsystem which can be embedded in the common computers. Platforms with this subsystem can get assistance for security in all kinds of levels: hardware, OS kernel and application[5].

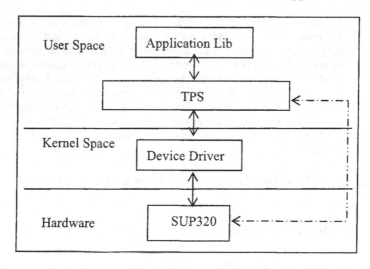

Figure 1. Framework of Secure Subsystem

The subsystem is composed of physical chip SUP320 including outside circuitry, device driver, Trust Platform Service (TPS) and libraries assisting for applications. Among those four levers, SUP320, a system on chip is the core of the subsystem. SUP320 has two kinds of important functions: accelerating security arithmetic and intercommunicating with host according to a robust protocol (explained in the broken line)[6][7]. This paper firstly presents the hardware architecture of SUP320; then describes the firmware modules in detail; following that, introduces the method for low power; Finally, tests the chip's performance.

2.   HARDWARE ARCHITECTURE

The SUP320 was implemented by 0.18um 1P6M CMOS technology, Its die size is 4.9×4.9 mm^2, The cost of power is about 0.6W. Its hardware architecture is presented in figure 2.

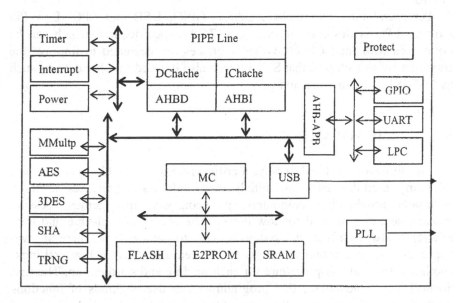

Figure 2. Hardware Architecture of SUP320

The core of SUP320 is a 32 bit RISC processor named "TengYue-1", which is designed by ourselves. The processor works well under frequency of 200 MHz, with a five-lever pipeline, an independent data cache and an instruction cache. Cache sizes are both 2K bytes.

MMultp, AES, DES and SHA are co-processors, which accelerate arithmetic of RSA/ECC, AES, DES and SHA. Because the operations of modular multiplication and modular power are frequently used in public-key arithmetic of RSA and ECC, we design a co-processor "MMultp" to speedup operation of modular multiplication, then speedup operation of modular power based on that. It is easy to realize the arithmetic of RSA (the length of key can be 512bit, 1024bit and 2048bit) and ECC (the length of key can be 160 bit and 192 bit) by different procedures[8][9][10].

TRNG is a true random numbers generator which produces numbers through noises in physical matter. These true random numbers are used in the module of communication protocol and public-key arithmetic to get big modular number pairs.

There are three kinds of memories on the chip: FLASH, EEPROM and SRAM. The size of flash is 128K bytes, in which the whole firmware is stored. The size of EEPROM is 128K bytes too, in which all kinds of keys and privacy data are stored. The SRAM's size is 16K bytes. It acts as a work room for the system on chip. These memories are all managed by a memory controller.

Peripheral interfaces include UART, GPIO, USB and LPC (Low Pin Count). LPC is designed according to TCPA specification, which can be connected with Intel CPU[4]. Other interfaces are designed to improve the flexibility of this chip so that SUP320 can also be used in other devices such as USB-KEY, secure disk and etc.

3.   FIRMWARE

The firmware of SUP320 plays a critical role in the SOC. It is composed of many modules such as initialization, self test, interfaces abstract, arithmetic accelerating, community protocol, keys management, session management and method for low power. The ability of real-time and high efficiency always affects the system on chip to great extent. There are two approaches to this problem: one is to clip real-time OS such as RtLinux; another is to code sub-procedure for each module and set them up. The first method has an advantage that programmer can use all kinds of functions supported by OS and ignore the work of task scheduling, but the code size of OS is often too big. The second method has advantage inversely. Our chip would be used in a complex platform, efficiency is very important to the host. And the size of memory on chip is limited. For these reasons, we adopt the second method to organize the firmware. The firmware's architecture is presented in figure 3.

The main job of SUP320 is to wait for commands and do them, so the module of communication protocol is the schedule center of firmware. Command issued by outer entity enters the chip through the module of interfaces abstract, which is aroused by interrupts from peripheral interfaces. The module of communication protocol checks the integrity of command and explains it according to TCPA specification. After that, it arouses other modules to run. The result of operation is also sealed and sent out by it. In the following section, we give more details of some critical modules.

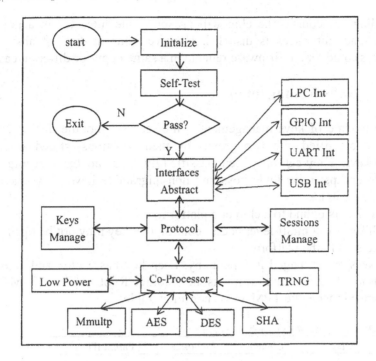

Figure 3. Firmware Modules

3.1 Interfaces Abstract

In operating system, all kinds of interfaces to peripheral devices are described abstractly as file interface. Sending and receiving data from device by applications are just like writing and reading data from a special file. The module of interfaces abstract uses a kind of data structure to describe communicating data in peripheral interfaces. The data structure is presented in figure 4.

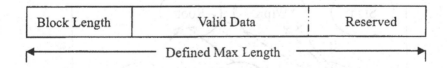

Figure 4. Data Structure Described In Interface Abstract Module

SUP320 has lots of peripheral interfaces, so it can be used in devices such as USB-KEY and secure disk. The module of interfaces abstract lets system on chip pay more attention to data processing but not data receiving or sending. Of course, the chip will use only one interface in a practical device, other interfaces is disabled. However, our method of abstracting interfaces could let the firmware reliable in a same copy for different cases.

3.2   Keys Management

The module of keys management is the securest module in SOC. Keys can been classified into two groups by their functions: stored keys and signing keys. They can also been classified by their capability in migration into two groups: migratable keys and non-migratable keys. Keys have the following attributes:

1. Some key is bound to a chip or a platform;
2. Each key has a multi-lever access control, one key may not be open to all processors in the platform;
3. All keys are managed in hiberarchy. Each key has a blob and naturally leads to a tree. The root of tree is the "Storage Root Key"(SRK) which is generated inside the TPM and is non-migratable.

Table 1. Data Structure of Key Blob

ID	ClassID	Content	Authorize	Parent ID	Next ID

To describe key blob in the key tree, a data structure presented in table 1 is adopted. All keys are managed in a thread tree (figure 5).

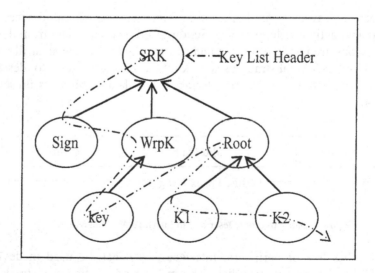

Figure 5. Key Tree

The key tree has a sub-tree marked "root". It managed all the migratable keys in the chip. The "root" key is assigned by user's command, for he always wants to set a "password" to protect his privacy.

The module of keys management is composed of three parts: keys generator, keys storage management and keys import and export. In asymmetrical arithmetic, the true random number generated by TRNG is used to produce big modular number pairs, and the user's public key and private key are produced by procedure according to them. In symmetrical arithmetic, the true random number is directly used for key. Key is managed by sub-module of keys storage management as soon as it is produced. Only public key and key of symmetrical arithmetic which are wrapped by parent key can be exported from chip. SOC provides two basic functions: Export_key and Load_key. In theory, SUP320 can produce unlimited keys through its capability of key generating. But the size of memory on chip is limited, we just design the chip as a portal for keys management. The host must use the two functions to maintain consistency of key view inside and outside the chip.

3.3   Session Management

The module of session management maintains the session information of processes in platform. Any process that wants to intercommunicate with SUP320 must start a session at the very start. A session is discarded when the conversation ends. But we must notice that the number of sessions is changing momentarily. There are lots of processes using the chip in the host, the capability of real-time must be considered. In this project, a list is adopted to manage the session handles. A new session is put ahead of the list when it is created, and a time-out session or an unused session is deleted from the list soon. When looking for a session which belongs to a special process, system searches the list from the head to the end. Other information belonging to a session is saved with the session handle.

3.4   Communication Protocol

The module of communication protocol does two kinds of things. One is to restrict SUP320 to explain input command in specific format. Another is to perform access control on the command according to current session, function and target key.

Different commands have different parameters. We don't intend to describe the format of each command. But each command includes fields such as TAG, Command ID, session information, key information, authorization data, nonce and etc. Table 2 shows most fields of a command.

Table 2. Fields of Command

TCG_TAG	ParamSize	CMD_ID	KeyHandle	InArg	Nonce	...

SUP320 provides two protocols for authorizing the use of entities without revealing the authorization data on the network or the connection to the SUP320. The first protocol is the "Object-Independent Authorization Protocol" (OI-AP), which allows the exchange of nonces with a specific SUP320. Once an OI-AP session is established, its nonces can be used to authorize any entity managed by the SUP320. The session can live indefinitely until either party request the session termination. The TPM_OIAP function starts an OI-AP session. The second protocol is the "Object Specific Authorization Protocol" (OS-AP)". The OS-AP allows establishment of an authentication session for a single entity. The session creates nonces that can authorize multiple commands without additional session establishment overhead, but is bound to a specific entity. The TPM_OSAP command starts the OS-AP session.

To depict problem easily, we suggest: inParamDigest is the result of the following calculation: SHA1(ordinal, inArg); outParamDigest is the result of the following calculation: SHA1(returnCode, ordinal, outArg); inAuthSetupParams refers to the following parameters, in this order: auth Handle, authLastNonceEven,nonceOdd, continueAuthSession; OutAuthSetupParams refers to the following parameters, in this order: auth Handle, nonceEven, nonceOdd, continueAuthSession. Then steps of OI-AP list below:

1. The caller sends command TPM_OIAP to start a conversation with SUP320;
2. SUP320 creates a new session, gets a new authHandle, associates session with authHandle. Gets a true random number to use for authLastNonceEven, saves authLastNonceEven with authHandle. Returns both authLastNonceEven and authHanle to the caller.
3. The caller receives anthHandle and authLastNonceEven and saves them. Gets a true random number to use for nonceOdd, computes inAuth = HMAC(Key.usageAuth, inParamDigest, inAuthSetupParams). Saves nonceOdd with authHandle, sends a command TPM_example, whose message include nonceOdd, authHandle and inAuth.
4. SUP320 receives command TPM_example, verifies authHandle is valid, retrieves authLastNonceEven from internal session storage, computes HM = HMAC (key.usageAuth, inParamDigest, inAuthSetupParams). Compares HM to inAuth, if anything is ok, executes TPM_Example and creates returnCode. Generates nonceEven to replace authLastNonceEven in session. Sets resAuth = HMAC (key.usageAuth,outParamDigest, outAuthSetupParams). Returns output parameters, message returned

includes nonceEven, resAuth and continueAuthSession. If continueAuthSession is FALSE, then destroys the session.

5. The caller saves nonceEven, HM = HMAC (key.usageAuth, outParamDigest, outAuthSetupParams). compares HM to resAuth. This verifies returnCode and output parameters.

6. The caller can use key authHandle to a new key, just follows step 3 to step 5 until the command of Terminate_Session is executed.

The difference of OS-AP and OA-IP is that: OS-AP names the entity to which the authorization session should be bound. More detail information can be found in [1].

3.5 Self Test

The task of self test module is to test the state of hardware components according to the configure parameters when the chip initializes. System on chip reports error information and exits once it finds any component is bad. To test a component effectually, we choose typical test programs for different components. Get test results after execution completed, then we can judge the state of component by the results.

4. METHOD FOR LOW POWER

The method for low power is paid more attention to the embedded system than any other problem. Although SUP320 has four co-processors, it nearly do not use two of them at the same time. So we can disable some co-processor to reduce the cost of chip power when it is not been used. The module of Low Power just does this kind of things, it manages all the states of co-processors and enables or disables the co-processors on behalf of other modules. To simplify the problem, we do not put the true number generator into consideration, for the true number is frequently used in module of communication protocol. In the following, a sub-routine of key-pair generating in RSA arithmetic is used as an example:

1. After SUP320 having done self test, all co-processors are disabled;

2. SUP320 receives the command of Key-Pair generating, procedure of RSA initialization calls module for low power, who enables the co-processor of MMultp;

3. When the operation finished, module for low power disables the co-processor of MMultp in the end of procedure of RSA;

4. SUP320 stores the keys and returns result to the caller.

The method is simple, but test result indicates that the cost of power is almost reduced by 50%.

5.   PERFORMANCE

In a Pentium 2.4GHZ, Windows XP installed machine, SUP320 is tested by USB interface. We did some typical operations and compare efficiency of it with that of a common smart card, Compare results list in table 3.

Table 3. Performance Test

Arithmetic	KeySize	Function	SmartCard	SUP320
SHA	160	HASH	N.A.	500M bps
3DES	3*64	Encrypt/Decrypt	100Mbps	210Mbps
AES	128	Encrypt/Decrypt	N.A.	500M bps
RSA	1024bit	Sign	6/s	300/s
	$1024bit(e=2^{16}+1)$	Authentication	24/s	28000/s
ECC	160bitGF(p)	Sign	N.A.	1200/s
	160bitGF(p)	Authentication	N.A.	600/s

* The symbol "N.A." denotes the device has not this kind of function.

It is obvious that chip of SUP320 is better than a common smart card in both security and efficiency. It is a good choice to build up trust computing platform.

6.   CONCLUSION

Building up a secure subsystem based on a physic chip, the whole platform gets security assistance in all levers of hardware, OS kernel and application. We can build a secure system out of insecure environment. Chip of SUP320 is designed by SOC technology, which can bind data and programs together in one chip. It can be used in many fields such as TPM, PKI and etc. The hardware architecture, software modules and method for low power is recommendable to design the system on chip. In the following days, we plan to consider the problems on cooperation of chip and platforms such as PC, PDA etc.

ACKNOWLEDGMENTS

The authors are extremely grateful to the members in trust computing group for their effort on the specifications. This research has been supported in part by The Chinese National Science Foundation(NSF 90104025).

REFERENCES

1. Trusted Computing Platform Alliance (TCPA), Main specification, February 2002. Version 1.1b.
2. Trusted Computing Platform Alliance (TCPA), PC Specific Implementation Specification version 1.0.
3. Trusted Computing Platform Alliance (TCPA), Trusted platform module protection profile, July 2002. Version 1.9.7.
4. Intel Low Pin Count (LPC) interface Specification Revision 1.1.
5. J.E.Dobson and B.Randell, Building Reliable Secure Computing Systems Out of Unreliable UnSecure Compinents, IEEE July 2003.
6. Ross Anderson, TCPApalladium frequently asked questions, http://www.cl.cam.ac.uk/users/rja14/tcpafaq.html accessed 13 March 2003.
7. W. A Arbaugh, D J Farber, and J. M Smith. A secure and reliable bootstrap architecture, In Proceedings 1997 IEEE Symposium on Security and Privacy, pages 65-71, May 1997.
8. Jean-Francois, Design of an Efficent Public-key Cryptographic Library for RISC-based smart cards. Ph.D. Thesis, University Catholique de Louvain,May 1998.
9. Koc,C.K,Acar,T., Burton S.kaliski Jr, Analyzing and Comparing Montgomery Multiplication Algorithms, IEEE Micro 16(3):26-33, june 1996.
10. Tung, C., "Signed-Digit Division Using Combinational Arithmetic," IEEE Trans. On Comp., vol. C-19, no. 8, pp. 746-748, Aug, 1970.

MATHEMATICAL MODELS OF IP TRACEBACK METHODS AND THEIR VERIFICATION

Keisuke Ohmori[1], Ayako Suzuki[1], Manabu Ohmuro[1], Toshifumi Kai[2], Mariko Kawabata[1], Ryu Matushima[1] and Shigeru Nishiyama[1]

[1]*NTT Advanced Technology Corp. Systems Development Unit,1-19-3, Nakacho, Musashino-shi, Tokyo, 180-0006, Japan;* [2]*Matsushita Electric Works, Ltd. Systems Technology Reserch Laboratory, 4-8-2, Shiba, Minato-ku, Tokyo 108-0014, Japan*

Abstract: IP traceback is a technology for finding distributed-denial-of-service (DDoS) attackers. Various IP traceback methods have been proposed. When a new method is proposed, a performance comparison with the conventional methods is required. In this paper, mathematical models of ICMP, probabilistic packet marking, hash-based, and Kai's improved ICMP method are proposed. The mathematical models proposed can be applied to arbitrary network topologies, and are applicable for evaluating the performance of a new traceback. The mathematical models are verified by comparing the theoretical values with actual measurements of a network of about 600 nodes.

Key words: ICMP traceback,; Probabilistic packet marking traceback; Hash-based IP traceback; Mathematical model

1. INTRODUCTION

Distributed-denial-of-service attacks (DDoS attack) cause serious damage to the Internet community; programs for implementing such attacks are typically propagated using worms or viruses. Research and development to prevent such attacks is necessary.

IP traceback can look for the attack routes, even if the IP address of the attacker is forged. It is one technology that may be employed to defend a computer system from DDOS attacks. ICMP traceback[1], probabilistic

packet marking traceback[2], and hash-based traceback[3] are typical IP
traceback methods; new traceback methods are also being proposed.

When a new traceback method is proposed, we need to compare the
performance with the conventional IP traceback methods. Ideally, one
would install the conventional IP traceback systems and evaluate the
performance; however, the systems are difficult to install. Therefore,
performance estimation by mathematical modeling becomes desirable.

The conventional mathematical models[4] apply to simple network
topologies such as linear and binary trees. They are not applicable to
arbitrary network topologies.

In this paper, we propose mathematical models of typical traceback
methods: ICMP traceback (**iTrace**), probabilistic packet marking traceback
(**PPM**), and hash-based traceback. We also propose a mathematical model
for improved ICMP traceback method, which does not use probabilistic
packet sampling. These models can be applied to arbitrary network
topologies and we show the validity of the mathematical models by
comparing them with actual measurements of a large-scale verification
network.

This paper is organized as follows. We propose mathematical models of
the IP traceback methods in Section 2. We present the verification method
and a verification network in Section 3. We compare the theoretical values
with actual measurements in Section 4. Finally we summarize our results
and present areas for future research in Section 5.

2.   THE MATHEMATICAL MODELS

2.1   Summary of IP traceback methods

First, we present an overview of IP traceback. An IP traceback looks for
DDoS attackers by examining the flow of attack packets. An agent of the IP
traceback is sent to each router. It generates traceback information, which
includes the packets that pass through the router. This traceback information
is sent to a collector and is used for the traceback.

An example of a traceback is shown in Fig. 1. V is a victim, and A1, A2,
A3, A4 are attackers.The attackers A1, A2, A3, A4 attack the victim V
through routers. For example, attack packets from the attacker A1 reach the
victim through the edge e6, e3, and e1. Therefore, a trace back to the
attacker A1 becomes possible when traceback information about e1, e3, and
e6 is generated.

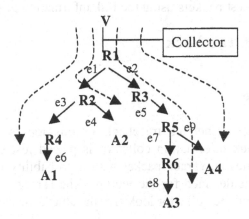

Figure 1. Example of IP traceback

IP traceback methods may be divided into those that use probabilistic packet sampling, such as ICMP and probabilistic packet marking, and those that do not, such as hash-based traceback and improved ICMP traceback.

In ICMP and probabilistic packet marking, traceback information is generated probabilistically for packets, both normal and attack packets. Therefore, the discovery probability of the attackers can be calculated from the generation probability of the traceback information about each edge of the attack routes. For example, the discovery probability Pr (A1 ∩ A2 ∩ A3 ∩ A4) of the four attackers A1, A2, A3, A4 can be calculated with the Eq. (1), using the probability that traceback information is generated for the edge ei.

$$Pr(A1 \cap A2 \cap A3 \cap A4) = \prod_{i=1}^{9} Pr(ei) \tag{1}$$

The hash-based traceback and improved ICMP traceback methods are not probabilistic packet sampling methods. They have no collector; a manager controls the agents on the routers.

In hash-based traceback, an agent of each router calculates the hash value of every packet and registers it in a hash table. The manager asks agents whether they routed an attack packet using the hash value of the attack packet. The attackers can be found using the answers from the agents.

In improved ICMP traceback, the manager sends a request packet asking the router to inform it if it routes an attack packet. The agent informs the manager when the attack packet is routed. The manager decides where to send the next request packets using the link information of the routers.

2.2   ICMP traceback

2.2.1   Outline

An agent which generates traceback information is installed in each router. A traceback information collector is placed just before the victim. The agent generates an iTrace packet with a probability p (usually, one in 20000) for packets destined for the victim. The iTrace packet includes the original packet, and the collector looks for the attackers.

2.2.2   Implementation

Implementation is based on the Internet-Draft[1]. In a normal ICMP traceback, a single iTrace packet including an attack packet means that the attack packet was routed down the edge. To reduce false positives, it was decided that two iTrace packets must include an attack packet before attributing the attack to the edge.

2.2.3   Mathematical model

We find the probability that the agent of a router generates two iTrace packets including the attack packet on an edge. Let p be the probability of generating an iTrace packet and N be the number of attack packets arriving on edge ei; then the probability of generating two or more iTrace packets becomes

$$Pr(ei) = F(N) = 1 - (Np(1-p)^{N-1} + (1-p)^{N}) \qquad (2)$$

Here $(1-p)^{N}$ is the probability that no iTrace packets are generated and $Np(1-p)^{N-1}$ is the probability that only one iTrace packet is generated. We can calculate the discovery probability $Pr(\prod Aj)$ of all the attackers by using the iTrace packet generation probability on each edge ei of the attack routes.

$$Pr(\prod_{j=1}^{n} Aj) = \prod_{i=1}^{m} Pr(ei) \qquad (3)$$

Changing the number of packets N in Eq. (2) and Eq. (3) allows us to calculate the discovery probability for different attack scenarios. Traceback time is calculated from the number of packets found with the formula Eq. (3).

For example, we may apply Eq. (2) and Eq. (3) to the scenario illustrated in Fig. 1. In Fig. 1, the attackers A1, A2, A3, A4 carry out a DDoS attack; each sends *a* attack packets per second. We may calculate the probability that the attackers are discovered after *t* seconds as shown in Table 1. Here the number of packets on edges e1, e2, e5 is twice that of the other edges because two edges join into one. For example, suppose that each attacker sends 1000 attack packets per second. In this case, the traceback takes 96 seconds for the discovery probability to reach 95% for A1, A2, A3, and A4.

Table 1. Example for how to calculate the discovery probability of the attackers in ICMP traceback

Edge	Number of attack packets after t seconds	Probability that two or more iTrace packets are generated at each edge
e1, e2, e5	2at	$F(2at)$
e3, e4, e6, e7, e8	at	$F(at)$
the discovery probability of attackers A1, A2, A3, A4		$F(2at)^3*F(at)^5$

2.3 Probabilistic packet marking traceback

2.3.1 Outline

The agent, which marks routed packets, is installed in each router. The collector, which collects marked packets, is arranged just before the victim. A hash value for a packet is stored at each router with probability p (usually, 1/20). The collector can look for the attackers using marked packets sent to the victim.

2.3.2 Implementation

Implementation is based on Song and Perrig's AMS-II(Advanced and Authenticated Marking Scheme-II). The probabilistic packet marking traceback evaluated here divides the 64-bit hash value into 8 individual fragments; one of these is chosen at random and marked. The collector considers an attack packet to have been routed when the 64-bit hash value (16 marked packets) arrives twice.

2.3.3   Mathematical model

We assume that the d individual routers are found in a direct route between the attacker and the victim. First, a router Ri marks a packet, and the probability that the other routers do not rewrite the marked packet is calculated.

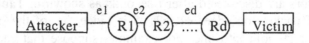

Figure 2. For mathematical model computation of PPM

For example, in Fig. 2, if router R1 marks one packet, the probability that it is not marked by other routers is $p(1-p)^{d-1}$. The hash value is divided into 8 individual fragments and one of those is sent at random. Therefore, the generation probability of a marked packet is p/8. One mark is generated, and the probability F_d, that the mark is not rewritten by other routers, becomes

$$F_d = p(1-p)^{d-1}/8. \tag{4}$$

The probability Pr(ei), that two or more marked packets arrive on edge ei as set of eight individual fragments is

$$Pr(ei) = G(d, N) = (1-(N*F_d(1-F_d)^{N-1}+(1-F_d)^N))^8 \tag{5}$$

Here N is the number of packets, $N*F_d(1-F_d)^{N-1}$ is the arrival probability of one marked packet, and $(1-F_d)^N$ is the probability that no marked packet reaches the collector. By deducting the two values from 1, the probability that two or more marked packets arrive at the collector can be calculated. It is raised to the eighth power because eight fragments are necessary for the traceback. The discovery probability of the attackers and traceback time can be found with Eq. (3) in the same way as for ICMP traceback.

We may apply Eq. (3) and Eq. (5) to the scenario illustrated in Fig. 1. In an IP marking system, there are routers which may rewrite marked packets between a router to mark and the victim, unlike the ICMP system. Let a be the number of attack packets per second from A1, A2, A3, and A4. The discovery probabilities of the attackers after t seconds are shown in the Table 2.

For example, suppose that each attacker sends 25 packets / sec; then the traceback time takes 65 seconds for the discovery probability to reach 95% for A1, A2, A3, and A4.

Table 2. Example for how to calculate the discovery probability of the attackers in probabilistic packet marking traceback

Edge	d	Number of routed packets after t seconds	Probability of generating two or more marked packets on each edge
e1,e2	1	2at	G(1,2at)
e5	2	2at	G(2,2at)
e3,e4	2	at	G(2,at)
e6,e7,e9	3	at	G(3,at)
e8	4	at	G(4,at)
Discovery probability of A1,A2,A3,A4		$G(1,2at)^2*G(2,2at)*G(2,at)^2*G(3,at)^3*G(4,at)$	

2.4   Hash-based traceback

2.4.1   Outline

The agent of each router registers the hash value of every packet. The manager asks the agents whether attack packets with the same hash value were routed through each router. An example of hash-based traceback is shown in Fig.3.   In this example, the manager is tracing A3. The agent of each router returns the answer to manager's inquiry by yes or no.

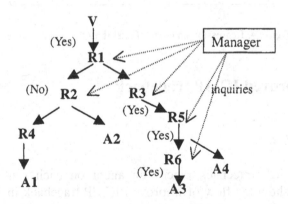

Figure 3. Example of Hash-based traceback

2.4.2   Implementation

We use the system of BBN technologies. Because of hash collisions, we let the hash table size be 14 bits in the evaluated system.

2.4.3　Mathematical model

The traceback time depends on the total number of inquiries from the manager. The number of inquiries increases very much as the attackers increase. It is difficult to derive the mathematical model because the traceback is done by the parallel processing. Therefore, the regression analysis was done based on the measurement data this time.

The measurement data and regression analysis are shown in Fig.4. The measurement data was obtained by the verification item shown in paragraph 4.2 . The number of attackers is changed from 1 to 100.

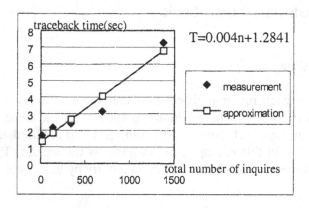

Figure 4. Regression analysis of hash-based traceback time

2.5　Improved ICMP traceback

2.5.1　Outline

Improved ICMP traceback places an agent on each router and has a manager. We show the flow of improved ICMP traceback in Fig. 5. First, the manager sends a traceback request to an agent on the router just before the victim. The traceback request has information about the attack packet. When the agent detects the requested attack packet, it generates a uTrace packet, which includes link information, and sends the uTrace packet to the manager. The manager receives the uTrace packet and sends a traceback request to the routers which link to the previous router. The manager then repeats these transactions.

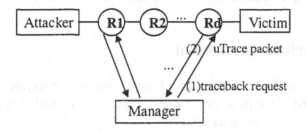

Figure 5. The flow of improved ICMP traceback

2.5.2 Implementation

An agent usually generates an uTrace packet when one attack packet comes on an edge. An exception is the edge just before an attacker. In this case, the agent generates an uTrace packet when two attack packets come on the edge to prevent false positives.

2.5.3 Mathematical model

The traceback time T of an attack route is the following.

$$T = \sum T_{wait(i)} \tag{6}$$

Here $T_{wait(i)}$ is the waiting time that an agent waits for an attack packet on an edge ei. The traceback time is the sum of the waiting times of each router on the attack route. When there are many attack routes, the traceback time is the one route with the longest traceback time.

In Fig.1, when A1, A2, A3, A4 attack with 50 packets per second, the traceback of the attack route from A3 takes the most time. The average waiting time of an attack packet on edges e2, e5, e7 and e8 are 100ms, 100ms, 200ms, 200ms × 2, respectively. The average traceback time is 800ms.

3.   VERIFICATION METHOD AND NETWORK

3.1   Verification method

In order to verify our mathematical model against actual measured values, we constructed a verification network. The network had 600 nodes. The network has the following parameters:

- The number of the hops between the victim and the attacker
- The number of attackers, arranged at random.

In the measurement of the verification network, the traceback time was measured when the false negative rate and the false positive rate were within 5% of each other.

3.2   Verification network

The specification of the verification network is shown in the Table 3.

Table 3. The specification of verification network

Item	Specification
Network scale	300 servers, 600 nodes
OS	Linux
Network speed	100Mbps
The number of DoS/DDoS attackers	at most 100
Attack packet amount per a machine	at most 25000 packet/sec
Network topology	a mesh at the core, trees at the edges

When a large-scale network is constructed, we must consider the cost, the setting, securing a power supply, and the problem of heat in the room. In this research, these problems are solved by virtual OS technology. We selected UML (User Mode Linux)[5] because the specified OS is Linux. We assigned 32 MB to each virtual OS. The maximum number of virtual servers is six. We chose not to use the virtual OS for the router because of the limitation of the input and output interface speed of the PC. The structure of the verification network is shown in Fig. 6. We had 380 servers and clients and 110 routers. Each router was a PC router using zebra[6]. The number in parenthesis is the actual number of PCs.

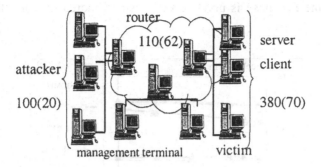

Figure 6. The structure of the verification network

4.   EVALUATION

4.1   The number of the hops

We measured the traceback time of a linear topology by changing the number of hops; we used the values 1, 3, 5, 10, 15, and 20. The measurement conditions are shown in Table 4. The number of trials was decided from both the standard deviation in measurement values and the test efficiency.

Table 4. The measurement condition

Traceback method	Number of hops	Number of attack packets per second	Number of trials
ICMP	1,3,5,10,15,20	1250pps	10
PPM	"	50pps	100
Hash	"	50pps	5
Improved ICMP	"	50pps	60

We evaluate the relationship between the number of the hops and traceback time. Actual measurement values and the theoretical values from the mathematical model are shown in the Fig. 7. Character M and T in parentheses mean the measurement values and the theoretical values respectively. The theoretical value is about the same as the actual measurement value. The hash-based and improved ICMP traceback are different from tracebacks where traceback information is generated probabilistically, such as ICMP and PPM traceback. They are very fast, because they can trace back with the arrival of a single attack packet.

Traceback time for these is under 2 seconds and increases with the number of the hops.

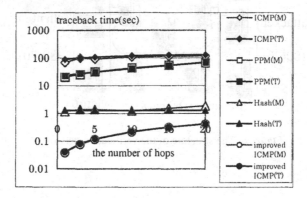

Figure 7. The relation between the number of hops and the traceback time

4.2   The number of attackers

We also measured the change in traceback times in relation to the number of attackers. The numbers of attackers were 1, 10, 20, 50, and 100. We fixed the total number of attack packets sent to the victim. The measurement conditions are shown in Table 5; the topology of the verification network is shown in Fig. 8.

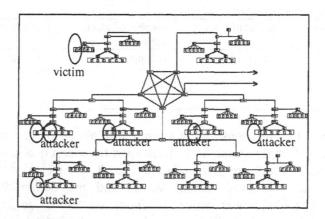

Figure 8. Topology of the verification network

Table 5. The measurement condition

Traceback method	Number off attackers	Total number of attack packets per second	Number of trials
ICMP	1,10,20,50,100	25000pps	10
PPM	"	1000pps	60
Hash	1,10,25,50,100	50pps	5
Improved ICMP	10,20,50,100	100pps	60

The actual measurement values and the theoretical values from the mathematical model are shown in Fig. 9. Character M and T in parentheses mean the measurement values and the theoretical values respectively. The theoretical values are about the same as the actual measurement values. The traceback time for ICMP traceback with 100 attackers is not shown because it took longer than the measurement time limit of 10 minutes.

In the Hash-based and improved ICMP traceback, traceback times were almost the same even when the number of attackers changed, because the maximum number of hops is about the same even if the number of attackers is different.

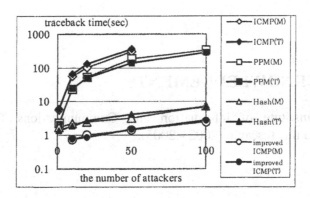

Figure 9. The relation between the number of attackers and the traceback time

5.   SUMMARY AND FUTURE RESEARCH

• Mathematical models of IP traceback

We proposed mathematical models of ICMP, probabilistic packet marking, hash-based and improved ICMP traceback methods. In improved ICMP traceback, the manager sends a request packet to inform it if an attack packet is routed through each router. The agent informs the manager when the attack packet gets routed. We constructed a verification network to verify the mathematical models. We evaluated the relationship between the number of hops and traceback time and the relationship between the number of attackers and traceback time. We confirmed that the theoretical values of the mathematical models are almost same as the values actually measured, and the improved ICMP performance is about the same as hash-based traceback.

The models can be applied to any network topology; therefore, the models can be used for the performance comparison of a new traceback model. The models can also be used to predict the performance of a typical traceback.

- The large-scale verification network

The verification network had 600 nodes made by 152 PCs. We used a virtual OS to increase the number of "machines." In order to prevent performance decline, we chose to use PC routers rather than virtualize.

- Future research

This time, we verified the mathematical models on a verification network with one autonomous system (AS). We plan to evaluate them on a verification network which has multiple ASs.

ACKNOWLEDGEMENTS

National institute of information and Communications Technology (NICT) funded this research (2002 - 2005).

REFERENCES

1. Steven M. Bellovin, "ICMP Traceback Message", Internet Draft, Oct. 2001,
   http://mark.doll.name/i-d/itrace/obsolete
2. Dawn Xiaodon Song, Adrian Perrig, "Advanced and Authenticated Marking Schemes for IP Traceback", IEEE INFOCOM 2001,
      http://vip.poly.edu/kulesh/forensics/docs/advancedmarking.pdf
3. Alex C. Snoeren et al., "Hash-Based IP Traceback", Proc. of the ACM SIGCOMM conference 2001, San Diego, CA, Computer Communication Review Vol. 31, No 4, October 2001. http://nms.lcs.mit.edu/~snoeren/papers/spie-sigcomm.pdf

4. Vadim Kuznetsov, Andrei Simkin, Helena Sandstrom, "An evaluation of different IP traceback approaches", ICICS, 2002, 37-48
5. The User-mode Linux Kernel Home Page. http://user-mode-linux.sourceforge.net
6. Zebra Home Page, http://www.zebra.org

TRANSFERABLE E-CASH REVISIT

Joseph K. Liu[1], Sandy H. Wong[2]*, Duncan S. Wong[3]

[1]Department of Information Engineering
The Chinese University of Hong Kong
Shatin, Hong Kong
ksliu@ie.cuhk.edu.hk

[2]Wireless Technology
Hong Kong Applied Science and Technology
Research Institute Company Limited,
Hong Kong
sandy@astri.org

[3]Department of Computer Science
City University of Hong Kong
Kowloon, Hong Kong
duncan@cityu.edu.hk

* The work described in this paper was done when this author was at the Department of Information
Engineering, The Chinese University of Hong Kong

Abstract: Incorporating the property of transferability into an offline electronic cash (e-cash) system has turned out to be no easy matter. Most of the e-cash systems proposed so far do not provide transferability. Those who support transferability are either inefficient or requiring to be online with the help of a trustee. In this paper we present an efficient offline transferable e-cash system. The computational complexity of our system is as light as a single term e-cash system [13]. Besides, no trustee is involved in the transfer protocol. In addition to it, we propose two e-check systems constructed using similar techniques to our e-cash system. One is as efficient as a single term e-cash system and supports partial unlinkability. The other one provides complete unlinkability with a more complex setting.

Key words: electronic payment systems, secure electronic commerce, transferable e-cash, e-check

1.   INTRODUCTION

An electronic cash (e-cash) system provides a digital way to mint and transfer money, or so-called e-cash/e-coins. According to [17], an ideal e-cash system consists of six properties: independence, security, untraceability, offline payment, divisibility and transferability. Independence implies that the security of the e-cash system does not depend on any physical location, medium, time or users. Security means that an e-cash cannot be forged or double spent without being detected. Untraceability refers to the maintenance of the anonymity of any honest user. Offline payment does not need the bank to be involved during the payment process conducted by a customer and a merchant. Divisibility refers to the ability to divide an e-coin into smaller pieces provided that the total amount of those pieces equals the value of the original e-coin. Transferability allows a user to spend an e-coin received in a payment to another user immediately without having contact the bank.

Most of the e-cash systems [12,13,4,16,7,20] focus only on the first five properties but not on the last one: transferability. The first transferable e-cash system was proposed by Okamoto, et al. [17]. It is based on the costly cut-and-choose methodology. Chaum, et al. [11] outlined a method to extend [22] for transferability. Their idea is similar to ours. In this paper, we actually build a transferable one based on an entirely different and efficient scheme. In addition, we also illustrate how it can further be extended to construct e-check systems. Other transferable e-cash systems include the one proposed by Pagnia, et al. [18] and another one proposed by Anand, et al. [1]. However, this first one requires a trusted third party in the system to maintain users' anonymity while the second one requires the payment process to be online. Jeong, et al. also proposed a transferable cash system in [15] using group signatures [6,5,2]. However, the system needs an additional third party – the group manager. It can recover the identity of any group member and therefore it has to be trusted by all the users.

In addition to e-cash, a more convenient means of payment is the electronic check (e-check). An e-check can be used for only once but can be used for any amount up to a maximum value and then be returned to the bank for a refund of the unused part. Therefore an extra protocol, the refund protocol is needed. It is convenient because a customer can 'withdraw' some token with the bank and decides how much it wants to use later on. The leftover can be redeemed afterwards.

E-check was first proposed by Chaum, et al. in 1988 [8,10]. In [8], an online e-check system was proposed. Although [10,14] proposed offline e-check systems, they use the cut-and-choose methodology which appears to be quite inefficient in practice. The systems proposed in [3,21] avoid using the cut-and-choose technique. However, [21] requires a trustee which knows the owner of each e-coin in the system even without double-spending. In this paper, we propose two e-check schemes by direct extension from a e-cash scheme [13]. The second scheme which provides complete linkability is as efficient as that in [3] in terms of computational complexity.

In this paper, we propose an efficient offline transferable e-cash system. It extends directly from Ferguson's single term e-cash scheme [13] which is described in Section 2. In our proposed system, only users and a bank is present. A *shop* is eliminated because a payment to a shop can be regarded as a transfer of an e-coin from a user to anther user. The scheme is untraceable but secure. Moreover, we also propose two offline e-check systems which are as efficient as the one in [3]. One is highly efficient and supports partial unlinkability. The other one is completely unlinkable with a more complex setting.

The rest of the paper is organized as follows. In Sec. 2, the Single Term offline e-cash scheme proposed by Ferguson [13] is reviewed. In Sec. 3, we describe our proposed transferable e-cash scheme. This is followed by the proposed two e-check schemes in Sec. 4 and conclude the paper in Sec. 5.

2. FERGUSON'S SINGLE TERM OFF-LINE COINS [13]

It is an offline untraceable e-cash system without providing transferability. The scheme is efficient and does not use the cut-and-choose methodology. Here we give a brief review of the scheme.

Preliminaries. Let $\{0,1\}^*$ denote the set of finite binary strings. To denote that an element a is chosen uniformly at random from a finite set A, we write $a \in_R A$. Let n be the public RSA modulus [19] of the bank and v be its public exponent. It is required that v is a reasonably large prime (say 128 bits). Let g_a, g_b, g_c, h_b, h_c be publicly known numbers such that $g_a, g_b, g_c \in Z_n^*$ have large order and h_b, h_c are of order n in $GF(p)$, where $p-1$ is a multiple of n. Let U be an identity which is the concatenation of the user's identity and a unique coin number so that U is distinct for each e-coin. Let $f_1 : \{0,1\}^* \to Z_v$ and $f_2 : \{0,1\}^* \to Z_n^*$ be cryptographic hash functions.

2.1. Withdrawal Protocol

The withdrawal protocol consists of three parallel runs of the randomized blinded RSA signature scheme [9,13]. It is proceeded as follows.

1. The user picks $c_1, a_1, b_1 \in_R Z_n^*$, $\sigma, r, \phi \in_R Z_v$, $\gamma, \alpha, \beta \in_R Z_n$, and computes

$$G_c = \gamma^v c_1 g_c{}^\sigma \bmod n,$$

$$G_a = \alpha^v a_1 g_a{}^r \bmod n,$$

$$G_b = \beta^v b_1 g_b{}^\phi \bmod n.$$

2. It sends $M_1 = (U, G_c, G_a, G_b)$ to the bank. For simplicity, we omit the notation of modular reduction in the rest of the paper when it gets clear from its context.

3. The bank picks $c_2, a_2, b_2 \in_R Z_n^*$ and sends $M_2 = (h_c{}^{c_2} \bmod p, a_2, h_b{}^{b_2} \bmod p)$ to the user.

4. The user picks $t_1 \in_R Z_v^*$ and computes

$$e_c = f_1(h_c{}^{c_1 c_2}) - \sigma \bmod v,$$

$$e_b = f_1(h_b{}^{b_1 b_2}) - \phi \bmod v,$$

$$a = (a_1 a_2 f_2(e_c, e_b))^{t_1} \bmod n,$$

$$e_a = \frac{1}{t_1} f_1(a) - r \bmod v.$$

5. It then sends $M_3 = (e_c, e_a, e_b)$ to the bank[10].

6. The user also signs (M_1, M_2, M_3) and sends the signature to the bank[11].

7. The bank computes

$$\overline{C} = G_c c_2 g_c{}^{e_c},$$

$$\overline{A} = G_a a_2 f_2(e_c, e_b) g_a{}^{e_a},$$

$$\overline{B} = G_b b_2 g_b{}^{e_b}$$

---

[10] Note that the exponents e_c, e_b and e_a are computed modulo v. Certain corrections in the final signature (S_a, S_b) are needed to make the blinding perfect. This is done by multiplying the final signature by a suitable powers of g_c, g_a and g_b [13]. Corrections are not shown in this paper.

[11] This corresponds to a signature of the user for all the data in the first three transmissions. It is used to protect the user against framing by the bank. We refer readers to [13] for detail.

and selects $t_2 \in_R Z_v^*$. It sends $(c_2, b_2, t_2, (\overline{C}^{t_2}\overline{A})^{1/v}, (\overline{C}^U\overline{B})^{1/v})$ to the user.

8. The user computes

$$c = c_1 c_2,$$

$$b = b_1 b_2,$$

$$t = t_1 t_2 \bmod v,$$

$$C = c g_c^{f_1(h_c^c)},$$

$$A = a g_a^{f_1(a)},$$

$$B = b g_b^{f_1(h_b^b)},$$

$$S_a = (\frac{(\overline{C}^{t_2}\overline{A})^{1/v}}{\gamma^{t_2}\alpha})^{t_1},$$

$$S_b = \frac{(\overline{C}^U\overline{B})^{1/v}}{\gamma^U \beta}$$

and checks whether $S_a^{\,v} = C^t A$ and $S_b^{\,v} = C^U B$. If these two equalities hold, it accepts.

The user stores (a,b,c,t,S_a,S_b) as an e-coin. (a,b,c) are the *base numbers* of the coin.

2.2. Payment Protocol

To spend an e-coin (a,b,c,t,S_a,S_b), the user executes the following protocol with the shop.

9. The user sends (a,b,c) to the shop.

10. The shop randomly chooses a challenge x and sends it to the user.

11. The user computes and sends $r = tx + U$ and $S = (S_a)^x (S_b)$ to the shop.

12. The shop computes $C = c g_c^{f_1(h_c^c)}$, $A = a g_a^{f_1(a)}$, $B = b g_b^{f_1(h_b^b)}$ and checks if $S^v = C^r A^x B$. If the equality holds, the shop accepts the coin and stores (a,b,c,x,r,S). Otherwise, it rejects.

(x, r, S) is a proof of the user's ownership to the e-coin with base number (a, b, c). Obviously, the user can only provide one proof in order to prevent from revealing its identity.

2.3. Deposit Protocol

To deposit an e-coin, it sends (a, b, c, x, r, S) to the bank. The bank verifies the coin by following the steps below.

13.   Compute $C = cg_c^{f_1(h_c^c)}$, $A = ag_a^{f_1(a)}$ and $B = bg_b^{f_1(h_b^b)}$.

14.   Check if $S^v = C^r A^x B$. If it is false, the bank rejects the deposit.

15.   Otherwise, it checks if (a, b, c) are already existed in its database. If yes, the bank rejects the deposit. Otherwise it accepts and stores (a, b, c) in its database and it credits the shop.

Double-spending is detected if the bank finds the same triple (a, b, c) are already in its database. If the corresponding (x, r, S) are the same as the ones stored in the database, the bank concludes that the shop is cheating. Otherwise, it concludes that the user double spends the coin. The identity of the user, U can be obtained easily by solving the two linear equations.

3. OUR PROPOSED TRANSFERABLE E-CASH

Let the bank issue e-coins with N different denominations. For the i-th denomination, the bank has a distinct and reasonably large prime v_i be the corresponding public exponent. Define a *zero-value* coin with the corresponding public exponent v_0 as a distinct large prime. A zero-value coin is an e-coin which is worth nothing. It preserves all the properties of a non-transferable e-coin. Essentially, if the zero-value coin is double spent, the identity of the user would be revealed. For distinction, we call other nonzero-value coins as positive-value coins. Note that coins with various denominations are sharing the same public RSA modulus n.

How It Works. Each user obtains a number of zero-value coins from the bank using the withdrawal protocol described below during the system setup. When an owner, Alice, of a positive-value coin transfers the coin to a user, Bob, she carries out the payment protocol described below which is similar to the original payment protocol reviewed in Sec. 2.2. That is, Bob obtains the coin's base numbers and a proof of Alice's ownership. When Bob wants to transfer this coin to another user, Carol, Bob has to send, through a transfer protocol, the coin's base numbers and the proof of Alice's ownership appended with the base numbers of one of his zero-value coins and a proof of his ownership to Carol. Now when Carol wants to transfer the coin to another user, Daniel, Carol sends, through the transfer protocol, the coin's base numbers, the proof of Alice's ownership, the base numbers of Bob's zero-value coin, the proof of Bob's ownership, appended with the base numbers of one of her zero-value coins and a proof of his ownership to Daniel. The procedure repeats until the final receiver of the coin decides to deposit it.

We refer to this transfer mechanism as a 'transfer-chain'. We will see shortly that this 'chain' is linked by a special relation between the proof of the sender's ownership of the coin and the base numbers of a receiver's zero-value coin. As long as a user provides only one proof of its ownership to a zero-value coin, the user's identity would not be compromised. This implies that each zero-value coin can only be appeared in at most one transfer-chain. Using twice or more will result in identity revocation.

When a user receives a transfer-chain, it can only transfer the chain to one user. Transferring to more than one user is equivalent to double-spending. On the other side, multiple transfer-chains can be combined into one transfer chain when they are transferring to one single user. That is, multiple coins can be transferred to a receiver in one run of the transfer protocol. This is accomplished by building up many-to-one relation of the proofs of multiple senders' ownerships of several coins and the base numbers of a receiver's zero-value coin.

In the system, a payment process is considered as a transfer of some e-coins from a user to a shop. It has no difference from a transfer process and a shop has no difference from a conventional user. Hence there are only users and a bank in the system and the payment protocol is replaced by a transfer protocol.

In the following, we describe the three protocols of our scheme, namely the withdrawal protocol, the transfer protocol and the deposit protocol.

3.1. Withdrawal Protocol

This protocol is executed when a user withdraws a coin from the bank, no matter the coin is a zero-valued or a positive-valued. The corresponding public exponent is used for each denomination of the coins. The protocol is similar to Ferguson's described in Sec. 2.1. For a user i, a zero-value coin and a positive-value coin are denoted as Z_i and P_i, respectively.

3.2. Transfer Protocol

As explained before, a transfer-chain is formed when a coin is transferred. Without loss of generality, let the transfer start from user 1, then to user 2, and so on. That is, user 1 withdraws a positive-value coin $P_1 = (a_1, b_1, c_1, t_1, S_{a_1}, S_{b_1})$ from the bank with corresponding identity U_1. It is later transferred to user 2 and then to user 3, and so on. Let the zero-value coin of user k, for $k > 1$, be $Z_k = (a_k, b_k, c_k, t_k, S_{a_k}, S_{b_k})$ with corresponding identities U_k. In this section, we will see that a transferred coin is derived directly from the concept of transfer-chain. Below is the structure of a transferred coin $Coin_k$ when it is transferred from user 1 all the way to user k, for $k > 1$.

Structure of a Transferred Coin. Suppose the value of P_1 is d-th denomination which corresponds to the public exponent v_d, where $1 \leq d \leq N$. After the coin has been transferred for $k - 1$ times for $k > 1$, user k has the coin and the following components constitute the transferred coin $Coin_k$.

$$S_k = s_1 \| s_2 \| \cdots \| s_{k-1}$$
$$A_k = a_1 \| a_2 \| \cdots \| a_{k-1} \| a_k$$
$$B_k = b_1 \| b_2 \| \cdots \| b_{k-1} \| b_k$$
$$C_k = c_1 \| c_2 \| \cdots \| c_{k-1} \| c_k$$
$$R_k = r_1 \| r_2 \| \cdots \| r_{k-1}$$

where

- $s_i = (S_{a_i})^{x_i} S_{b_i}$, $1 \leq i \leq k-1$, $x_i = H(a_{i+1}, b_{i+1}, c_{i+1})$ and H is some appropriate cryptographic hash functions. Hence $(a_1, b_1, c_1, S_{a_1}, S_{b_1})$ are from P_1 and $(a_j, b_j, c_j, S_{a_j}, S_{b_j})$ are from Z_j for $j > 1$.

- $r_i = t_i x_i + U_i$, $1 \leq i \leq k-1$.

For the boundary case (when $k = 1$), we define $Coin_1 = (S_1, A_1, B_1, C_1, R_1)$, for $A_1 = a_1$, $B_1 = b_1$, $C_1 = c_1$ and $S_1 = R_1 = \lambda$, where λ represents empty content.

Validation of a Transferred Coin. We describe the validation of a transferred coin $Coin_k$ as a function, $valid(Coin_k)$ which outputs accept if the coin is valid, otherwise, it outputs reject:

$valid = $ "On input $Coin_k = (S_k, A_k, B_k, C_k, R_k)$, for any $k \geq 1$,

1. Compute $A_i = a_i g_a^{f_1(a_i)}$, $B_i = b_i g_b^{f_1(h_b^{b_i})}$, $C_i = c_i g_c^{f_1(h_c^{c_i})}$

   and $x_i = H(a_{i+1}, b_{i+1}, c_{i+1})$, for $1 \leq i \leq k-1$.

2. Check whether

$$s_1^{v_d} = C_1^{r_1} A_1^{x_1} B_1,$$

$$s_i^{v_0} = C_i^{r_i} A_i^{x_i} B_i, \text{for} 2 \leq i \leq k-1$$

3. Output accept if all the equalities hold, otherwise output reject."

The Protocol. When user k (for any $k \geq 1$) transfers $Coin_k$ to user $k+1$, they execute the following transfer protocol. Here we assume that user $k+1$ has an unused (fresh) zero-value coin Z_{k+1} with base numbers $(a_{k+1}, b_{k+1}, c_{k+1})$.

1. User k sends $Coin_k = (S_k, A_k, B_k, C_k, R_k)$ to user $k+1$.

2. User $k+1$ executes $valid(Coin_k)$ to validate the coin. It continues if the function output accept. Otherwise, it halts with failure.

3. User $k+1$ computes and sends $x_k = H(a_{k+1}, b_{k+1}, c_{k+1})$ to user k.

4. User k computes $\begin{aligned} r_k &= t_k x_k + U_k \\ s_k &= (S_{a_k})^{x_k} S_{b_k} \end{aligned}$ and sends r_k, s_k to user $k+1$.

5. User $k+1$ computes

$$A_k = a_k g_a^{f_1(a_k)}$$

$$B_k = b_k g_b^{f_1(h_b^{b_k})}$$

$$C_k = c_k g_c^{f_1(h_c^{c_k})}$$

and checks whether

$$s_1^{v_d} = C_1^{r_1} A_1^{x_1} B_1, \text{if} k = 1$$

$$s_k^{v_0} = C_k^{r_k} A_k^{x_k} B_k, \text{if} k > 1.$$

6. It continues if the equality holds. Otherwise, it halts with failure.

7. User $k+1$ constructs

$$A_{k+1} = A_k \| a_{k+1}$$
$$B_{k+1} = B_k \| b_{k+1}$$
$$C_{k+1} = C_k \| c_{k+1}$$
$$R_{k+1} = R_k \| r_k$$
$$S_{k+1} = S_k \| s_k$$

and stores them as the new transferred coin $Coin_{k+1}$.

3.3. Deposit Protocol

The Deposit Protocol is straightforward. When user k deposits $Coin_k$ to the bank, the bank executes $valid(Coin_k)$ to validate the coin. Then it checks if (a_1, b_1, c_1) are already in its database. If not, the bank stores $Coin_k$ and credits user k. Otherwise, it means someone has double spent the coin.

Detection of Double-Spending. We use the following example to illustrate the detection mechanism of double-spending. Suppose user 1 withdraws a coin (U_1) from the bank and transfers to user 2 (U_2) and so on, until it reaches user 6 (U_6). User 6 deposits the coin. Also suppose that user 3 (U_3) and user 5 (U_5) double spend the coin. Their double-spent coins are finally transferred to user $6'$ and user $7''$, respectively, and then deposited to the bank. Hence the bank has three copies of the coin with the same initial base numbers (a_1, b_1, c_1). Let the transferred coin deposited by user 6, user $6'$ and user $7''$ be $Coin_6$, $Coin_{6'}$ and $Coin_{7''}$, respectively. The bank finds

- $\{(a_1, b_1, c_1, r_1), \cdots, (a_3, b_3, c_3, r_3), \cdots, (a_5, b_5, c_5, r_5), \cdots\}$ from $Coin_6$;

- $\{(a_1, b_1, c_1, r_1), \cdots, (a_3, b_3, c_3, r_{3'}), \cdots\}$ from $Coin_{6'}$; and

- $\{(a_1, b_1, c_1, r_1), \cdots, (a_5, b_5, c_5, r_{5'}), \cdots\}$ from $Coin_{7''}$.

From $Coin_6$ and $Coin_{6'}$, the bank finds the double spender to be user 3 and from $Coin_6$ and $Coin_{7''}$, the bank finds the double spender to be user 5. Their identities are easily obtained by solving the corresponding linear equations. For example, U_3 is obtained by computing $x_3 = H(a_4, b_4, c_4)$ and $x_{3'} = H(a_{4'}, b_{4'}, c_{4'})$ and solving the following equations:

$$r_3 = t_3 x_3 + U_3$$
$$r_{3'} = t_3 x_{3'} + U_3$$

Also, it is important to see that the identity of other honest users would not be revealed by the bank.

4. OUR PROPOSED E-CHECKS

In this section, we present two e-check systems. The first one is highly efficient and supports partial unlinkability. The second one support complete unlinkability with a more complex setting.

4.1. E-Check I

We base on Ferguson's e-cash system again and therefore use the same notations as before. In this e-check system, there is a list of reasonable large prime numbers $(v_1, \cdots, v_k)$ as public exponents of the bank with v_i corresponding the value of $\$2^{i-1}$. Define that multiplying any set of v_i, $1 \le i \le k$, represents to the sum of their corresponding values. v_d denotes the public exponent of the bank representing $\$d$ such that

$$v_d = \prod_{i=1}^{k} v_i <d>_i \tag{1}$$

where $<d>_i$ denotes the value of the i-th least significant bit of d. For example, $<6>_1 = 0, <6>_2 = 1, <6>_3 = 1$. In this way, we can represent any amount up to $\$2^k - 1$.

4.1.1. Withdrawal Protocol

Without loss of generality, suppose a user wants to withdraw an e-check of $\$2^k - 1$ as its maximum value. The withdrawal protocol is the same as Ferguson's one (Sec. 2.1) by setting the public exponent to $v = v_1 \cdot v_2 \cdots v_k$. Note that the maximum value of the e-check must be in the form of $\$2^i - 1$, for any $i > 1$. That is, all the bits of the maximum value of the e-check should be 1 in its binary representation. This ensures that the devaluation of v_d (first step of the Payment Protocol below) is always computable. Let the e-check be denoted as $K = (a, b, c, t, S_a, S_b)$ where (a, b, c) are the base numbers of the check.

4.1.2. Payment Protocol

Suppose the user wants to spend $\$d$ to the shop, where $1 \le d \le 2^k - 1$. The corresponding public exponent of the bank is v_d which can be publicly computed using equation (1). In the first step of the protocol, the user 'devalues' the check from $\$2^k - 1$ to $\$d$. The protocol proceeds as follow.

1. The user computes $\overline{v_d} = v_1 \cdots v_k \, div \, v_d$, and

$$S'_a = (S_a)^{\overline{v_d}} \qquad S'_b = (S_b)^{\overline{v_d}}.$$

2. Note: div is normal division without taking modulo.
3. The user then sends the base numbers of the check (a, b, c) to the shop.
4. The shop randomly picks a challenge x and sends it to the user.
5. The user computes $r = tx + U$, $S = (S'_a)^x (S'_b)$ and sends r, S to the shop.
6. The shop computes $C = c g_c^{\, f_1(h_c{}^c)}$, $A = a g_a^{\, f_1(a)}$, $B = b g_b^{\, f_1(h_b{}^b)}$

and checks whether $S^{v_d} = C^r A^x B$. If the equality holds, the shop accepts and stores (a, b, c, x, r, S, d). Otherwise, it rejects.

4.1.3. Deposit and Refund Protocols

The deposit protocol of our e-check system is the same as Ferguson's deposit protocol reviewed in Sec. 2.3, with the public exponent $v = v_d$.

The user can refund the remaining $\$2^k - 1 - d$ from the bank by executing a refund protocol. The protocol is almost the same as the deposit protocol, except the checking of double spending. In the refund protocol, the user sends the used check-tuple (a, b, c, x, r, S, d) to the bank. The bank verifies user's ownership of the e-check by first carries out the steps similar to the payment protocol, namely it sends a challenge x' and obtains a response pair (r', S'). Then it checks if the base numbers (a, b, c) are already in its database. If it exists and the amount is d, the bank refunds the remaining $\$2^k - 1 - d$ to the user and updates its database to record that the e-check has already been refunded.

Note that this part is not anonymous. The bank knows the identity of the user who asks for refund. The bank can also link the e-check which has already spent by the user in earlier time.

4.2. E-Check II

The e-check system proposed in last section is linkable at the refund stage. In this section, we propose another scheme which is completely unlinkable. In this scheme, the bank has only one public exponent v. Instead, we use different elements $g_{a_i} \in Z_n^*, 1 \leq i \leq k$ of large order to represent different values of the e-check. Like the representation system in E-Check I, we use g_{a_i} to represent $\$2^{i-1}$. In this way, with k consecutive elements, the e-check has a maximum value of $\$2^k - 1$. We further use g_{a_0} to prevent a user from using the e-check twice or more. Thus g_{a_0} is included in the payment of an e-check regardless of the payment amount.

E-Check II is similar to Ferguson's e-cash system. However, there are $k+1$ signatures in each e-check if its maximum value is $\$2^k - 1$, one is for embedding the identity of the user to prevent double-spending while the others are for composing the value of the e-check.

4.2.1. Withdrawal Protocol

Without loss of generality, we assume a user wants to withdraw an e-check of $\$2^k - 1$. We follow the notations of Sec. 2.1. Let $g_{a_0}, g_{a_1}, \cdots, g_{a_k}, g_b, g_c$ be public where $g_{a_0}, g_{a_1}, \cdots, g_{a_k}, g_b, g_c$ are of large order in Z_n^*. The Withdrawal Protocol proceeds as follows.

1. The user picks $b_1, c_1, a_{1_0}, a_{1_1}, \cdots, a_{1_k} \in_R Z_n^*$, $\sigma, \phi, r_0, r_1, \cdots, r_k \in_R Z_v$ and $\gamma, \beta, \alpha_0, \alpha_1, \cdots, \alpha_k \in_R Z_n$. It then computes

$$G_b = \beta^v b_1 g_b^{\phi}$$

$$G_c = \gamma^v c_1 g_c^{\sigma}$$

$$G_{a_i} = \alpha_i^v a_{1_i} g_{a_i}^{r_i}, \text{for } i = 0, \cdots, k$$

2. and sends $M_1 = (U, G_b, G_c, G_{a_0}, \cdots, G_{a_k})$ to the bank.

3. The bank picks $b_2, c_2, a_{2_0}, a_{2_1}, \cdots, a_{2_k} \in_R Z_n^*$ and sends $M_2 = (h_b^{b_2}, h_c^{c_2}, a_{2_0}, a_{2_1}, \cdots, a_{2_k})$ to the user.

4. The user picks $t_{1_0}, t_{1_1}, \cdots, t_{1_k} \in_R Z_v^*$, computes

$$e_b = f_1(h_b^{b_1 b_2}) - \phi \bmod v$$

$$e_c = f_1(h_c^{c_1 c_2}) - \sigma \bmod v$$

$$a_i = (a_{1_i} a_{2_i} f_2(i, e_c, e_b))^{t_{1_i}}, \text{for } i = 0, \cdots, k$$

$$e_{a_i} = \frac{1}{t_{1_i}} f_1(a_i) - r_i \bmod v, \text{for } i = 0, \cdots, k$$

and sends $M_3 = (e_b, e_c, e_{a_0}, e_{a_1}, \cdots, e_{a_k})$ to the bank.

5. The user also signs (M_1, M_2, M_3) and sends the signature to the bank. (Note: refer to Sec. 2.1 for discussions).

6. The bank computes

$$\overline{C} = G_c c_2 g_c^{e_c}$$

$$\overline{B} = G_b b_2 g_b^{e_b}$$

$$\overline{A_i} = G_{a_i} a_{2_i} f_2(i, e_c, e_b) g_{a_i}^{e_{a_i}}, \text{for } i = 0, \cdots, k$$

7. The bank selects $t_{2_0}, t_{2_1}, \cdots, t_{2_k} \in_R Z_v^*$ and sends $c_2, b_2, \{t_{2_i}\}_{0 \le i \le k}, \{(\overline{C}^{t_{2_i}} \overline{A_i})^{1/v}\}_{0 \le i \le k}, (\overline{C}^U \overline{B})^{1/v}$ to the user.

8. The user computes

$$c = c_1 c_2$$

$$b = b_1 b_2$$

$$t_i = t_{1_i} t_{2_i} \bmod v, \text{for } i = 0, \cdots, k$$

$$B = b g_b^{f_1(h_b^b)}$$

$$C = c g_c^{f_1(h_c^c)}$$

$$A_i = a_i g_{a_i}^{f_1(a_i)}, \text{for } i = 0, \cdots, k$$

$$S_b = \frac{(\overline{C}^U \overline{B})^{1/v}}{\gamma^U \beta}$$

$$S_i = (\frac{\overline{C}^{t_{2_i}} \overline{A_i})^{1/v}}{\gamma^{t_{2_i}} \alpha_i})^{t_{1_i}}, \text{for } i = 0, \cdots, k$$

and checks whether $S_b^v = C^U B$ and $S_i^v = C^{t_i} A_i$, for $i = 0, \cdots, k$. If all the equalities hold, he accepts.

The user stores $(a_0, \cdots, a_k, b, c, t_0, \cdots, t_k, S_b, S_0, S_1, \cdots, S_k)$ for the payment of the e-check.

4.2.2. Payment Protocol

Without loss of generality, suppose the user wants to spend $\$2^j - 1$, for some $1 \le j \le k$, to the shop. The payment protocol proceeds as follows.

1. The user sends $b, c, a_0, \cdots, a_j$ to the shop.

2. The shop selects a challenge number x and sends it to the user.

3. The user computes $r_i = t_i x + U$ and $S'_i = (S_b)(S_i)^x$, and sends (r_i, S'_i), $0 \le i \le j$, to the shop.

4. The shop computes

$$C = c g_c^{f_1(h_c^c)}$$

$$B = b g_b^{f_1(h_b^b)}$$

$$A_i = a_i g_{a_i}^{f_1(a_i)}, 0 \le i \le j,$$

and checks whether $S'^v_i = C^{r_i} A_i^x B$ for $0 \le i \le j$. If all the equalities hold, the shop accepts and stores $(a_0, \cdots, a_j, b, c, x, r_0, \cdots, r_j, S'_0, \cdots, S'_j)$. Otherwise, it rejects.

4.2.3. Deposit Protocol

The deposit protocol is constructed in its natural way. When the shop deposits the e-check, it sends the check-tuple

$$(a_0, \cdots, a_j, b, c, x, r_0, \cdots, r_j, S'_0, \cdots, S'_j)$$

to the bank. The bank verifies of the tuple as follows.

1. Compute $C = c g_c^{f(h_c^c)}$, $B = b g_b^{f(h_b^b)}$, $A_i = a_i g_{a_i}^{f(a_i)}$, for $i = 0, \cdots, j$.

2. Check whether $S'^v_i = C^{r_i} A_i^x B$, $0 \le i \le j$. If not all equal, the bank rejects the deposit.

3. Check whether the same values of (a_0, b, c) already exist in its database. If yes, the bank rejects the deposit and the double-spender can easily be found. Otherwise, it accepts and credits the shop.

4.2.4. Refund Protocol

If the user wants to refund the remaining amount of the e-check, that is, $\$2^k - 1 - (2^j - 1) = \$2^k - 2^j$, he has to inform the bank his account number and his identity U for the refund purpose and execute the following steps.

1. The user sends $U, a_{j+1}, \cdots, a_k$ and $t_{j+1}, \cdots, t_k$ to the bank.

2. The bank retrieves $\overline{B}, \overline{C}$ from the withdrawal record.

3. The bank checks if any of $a_{j+1}, \cdots, a_k$ are already in the database. If yes, it rejects. Otherwise, the bank selects a challenge number x and sends it to the user. The bank also computes $r_i = t_i x + U$, for $j + 1 \leq i \leq k$.

4. The user computes $r_i = t_i x + U$ and $S''_i = (S_b)(S_i)^x \gamma^{r_i} \beta$, and sends S''_i, $j + 1 \leq i \leq k$, to the bank.

5. The bank computes $A_i = a_i g_{a_i}^{f(a_i)}$ and checks whether $S''_i{}^v = \overline{C}^{r_i} A_i{}^x \overline{B}$ for $j + 1 \leq i \leq k$. If not all of them are equal, the bank rejects. Otherwise, the bank records that the e-check has been refunded in its database and refunds $\$2^k - 2^j$ to the user.

Unlike E-Check I, in this e-check system, the bank is unable to link the refunded e-check with the e-check that the user has already spent to the shop.

5. CONCLUSION

We have proposed an off-line transferable e-cash system. Unlike [15], we do not require any group manager or trustee. Our scheme does not use cut-and-choose technique, thus more efficient than those using cut-and-choose such as [17]. In addition, we have also proposed two e-check systems. One is almost as efficient as a single term e-cash such as [13] with partial unlinkability only. The other one provides complete unlinkability with a more complex setting.

We do not address divisibility in our transferable e-cash system. We may consider divisibility to be less important in practice as this can be easily be solved as in the world of physical cash. That is, using various denominations and conducting changes in transactions as the coins are transferable. Hence we consider that with transferability, divisibility becomes a less important property of e-cash.

REFERENCE

[1] R. Sai Anand and C.E. Veni Madhavan. An Online, Transferable E-Cash Payment System. *INDOCRYPT 2000*, LNCS 1977, pp. 93-103. Springer-Verlag, 2000.

[2] G. Ateniese, J. Camenisch, M. Joye and G. Tsudik. A Practical and Provably Secure Coalition-Resistant Group Signature Scheme. *CRYPTO 2000*, LNCS 1880, pp. 255-270. Springer-Verlag, 2000.

[3] S. Brands. An Efficient Off-line Electronic Cash System based on the Representation Problem. *Technical Report CS-R9323*, CWI, April, 1993.

[4] S. Brands. Untraceable off-line cash in wallets with observers. *Proc. CRYPTO 93*, LNCS 0773, pp. 302-318. Springer-Verlag, 1993.

[5] J. Camenisch and M. Michels. A group signature schemes with improved efficiency. *Proc. ASIACRYPT 98*, LNCS 1514, pp. 160-174. Springer-Verlag, 1998.

[6] J. Camenisch and M. Stadler. Efficient group signature scheme for large groups. *Proc. CRYPTO 97*, LNCS 1296, pp. 410-424. Springer-Verlag, 1997.

[7] A. Chan, Y. Frankel and Y. Tsiounis. Easy Come - Easy Go Divisible Cash *EUROCRYPT 98*, LNCS 1403, pp. 561-575. Springer-Verlag, 1998.

[8] D. Chaum. Online Cash Checks. *Proc. EUROCRYPT 89*, LNCS 0403, pp. 288-293. Springer-Verlag, 1990.

[9] D. Chaum. Randomized Blind Signature. manuscript., 1992.

[10] D. Chaum, B. den Boer, E. van Heyst, S. Mjolsnes and A. Steenbeek. Efficient offline electronic checks. *Proc. EUROCRYPT 89*, LNCS 0403, pp. 294-301. Springer-Verlag, 1990.

[11] D. Chaum, T.P. Pedersen. Transferred Cash Grows in Size *Proc. EUROCRYPT 92*, LNCS 0658, pp. 390-407. Springer-Verlag, 1992.

[12] D. Chaum, A. Fiat and M. Naor. Untraceable electronic cash. *Proc. CRYPTO 88*, LNCS 0403, pp. 319-327. Springer-Verlag, 1990.

[13] N. Ferguson. Single Term Off-line Coins. *Proc. EUROCRYPT 93*, LNCS 0765, pp. 318-328. Springer-Verlag, 1993.

[14] R. Hirschfeld. Making Electronic Refunds Safer. *Proc. EUROCRYPT 92*, LNCS 740, pp. 106–112. Springer-Verlag, 1993.

[15] I.R. Jeong, D.H. Lee and J.I. Lim. Efficient Transferable Cash with Group Signatures. *ISC 2001*, LNCS 2200, pp. 462–474. Springer-Verlag, 2001.

[16] T. Okamoto. An efficient divisible electronic cash scheme. *Proc. CRYPTO 95*, LNCS 0963, pp. 438-451. Springer-Verlag, 1995.

[17] T. Okamoto and K. Ohta. Universal electronic cash. *Proc. EUROCRYPT 91*, LNCS 0547, pp. 324-337. Springer-Verlag, 1991.

[18] H. Pagnia and R. Jansen. Towards Multiple-payment Schemes for Digital Money. *Financial Cryptography 1997*, LNCS 1318, pp. 203-216. Springer-Verlag, 1997.

[19] R. Rivest, A. Shamir and L Adleman. A method for obtaining digital signatures and public-key cryptosystems. *Communications of the ACM*, 21:120-126, Feb 1978.

[20] T. Sander and A. Ta-Shma. Auditable, Anonymous Electronic Cash. *Proc. CRYPTO 1999*, LNCS 1666, pp. 555-572. Springer-Verlag, 1999.

[21] A. de Solages and J. Traore. An Efficient Fair Offline Electronic Cash System with Extensions to Checks and Wallets with Observers. *Financial Cryptography 1998*, LNCS 1465, pp. 275-295. Springer-Verlag, 1998.

[22] Hans van Antwerpen. Electronic Cash. *Master's Thesis*, CWI, 1990.

A LICENSE TRANSFER SYSTEM FOR SUPPORTING CONTENT PORTABILITY IN DIGITAL RIGHTS MANAGEMENT*

Qiong Liu, Reihaneh Safavi-Naini and Nicholas Paul Sheppard
Center for Information Security, School of Information Technology and Computer Science, University of Wollongong, NSW 2522, Australia

Abstract: Current digital rights management systems typically bind the right to use content to a particular device whose location cannot be changed easily. Users may find it difficult to acquire a new license for each device. We propose a license transfer system that allows the user to share a license between devices and uses transaction track files to ensure that only one device can use the license at a time. We analyze the security properties of the proposed system.

Key words: digital rights management; content portability; license transfer.

1. INTRODUCTION

Digital rights management (DRM) systems are used to prevent unauthorized access to digital content and to manage content usage rights. Rather than trading the content as in traditional physical methods of distribution, the subject of trade is a license that grants certain rights over the content. Typically, the content, which is sold by the content provider, is encrypted with cryptographic algorithms to protect it from illegal copying and consumption. Protected content can be obtained by the user through various delivery channels. However, without possession of a valid license, the content cannot be decrypted.[1] To access protected content, the user's

* This work was funded by the Smart Internet Technology Cooperative Research Centre, Australia.

device needs to contact the license server, which is used by the license issuer to issue licenses, to acquire a valid license.

We define content portability as the ability to access the same digital content on multiple suitable devices. Traditional content distribution models bind the right to use the content to the physical object that carries the content, such as a CD, which can be easily moved to wherever the owner wishes to go. One of the problems in current DRM implementations is that most solutions offer licenses that are bound to a playback device, which is often a desktop computer whose physical location cannot be changed easily. Usually digital content is locked to the device that it was downloaded to. Having purchased the right to use the content for one device, the user still needs to acquire a new license to access the same content on a different device,[2-3] which discourages the use of DRM. It would be welcomed by the users if the usage rights can be shared among multiple devices, so that the users do not need to re-acquire the rights when they have a new device, or when they want to use the content at different physical places.

In this paper we propose a license transfer system that allows the user to share a license among multiple devices while ensuring that only one device can use the license at a time. We present surveys of existing DRM models that support content portability in Section 2. Section 3 describes our proposed license transfer system. We analyze the security properties of the proposed system in Section 4. Section 5 describes our demonstration system that supports content portability using the proposed license transfer protocol. Finally, we give conclusions in Section 6.

2.   RELATED WORKS

Several existing DRM systems or proposed schemes provide a limited degree of content portability. We identified three DRM models as follows.

2.1   Rights Locker Model

A rights locker[4] is a storage system that contains the digital rights purchased by a user. It is implemented as a central server to facilitate consumers' access to their rights anywhere anytime.

In a DRM rights locker model, permissions to use content are no longer bound to a particular device, but to the consumer himself. Each consumer has an account with the locker service that allows him to redeem his content usage rights from multiple locations using any DRM enabled device. Every time the user wants to access the content on his device, he needs to log on to his locker account via a web browser by typing in his username and password. After the central server has verified the authenticity of the request, it grants the access rights for the content to the user. Digital World Services[5] and MP3.com[6] use such a model.

The rights locker model requires Internet connections between users' devices and the central server. Users, especially those who do not have permanent Internet access, may find it difficult to access their rights using this model. Another drawback of this approach is that the central server can easily become a bottleneck. There may be a single point of failure: if the server were crashed or compromised by an attacker, e.g. under denial of service attacks, all the consumers would not be able to access their rights. Moreover, cracking a DRM rights locker would violate user's privacy[4].

2.2 Rights-Sharing Model

In the rights-sharing model, content can be shared among a collection of devices, which is called an authorized domain. Usually, a domain represents a set of devices belonging to a consumer. The content can only be accessed in the domain for which it has been authorized. Several schemes have been proposed to address the need for content protection in the authorized domain, such as the Family Domain concept[7], the xCP Cluster Protocol[8] and OMA DRM domain sharing[9]. The disadvantage of such model is that it introduces the overhead of setting up the domain prior to using it. Only when the domain is formed and the devices are enrolled in the domain, can content be sharing among devices in the domain. If user wants to access the content using devices that are not part of the domain, he is required to register each of these devices as a member of the domain.

2.3 Rights Transfer Model

The rights transfer model allows the consumer to transfer the rights to use the content to machines on the condition that the original copy of the content cannot be used. A few content protection schemes use this model.

FlexiToken[10] proposed by NTT Laboratories is a generic copy prevention scheme for trading digital rights. In this scheme, a digital right is represented using two types of information: the rights description object and the token object. The token object represents the "genuineness" of the rights object and is stored and circulated using tamper-proof devices such as smart cards. The rights object can be held in any storage medium, but to redeem the rights, the user must present the token of the rights to the service provider. The rights transfer protocol proposed in this scheme takes place between two smart cards of the user, using public key cryptography. The token object must be deleted from the original card after the rights transfer procedure.

FlexiToken assumes that neither participant flees from the other, i.e. the sender will delete the token object after it receives the receipt from the receiver. However, this assumption may be violated if the operation of the rights transfer protocol is interrupted either intentionally or accidentally. For example, a dishonest user may cut power from the sender before it deletes the token object.

Aura and Gollmann[11] proposed a simple license transfer procedure for transferring software licenses between smart cards. Since this protocol is based on the assumption that there is no communication interruption between two the cards, it shares the same problem as FlexiToken.

The license transfer model has several advantages over the rights locker model and the license-sharing model: First, in the license transfer model, the digital rights portability among various devices does not depend on the availability of the central server. And it does not require Internet connection if licenses are transferred using local area network (LAN) or between PC and PDA. Second, the license transfer model is a low cost solution because it does not require domain set-up and management process, as the license-sharing model does. We consider using the license transfer model to support content portability in DRM.

3.   ROPOSED SOLUTION

We assume that encrypted content can be obtained by the user in any way the user wishes. Access to the content will be provided if the device has a valid license and can extract the content key from the license. Suppose that the user has purchased a license for a content file using one of his devices. Encrypted content is allowed to be copied to any device. To access the content on a particular device, the user needs to transfer the license from the original device to this machine. We require that at any one time, a license should only be used by one device to consume the digital content. Rogue users may try to make copies of one purchased license, then exchange or resell the license copies, thus allowing many devices to consume the digital content at the same time with the cost of a single license. In our system, the management of a single copy is done by the player software by using a transaction flag to record the state of the license: there are four different transaction flags, only when the license has an 'Active' flag, can the content be played. When the transaction flag changes to one other than 'Active', playing stops. Each device keeps a transaction track file that records the current state of each license stored on the device. Only the player can read and update transaction flags. In our proposed license transfer protocol, suppose that license L is for user U to use content C on device D_1, L's usage rules allow a new license to be created for the same user U to use the same content C on another device D_2, subject to the condition that after transfer of the new license from D_1 to D_2, L becomes invalid.

3.1   System Overview

Figure 1 shows the components of our system: The license database is a conceptual database, such as a file directory, on the user's device, which stores all the licenses that the user has purchased. The transaction track file is a digital data file that records the current state of these licenses. The digital library is a digital content repository on the user's device that stores protected content files obtained by the user. To decrypt and use the protected content, there must be a valid license in the license database and the transaction flag for that license must be 'Active' in the transaction track file. The player is a content viewer responsible for content decryption and playback, and for providing an interface with which the user can request/transfer a license from/to another device through a network.

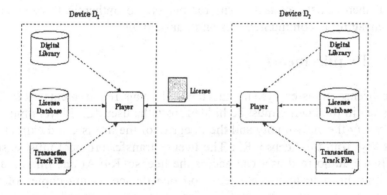

Figure 1. Components of license transfer system

Following is a use scenario of our system: The user has acquired a license from the license server and stored the license on his home PC. If he wants to consume the content on multiple devices, he must transfer the license to the appropriate device. Transfer can be through mobile phones or other handheld devices with wireless connections. The user carries the mobile phone wherever he goes. Using the mobile phone to transfer licenses to a DRM-enabled device is a convenient solution to the above scenario.

3.2   Assumptions

In DRM, end users cannot be assumed to be trusted. Player software executes in a hostile environment, taking control of how the content is used, by whom and under which conditions. Rogue users may try to modify the player software to circumvent the enforcement of usage rules.

We assume the player is trusted. The trusted player is responsible for rendering digital content while enforcing the permissions and constraints associated with the content. By "trusted", we mean trusted by the content provider. The integrity of the trusted player can be protected by using a trusted computing platform[12], for example, Microsoft's Next-Generation Secure Computing Base[13] (NGSCB). If a trusted computing platform is not available, there are techniques such as encryption or code obfuscation[14] exist to make it hard for rogue users to tamper with the player.

We assume each device has a certified public/private key pair. The corresponding private key and the transaction track file stored on the device are only accessible to the trusted player.

We assume that license transfers take place between two trusted players and that there exists a mechanism for players to authenticate each other before engaging in communications and transactions.

3.3   Requirements

Our license transfer system has the following requirements: R1: The content key in the license must be hidden from the user. R2: Players must be able to verify the authenticity and the integrity of the license, and extract the content key from the license. R3: The license transfer protocol must ensure that only the authorized user can access the license. R4: At one time, a user should only be able to use a license on one device. Using copies of the license on other devices does not give access to the protected content. R5: The license transfer protocol must satisfy the atomicity property, which is to ensure that exactly one device (D_1 or D_2) has a valid copy of the license at the end of the transfer procedure regardless of any communication failure between two devices.

3.4   License Transfer Protocol

In our system, the license transfer protocol is run between two trusted player programs. Consider the scenario there are two devices D_1 and D_2 with trusted players P_1 and P_2 respectively. Suppose the user uses device D_2 to request the license L stored on device D_1. The user should first authenticate himself to P_2. P_1 and P_2 will perform a mutual authentication protocol to authenticate their mutual identities. P_1 only allows transfer of the license if the identity of the user matches the identity of the licensee stored in the license on D_1.

We use a flag mechanism to define license status. Each license is associated with a transaction flag that describes the status of the license. There are four different transaction flags: Active, Deactivated, Request and Recover. The meaning of these flags is as follows:

- Active: the player can use the license to decrypt the content.
- Deactivated: the license is deactivated, so the player cannot use it to decrypt the content.
- Request: the license is not physically stored on the device. The player can request it to be transferred from another device.
- Recover: the license is physically stored on the device. The player can request its status to be changed from 'Deactivated' to 'Active'.

We assume P_1 and P_2 share an authenticated session key K, which can be obtained by executing an authenticated key establishment protocol.[15] In the following license transfer protocol, $Req(P_1, P_2, L)$ denotes the license request that P_2 sends to P_1 for license L. ID_L is L's identifier. Flg denotes the message type, which is to show that the message is for license request or license recovery. PK_1/SK_1 and PK_2/SK_2 are public/private key pairs of device D_1 and D_2 respectively. Let $ENC_K(X)$ denote encryption of a message X using a symmetric key K. T is the timeout value of the protocol. The license transfer protocol is described as follows:

Step1: P_2 sends to P_1: $ENC_K(Req(P_1, P_2, L))$, P_2 writes (ID_L, 'Flag=Request')

where $Req(P_1, P_2, L) = P_1, P_2, ID_L, Flg$.

Step2: If Flg = 'Request' and the transaction flag for L on D_1 is 'Active' or 'Deactivated by P_2', P_1 sends to P_2: $ENC_K(L)$, P_1 writes (ID_L, 'Flag=Deactivated by P_2'); else P_1 quits.

Step3: If L is valid, P_2 stores L and writes (ID_L, 'Flag=Active'); If L is invalid or P_2 does not receive L within time T, P_2 quits.

In step 1, P_2 sends a license request $Req(P_1, P_2, L)$ to P_1, which is encrypted using the shared session key K. This encryption provides privacy for the user against eavesdropping and also possible theft and misuse of the license. The message type Flg in $Req(P_1, P_2, L)$ is 'Request'. At the same time, P_2 writes (ID_L, 'Flag=Request') as the entry for L in the transaction track file on D_2, which indicates that L is being requested by P_2.

In step 2, P_1 uses the session key K to decrypt the license request $Req(P_1, P_2, L)$ and checks the message type Flg in $Req(P_1, P_2, L)$. If Flg is 'Request', P_1 checks transaction flag for L on D_1. If L has the 'Active' flag, P_1 sends license L to P_2 and updates L's transaction flag on D_1 from 'Active' to 'Deactivated by P_2'. If P_1 finds that L was deactivated by P_2, which indicates that P_2 failed to receive the license in the previous license transfer procedure, P_1 will send L to P_2 again. Once L is deactivated, it cannot be used by P_1 anymore, although L is still physically kept on device D_1, i.e. P_1 will refuse to use L to decrypt the content if P_1 finds that L is marked as deactivated in the transaction track file. If P_1 finds that L does not have the 'Active' flag or was deactivated by other players, P_1 quits.

In step 3, P_2 receives L from P_1 and verifies L's integrity and authenticity using the public key of the license issuer. If this succeeds, P_2 stores L and sets the transaction flag for L as 'Active', i.e. L's entry in the D_2's transaction track file is changed from (ID_L, 'Flag=Request') to (ID_L, 'Flag=Active'). If the license verification fails or P_2 does not receive L from P_1 within time T after sending the license request, P_2 quits. To get the license L, P_2 needs to request the license again, starting from step 1.

The license recovery protocol is similar to the above approach. A license recovery scenario is that both D_1 and D_2 have a copy of the license L. The transaction flag for L is 'Active' on device D_2, but is 'Deactivated by P_2' on device D_1. P_1 requests L's transaction flag on D_1 to be set to 'Active'. In this procedure, P_1 sends the license recovery request to P_2 in which the message type Flg is 'Recover'. At the same time, P_1 writes (ID_L, 'Flag=Recover') as the entry for L in the transaction track file on D_1. The transaction flag 'Recover' indicates that L is physically stored on D_1 but cannot be used and P_1 requests reactivation of L. After P_2 receives and verifies the license recovery request, it sets the transaction flag for L on D_2 from 'Active' to 'Deactivated by P_1', so P_2 will not be able to use the license L. On receipt of the respond message from P_2, P_1 updates L's transaction flag on D_1, changing it to 'Active', so L can only be used by P_1 to decrypt the content. If P_1 does not receive the response from P_2 within time T, P_1 quits. To activate L on D_1, P_1 needs to send the license recovery request to P_2 again.

Using transaction flags rather than deletion of licenses guarantees atomicity property, i.e. at any one time, exactly one device can use the license to get access to the content. In comparison with the FlexiToken scheme, our license transfer protocol is robust against communication interruption between two devices.

3.5   Content Key Management in License Transfers

In current DRM implementations, a license includes the content identifier, the identity of the licensee, usage rules, and the encrypted content key. The content key is usually encrypted with the public key of the user's device. Only the device that possesses the correct private key is able to decrypt the encrypted content key and gain access to the content. The license is usually digitally signed by the license issuer to enable its integrity and authenticity to be verified. A problem with this in the license transfer system is that the content key, which is encrypted with the original device's public key, in the transferred license cannot be accessed by the receiving device. We need to design a mechanism that protects the license in such a way that enables the content key to be successfully extracted by the target device in license transfer while ensuring that the license integrity can be verified. We consider using broadcast encryption and point-to-point encryption for content key encryption. Let K_C denote the content key and $PENC_{PK}(X)$ denote encryption of a message X using a public key PK. Let PK_1/SK_1, PK_2/SK_2, ... PK_n/SK_n be public/private key pairs of devices D_1, D_2, ...D_n respectively.

3.5.1   Broadcast Encryption

In broadcast encryption, the user needs to register all the devices he intends to use with the content provider. During license transfer, the sender does not need to modify the original license. Only legitimate devices can access the content key after receiving the license. We consider two types of broadcast encryption:

Public key broadcast encryption[16]: The content provider generates a public key (PK_G) for the group of devices that the user will be using. Each legitimate device has a different private key (SK_1, SK_2, ... SK_n) stored in a secure storage on the device. The license contains the encrypted content key $PENC_{PKG}(K_C)$. Player P_i (i = 1, 2, ... n) on device D_i uses D_i's private key SK_i to decrypt the content key K_C and then uses K_C to decrypt the content.

Using a public key infrastructure[17]: The content key in the license is encrypted with each device's public key: $PENC_{PK1}(K_C)PENC_{PK2}(K_C)$ ... $PENC_{PKn}(K_C)$. When player P_i (i = 1, 2, ... n) on device D_i receives the license, it uses D_i's private key SK_i to decrypt $PENC_{PKi}(K_C)$ and then uses K_C to decrypt the content.

The disadvantage of using broadcast encryption is that new devices must be registered to the content provider. When the user replaces old devices with new ones, he wants to continue to use the content he has purchased. The new devices must receive a private key. If a device is compromised, the content provider must change the public key and update the private keys of all devices. Thus, the content provider will have to save and periodically update a record of the user and the device set. Moreover, if the user wants to subscribe content from different content providers, the user has to register his devices with each content provider, which is inconvenient for the user.

3.5.2   Point-to-Point Encryption

Suppose that the license issuer's public key and private key pair is PK_{LI}/SK_{LI}. When the license is distributed to device D_1, the license issuer encrypts the content key K_C with D_1's public key, i.e. $PENC_{PK_1}(K_C)$. In the license transfer procedure from D_1 to D_2, player P_1 on device D_1 needs to use D_1's private key SK_1 to decrypt the encrypted content key first and then re-encrypt the content key using D_2's public key, i.e. $PENC_{PK_2}(K_C)$, so that player P_2 on device D_2 can decrypt the content key using D_2's private key SK_2.

A problem with this scenario is that the license integrity cannot be verified, because the encrypted content key in the license has changed from $PENC_{PK_1}(K_C)$ to $PENC_{PK_2}(K_C)$. When player P_2 on device D_2 checks the integrity of the license using the license issuer's public key, the verification will fail.

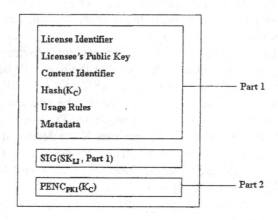

Figure 2. License format on device D_1

To solve the above problem, we propose a license format as shown in Figure 2. A license is split into two parts: the first part of the license includes the license identifier, licensee's identity, the content identifier, the hash value of the content key, usage rules and other related licensing information. The license identifier is used to link the license with its entry in the transaction track file. The licensee's identity shows the identity of the user to whom the rights are granted, using the user's public key. The content identifier, which may be in the form of a Uniform Resource Identifier[18] (URI) or a Uniform Resource Locator[19] (URL), is used to uniquely identify the protected content file associated with this license. The first part is digitally signed by the license issuer to enable its integrity and authenticity to be verified. We use SIG(SK$_{LI}$, Part 1) to denote the license issuers signature on the first part of the license. The second part includes the content key encrypted with the public key of the device. The reason for constructing the license in this way is to prevent usage rules from unauthorized modification and to ensure that the license issuer's signature can be verified when the content key is encrypted with another device's public key during license transfer.

One may ask: what happens in the case of a dispute when the user claims that the license issuer put the wrong content key in the license? To avoid such dispute, the hash function has to be one-way and collision-free, so it would be infeasible for the license issuer to generate two content keys with the same hash value. When the player receives the license, it checks the signature of the first part of the license. If the verification succeeds, the player checks the transaction flag for the license in the transaction track file according to the license identifier specified in the license. If the transaction flag for the license is 'Active', the player verifies the user's identity, checks the content identifier, decrypts the encrypted content key using the device's private key and passes the resulting content key to the hash function. If the computed result is the same as the hash value contained in the license, the player will accept the license. Otherwise, the license will be rejected and the player will contact the license server for license reissue. If the license is accepted but the key cannot be used to decrypt the content, the license issuer needs to reissue a license that contains the correct content key.

3.6   Transaction Track File

Each device has a transaction track file that records the current state of each license stored on the device. There are two fields in a track entry: the license identifier and the transaction flag. If the license identifier in the track entry matches that in a license, the track entry is for the corresponding license. Each time the user wants to use the license to access the content, the player checks the transaction flag of the license. Access to the content can be granted only when the transaction flag for the license is 'Active'.

There is only one entry in the track file for a specific license. When a license is delivered to the user's device for the first time, the player adds an entry for the license to the track file after the license integrity is verified. The transaction flag for the license is set to 'Active'. When a license transfer happens, the player reads the license identifier specified in the transferred license and checks the track entry for the license according to the identifier. The player will update the transaction flag for the license after the license has been transferred to another device.

To prevent track entries from unauthorized manipulation, the transaction track file is only accessible to the trusted player. Rogue users may take a snapshot of the hard disk drive, perform one or more license transfers to another device and then restore the transaction track file to the state before license transfers happen. To prevent this attack, software tamper resistant techniques need to be deployed. We do not discuss this topic here, because it is out of the scope of this paper.

4.   SECURITY ANALYSIS

This section analyzes the security properties of our system according to the requirements listed in 3.3.

Requirement R1 is satisfied, because the content key in the license is encrypted using the device's public key (or the group key of authorized devices). The corresponding private key of the device is only accessible to the trusted player, so the user cannot decrypt the content key.

Requirement R2 is satisfied. The license is digitally signed by the license issuer, so the integrity and authenticity of the license can be verified by the trusted player using the license issuer's public key. As discussed in 3.5, the proposed license format ensures that the license issuer's digital signature will work properly when a license transfer happens using point-to-point encryption. The content key in the license is encrypted using the device's public key (or the group public key of authorized devices), which can be decrypted by the trusted player using the device's private key.

Requirement R3 is satisfied. The license is transferred through a secure communication channel between two devices. The shared session key of the two trusted players is unknown to any third party, so rogue users cannot intercept and get access to the license during license transfers. Moreover, license transfers can only be allowed if the license requestor is the licensee.

Requirement R4 is satisfied. Unauthorized devices will not be able to use copied licenses to consume the protected content, because the content key in the license is encrypted using the authorized device's public key (or the group public key of authorized devices). Only the authorized device has the knowledge of its own private key and hence can decrypt the encrypted content key, which then can be used to decrypt the content.

Requirement R5 is satisfied. After a license transfer procedure takes place, exactly one device has the license with 'Active' flag. This property is analyzed on a case-by-case basis, as follows:

Case 1: There is no communication problem between P_1 and P_2. Exchanged messages are not disrupted by an attacker. The protocol runs successfully. At the end of the license transfer, only P_2 gets the license with 'Active' flag.

Case 2: P_1 fails to receive a valid license request from P_2 in step 2. L is still kept on device D_1. P_2 does not get the license. The transaction flag for L on D_1 is kept unchanged, which is still 'Active'.

Case 3: P_2 fails to receive the license from P_1 in step 3. The protocol aborts after timeout. The transaction flag for L in the transaction track file on D_1 is marked as 'Deactivated by P_2'. However, P_2 can get the license from P_1 through a negotiation procedure, i.e. P_2 re-starts the license transfer protocol by sending the license request $Req(P_1, P_2, L)$ to P_1 again. The message flag in the license request shows that the current transaction flag for L on D_2's transaction track file (i.e. 'Request'). From the message flag and L's transaction flag on D_1, P_1 gets to know that P_2 did not receive the license previously. Since L is still physically stored on D_1, P_1 sends L to P_2 again. Finally, P_2 gets the license L and changes the transaction flag for L on D_2 from to 'Request' to 'Active'.

Case 4: The trusted player P_3 on the third device D_3 initiates the license transfer protocol, requesting L stored on D_1. Upon on the receipt of the license request, P_1 checks L's transaction flag on D_1 that indicates that L was deactivated by P_2 due to the previous license transfer procedure that transfers L from D_1 to D_2. The deactivating device is D_2 rather than D_3, which tells P_1 that this is not a negotiation procedure due to the previous communication failure between P_1 and P_3. Therefore, P_1 refuses the license request from P_3 by terminating the license transfer protocol.

In conclusion, our license transfer system satisfies all the requirements listed in 3.3. Our license transfer protocol is robust against intentional or unintentional communication failures of the protocol. The license is not bound to a particular device so it can be easily transferred within a household if required. Finally, our scheme is an offline solution, so the user does not need to be connected on the Internet when he wants to share a license between devices.

5.   IMPLEMENTATION

We have developed a functional DRM demonstration system that supports content portability among different devices using the license transfer mechanism defined in this paper. The player is implemented in Java, which allows the user to access protected content using the licenses stored on the device and supports transfers of licenses from one device to another using the license transfer protocol. In our system, licenses are generated based on the MPEG-21 REL data model. We use the Java cryptography extension library provided by Cryptix Foundation Limited to implement cryptographic operations, such as encryption, hashing and digital signature.

Our implementation of license transfer includes a client program and a server program. The client requests licenses to be transferred from the server. We use the following software modules, as shown in Figure 3.

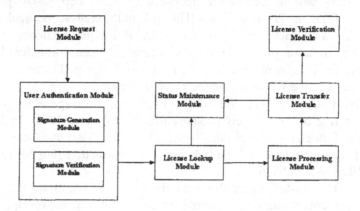

Figure 3. Software modules of license transfer

The license request module connects to the server and creates a secure communication channel between the client and the server using secure socket layer. The user authentication module uses a "challenge-response" mechanism to authenticate the user's identity. It includes signature generation and verification modules. The signature generation module takes the challenge and the user's private key as input and generates a digital signature using the user's private key. The signature verification module verifies the user's signature on the challenge using the user's public key. We implemented these modules using RSA digital signature scheme. The license lookup module is called by the server. It takes each license stored on the server machine and the user's public key as input and checks if the identity of the user matches the identity of the licensee (i.e. the public key of the key holder) in the license. It also consults the status maintenance module to check the status of the license. The license processing module is called by the server. It processes the requested license to make the content key inside the license accessible to the client machine. It uses the private key of the server machine to decrypt the content key in the requested license, and re-encrypts the content key using the public key of the client machine. The license transfer module accepts the output of the license processing module and transfers the license to the client through the secure communication channel created by the license request module. The license verification module is called by the client. It verifies the integrity and authenticity of the license by checking the digital signature on the license using the license issuer's public key. The status maintenance module maintains status of all the licenses stored on the device. It updates transaction flags for the transferred licenses in the transaction track file.

6.   CONCLUDING REMARKS

This paper describes the requirements for a digital rights management system that supports content portability. We propose a license transfer system that allows the user to share a license between multiple devices, which satisfies these requirements. We describe the license transfer protocol and analyze content key management problem in license transfers. We propose a new license format by which the integrity of the license can be verified. We use a transaction track file to ensure that at one time, exactly one device can use the license to decrypt content. We evaluate the security properties of our solution. The development of our DRM test bed shows a working implementation of possible DRM services that could be offered to the customers. Through implementation, we get better understanding about how DRM components work together to provide a secure DRM solution. We believe that DRM systems can be widely used only if customers find it easy to use.

Currently, our system only supports transfers of licenses that grant stateless rights to the user. Stateless rights are usage rights for which the device does not have to maintain state information. Permissions that require maintenance of state by the device, for example a limited number of plays or a maximum period of metered usage time, are considered stateful rights. To make our license transfer system correctly enforce stateful rights expressed in the license, we need to design a mechanism that keeps track of the uses of the associated content.

In the future, we will work on the loan application to make our system support the loan right. The loan right represents the right to lend the content to another user for a specific period of time. While the content is on loan, the original copy of the content cannot be used. At the end of the loan period, the loaner copy deactivates and the original copy reactivates. Currently, our system supports license transfers for a single user between different devices. To support the loan application, the license transfer protocol needs to handle license transfers between different users and take the loan period into account, which provides a challenge for us to conduct future work on this field.

REFERENCES

1. Q. Liu, R. Safavi-Naini, and N. P. Sheppard, "Digital rights management for content distribution," In Proceedings of the Australasian information security workshop conference on ACSW frontiers 2003, pp. 49-58, Australian.
2. D. K. Mulligan, J. Han, and A. J. Burstein, "How DRM-based content delivery systems disrupt expectations of 'personal use'," In Proceedings of the 2003 ACM workshop on Digital rights management, pp. 77-89, ACM Press, 2003.
3. W. Eric, "Protecting digital music using DRM: example of a mobile service over WLAN (version 1.0)," 2002.
4. J. Feigenbaum, M. J. Freedman, T. Sander, and A. Shostack, "Privacy engineering for digital rights management systems," In Digital Rights Management Workshop, pp. 76-105, 2001.
5. Digital World Services, "Digital world services launches rights locker and content manager," http://www.dwsco.com/, 2001.
6. Cyber Patrol, "Free MP3.com technology takes cue from open-source movement,"http://news.com.com/2100-1023-251014.html?legacy=cnet, 2001.
7. T. S. Messerges and E. A. Dabbish, "Digital rights management in a 3G mobile phone and beyond," In Proceedings of the 2003 ACM workshop on Digital rights management, pp. 27-38, ACM Press, 2003.
8. F. Pestoni, J. B. Lotspiech, and S. Nusser, "xCP: Peer-to-Peer Content Protection," IEEE Signal Processing Magazine, 2004.
9. Open Mobile Alliance, "OMA DRM specification v2.0. draft version," http://member.openmobilealliance.org/ftp/Publicdocuments/BAC/DLDRM/, Jan. 2004.
10. M. Terada, H. Kuno, M. Hanadate, and K. Fujimura, "Copy prevention scheme for rights trading infrastructure".
11. T. Aura and D. Gollmann, "Software license management with smart cards," USENIX Workshop on Smartcard Technology, May 1999.
12. P. England, B. Lampson, J. Manferdelli, M. Peinado, and B. Willman, "A Trusted Open Platform," IEEE Computer Security, 2003.
13. Microsoft Corporation, "Next-generation secure computing base: the road to security," http://www.microsoft.com/ngscb, 2003.
14. M. Sosonkin, G. Naumovich and Nasir Memon, "Obfuscation of design intent in object-oriented applications," In Proceedings of the 2003 ACM workshop on Digital rights management, pp. 142 – 153, ACM Press, 2003.
15. W. Stallings, "Cryptography and network security: principles and practice," Prentice-Hall, Inc., Upper Saddle River, NJ 07458, USA, third edition, 2002.
16. D. Boneh and M. K. Franklin, "An efficient public key traitor tracing scheme," In Proceedings of the 19th Annual International Cryptology Conference on Advances in Cryptology, pp. 338-353, Springer-Verlag, 1999.
17. R. Safavi-Naini and H. Wang, "Broadcast authentication for group communication," Theoretical Computer Science, 269(1-2): 1-21, October 2001.
18. The Internet Society, "Uniform Resource Identifiers (URI): Generic Syntax. http://www.ietf.org/rfc/rfc2396.txt, 1998.
19. The Internet Society, "Registration Procedures for URL Scheme Names", http://www.ietf.org/rfc/rfc2717.txt, 1999.

SECURE HUMAN COMMUNICATIONS BASED ON BIOMETRICS SIGNALS

Yongdong Wu[1], Feng Bao[1] and Robert H. Deng[2]
[1]*Institiue for Infocomm Research* (**I²R**), *Singapore,* {*wydong,baofeng*}*@i2r.a-star.edu.sg;*
[2]*School of information Systems, Singapore Management Univeristy, robertdeng@smu.edu.sg*

Abstract: User authentication is the first and probably the most challenging step in achieving secure person-to-person communications. Most of the existing authentication schemes require communicating parties either share a secret/password or know each other's public key. In this paper we suggest a novel user authentication scheme that is easy to use and overcomes the requirements of sharing password or public keys. Our scheme allows two human users to perform mutual authentication and have secure communications over an open channel by exchanging biometrics signals (e. g., voice or video signals). In addition to user authentication, our scheme establishes a secret session key between two users by cryptographically binding biometrics signals with users's Diffie-Hellman public values. Under the assumption that the two communicating persons are familiar with each other's biometrics signals, we show that the scheme is secure against various attacks, including the man-in-the-middle attack. The proposed scheme is highly suitable for applications such as Voice-over-IP.

1.   INTRODUCTION

The explosive growth of computer systems and their applications has considerably increased the dependence of both organizations and individuals on the information communicated using the Internet. However, the Internet is an interconnection of open public networks. Without security measures, communications over the Internet, such as Voice-over-IP (VOIP) and video conferences, can be eavesdropped without much difficulty. This in turn has led to a heightened effort to protect data from disclosure and to guarantee the integrity of data and messages communicated over open networks. User authentication is the first and probably the most challenging step in achieving secure communications in the Internet.

To date, the most pervasive user authentication schemes are based on cryptographic techniques which require that the parties either share a secret key (e.g., a password)[1] or know each other's public key[2]. Although password based authentication protocols are widely used, there are many potential difficulties for a human user to share passwords with a large number of remote users. First of all, establishing a shared password between two users requires a secure secret distribution mechanism to be in place. This is very challenging. Second and more importantly, human users are not good at remember passwords of good quality, not to mention remembering multiple passwords shared with many remote users. Public key based authentication protocols require users to know each other's public key in authenticated manners in the form of public key certificates. This turn requires the existence of a public key infrastructure in the Internet, an impossible task at least in the near to medium terms[3].

In this paper our focus is on human user authentication in person-to-person communications in an open environment such as the Internet. In this case, it is much more convenient and natural for human users to authenticate each other using biometrics techniques.

Most of the existing research on biometrics based user authentication techniques allows a human user to authenticate himself or herself to a local machine. Little effort has been spent to study biometrics based methods which perform authentication between two remote human users. To our knowledge, the only work related to our effort is the Pretty Good Privacy Phone or PGPfone[4]. PGPfone implements an authentication protocol based on the exchange of voice signals. However, PGPfone is vulnerable to replay attack. If an attacker is able to collect sound samples of all the 256 octets by, for example, eavesdropping on someone's phone calls, the attacker is able to impersonate the victim at will.

As in PGPfone, our scheme requires that communicating users be able to identify each other based on the other party's biometrics signals (such as acoustic waves or face expression). Based on the exchange of biometrics signals, the proposed scheme not only authenticates remote human users but also enables them to have secure communications over open channels. Specifically, to achieve authentication and agreement of a secret session key, the Diffie-Hellman public key values are cryptographically committed or bound with biometrics signals such that the trust on the biometric information is extended to the Diffie-Hellman public values. The trusted Diffie-Hellman public values are then used to perform the Diffie-Hellman Key Exchange Protocol so as to defeat the man-in-the-middle attack. Since our scheme does not require users to share any password or know each other's public key in advance, it is attractive for applications such as secure VOIP or secure video conferences.

The reminder of the paper is organized as follows. Section 2 addresses the primaries for clarity. Section 3 elaborates the proposed scheme and its variant. Section 4 discusses the availability and security. Section 5 contains our concluding remarks.

2.   PRILIMINARIES

2.1   Notations

A: shorthand notation for Alice (or her communication device) who initiates the communication unless stated otherwise. Preminatary

B: shorthand notation for Bob (or his communication device) who responses to Alice's communication request.

C: shorthand notation for Clark who tries to attack the communications between Alice and Bob.

C_X: a challenge biometrics signal. Without loss of generality, we will use voice signals as the representative biometrics signals throughout the paper. Thus, C_X is the acoustic wave or digital representation of a challenge statement spoken by user **X** (either **A** or **B**); whether it is the acoustic wave or the digital representation should be clear from the context of discussion.

R_Y: an acoustic wave or digital representation of a response statement spoken by user **Y** in reply to C_X.

$R_Y \sim C_X$: The response R_Y matches challenge C_X. For instance, the content of R_Y is the same/similar to that of C_X, or R_Y is a correct answer to C_X.

$|C_X|$: the time duration of C_X.

$|R_Y|$: the time duration of R_Y.

$e(K, m)$: encryption of message m with a symmetric key cryptosystem (e. g., AES) using a secret key K.

$d(K, c)$: decryption of a ciphertext c with a symmetric key cryptosystem using a secret key K.

$h(\cdot)$: a one-way hash function (e.g., SHA-1).

T: the required minimum time duration (e. g., 10 seconds) of any statement spoken by a user.

δ: a threshold value which is much less than T, (e.g. $\delta = 0.1T$). The value of δ (or equivalently that of T) plays an important role in deciding the security strength of the protocol (refer to Eq.(1)).

To keep our notation compact, only residue modulo is shown in the following. That is, we will write $g^x \bmod p$, $g^y \bmod p$ and $g^{xy} \bmod p$ simply as g^x, g^y and g^{xy} respectively, where p is a predefined large prime.

2.2   System Architecture

The system architect for person-to-person communications between two remote users, Alice and Bob, is depicted in Figure 1. We assume that Alice and Bob are aware that they will have an authenticated and confidential communication session and Alice will start the present secure protocol. This awareness assumption can be satisfied easily via any non-secure channel. The transmission channel includes but is not limited to any communication systems or media such as computer networks, public telephone switching networks and radio links.

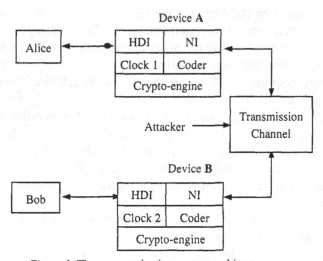

Figure 1. The communication system architecture.

Alice and Bob communicate with each other by interfacing with Device **A** and Device **B**, respectively. Device **A** (or Device **B**) accepts audio input from Alice (Bob) and outputs Bob's (Alice's) audio signal to Alice (Bob). The signals are sent and received via the Network Interface (NI). Each device has a clock for timing purpose, a coder performing audio encoding/decoding operations, and a crypto-engine executing the Diffie-Hellman and symmetric key cryptosystem operations. We assume that the Diffie-Hellman parameters, g and p, are negotiated on-line or hard coded in the software. Without loss of generality, Alice is assumed to be the initiator and Bob is the responder of a communication session.

2.3   Assumptions

The attacker Clark sits in the middle of the channel between Alice and Bob. He is able to perform both passive (eavesdropping) and active (message tampering, delay, replay). He may know biometrics data of Alice and Bob recorded from their past conversations. Clark may have much more powerful resources (e.g. super-computers and large storage devices) than Alice and Bob. The only restriction is that Clark is not able to mimic the natural speech of Alice or Bob in real time.

Alice and Bob neither share any secret data (e.g., password) nor have each other's public key. In order to achieve user authentication, we make the following assumptions:

S1: Alice and Bob are familiar with each other's voice (biometrics characteristics in general) and able to recognize each other by listening each other's speech. This assumption is reasonable and practical since there are generally no confidential topics between two strangers unless there is the involvement of a trusted third party.

S2: It is difficult for a human being to mimic the dynamic biometrics features of others in real time without being detected.

S3 It is difficult for a machine to mimic the dynamic biometrics features of a human being without being detected. Text-To-Speech (TTS) technology targets for creation of audible speech from computer readable text. A high quality TTS has to select text units from large speech databases in an optimum way[5]. To make use of TTS, an attacker needs to organize a database of large samples. On the other hand, although speech syntheses technology has made significant advancement in minimizing audible signal discontinuities between two successive concatenated units, and prosodic variation, it is still not satisfactory to mimic natural speech[6]. For example, in the TTS demo[7] of Microsoft Research, the speech is not nature although each word or short phrase is pronounced accurately, such that it is easy to distinguish the voice of a machine from that of a natural human. Similarly, the concatenation artifacts of TTS from AT&T[8] can be detected easily. In other words, presently, synthesized speech is still distinguishable from human speech after many years of research and development.

S4: Each participant can speak fresh sentences whose durations are sufficient long (e.g. at least T).

S5: The RTT (round-trip-time) of the communication channel can be estimated (e.g., command `ping www.yahoo.com`). It is required that $RTT \ll T$. This requirement must be met in order for the conversations between the communicating parties to be audible.

3.   AUTHENTICATION PROTOCOLS BASED ON BIOMETRICS SIGNALS

In this section, we present authentication protocols based on the exchange of users' biometrics signals. The protocols are designed to perform mutual authentication between two parties called Alice and Bob and at the same time allow them to share a secret session key for securing their subsequent communications.

3.1   Basic Idea

To start a secure communication session with our proposed scheme, Alice initiates the session by speaking a challenge statement, such as

"This is Alice! The time is 21 minutes passed 9am. How was your mid-term examination, Bob? "

Bob receives and listens to Alice's challenge, and makes sure that the message is indeed spoken by Alice. He then speaks a response statement, such as

"Hi, Alice! Bob's here. My mid-term exam was not very good. But thank God, it was over! ".

Upon hearing Bob's response, Alice decides whether the response is spoken by Bob and whether it is related to her challenge. If the answer is positive, Alice authenticates Bob. Bob can authenticate Alice in the same way.

In order to establish a secret session key during the above authentication process, we incorporate the Diffie-Hellman key exchange into our scheme. By cryptographically binding biometric signals with Diffie-Hellman public values, the proposed scheme is protected against the man-in-the-middle attack. The above conceptual description seems very simple, the scheme is more complicated. To demonstrate the above concept, we present two protocols, a sequential protocol and a parallel protocol in the following.

3.2   A Sequential Protocol

The authentication protocol consists of three phases: Authentication of Bob, Authentication of Alice. Additionally, a Key Confirmation will be executed so as to guarantee that both share the same session key.

3.2.1   Authentication of Bob

This phase, shown in Figure 2, allows Alice to authenticate Bob and

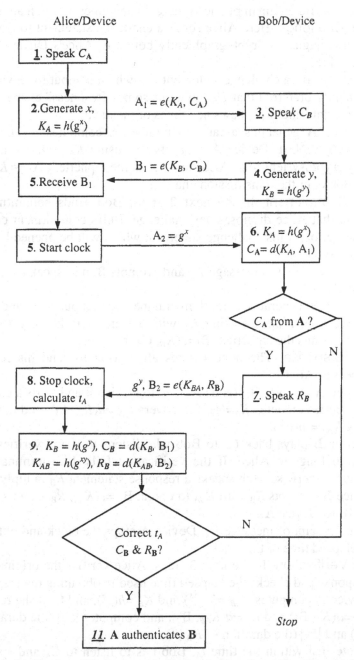

Figure 2. Authentication of Bob. An underlined step is performed by Alice or Bob, while other steps are executed by devices. $R_B \sim C_A$ means that the reply R_B matches the challenge C_A. For instance, the content of R_B is the same/similar to that of C_A, or RB is a correct answer to C_A.

obtain Bob's Diffie-Hellman public value g^y in an authenticated manner.

- **Alice's Challenge.** Here Alice sends a challenge statement to Bob. This biometrics signal is cryptographically bound to Alice's Diffle-Hellman public value.

  (1) Alice speaks a challenge statement C_A which is input to Device **A**. It is highly preferred that C_A contains some "freshness" elements such as the date and time, news headlines of the day.

  (2) Device **A** generates a random number x, computes g^x and a key $K_A = h(g^x)$. Next, Device **A** encrypts C_A using K_A with a symmetric key cryptosystem (e.g. AES) and sends the ciphertext $A_1 = e(K_A, C_A)$ to Bob over the transmission channel.

- **Bob's Commitment.** In the next 2 steps, Bob sends commitment to Alice so that Alice discloses her challenge. Bob's commitment contains the encryption of his challenge statement which will be opened by Alice at a later stage.

  (3) Device **B** receives message A_1 and prompts Bob to speak a challenge statement C_B.

  (4) Device **B** generates a random number y, computes g^y and a key $K_B = h(g^y)$, encrypts C_B using K_B with a symmetric key cryptosystem and transmits the ciphertext $B_1 = e(K_B, C_B)$ to Alice.

- **Bob's Response.** The next 4 steps allow Bob to send his response statement to Alice.

  (5) Device **A** receives B_1, sends $A_2 = g^x$ to Device **B** and starts a clock.

  (6) Device **B** computes $K_A = h(g^x)$, recovers $C_A = d(K_A, A_1)$, and computes a key $K_{BA} = h(g^{xy})$.

  (7) Device **B** plays back C_A to Bob who listens to it and verifies if the voice belongs to Alice. If the verification fails, Bob terminates the session; otherwise, Bob speaks a response statement R_B in reply to C_A. Device **B** encrypts R_B with K_{BA} to obtain $B_2 = e(K_{BA}, R_B)$, and sends g^y and B_2 to Device **A**.

  (8) Upon receipt of message B_2, Device **A** stops the clock and obtains t_A the elapsed time of the clock.

- **Alice's Verification.** In the next 3 steps, Alice verifies the originality of the response and checks the elapsed time used in obtaining the response.

  (9) Device A computes $K_{AB} = h(g^{yx})$ and $K_B = h(g^y)$, and then she recovers $C_B = d(K_B, B_1)$ and $R_B = d(K_{AB}, B_2)$, and computes $|C_A|$ (the duration of C_A) and $|R_B|$ (the duration of R_B).

  (10) Note that within the time t_A, Bob has to listen to C_A and speaks a response R_B. Hence, $t_A \geq (|C_A| + |R_B| + \Delta_B)$, where Δ_B is the delay due to transmitting messages A_2 and B_2, and processing time introduced by Device **B** in steps (6) and (7). Δ_B can be estimated by device A. Then, Device **A** calculates

$$\delta_A = t_A - (|C_A| + |R_B| + \Delta_B) \tag{1}$$

If $\delta_A > \delta$, Device A terminates the session; otherwise, Alice listens and verifies R_B. If Alice recognizes that either R_B is not in Bob's voice or R_B is not a reply to C_A, she stops the session.

(11) Alice concludes that g^y comes from Bob and authenticates Bob.

3.2.2 Authentication of Alice

To provide mutual authentication and key agreement, Bob will proceed to authenticate Alice and obtain Alice's Diffie-Hellman public value g^x in an authenticated manner. The process is similar to that given above with the exception that Bob is the initiator and Alice is the responder. Note that Bob's challenge statement C_B was sent to Alice in step (4). This is done intentionally so as to prevent the man-in-the-middle attack during the process of authenticating Alice.

After both Alice and Bob have obtained the each other's authenticated Diffie-Hellman public key values, they are confident that the agreed Diffie-Hellman key K_{AB} is shared only among them. After mutual authentication, Alice and Bob can confirm their shared key easily.

3.3 A Parallel Protocol

A careful reader might have noticed that certain steps in Figure 2 can be executed in parallel so as to speed up the protocol. Figure 3 depicts the flow chart of the parallel protocol which has the same phases as those of the sequential protocol.

- Challenges
  (1) Alice starts the session by speaking a challenge statement C_A.
  (2) Device **A** generates a random x, computes a key $K_A = h(g^x)$, encrypts C_A as $A_1 = e(K_A, C_A)$ and sends the ciphertext A_1 to Bob.
  (3) After receiving message A_1, Bob speaks a challenge statement C_B.
  (4) Device **B** generates a random number y, computes a key $K_B = h(g^y)$, encrypts C_B as $B_1 = e(K_B, C_B)$ and sends the ciphertext B_1 to Alice.
  (5) After receiving B_1, Device **A** sends $A_2 = g^x$ to Bob and starts clock 1.
  (6) Upon receipt of message A_2, Device **B** starts its clock 2 and sends message $B_2 = g^y$ to Alice.
- Responses
  (7) After receiving message B_2, Device **A** computes $K_B = h(g^y)$, recovers Bob's challenge message $C_B = d(K_B, B_1)$. On the other hand, Device **B** computes $K_A = h(g^x)$, recovers Alice's challenge as $C_A = d(K_A, A_1)$.
  (8) Alice listens C_B and stops the protocol if she believes that C_B is not in Bob's voice; Bob listens to C_A and terminates the protocol if he

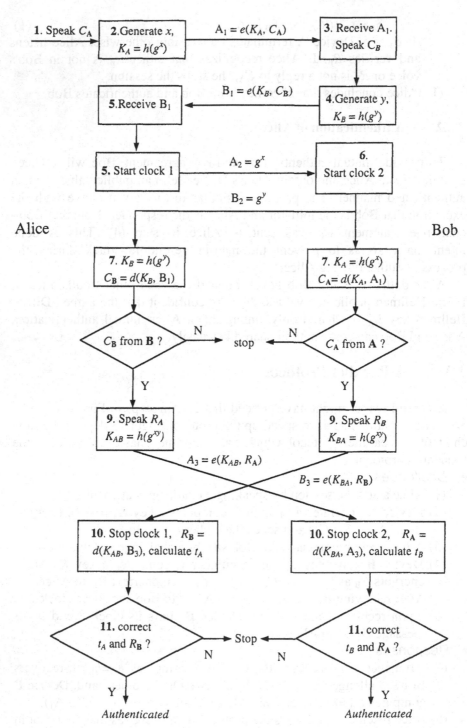

Figure 3. Parallel protocol.

doubts on the originality of C_A.

(9) Alice speaks a response R_A to the challenge C_B. Device **A** computes $K_{AB}=h(g^{yx})$, encrypts R_A and sends the ciphertext $A_3=e(K_{AB}, R_A)$ to Bob. Meanwhile, Bob speaks R_B in reply to C_A. Device **B** computes a key $K_{BA}=h(g^{xy})$, and sends the ciphertext $B_3=e(K_{BA}, R_B)$ to Alice.

(10) After receiving message B_3, Device **A** stops clock 1, recovers Bob's response $R_B=d(K_{AB}, B_3)$ and calculates the elapsed time t_A. After receiving message A_3, Device **B** stops clock 2, recovers Alice's response $R_A =d(K_{BA}, A_3)$, and calculates the elapsed time t_B.

- Verifications

(11) Device **A** calculates δ_A as
$$\delta_A= t_A - (|C_A| + |R_B|+ \Delta_B)$$
where Δ_B is the delay due to transmitting messages A_2 and B_3, and processing interval introduced by Device **B** in steps (7)-(9). Device **A** terminates the session if $\delta_A >\delta$.

Simultaneously, Device **B** calculates δ_B as
$$\delta_B= t_B - (|C_B| + |R_A|+ \Delta_A),$$
where Δ_A is the delay due to transmitting messages B_2 and A_3, and processing interval introduced by Device **A** in steps (7)-(9). Device **B** terminates the session if $\delta_B > \delta$. Alice listens and verifies R_B. She stops the session if she is not convinced that R_B is Bob's response to C_A. Bob listens and verifies R_A. He stops the session if he is not convinced that R_A is Alice's response to C_B.

3.4   Variant

An alternative approach in the protocol is that the symmetric key cryptosystem for messages C_A and C_B can be replaced by a cryptographic commitment function. For example, the commitment function is using a cryptographic one-way function $h(\cdot)$. To commit to an item m, the committing party computes the commitment $h(k \| m)$, where k is a secret key and $\|$ is the concatenation. To verify the commitment, the verifying party must have k and m, compute $h(k \| m)$ and compare it with the commitment. In other word, A_1 can be replaced with $h(K_A \| C_A)$, then C_A will be transmitted along with A_2. Similarly, the parallel protocol can be implemented with the commitment variant too.

4.   DISCUSSION

4.1   Availability

In the present protocols, time restriction plays an important role for the availability. The scheme requires the responsor produce a related response in real time, otherwise, the authentication fails. To relieve this burden, the responsor may merely repeat the challenge in his/her own voice. Here, the challenge can be prepared in advance and has no impact on the availability.

Another factor related to the availability is the variability of the network delay T. An inappropriate parameter T may disable to set up an authenticated channel. Thus, the proposed scheme is applicable such as in VOIP where the quality of the service itself is required to be high.

Despite the proposed protocols may reject some genuine communication, no forgery is possible. In other words, although false rejection ratio FRR$\neq$0 due to network traffic, FAR (false acceptance ratio) is negligible. From the viewpoint of security, FAR is much more important than FRR.

4.2   Impersonating Bob

In the proposed protocols, an important condition for Alice (Bob) to authenticate Bob (Alice) is that Bob's (Alice's) response to her (his) challenge must arrive within a defined time interval. Therefore, if Clark can obtain the correct answer in the voice of Bob (or Alice) in the predefined time interval, he can impersonate Alice (Bob) successfully. To this end, Clark may adopt one of the following three methods to provide the response in the voice of the impersonated party.

- Clark replays recorded speeches of the impersonated party.
- Clark or his device responses to the challenge by emulating the speech of the impersonated party.
- Clark lures the impersonated party to answer the challenge.

The first two methods are not possible based on security assumptions S1-S3. To defend against Clark's attack using the third method, it is crucial to check the lengths of the elapsed time of the clocks.

Assume that an attacker would like to impersonate Bob, Figure 4 illustrates a possible way to lure Bob to respond with R_B. To this end, Clark performs a man-in-the-middle attack shown in Figure 4 so as to obtain C_A and R_B. In this simulated attack, Alice proceeds in the same way as that shown in Figure 2. For the sake of simplicity, we will only show the main steps which are related to the attack.

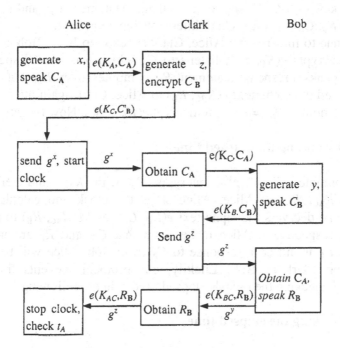

Figure 4. Impersonate Bob. Clark shares a channel with Alice and Bob simultaneously so that he can eavesdrop messages between them.

4.2.1 Obtaining Alice's challenge C_A

In Figure 4, Alice starts a session and sends the ciphertext $A_1 = e(K_A, C_A)$ to Bob. After intercepting the ciphertext A_1, Clark generates a number z, computes a key $K_C = h(g^z)$. Based on Assumption S2, Clark can not mimic Bob to speak a challenge, but he may reuse Bob's recorded speech. Clark encrypts an old statement C'_B spoken by Bob, and sends to Alice the ciphertext $e(K_C, C'_B)$.

Alice receives the ciphertext, replies with g^x and starts a clock. Clark derives a key $K_A = h(g^x)$ and recovers $C_A = d(K_A, C_A)$.

4.2.2 Obtaining Bob's response R_B

Because Clark can not mimic Bob's voice to produce an appropriate response R_B, he has to lure Bob to respond to Alice's challenge C_A. To this end, he impersonates Alice and starts a new session with Bob by sending Bob $e(K_C, C_A)$. Next, upon receipt of $e(K_C, C_A)$, Bob generates a random y, computes a key $K_B = h(g^y)$, speaks a challenge statement C_B, and sends the ciphertext $e(K_B, C_B)$ to Alice. Clark intercepts the ciphertext.

To continue to impersonate Alice, Clark sends g^z to Bob. Bob computes $K_C = h(g^z)$, decrypts $e(K_C, C_A)$, listens to C_A which was indeed spoken by Alice. Bob speaks a response statement R_B, computes a key $K_{BC} = h(g^{zy})$, and transmits g^y and the ciphertext $e(K_{BC}, R_B)$ to Alice. Clark again intercepts the ciphertext, computes $K_{BC} = h(g^{yz})$ to decrypt $e(K_{BC}, R_B)$. Now he gets R_B !

4.2.3   Calculating the elapsed time

Clark computes $K_{AC} = h(g^{xz})$, encrypts R_B with K_{AC}, and sends the ciphertext $e(K_{AC}, R_B)$ to Alice. Alice stops the clock and calculates the elapsed time t_A, decrypts the ciphertext $e(K_C, C'_B)$ and $e(K_{AC}, R_B)$ to recover C'_B and R_B, respectively. Alice makes sure that C'_B and R_B are in Bob's voice. Since R_B is indeed a response to C_A from Bob, Alice will be fooled into believing Clark as Bob! Luckily, our protocol prevents this from happening by checking the clock's elapsed time t_A in the following.

4.2.4   Checking the elapsed time

Consider the man-in-the middle attack shown in Figure 4. Within the time interval t_A, Bob has to speak his challenge statement C_B, listens to C_A, and speaks the response R_B; therefore, $t_A \geq |C_B| + |C_A| + |R_B| + \Delta_b$, where Δ_b is the time used in computation and transmission. With reference to Eq.(1), Alice checks $\delta_A = t_A - (|C_A| + |R_B| + \Delta_B) \geq |C_B| \geq T > \delta$.

Therefore, by checking the value of t_A, Alice detects the man-in-the-middle attack and stops the session.

4.3   Impersonating Alice

The other kind of possible attack is to impersonate the initiator Alice. To this end, Clark has to obtain the original challenge C_B and then Alice's respond R_A to Bob's challenge C_B. Figure 5 shows the second scenario of the man-in-the-middle attack, where Clark impersonates Alice to Bob. Therefore, Clark must start the communication with Bob at first.

4.3.1   Obtaining Bob's challenge C_B

Clark generates z, computes a key $K_C = h(g^z)$, and encrypts C'_A - an old statement from Alice. Clark starts the impersonation by sending the ciphertext $e(K_C, C'_A)$ to Bob.

Upon receipt of Clark' message, Bob generates y, and a key $K_B = h(g^y)$. He then speaks a reply C_B, and transmits the ciphertext $e(K_B, C_B)$ to Alice. Clark sends g^z to Bob. Bob derives $K_C = h(g^z)$, decrypts $e(K_C, C'_A)$ with K_C to recover C'_A. Bob listens to C'_A and believes that C'_A was indeed spoken by Alice. He then speaks a response statement R_B, derives a key $K_{BC} = h(g^{zy})$, and transmits g^y and the ciphertext $e(K_{BC}, R_B)$ to Alice. Bob then starts a clock.

Next, upon interception of $e(K_{BC}, R_B)$ and g^y, Clark derives $K_B = h(g^y)$ and $K_{BC} = h(g^{zy})$, decrypts $e(K_B, C_B)$ to recover C_B!

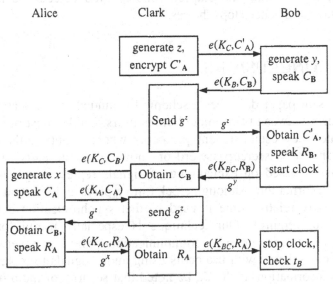

Figure 5. Impersonate Alice.

4.3.2 Obtaining Alice's response R_A

Since Clark is not able to reply C_B in Alice's voice, he starts a session with Alice by sending $e(K_C, C_B)$ to Alice. Upon receipt of $e(K_C, C_B)$, Alice generates x, and computes a key $K_A = h(g^x)$. She speaks a challenge C_A, and sends the ciphertext $e(K_A, C_A)$ to Bob.

Clark intercepts the ciphertext and sends g^z to Alice. Alice derives a key $K_C = h(g^z)$, decrypts $e(K_C, C_B)$ to recover C_B. She listens to C_B and believes that it is in Bob's voice.

Alice speaks a response statement R_A, computes $K_{AC} = h(g^{xz})$, sends g^x and the ciphertext $e(K_{AC}, R_A)$ to Bob. Clark intercepts the message from Alice and decrypts the ciphertext to obtain R_A!

4.3.3   Checking the elapsed time

After obtaining R_A, Clark sends the ciphertext $e(K_{BC}, R_A)$ to Bob. Bob receives $e(K_{BC}, R_A)$, stops the clock and calculates the elapsed time t_B. Without checking the elapsed time t_B, Bob would have been fooled into believing that he is talking to Alice since R_A is Alice's reply to C_B. However, within interval t_B, Alice has to speak C_A, listen to C_B and speak R_A, thus,

$$\delta_B = t_B - (|C_B| + |R_A| + \Delta_A) \geq |C_A| \geq T > \delta.$$

Therefore, by checking the elapsed time t_B, Bob detects the man-in-the-middle attack and hence stops the session.

5.   CONCLUSIONS

The present paper describes a scheme for mutual authentication and key establishment between two remote human users. Unlike most of the existing authenticated key establishment protocols where remote authentication is based on sharing a secret/password or knowing remote party's public key, our scheme is based on exchanging of signals representing remote user's biometrics information. Although clock timing plays an important role in our protocols, only relative time is used so that synchronization between two parties is not required. Our technique is especially useful for securing telephony or videoconference communications over open networks. We illustrated our scheme with protocols using audio signals to represent users' biometrics information. It should be noted that security of the protocols can be improved with additional biometrics information such as facial image and mouth movement. Such additional information adds few burdens to the human users, but greatly increases the difficulty of attacking the protocols.

REFERENCE

1. D. Otway and O. Rees, "Efficient and timely authentication", Operating Systems Review, Vol. 21, No. 1, pp. 8-10, 1987.
2. C. Kaufman, R. Perlman, and M. Speciner, Network Security - Private Communication in A Public World, PTR Prentice Hall, Englewoor Cliffs, NJ, 1995.
3. R. H. Deng, J. Zhou and F. Bao, "Defending against redirect attacks in mobile IP," ACM Conference on Computer and Communications Security, pp. 59-67, 2002.
4. P. Zimmermann, PGPfone Owner's Manual, Version 1.0 beta 5, 5 January 1996, http://web.mit.edu/network/pgpfone/manual.
5. Hunt and A. Black, "Unit selection in a concatenative speech synthesis system using large speech database," IEEE International Conf. on Acoustics, Speech, and Signal Processing (ICASSP), pp.373-376, 1996.

6. Matthias Jilka, Ann K. Syrdal, Alistair D. Conkie and David A. Kapilow, "Effects on TTS Quality of Methods of Realizing Natural Prosodic Variations," 15th International Congress of Phonics Science (ICPhS) 2003.
7. TTS, `https://research.microsoft.com/speech/tts/TTS.dll?TTS`
8. AT&T,   TTS: Synthesis   of   Audible   Speech   from   Text, `http://www.research.att.com/projects/tts/`

DISTANCE-BOUNDING PROOF OF KNOWLEDGE TO AVOID REAL-TIME ATTACKS[12]

Laurent Bussard and Walid Bagga
Institut Eurécom, Corporate Communications, 2229 Route des Crêtes, BP 193,
06904 Sophia Antipolis (France)
{bussard,bagga}@eurecom.fr

Abstract: Traditional authentication is based on proving the knowledge of a private key corresponding to a given public key. In some situations, especially in the context of pervasive computing, it is additionally required to verify the physical proximity of the authenticated party in order to avoid a set of real-time attacks. Brands and Chaum proposed distance-bounding protocols as a way to compute a practical upper bound on the distance between a prover and a verifier during an authentication process. Their protocol prevents frauds where an intruder sits between a legitimate prover and a verifier and succeeds to perform the distance-bounding process. However, frauds where a malicious prover and an intruder collaborate to cheat a verifier have been left as an open issue. In this paper, we provide a solution preventing both types of attacks.

Keywords: Real-time attack, distance-bounding, authentication, proof of knowledge

INTRODUCTION

The impressive development in the areas of web technologies, wireless networks, mobile computing, and embedded systems in the past decade has lead to an increasing interest in the topics of pervasive computing and open environments computing. In these contexts, authentication of communicating parties is considered as a major security requirement. As described in [8], a careful authentication may require the verification of the physical proximity of the authenticated party in order to prevent some real-time attacks. A

[12] The work reported in this paper is supported by the IST PRIME project and by Institut Eurécom; however, it represents the view of the authors only.

typical example is applications where digital authentication is required to access a building.

In the scenario depicted in Figure 1-a, a researcher carries around a mobile device (a mobile phone with extended functionalities or a PDA enhanced with communication capabilities) that takes care of computing, storage and communication on his behalf within the laboratory environment. Whenever the researcher approaches the door of a confidential research area, a communication is established between his mobile device and a lock device installed at the door. If the researcher is authorized to access the research area, the door is unlocked. Whenever the combination of the physical proximity and the cryptographic identification is not carefully addressed, some frauds could be performed such as the one depicted in Figure 1-b. In this fraud, a distant researcher (prover) that is allowed to access the confidential research area helps a friend (intruder) that is close by to access the area. For instance, a radio link could be used to establish the communication between the prover and the intruder.

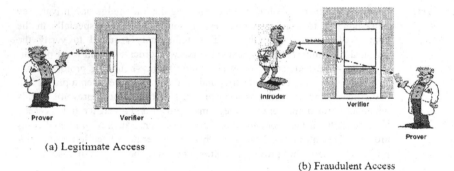

(a) Legitimate Access

(b) Fraudulent Access

Figure 1.   Access to a Confidential Research Area

The scenario described above falls under a quite recurring family of security protocols where a prover tries to convince a verifier of some assertion related to his private key. In order to address this problem, Brands and Chaum introduced the concept of distance-bounding protocols in [4]. Such protocols allow determining a practical upper bound on the distance between two communicating entities. This is performed by timing the delay between sending out a challenge bit and receiving back the corresponding response bit where the number of challenge-response interactions is determined by a system-specific security parameter. This approach is feasible if, on one hand, the protocol uses very short messages (one bit) on a dedicated communication channel (e.g. wire, IR) and if, on the other hand, no computation is required during each exchange of challenge-response bits (logical operations on the challenge bit). These conditions allow having round-trip times of few-nanoseconds.

The protocols given in [4] allow preventing *mafia frauds* where an intruder sits between a legitimate prover and a verifier and succeeds to perform the distance-bounding process. In this paper, we provide an extension of such protocols. Our solution allows preventing *terrorist frauds* [8] that have not been addressed so far. In these frauds the prover and the intruder collaborate to cheat the verifier. Note that even if the prover is willing to help the intruder to cheat the verifier, we assume that he never discloses his valuable private key. The key idea in our solution consists of linking the private key of the prover to the bits used during the distance-bounding process. This relies on an adequate combination of the distance-bounding protocol with a bit commitment scheme and a zero-knowledge proof of knowledge protocol [12].

The remainder of this paper is organized as follows. In Section 1, we define the frauds being addressed and give some related work. In Section 2, we draw a general scheme for distance-bounding proof of knowledge protocols, while we give a description of our protocol in Section 3. In Section 4, we analyze the security properties of the proposed protocol. At the end, we conclude and describe further work.

1. PROBLEM STATEMENT

In this section, we provide the definitions of the three attacks we tackle in this paper, namely *distance fraud*, *mafia fraud*, and *terrorist fraud*. Next, we present related work and we show why the existing approaches are not satisfactory.

1.1 Definitions

Distance-bounding protocols have to take into account the three real-time frauds that are depicted in Figure 2. These frauds can be applied in zero-knowledge or minimal disclosure identification schemes. The first fraud is called the *distance fraud* and is defined in the following (Figure 1-a).

DEFINITION 1 (DISTANCE FRAUD) *In the distance fraud two parties are involved, one of them (V the verifier) is not aware of the fraud is going on, the other one (P the fraudulent prover) performs the fraud. The fraud enables P to convince V of a wrong statement related to its physical distance to prover V.*

The distance fraud has been addressed in [4]. This fraud consists of the following: if there's no relationship between the challenge bits and the response bits during the distance-bounding protocol and if the prover P is able to know at what times the challenge bits are sent by the verifier V, he can make V compute a wrong upper bound on his physical distance to V by sending out the response bits at the correct time before receiving the challenge bit, regardless of his physical distance to V.

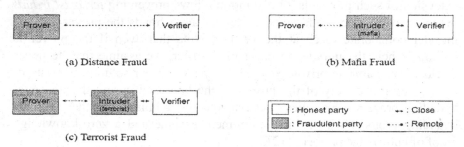

(a) Distance Fraud   (b) Mafia Fraud

(c) Terrorist Fraud

Figure 2.   Three Real-Time Frauds

The second fraud is called the *mafia fraud* and is defined in the following (Figure 1-b).

DEFINITION 2 (MAFIA FRAUD) *In the mafia fraud three parties are involved, two of them (P the honest prover and V the verifier) are not aware of the fraud is going on, the third party (I the intruder) performs the fraud. The fraud enables I to convince V of an assertion related to the private key of P.*

The mafia fraud has been first described in [8]. In this fraud, the intruder I is usually modeled as a couple $\{\overline{P}, \overline{V}\}$ where $\overline{P}$ is a dishonest prover interacting with the honest verifier V and where $\overline{V}$ is a dishonest verifier interacting with the honest prover P. Thanks to the collaboration of $\overline{V}$, the fraud enables $\overline{P}$ to convince V of an assertion related to the private key of P. The assertion is that the prover is within a certain physical distance. This fraud was also called *Mig-in-the-middle attack* in [2].

The third fraud is called the *terrorist fraud* and is defined in the following (Figure 1-c).

DEFINITION 3 (TERRORIST FRAUD) *In the terrorist fraud three parties are involved, one of them (V the verifier) is not aware of the fraud is going on, the two others (P the dishonest prover and I the intruder or terrorist) collaborate to perform the fraud. Thanks to the help of P, the fraud enables I to convince V of an assertion related to the private key of P.*

The terrorist fraud has been first described in [8]. In this fraud, the prover and the intruder collaborate to perform the fraud whereas in the mafia fraud the intruder is the only entity that performs the fraud. Note that the prevention of terrorist frauds implies the prevention of mafia frauds.

1.2   Related Work

In this section we review different techniques that have been proposed and show why they are not sufficient when it is necessary to verify that some entity knowing a private key is indeed physically present.

Constrained Channel [13] aims at exchanging some secret between two physical entities and thus ensures the proximity of two devices. An obvious implementation is to have a physical contact [16] between the two artifacts. This scheme works only when the attacker is not physically present. It can protect a system only against distance frauds.

Context Sharing is a straightforward extension of constrained channels where some contextual data is used to initiate the key exchange. For instance, in [10], the pairing mechanism is done by shaking artifacts together in order to create a common movement pattern that is subsequently used to bootstrap the security of communications. This approach prevents distance frauds and can partially avoid mafia frauds when the context is difficult to reproduce.

Isolation [3] is a widely deployed solution to check whether a physical entity holds a secret. The device is isolated in a Faraday cage during a challenge-response protocol. This solution prevents distance frauds, mafia frauds as well as terrorist frauds. However, it is difficult to deploy, it is not user-friendly, and does not allow mutual authentication.

Unforgeable Channel aims at using communication channels that are difficult to record and reconstruct without knowing some secret. For instance, channel hopping [1] or radio frequency watermarking [11] makes it difficult to transfer data necessary to create the signal in another place. This scheme protects against distance frauds and the solution proposed in [1] can prevent mafia frauds as well when it is not possible to identify communication sources. Quantum cryptography can also be envisioned as an unforgeable channel.

Time of Flight relies on the speed of sound and/or light. Sound and especially ultra-sound [15] is interesting to measure distance since it is slow enough to authorize computation without reducing the accuracy of the measure. Sound-based approaches cannot protect against physically present attackers and thus can only prevent distance frauds. Some works also rely on the speed of light when measuring the round trip time of a message to evaluate the distance to the prover. However, one meter accuracy implies responding within few nanoseconds and thus it cannot be done through standard communication channels and cannot use cryptography [17]. Such schemes prevent distance frauds and the solution proposed in [4] prevents mafia frauds as well.

As shown above, only *isolation* allows preventing distance, mafia and terrorist frauds all together. In this paper, we focus on *distance-bounding protocols* and propose a solution that prevents the three real-time attacks. In contrast with *isolation*, our approach is easy to deploy and allows mutual authentication.

2.   THE GENERAL SCHEME

In this section, we present a general scheme (denoted **DBPK**) containing the basic building blocks of *distance-bounding proof of knowledge* protocols.

2.1   Description

The **DBPK** scheme is depicted in Table 1. It relies on a set of global settings that have to be performed before the execution of any interaction between the prover and the verifier. Besides the cryptosystem's public parameters, these global settings allow the prover to have a valuable private key and a certificate on the corresponding public key. That is, before any interaction with the verifier, the prover holds a private key x which importance, by assumption, is so high that the prover should not reveal it to any other party. In addition, the prover holds a certificate (generated by a globally trusted authority) on its public key $y = \Gamma(x)$.

The first stage of the **DBPK** protocol is called the *Bit Commitment* stage. During this stage the prover first picks a random one-time key $k \in_R K$ and uses it to encrypt its private key x according to a publicly known symmetric key encryption algorithm E. This leads to the ciphertext $e = E_k(x)$. Once the encryption performed, the prover commits to each bit of both k and e according to a secure bit commitment scheme *commit*. For each bit $k[i]$ (resp. $e[i]$), a string v_i (resp. v'_i) is randomly chosen by the prover to construct the commitment blob $c_{(k,i)}$ (resp. $c_{(e,i)}$).

Once the *Bit Commitments* stage is completed, the actual distance-bounding interactions are executed during the *Distance-Bounding* stage. Basically, N interactions are performed between the prover and the verifier. In the ith interaction, the prover releases either $k[i]$ or $e[i]$ depending on whether the challenge bit is equal to 0 or to 1. Note that $k[i]$ (resp. $e[i]$) denotes the ith bit in the binary representation of k (resp. e) where $k[0]$ (resp. $e[0]$) is the least significant bit of k (resp. e). During each bit exchange, the round trip time (few nanoseconds) is measured in order to verify the distance to the prover.

$$\begin{array}{cc}
\mathbf{P} & \mathbf{V} \\
\text{Prover} & \text{Verifier} \\
(\text{private key } x) & (P\text{'s public key } y) \\
(\text{public key } y = \Gamma(x)) &
\end{array}$$

Bit Commitments

secret key $k \in_R \mathcal{K}$

$N = \lceil log_2(|\mathcal{K}|) \rceil$, $\mathcal{M} = \{0, \ldots, N-1\}$

$e = \mathcal{E}_k(x) \in \{0,1\}^N$

for all $i \in \mathcal{M}$ $v_i, v_i' \in_R \{0,1\}^*$

for all $i \in \mathcal{M}$ $c_{(k,i)} = commit(k[i], v_i)$

for all $i \in \mathcal{M}$ $c_{(e,i)} = commit(e[i], v_i')$

$$\text{for all } i \in \mathcal{M} \quad c_{(k,i)}, c_{(e,i)} \longrightarrow$$

Distance-Bounding (for all $i \in \mathcal{M}$)

$$a_i \in_R \{0,1\} \longleftarrow$$

$b_i = k[i]$ if $\delta(a_i) = 0$

$b_i = e[i]$ if $\delta(a_i) = 1$

$$b_i \in \{0,1\} \longrightarrow$$

Commitment Opening

for all $i \in \mathcal{M}$

$$v_i \text{ (if } \delta(a_i) = 0) \quad v_i' \text{ (if } \delta(a_i) = 1) \longrightarrow$$

$$c_{(k,i)} \stackrel{?}{=} commit(b_i, v_i) \quad \text{if } \delta(a_i) = 0$$

$$c_{(e,i)} \stackrel{?}{=} commit(b_i, v_i') \quad \text{if } \delta(a_i) = 1$$

Proof of knowledge

$$\{c_{(k,i)}, c_{(e,i)}\}_{0 \le i \le N-1} \rightarrow z = \Omega(x, v)$$

$$PK[(\alpha, \beta) : z = \Omega(\alpha, \beta) \wedge y = \Gamma(\alpha)] \longleftarrow$$

Figure 3.   A general scheme for $DBPK[\alpha : y = \Gamma(\alpha)]$

After the execution of the N successful challenge-response bit exchanges, the prover opens the commitments on the released bits of k and e. The *Commitment Opening* stage consists of sending the string v_i if $k[i]$ has been released and v_i' otherwise. Only half of the bits of k and e are released to the verifier. This must not allow the verifier to get any significant information about the valuable private key x. In the case where the verification of $c_{(k,i)}$ (resp. $c_{(e,i)}$) fails, the verifier sends back an error notification of the form $error_k(i)$ (resp. $error_e(i)$).

The last step in the **DBPK** protocol is the *Proof of Knowledge* stage. During this stage, the prover convinces the verifier in a zero-knowledge interaction that he is the party who performed the three previously described stages. That is, the prover proves that he has generated the different commitments, that the generated commitments correspond to a unique private key, and that this private key corresponds to the public key y that is used by the verifier to authenticate the prover. Before the proof of knowledge process can be performed, the verifier must compute a one way function on the private key x: $z = \Omega(x, y)$ where v and x are known only by the prover. As z depends on and only on the commitments on the bits of k and e, it may even be computed just after the *Bit Commitments* stage. The proof of knowledge we use is denoted $PK[(\alpha, \beta) : z = \Omega(\alpha, \beta) \wedge y = \Gamma(\alpha)]$ where the Greek small letters denote the quantity the knowledge of which is being proved, while all other parameters are known to the verifier. The functions Ω, Γ, δ, E, and *commit* are adequately chosen to meet our security requirements, namely the prevention of the distance, mafia, and terrorist frauds.

The distance-bounding proof of knowledge of a secret x such that $y=\Gamma(x)$ is denoted $DBPK[\alpha : y = \Gamma(\alpha)]$.

3.   OUR PROTOCOL

This section presents a concrete distance-bounding proof of knowledge protocol that consists of exactly the same building blocks of the **DBPK** protocol. The protocol will be denoted **DBPK-Log** $= DBPK[\alpha : y = g^{\alpha}]$.

3.1   Global Settings

We first describe the global settings on which relies the **DBPK-Log** protocol. These settings consist of two main phases: *Initialization* and *Registration*. In the *Initialization* stage, a trust authority (TA) provides the public parameters of the system.

Initialization:
TA sets up the system's global parameters
- TA chooses a large enough strong prime p, i.e. there exists a large enough prime q such that $p = 2q+1$
- TA chooses a generator g of Z_p^{*}
- TA chooses an element $h \in_R Z_p^{*}$

The randomly chosen element h will be used by the commitment scheme. The only requirement is that neither of the prover and the verifier knows $\log_g(h)$. This can be achieved either by letting the trusted authority generate this element, or by making the prover and the verifier jointly generate h. The two alternatives rely on the intractability of the *discrete logarithm* problem [14].

In the *Registration* stage, a user chooses a private key and registers at the trust authority so to get a certificate on the corresponding public key.

Registration:
The following steps are taken by P to get a certified public key corresponding to a valuable private key
- P selects an odd secret $x \in_R Z_{p-1} \setminus \{q\}$, then computes $y = g^x$. The public key of P is y and his private key is x
- P registers his public key with TA so TA publishes a certificate on this public key

Note that the two phases described above are executed only once. They allow generating the prover's public and private keys that will be used in the different subsequent distance-bounding proofs of knowledge.

3.2   Interactions

Our distance-bounding proof of knowledge protocol starts with the *Bit Commitments* stage where the prover P generates a random key k, and uses this key to encrypt the private key x. Then, P performs a secure commitment on each bit of the key k and encryption e.

Bit Commitments:
The following steps are performed
- Given a security parameter m', P picks at random $k \in_R \{0, ..., 2^N-1\}$, where $N = m'+m$ and $m = \lceil log_2(p) \rceil$.
- P computes $e \in \{0, ..., 2^N-1\}$ such that $e \equiv x-k \mod p-1$.
- For all $i \in \{0, ..., N-1\}$, P chooses $v_{k,i}, v_{e,i} \in_R Z_{p-1}$, computes $c_{k,i} = g^{k[i]} \cdot h^{v_{k,i}} \mod p$ and $c_{e,i} = g^{e[i]} \cdot h^{v_{e,i}} \mod p$, then sends $c_{k,i}$ as well as $c_{e,i}$ to V

Once the verifier V receives all the commitment blobs corresponding to the bits of k and e, the *Distance-Bounding* stage can start. Thus, a set of fast single bit challenge-response interactions is performed. A challenge corresponds to a bit chosen randomly by V while a response corresponds either to a bit of k or to a bit of e.

Distance-Bounding:
For all $i \in \{0, ..., N\text{-}1\}$,
- V sends a challenge bit $a_i \in_R \{0,1\}$ to P
- P immediately sends the response bit $b_i = \bar{a}_i k[i] + a_i e[i]$ to V

At the end of the *Distance-Bounding* stage, the verifier V is able to compute an upper bound on the distance to P. In order to be sure that P holds the secrets k and e, the prover P opens, during the *Commitment Opening* stage, the commitments on the bits of k and e that have been released during the *Distance-Bounding* stage.

Commitment Opening:
The commitments of the released bits are opened. If all the checks hold, all the bit commitments on k and e are accepted, otherwise they are rejected and an error message is sent back
- For all $i \in \{0, ..., N\text{-}1\}$, P sends $\bar{a}_i v_{k,i} + a_i v_{e,i}$ to V
- For all $i \in \{0, ..., N\text{-}1\}$, V performs the following verification:
$$\bar{a}_i c_{k,i} + a_i c_{e,i} \overset{?}{=} g^{\bar{a}_i k[i] + a_i e[i]} \cdot h^{\bar{a}_i v_{k,i} + a_i v_{e,i}} \bmod p$$

The proof of knowledge allows the verifier V to be sure that e is indeed the encryption of the private key x corresponding to the public key y of the prover. From the bit commitments, V can compute:

$$z = \prod_{i=0}^{N-1} (c_{k,i} \cdot c_{e,i})^{2^i} = g^{\sum_{i=0}^{N-1}\left(2^i \cdot k[i] + 2^i \cdot e[i]\right)} \cdot h^{\sum_{i=0}^{N-1}\left(2^i \cdot (v_{k,i} + v_{e,i})\right)}$$

$$= g^{k+e} \cdot h^v = g^x \cdot h^v \bmod p$$

Note that V is able to compute z as soon as all the commitments on the bits of k and e are received.

Proof of Knowledge:
Given $z = g^x \cdot h^v$, the following proof of knowledge is performed by P and V:
$PK[(\alpha, \beta) : z = g^\alpha h^\beta \wedge y = g^\alpha]$.

4.   SECURITY ANALYSIS

In this section, we discuss the relevant security properties of the **DBPK-Log** protocol. First, we show that our protocol prevents distance, mafia, and terrorist frauds. Next, the security properties of the encryption scheme that is used to hide the prover's private key are studied.

4.1 Preventing Distance, Mafia and Terrorist Frauds

The first security requirement for our distance-bounding proof of knowledge protocol is a correct computation of an upper bound on the distance between the prover and the verifier. This requirement is being already achieved in the **DBPK** general scheme according to the following.

Proposition 4.1 *If the **DBPK** protocol is performed correctly, then the distance fraud has a negligible probability of success.*

Proof: Assume that the prover P knows at what times the verifier V will send out bit challenges. In this case, he can convince V of being close by sending out the bit response b_i at the correct time before he receives the bit a_i. The probability that P sends correct responses to V before receiving the challenges is equal to

$$\prod_{i=1}^{N}(P[b_i = k[i]|\delta(a_i) = 0] + P[b_i = e[i]|\delta(a_i) = 1]) = 2^{-N}$$

In Proposition 4.1, the correct execution of the protocol means that each party performs exactly and correctly the actions specified in the different steps of the protocol.

The **DBPK-Log** protocol is an implementation of the **DBPK** protocol where the function δ corresponds to the identity function, i.e. $\forall i\ \delta(a_i) = a_i$. This leads to the following proposition.

Proposition 4.2 *If the **DBPK** protocol is performed correctly, then the distance fraud has a negligible probability of success.*

Respecting the notations of Section 2, we introduce the three following properties.

Property 4.1 *Let $\Gamma : \{0,1\}^* \rightarrow \{0,1\}^*$ be the function such that $y = \Gamma(x)$, then the following holds:*
- *Given y, it is hard to find x such that $y = \Gamma(x)$.*
- *It is hard to find $x \neq x'$ such that $\Gamma(x) = \Gamma(x')$.*

Property 4.2 *Let $\Omega : \{0,1\}^* \times \{0,1\}^* \rightarrow \{0,1\}^*$ be the function such that $z = \Omega(x,v)$, then the following holds*
- *Knowing z and Ω, it is hard to find (x,v).*
- *It is hard to find $(x,v) \neq (x',v')$ such that $\Omega(x,v) = \Omega(x',v')$.*

Property 4.3 *Let E be the function such that $e = E_k(x)$, then the following holds*
   (a) E is an encryption scheme: knowing e and E, it is hard to find x without knowing k; and given e and k, $x = D_k(e)$ is efficiently computable.

(b) *Given either k[i] or e[i] for all i $\in$ {0,...,N-1}, it is hard to find x.*

(c) *It is efficient to compute z = $\Omega(x,v)$ from the commitments on the bits of k and e.*

The second security requirement for distance-bounding proof of knowledge protocols consists in preventing terrorist frauds. This requirement can already be achieved in the **DBPK** general scheme according to the following proposition.

Proposition 4.3 *If Property 4.1, Property 4.2, and Property 4.3 are respected and if the **DBPK** protocol is performed correctly, then the terrorist fraud has a negligible probability of success.*

Proof: A successful execution of the Proof of Knowledge stage proves that the entity knowing the private key corresponding to the public key y have performed the *Bit Commitments* stage. Assume that the latter has been performed using k and e. Then, the probability for an intruder to perform the Distance-Bounding stage successfully using (k', e') $\neq$ (k,e) is equal to

$$\prod_{i=1}^{N} \left(P[k[i] = k'[i] | \delta(a_i) = 0] + P[e[i] = c'[i] | \delta(a_i) = 1] \right) = 2^{-N}$$

This shows that without knowing (k,e), i.e. without knowing $x = D_k(e)$, the probability of success of a terrorist fraud is negligible.

The **DBPK-Log** consists of the same building blocks than those of the **DBPK** protocol. Moreover, the three following statements hold

(1) The function $\Gamma : x \rightarrow g^x$ respects Property 4.1 thanks to the intractability of the discrete logarithm problem.

(2) The function $\Omega : (x,v) \rightarrow g^x \cdot h^v$ respects Property 4.2 thanks to the intractability of the *representation* problem.

(3) The one-time pad $E_k(x) = $ x-k mod p-1 respects Property 4.3 (see Section 4.2).

The properties listed above lead to the following.

Proposition 4.4 *If the **DBPK-Log** protocol is performed correctly, then the terrorist fraud has a negligible probability of success.*

Recall that the prevention of terrorist frauds makes the prevention of mafia frauds straightforward.

4.2   Encryption of the Private Key

Since some bits of the key k are revealed, it is straightforward that a one-time key is necessary. To be compliant with Property 4.3, we propose a dedicated one-time pad: $e = E_k(x) = x-k \mod p-1$ where $k \in \{0,..., 2^N-1\}$ is randomly chosen before each encryption. The parameter N is such that $N = m+m'$, where m is the number of bits of the private key x and m' is a system-specific security parameter. The prime number p is a strong prime, i.e. $p = 2q+1$ where q is an enough large prime. This scheme is compliant with the following:

- Property 4.3.a: with this encryption scheme, revealing e still ensures perfect secrecy of x: $P_{X|E}(X = x| E = e) = P_X(X = x) = 2^{-N}$ for all x,e
- Property 4.3.b: in the following, we show that the knowledge of either $k[i]$ or $e[i]$ for all $i \in \{0,...,N-1\}$, allows to retrieve information on x with probability less than $2^{-2m'}$. We basically study the impact of the security parameter m' on the probability of revealing information on x. Knowing b such that $b = x-b'$, information on x can be statistically obtained when a large enough number n of samples can be collected to have a sample mean $\overline{Y}_n$ close to the mean μ i.e. $|\overline{Y}_n - \mu| < p-1$. The *Central Limit Theorem* states that the sum of a large number of independent random variables has a distribution that is approximately normal. Let $Y_1, Y_2,..., Y_n$ be a sequence of independent and identically distributed random variables with mean μ and variance σ^2 and $\overline{Y}_n = (1/n)\sum_{i=1}^{n} Y_i$ then,

$$\frac{\sqrt{n}(\overline{Y}_n - \mu)}{\sigma} \to N(0,1), \quad \text{when } n \to \infty$$

That is

$$P\left\{-a \le \frac{\sqrt{n}(\overline{Y}_n - \mu)}{\sigma} \le a\right\} \to \frac{1}{\sqrt{2\pi}} \int_{-a}^{a} e^{-z^2/2} dz$$

Since the following holds

$$P\left\{-a \le \frac{\sqrt{n}(\overline{Y}_n - \mu)}{\sigma} \le a\right\} = P\left\{\mu - a\frac{\sigma}{\sqrt{n}} \le \overline{Y}_n \le \mu + a\frac{\sigma}{\sqrt{n}}\right\}$$

then, the probability of having a sample mean close to the mean is

$$P\left\{\mu - 2^{m-1} \le \overline{Y}_n \le \mu + 2^{m-1}\right\} \to \frac{1}{\sqrt{2\pi}} \int_{-2^{m-1}\frac{\sqrt{n}}{\sigma}}^{2^{m-1}\frac{\sqrt{n}}{\sigma}} e^{-z^2/2} dz$$

The mean is $\mu = x + 2^{N-1}$ and the variance is σ^2, where $\sigma = (2^N)^2/12$. Hence, $(2^{m-1}\sqrt{n})/\sigma = (6\sqrt{n})/2^{m+2m'} \gg 1$. In other words,

$$P\left\{\mu - 2^{m-1} \le \overline{Y}_n \le \mu + 2^{m-1}\right\} \cong \frac{12\sqrt{n}}{2^{m+2m'}} \cdot \frac{1}{\sqrt{2\pi}}$$

We study the number of samples necessary to get information on x:

$$\frac{12\sqrt{n_0}}{2^m} \cdot \frac{1}{\sqrt{2\pi}} = \frac{12\sqrt{n}}{2^{m+2m'}} \cdot \frac{1}{\sqrt{2\pi}} \Rightarrow \frac{n}{n_0} = 2^{4m'}$$

Here n_0 is the number of sample necessary when $m' = 0$, i.e. when the parameters x, k, and e are of equal length. In the worst case, n_0 is equal to 1 and the number of samples necessary to get information on x is $2^{4m'}$.

Note that the security parameter $m0$ ensures a probability of successful attack less than $2^{-4m'}$ at the expense of m' additional challenge-response bit exchanges. For instance, for $m = 1024$ bits and $m' = 50$ bits, the probability of retrieving information about x is less than 2^{-200} at the expense of around 5% of additional challenge-response bit exchanges.

- Property 4.3.c: it is possible to deduce a representation of z depending on x from commitments on bits of k and e (Section 3):

$$z = \prod_{i=0}^{N-1} (c_{k,i} \cdot c_{e,i})^{2^i} = g^x \cdot h^y \bmod p$$

CONCLUSION AND FURTHER WORK

In this paper, we addressed the problem of terrorist frauds in application scenarios where cryptographic authentication requires the physical proximity of the prover. Our solution consists in distance-bounding proof of knowledge protocols that extend Brands and Chaum's distance-bounding protocols [4]. We first presented a general scheme that shows the main building blocks of such protocols. We then presented a possible implementation of such protocols and analyzed its security properties. Even though we have not reached perfect secrecy, our solution remains secure in the statistical zero-knowledge security model.

The general scheme presented in this paper (**DBPK**) could be used with any public key scheme Γ if adequate commitment scheme *commit*, encryption method E, and representation function Ω exist. We proposed a solution relying on a public key scheme based on the discrete logarithm problem, bit commitment based on discrete logarithm, group addition one-time pad, and representation problem: **DBPK-Log** = $DBPK[\alpha : y = g^\alpha]$. This scheme could directly be used with ElGamal's and Schnorr's identification schemes that both rely on the discrete logarithm problem.

The integration of distance-bounding with Fiat-Shamir identification scheme [9] is not straightforward. The public key x is chosen in Z_n where $n = pq$ and the public key is $x2 \bmod n$. It is necessary to define $DBPK[\alpha : y = \alpha^2]$. Using the commitment scheme presented in this paper, the following proof of knowledge is required: $PK[\alpha,\beta : z = g^{\alpha} \cdot h^{\beta} \wedge y = \alpha^2]$. In other words, the parameter g has to be a generator of a cyclic group of order n.

We are also studying whether such a scheme can be used in a privacy preserving way. **DBPK** could be integrated in a group signature scheme, e.g. the initial one proposed in [7] would be: $DBPK[\alpha : \tilde{z} = \tilde{g}^{(\alpha^2)}]$; $PK[\beta : \tilde{z}\tilde{g} = \tilde{g}^{(\beta^2)}]$.This way, the verifier can verify that he is in front of a member of some group. However the verifier does not get any information on the identity of this group member. In this case, the encryption has to be done modulo n. A further step would be the integration of distance-bounding protocols in unlinkable and/or pseudonymous credentials schemes such as Idemix [6].

An alternative way to address terrorist frauds would be by combining trusted hardware with any protocol preventing mafia frauds [5]. In other words, a tamper-resistant hardware trusted by the verifier has to be used by the prover to execute the protocol. However, our approach is more general and easier to deploy since it neither relies on tamper-resistant hardware nor requires device certification process.

REFERENCES

[1] A. Alkassar and C. Stuble. Towards secure iff: preventing mafia fraud attacks. In *Proceedings of MILCOM 2002*, volume 2, pages 1139–1144, October 2002.

[2] Ross Anderson. *Security Engineering: A Guide to Building Dependable distributed Systems*. John Wiley and Sons, 2001.

[3] S. Bengio, G. Brassard, Y. Desmedt, C. Goutier, and J.J. Quisquater. Secure implementation of identification systems. *Journal of Cryptology*, 4(3):175–183, 1991.

[4] S. Brands and D. Chaum. Distance-bounding protocols (extended abstract). In *Proceedings of EUROCRYPT 93*, volume 765 of *LNCS*, pages 23–27. Springer-Verlag, May 1993.

[5] L. Bussard and Y. Roudier. Embedding distance-bounding protocols within intuitive interactions. In *Proceedings of Conference on Security in Pervasive Computing (SPC'2003)*, LNCS. Springer, 2003.

[6] J. Camenisch and A. Lysyanskaya. An efficient system for non-transferable anonymous credentials with optional anonymity revocation. *Lecture Notes in Computer Science*, 2045, 2001.

[7] J. L. Camenisch and M. A. Stadler. Efficient group signature schemes for large groups. In *Advances in Cryptology – CRYPTO '97 Proceedings*, volume 1294 of *LNCS*, pages 410–424. Springer-Verlag, 1997.

[8]   Yvo Desmedt. Major security problems with the 'unforgeable' (Feige)-Fiat-Shamir proofs of identity and how to overcome them. In *Proceedings of SecuriCom '88*, 1988.

[9]   A. Fiat and A. Shamir. How to prove yourself: Practical solutions to identification and signature problems. In *Advances in Cryptology—Crypto '86*, pages 186–194, New York, 1987. Springer-Verlag.

[10] L.E. Holmquist, F. Mattern, B. Schiele, P. Alahuhta, M. Beigl, and H-W. Gellersen. Smart-its friends: A technique for users to easily establish connections between smart artefacts. In *Proceedings of UbiComp 2001*, 2001.

[11] Yih-Chun Hu, A. Perrig, and D.B. Johnson. Packet leashes: a defense against wormhole attacks in wireless networks. In *Proceedings of INFOCOM 2003. Twenty-Second Annual Joint Conference of the IEEE Computer and Communications Societies*, volume 3, pages 1976–1986, March 2003.

[12] J.Pieprzyk, T.Hardjono, and J.Seberry. *Fundamentals of Computer Security*. Springer, 2003.

[13] T. Kindberg, K. Zhang, and N. Shankar. Context authentication using constrained channels. In *Proceedings of the IEEE Workshop on Mobile Computing Systems and Applications (WMCSA)*, pages 14–21, June 2002.

[14] Alfred J. Menezes, Scott A. Vanstone, and Paul C. Van Oorschot. *Handbook of Applied Cryptography*. CRC Press, Inc., 1996.

[15] N. Sastry, U. Shankar, and D. Wagner. Secure verification of location claims. In *Proceedings of the 2003 ACM workshop on Wireless security*, 2003.

[16] Frank Stajano and Ross J. Anderson. The resurrecting duckling: Security issues for adhoc wireless networks. In *Security Protocols Workshop*, pages 172–194, 1999.

[17] B. Waters and E. Felten. Proving the location of tamper-resistant devices. Technical report.

AN ADAPTIVE POLLING SCHEME FOR IEEE 802.11 WIRELESS LAN

Kyung-jun Kim, Hyun-sook Kim, Sang-don Lee, Ki-jun Han
Department of Computer Engineering, Kyungpook National University, San-kyuk Dong, Puk-gu, Daegu 702-701, Korea

Abstract: The point coordination function of IEEE 802.11 is defined to support time bounded traffic in wireless LANs. However, current WLAN standard does not consider the traffic characteristics such as time bound traffic, null poll, and priority. In the case of time bounded traffic, PCF must evolve to support degrees of priority. In this paper, we proposed an adaptive polling scheme to increase the performance of wireless LANs. Moreover, we focused to a polling schedule to serve voice traffic. The simulation results show that the proposed scheme is more efficient than using only standards.

Key words: WLAN, IEEE 802.11, Polling, PCF, Real-time

1. INTRODUCTION

The dramatic increasing of wireless devices has recently generated a lot of interest in wireless networks that support multimedia services, such as streaming video and audio that has critical quality of service requirements such latency, jitter and bandwidth [6].

IEEE802.11 WLAN standard protocol supports two kinds of access methods: distributed coordination function and PCF [1]. The DCF is designed for asynchronous data transmission by using CSMA/CA (carrier sense multiple access with collision avoidance) and must be implemented in all stations. On the other hand, the PCF is a polling-based scheme which is mainly intended for transmission of real-time traffic. This access method is optional and is based on polling controlled by a PC (point coordinator). Currently, the polling mechanism for IEEE 802.11 PCF mode is based on the round robin principle which polls every station in sequence regardless of whether it has packets to transmit or not. Real-time applications are characterized by QoS parameters such as delay, jitter, etc [8]. These applications generally have higher priority than non-real time traffic and hence they are allocated a significant portion of the bandwidth.

In our knowledge, since the round robin polling scheme does not consider traffic characteristics in each station, the PCF suffers from performance degradation such as delay since empty poll which a happen when the AP polls to stations in silence state [2, 3]. However, current WLAN standard does not consider the traffic characteristics such as time bound traffic, null poll, and priority. In the case of time bounded traffic, PCF must evolve into support priority and to reduce packet drop due to buffer overflow, traffic to the bottleneck node must be controlled. This can be achieved by decrease of the number of polling.

In this paper, we proposed dynamically adaptive polling scheme that reducing the unnecessary polling overhead for service real-time data in IEEE 802.11 WLAN efficiently. Thus, our scheme is to increase the bandwidth effectively without losing fairness at the MAC layer in wireless local area networks.

The remainder of this paper is organized as follows. In Section 2, we briefly describe the related works. In section 3 presents an adaptive polling scheme. In section 4, we evaluate the performance of the proposed scheme deriving the packet discard ratio and maximum number of real-time stations handled by PCF. Finally, section 5 concluded in this paper.

2.   RELATED WORKS

This section briefly summarizes the some of the features of the 802.11 WLAN sub-layer with the emphasis on the PCF mode of operation.

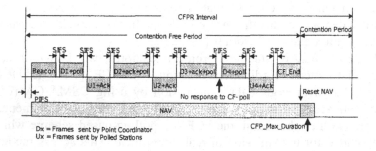

Figure 2. Example of PCF frame transfer

A typical medium access sequence during PCF is shown in Fig. 1. A station being polled is allowed to transmit a data frame. In case of an unsuccessful transmission, the station retransmits the frame after defer or during the next contention period. A PC [1] polls the stations in a round-robin fashion. A polled station always responds to the received a poll. If there is no pending transmission, the intended station responds to a null frame without payload. This frame is called empty poll. In the PCF mechanism, prior to all stations polling, if the *CFP_Max Duration* terminates, next polling sequence resumed at a station not called station. If transmission failed, the station retransmits the frame after delay or during the next contention period.

If contention traffic from the previous repetition interval carries over into the current interval, the *CFP_Max Duration* may be shortened. Also, if a station lasts longer than the remaining contention periods, the PC has to defer the start of its real time traffic transmission until the medium becomes free for a PIFS. This is because of bandwidth of shared medium. Therefore, recently, bandwidth has been a major design goal for wireless LAN.

In the last few years, many researchers actively explored advanced bandwidth reuse approaches for wireless LAN. A set of related research issues needs to be addressed before our approach called as adaptive polling scheme which is technically feasible and economically practical.

In [2, 3, 4] scheme, in the case of real time traffic, the existence of transmission delay and packet discard by empty poll is not assumed in order to decrease transmission delay. However, our adaptive polling scheme predicts an adaptive (optimal) priority for the current transmission, based on queue sizes in each station. Most of the known wireless MAC protocols are not specifically designed to support multimedia traffic QoS such as transmission delay, which severely impairs the system performance, can be averaged out.

Eustathia et al. [2] proposed a scheme called cyclic shift and station removal polling process (CSSR) in which the AP's polling list temporarily removes stations that enter silence state. However, when it leaves silence state, its voice packet may be discarded in the next round because it does not receive a poll in the maximum allowable delay.

O.Sharon et al. [4] proposed an Efficient Polling MAC Scheme in which stations are separated into two groups, active group and idle group, according to whether there are any pending data ready to be sent. a station in active group and a station in the idle group can simultaneously respond to the polling from the PC by using signals of different strengths.

In contrast, our polling scheme, each station has a priority and AP multi-polls in each station, which is dynamically assigned by the PC based on the queue size of each station.

3.   AN ADAPTIVE POLLING SCHEME

3.1   Overview

IEEE 802.11 PCF mode has been addressed many issues, very important one of these issues which is empty poll problem, this is happens during polling to stations under the silence state. For example, if any station leaves silence state, voice packet of the idle station may be discarded in the ongoing next round, since excess maximum delay for receiving a poll.

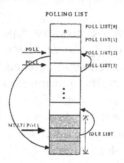

Figure 2. The polling list used for the re-setup

Therefore, an adaptive polling scheme for avoidance null-poll without transmitting delay is an important challenge to serve real-time traffic. When large frames are transmitted in a noisy channel, they accumulate bit errors that triggers retransmissions. These retransmissions consume valuable bandwidth and degrade the efficiency of the entire system. Thus, an adaptive polling scheme achieves two benefits: 1) the priority of a frame is guaranteed; and 2) avoid to waste bandwidth, which increases the throughput of the system. Fig. 2 shows basic operation of adaptive multicast polling list re-setup.

3.2   Management of Polling List

In this section, we propose an adaptive polling scheme in the IEEE 802.11 PCF to reduce polling overhead. It is occur because of managing the polling list based on the amount of packets at each station. A station with many data is expected to having a higher priority. For each intended station, its priority is given depending on an amount of payload, which is dynamically assigned by the PC based on the queue size of each station. The PC received a packet which includes the information of queue size and an amount of payload of each station. At this time, idle stations without data are assigned the lowest priority level. and then at one time they will be received a multicast poll instead of separated poll. For that reasons, a number of empty poll can be reduced as many as the number of idle stations, since using a multicast poll.

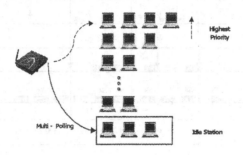

Figure 3. Multi-polling illustration

When an idle station receives the poll frame, it can transmit its packet after a designated constant time. We use silence detection mechanism to avoid empty-poll of the silence terminals.

Fig. 3 shows the illustration of the multi-polling. Frequently, any station within silence period will to be less polled. The use of silence detection can increase the number of voice stations supported by the network. The PC transmits sequentially a poll frame to each station based-on polling priority of polling list, and then the PC receives queue length information by a poll-feedback. Using poll-feedback, the PC allows priority to stations based on the queue length. In the Fig.4, we will describe how node determines that its priority given own.

3.3 An Adaptive Multicast Polling Scheme

We now describe an adaptive multicast polling scheme for idle station to avoid empty polling. As discussed above, the main objective of this scheme is to reduce delay depending on an empty poll and packet loss due to buffer overflow, as well as maintain the network bandwidth.

Our scheme can be expressed in an example of the management for polling list shown in Fig, 4. The arrow mark with small circle indicates multicast poll and general poll without circle, respectively. In Fig. 4, S_A, S_b, S_c, and S_d represents its station A, B, C, and D. The rectangular indicate pending packet. The multi column boxes represent polling list. The PHY show the sequence of packet transmission in physical layer.

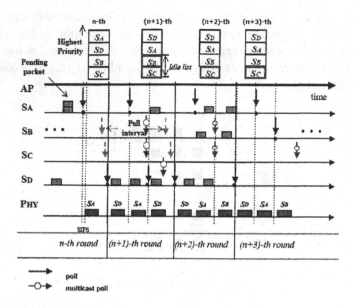

244   *Kyung-jun Kim, Hyun-sook Kim, Sang-don Lee, Ki-jun Han*

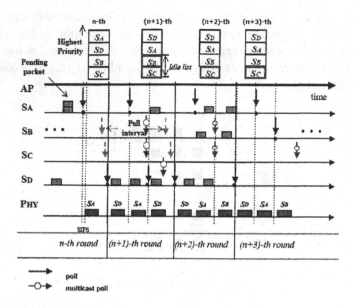

Figure 4. An example of the management for polling list

First, in the n*th* PCF round, S_B and S_C enter silence state following packet generation interval and have no more data to send (that is, because of the poll interval is longer than the packet generation interval), then S_B and S_c are added to the idle station group in the next (n+1)*th* round which will receive the multi-poll. If the idle station group has no data to send, it still leaves the idle station group.

```
PROCEDURE AP_POLLSEND (i) {
  for each station i in current round
  for (i = 0; i < n; i++)
    if poll was sent by AP
      return STA_POLLFEEDBACK()
}
PROCEDURE STA_POLLFEEDBACK( queue_length , i) {
  for each station i in current round
    for (i =0; i < n; i++)
      if queue_length i is not empty
        AP      queue_length of the pollable STA
}
PROCEDURE UPDATE_PRIORITY(i, poll_list[]) {
  if queue_length i > queue_length i-1
    poll_list[n]      station i
    poll_list[n+1]      station i-1
  else queue_length is empty ‖ poll_interval > packet generation interval
    idle list group      pollable STA
}
```

Figure 5. Basic operation of multicast polling scheme

In the (n+1)*th* PCF round, S_B leaves the idle station group because it has a packet to be sent when receiving the multi-poll at the same time. Assuming S_d transmits its packet in the (n+1)*th* PCF round, the PC gives the highest priority to S_d in the next PCF round since its queue is the longest of all stations. At time (n+2)*th*, S_b can transmit its packet. In other hand, S_c is remained idle station status. In the Fig. 5 shows an algorithm of basic operation of multicast polling scheme.

4.   ANALYSIS AND SIMULATION RESULTS

The system parameters for the simulation environment are listed in TABLE 1 as specified in the IEEE 802.11b standard. To simplify the simulation, the radio link propagation delay is assumed zero with no transmission errors. Fig. 6 shows the simulation model.

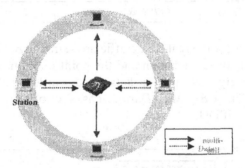

Figure 6. Simulation model

We consider on/off model of voice traffic. In the simulation, we assume that a voice packet is generated by exponential distribution [7]. While the previous packet has been transmitted, the older packet is discarded since a new packet is generated. In this model, the talk spurt period over silence period is 1.0 sec and 1.35 sec, respectively. The frame length of real-time traffic is set to 200 bytes considering the overheads of upper layer protocols.

Assuming the ratio of PCF duration within a super-frames, r, throughput of an adaptive polling scheme is approximately as follows,

$$r * F_p(n) + (1-r) * F_d(n) \tag{1}$$

Actually r can be dynamically modified by changing priority re-setup, *CFP_Duration*. For any given n, we want to set appropriate *CFP_Duration* to maximize the total throughput, as shown below [3].

$$Th = \left\lfloor r \mid Max\{r * F_p(n) + (1 - r) * F_d(n)\} \right\rfloor \qquad (2)$$

, where F_d is generally throughput, this parameter is function of n, denoted as F_d (n) and F_p is a function of the total number of associated stations T, the number of current active stations n, and r shown re-try limit.

Table 1 System parameters for simulation

Symbol	Meanings	Value
R	*Channel rate*	11Mbps
CW_{min}	*Minimum contention window*	31
CW_{max}	*Maximum contention window*	1023
T_{PIFS}	*PIFS time*	30us
T_{Rep}	*CFP repletion interval*	30ms
T_{MaxCFP}	*CFP_Max_Duration*	28ms

The parameters for the real-time traffic are summarized in TABLE 2. The maximum delay between a station and the point coordinator, D_{max} is set by 35ms [6, 7]. Namely, real-time packets are discarded if their waiting time exceeds 35ms. The *CFP_Max_Duration* is set to 28ms considering the maximum size of MPDU.

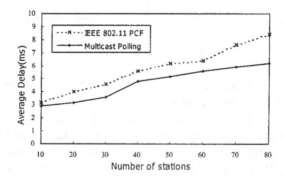

Figure 7. Average delay according to number of stations

In Fig. 7, the simulation shows the effects of changing the number of stations versus the average delay. If the number of node increases, the entire average delay increases. The average delay of proposed scheme is shorter than the original IEEE 802.11 scheme because the proposed scheme can reduce the amount of the empty polls.

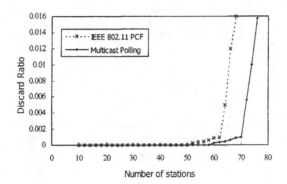

Figure 8. Discard ratio according to number of stations

In Fig. 8, we see that the performance of the proposed scheme for the packet discard ratio. Reflecting the end-to-end delay bound of real-time traffic, remaining time to service deadline between a station and the point coordinator is considered which is instead of end-to-end delay between two communicating stations. The discard ratio for real-time traffic using the proposed scheme stayed low. The maximum delay between a station and the point coordinator is set by 35ms [6].

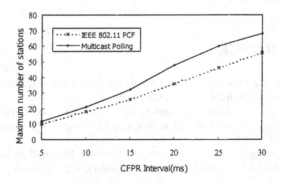

Figure 9. Maximum number of station according to CFPR interval

In Fig. 9, we see that the maximum number of supported stations, while the CFPR interval is increased. This is due to the reduction of delay and packet discard ratio with our scheme. Fig. 10 shows an increase of throughput of the proposed scheme in the same simulation setting.

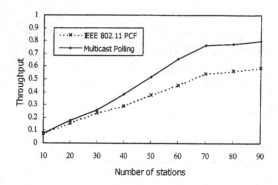

Figure 10. Throughput according to number of stations

5.   CONCLUSION

To reduce the number of empty poll in IEEE 802.11 PCF mode, this paper proposed a multicast poll scheme. Multicast poll scheme spreads a poll to the silence station group at the same time. Simulation studies revealed that our scheme could improve the average delay and packet discard ratio by preventing serious empty poll.

REFERENCE

1   IEEE: Draft and standard for wireless LAN Medium Access Control (MAC) and Physical Layer (PHY) Specification, IEEE 802.11, (May 1999).
2   Eustathia Ziouva and Theodore Antonakopoulos, A   dynamically adaptable polling scheme for voice support in IEEE 802.11 networks, *IEEE Computer Communications*, vol. 26, no.2, 129-142, (Feb. 2003).
3   Shou-Shin Lo, Guanling Lee and Wen-Tseun, An Efficient Multipolling Mechanism for IEEE 802.11 Wireless LANs, *IEEE Trans. on Computers*, vol, 52, no. 6, 764-768, (Jue. 2003).
4   Oran Sharon and Eitan Altman, An Efficient Polling MAC for Wireless LANs, *IEEE/ACM Trans. on Networking*, vol, 9, no. 4, 439-451, (Aug. 2001).
5   Moustafa A. Youssef and Arunchandar Vasan, Specification and Analysis of the DCF and PCF Protocols in the 802.11 Standard Using Systems of Communicating Machines, *Proc. of the 10th IEEE INCP*, (Nov. 2002).

6 R. S. Ranasinghe, L.L. H. Andrew and D. Everitt, Impact of polling strategy on capacity of 802.11 based wireless multimedia LANs, *IEEE Inter. Conf. on Network (ICON '99)*, 96-103, (Oct. 1999).

7 Sven Wietholter and Christan Hoene, Design and Verification of an IEEE 802.11 EDCF Simulation Model in ns-26, (Nov. 2003).

8   S. Agarwal and S.V.Krishnamurthy, Distributed Power Control in Ad Hoc Wireless Networks, *in Proceedings of IEEE PIMRC '01*, vol. 2, 59-66, (Oct. 2001).

THE PAIRING PROBLEM WITH USER INTERACTION

Thomas Peyrin
EPFL
Lausanne, Switzerland
thomas.peyrin@gmail.com

Serge Vaudenay
EPFL
Lausanne, Switzerland
serge.vaudenay@epfl.ch

Abstract   Bluetooth-like applications face the pairing problem: two devices want to establish a relationship between them without any prior private information. Hoepman studied the ephemeral pairing problem by regarding the human operator of the devices as a messenger in an authenticated and/or private low-bandwidth channel between the nodes. Here we study the pairing problem with user interaction in which the operator can participate by doing extra (simple) computations.

Keywords: Authentication, pairing, key exchange

1.   Introduction

A typical problem in wireless networks is that we do not know if two communicating devices are actually talking to each other. The pairing problem consists of securely establishing a private key between two or more specific physical nodes in the network. We assume that no secret information is shared between the nodes before the pairing. Furthermore, we want a high level of security and a minimal human interaction. Pairing between Bluetooth devices is a typical setting. In Hoe04, Hoepman studied the ephemeral pairing problem (denoted φKE): given a low bandwidth authentic and/or private communication channel between two nodes (called Alice and Bob), and a high bandwidth broadcast channel, can we establish a high-entropy shared secret session key without relying on any a priori shared secret information? The low bandwidth channel can be a (passive) human user who can read a PIN code on one device and write it on the other in a secure way. However there are many cases where this model is not sufficient: first, the standard Bluetooth pairing in which the user *generates* the PIN code; second, cases where the devices have no input keyboard or no output screen; third, when confidentiality (for instance) is guaranteed from the user to one device but not the other; etc. In this paper, we extend the model by introducing the user as a real participant who can further do simple computations. We call it the user-aided key exchange (UAKE) problem.

Gehrmann and Nyberg gave in GN01 two schemes. They also created a new scheme in GN04 using a MAC function and Jakobsson provided a variant of this scheme in Jak01. Those schemes are adapted to cases where one device has no input keyboard or no output screen.

The pairing problem is highly related to the authenticated key exchange problem (AKE): two users want to establish an authenticated high-entropy private key from scratch. Bellovin and Merritt BM92 gave a class of protocols called EKE (Encrypted Key Exchange) that solves the AKE problem using the assumption that the two peers already share a low-entropy password. EKE is basically an encrypted Diffie-Hellman DH76 key exchange. Jaspan Jas96 analyzed the Diffie-Hellman parameters in order to avoid partition attacks against EKE (in the case where the password is not ephemeral). Then Boyko *et al.* BMP00 specified a slightly different version of Diffie-Hellman based EKE called PAK (Password Authenticated Key exchange). MacKenzie Mac02 provided proofs in the Bellare-Rogaway model BR94. (A survey on authenticated key establishment protocols is available in BM03.) Note that in this paper, "EKE protocol" denotes independently the EKE or PAK protocol.

2.   The pairing problem models

2.1   The pairing problems

In the pairing problem, two nodes in a (wireless ad-hoc) network, that do not yet share any secret, want to establish a secure association. They may be able to exchange small amounts of information reliably and/or in a private way by being attended by a human operator. The ephemeral key exchange (φKE) problem considers the human operator as a simple messenger between the nodes. In this paper we consider the user-aided key exchange (UAKE) problem in which the operator really participates. The nodes can communicate through the insecure channel and the user can securely exchange small amounts of information with the nodes and perform simple operations. Protocols must be such that:

1 both nodes and the user are ensured that the secret is shared with the correct physical node

2 no other node learns any part of the shared secret

3 a user needs to perform only simple and intuitive steps

For the third requirement, we allow the following operations: pick a random string, compare two strings, copy a string, XOR two strings. Avoiding the (quite complicated) XOR will be addressed in Section 4.1. We further limit user channels to a small bandwidth. The second requirement will be made clear by formalizing the security model. Once achieved, the first requirement is satisfied by standard key confirmation techniques. Note that we do not consider denial of services attacks.

By directly introducing the user in the problem, we can consider many different situations that can be encountered in practice. For example, we can easily describe the Bluetooth pairing in many different scenarios such as devices with no output screen or no input keyboard, pairing in a hostile environment when anyone can look over the user's shoulder, etc.

2.2 The communication model

Two nodes Alice and Bob are connected through a high bandwidth channel network. The adversary Eve has full control over this channel. Both nodes however share with the user two communication channels (one in each direction) which can have specific security properties:

1 **confidentiality:** the sender is guaranteed that the messages she sends can not be read by anyone but the right receiver (Eve can not read it).

2 **integrity:** the receiver is guaranteed that the message he receives was actually sent as is (Eve can not modify it).

3 **authentication:** the receiver is guaranteed that the message he receives was actually sent by the right sender (Eve can not modify or insert a message in the channel but can delay or replay a message). (Note that our definition of authentication implicitly assumes integrity.)

These properties may hold in both directions, or only in one direction. In this paper, we will not consider the integrity property except in our final discussion in Section 4.1 to simplify the protocols. Note that lack of integrity protection in confidential channels means that it could be possible for Eve to replace a confidential z message by a message $z \oplus \delta$ with a δ of her choice. (This is typically the case when the confidential channel is implemented by a stream cipher, e.g. in Bluetooth.) We further assume independence between the channels in the sense that it is impossible for an adversary e.g. to take a message from a secure channel and to insert it into another.

We thus have 4 unidirectional channels that can have one of four attributes: AC (authenticated and confidential channel), A (authenticated channel), C (confidential channel) and 0 (no security property). Those channels represent all the interactions with the user. For example, a screen on a device represents a channel of type A from the device to the user who is watching the screen, a device holder typing a code on the device's keyboard in a private way represents a channel of type AC from the user to the device. Moreover, we can consider channels with an extremely low bandwidth (typically one bit) if we use a single light, or a single Boolean button for low cost devices.

2.3 The security model

We use the adversary model of Bellare *et al.* BPR00. Each participant p may engage in the protocol many times in a concurrent way. For each new protocol run where p is asked to play a role, a unique instance π_p^i is created. Eve has the entire control of the network and about who is running a new step of a protocol run. In a UAKE protocol with participants p, q, and r playing the role of Alice, Bob, and User respectively, we create new instances π_p^i, π_q^j, and π_r^k with input (p,q,r). π_p^i and π_q^j should terminate with a key. (The φKE protocol is similar: r is simply hidden.) The attack is formalized by giving access to oracles for the instances of the network to the adversary:

- Execute$(\pi_p^i,\pi_q^j,\pi_r^k)$: execute a complete protocol run with π_p^i, π_q^j, and π_r^k. This query models passive attacks.

- Corrupt(p,x): get all internal information about p and force its secret data (if any) to become x.

- Reveal(π_p^i): reveal the key generated by π_p^i to the adversary.

- Send(π_p^i,m): send a message m to the instance π_p^i and run a new step of the protocol.

- Test(π_p^i): this query can be called only once. A bit b is flipped at random a random key (if $b=0$) or the key from π_p^i (if $b=1$) is output.

Eve makes a Test query and tries to correctly guess the bit b. The attack is successful if p,q,r are not corrupted and if Test(π_p^i) or Test(π_q^j) led to the right guess for b. Thus we define the advantage of Eve attacking the protocol by $Adv_E = 2\Pr[\text{correct}] - 1$. Note that we can not send a Test(π_p^i) query if a Reveal(π_p^i) or Reveal(π_q^j) query has already been sent, otherwise finding the value of the bit b would be trivial. We do not consider long term passwords as in regular EKE schemes but rather ephemeral ones. So oracles Reveal(π_p^i) and Corrupt(p,x) are not relevant in our context.

3.   Key exchange with user interaction

3.1   The ephemeral pairing problem

In the original φKE problem, we have $2^4 = 16$ different possible configurations (2 channels and 4 possible security properties for each channel). We can represent each of those configurations by a 2×2 Boolean matrix: each row corresponds to a security property (A and C), and each column corresponds to a channel. For more readability, we represent the matrix by $M = [A \underset{b}{\overset{a}{\rightleftharpoons}} B]$ where $a, b \in \{0, A, C, AC\}$ are the columns of M. We denote φKE(M) the φKE problem with the configuration represented by the matrix M. If a secure protocol can be found for the φKE(M) problem, we say that φKE(M) is possible. Otherwise, we say that it is impossible. First of all, we can see that the φKE problem is symmetric: φKE(M) is equivalent to φKE(sym(M)) where sym(M) is the M matrix with the columns inverted. Furthermore, if a φKE(M_1) problem represented by the configuration matrix M_1 is possible, we can solve the problem with additional security properties by using the same protocol. We denote $M_1 \leq M_2$ for corresponding configuration matrices M_2.

FACT 1 *Let M_1 and M_2 be two φKE problem configuration matrices. If $M_1 \leq M_2$, any protocol which solves φKE(M_1) solves φKE(M_2) as well.*

FACT 2 *Let M be a φKE problem configuration matrix. φKE(M) is possible if and only if φKE(sym(M)) is possible.*

THEOREM 3 (HOE05) *φKE($A \underset{A}{\overset{C}{\rightleftharpoons}} B$) is impossible.*

Hoepman provided protocols for all minimal possible configurations. We can see that two types of protocols are used: we can try to make Alice and Bob share a low-entropy password and compute the EKE protocol with that password (see Figure 1 and Figure 2). The two devices can also try to run a Diffie-Hellman key exchange with commitment and authenticate with the low bandwidth channel (see Figure 3).

THEOREM 4 (HOE04) *φKE($A \xrightarrow{AC} B$) and φKE($A \underset{C}{\overset{C}{\rightleftharpoons}} B$) are possible by using the protocol from Figures 1 and 2 respectively. The advantage of an adversary which is limited to q oracles queries is at most the best advantage of an adversary to the EKE protocol with the same parameter q.*

THEOREM 5 (HOE04) *We consider a group G of order at least 2^{2s} in which the decisional Diffie-Hellman problem is hard. We consider five hash functions $h_1 : G \rightarrow G$, $h_2 : G \rightarrow \{0,1\}^i$, $h_3, h_4, h_5 : G \rightarrow \{0,1\}^\sigma$ such that $h_1(X)$ and $h_2(X)$ are independent for $X \in_U G$, h_2 is balanced, and h_3, h_4, and h_5 are* *(independent) pairwise independent random hash functions.* $\varphi KE(A \overset{A}{\underset{A}{\rightleftharpoons}} B)$ *is possible by using the protocol from Figure 3. The advantage of an adversary which is limited to q oracle queries is* $O(1 - e^{-q \cdot 2^{-t}}) + O(2^{-s})$.

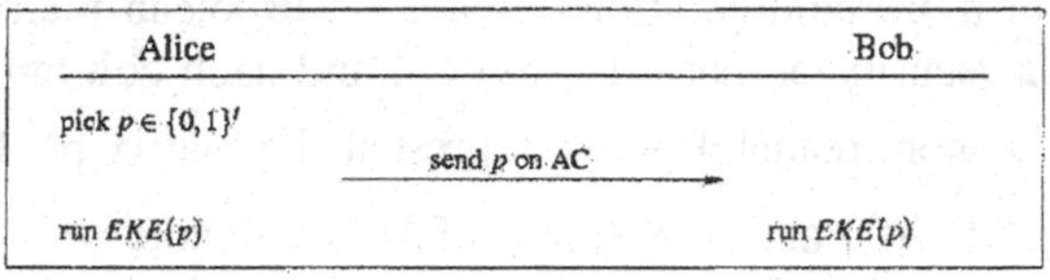

Alice	Bob
pick $p \in \{0,1\}^t$	
	send p on AC →
run $EKE(p)$	run $EKE(p)$

Figure 1. $\varphi KE(A \xrightarrow{AC} B)$

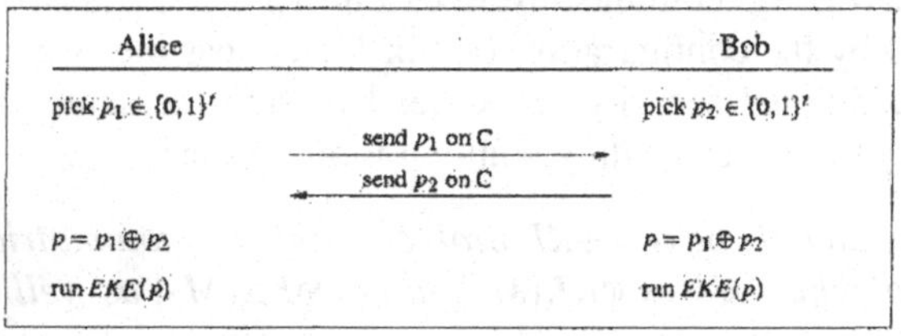

Alice	Bob
pick $p_1 \in \{0,1\}^t$	pick $p_2 \in \{0,1\}^t$
	send p_1 on C →
	← *send p_2 on C*
$p = p_1 \oplus p_2$	$p = p_1 \oplus p_2$
run $EKE(p)$	run $EKE(p)$

Figure 2. $\varphi KE(A \overset{C}{\underset{C}{\rightleftharpoons}} B)$

3.2   The user-aided key exchange problem

In the UAKE problem we have $4^4 = 256$ different possible configurations (4 channels and 4 different states for each channel). We can represent each of those configurations by a 2×4 matrix as in the φKE problem. For more readability, we represent the matrix by $M = [A \overset{a}{\underset{b}{\rightleftharpoons}} U \overset{d}{\underset{c}{\rightleftharpoons}} B]$ where $a, b, c, d \in \{0, A, C, AC\}$ correspond to the columns. We denote UAKE(M) the UAKE problem with the configuration represented by the matrix M. The UAKE problem is symmetric: UAKE(M) is the same problem as UAKE(sym(M)) where sym(M) is the M matrix with some columns inverted so that the role of Alice and Bob is exchanged.

FACT 6 *Let M_1 and M_2 be two UAKE problem configuration matrices. If $M_1 \leq M_2$, any protocol solving UAKE(M_1) solves UAKE(M_2) as well.*

FACT 7 *Let M be a UAKE problem configuration matrix. UAKE(M) is possible if and only if UAKE(sym(M)) is possible.*

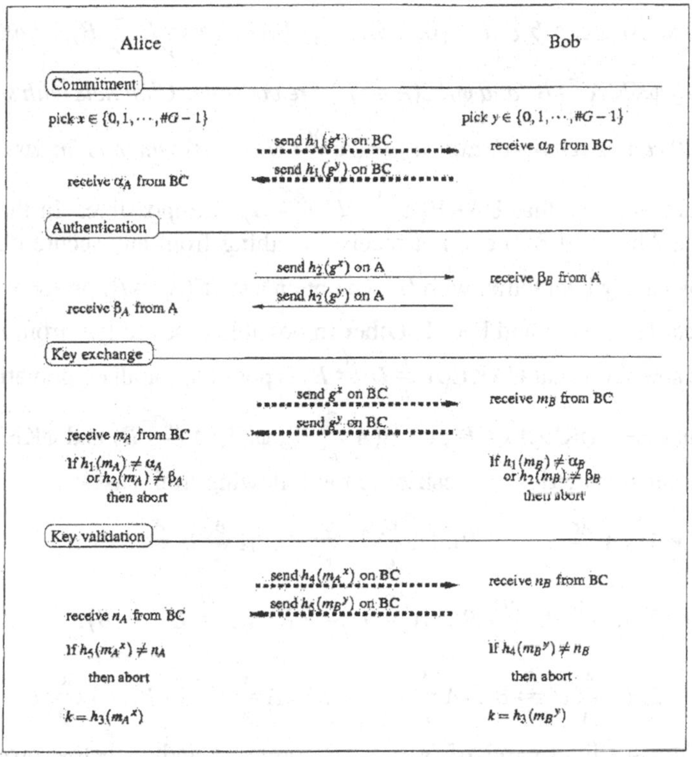

Figure 3. $\varphi KE(A \underset{A}{\overset{A}{\rightleftharpoons}} B)$

We consider two participants Alice and Bob of UAKE($A \underset{b}{\overset{a}{\rightleftharpoons}} U \underset{c}{\overset{d}{\rightleftharpoons}} B$) protocol. By simulating the interaction between Alice and User by a participant C. We obtain a protocol for $\varphi KE(C \underset{c}{\overset{d}{\rightleftharpoons}} B)$. We deduce:

FACT 8 *Let $a,b,c,d \in \{0,A,C,AC\}$. If UAKE($A \underset{b}{\overset{a}{\rightleftharpoons}} U \underset{c}{\overset{d}{\rightleftharpoons}} B$) is possible then $\varphi KE(A \underset{b}{\overset{a}{\rightleftharpoons}} B)$ and $\varphi KE(A \underset{c}{\overset{d}{\rightleftharpoons}} B)$ are also possible.*

Considering a messenger U who forwards messages, we obtain:

FACT 9 *Let $a,b \in \{0,A,C,AC\}$. If $\varphi KE(A \underset{b}{\overset{a}{\rightleftharpoons}} B)$ is possible then UAKE($A \underset{b}{\overset{a}{\rightleftharpoons}} U \underset{b}{\overset{a}{\rightleftharpoons}} B$) is also possible.*

THEOREM 10 *Let $a,b,c,d \in \{0,A,C,AC\}$. UAKE$(A \underset{b}{\overset{a}{\rightleftharpoons}} U \underset{c}{\overset{d}{\rightleftharpoons}} B)$ is possible if and only if $\varphi KE(A \underset{b}{\overset{a}{\rightleftharpoons}} B)$ and $\varphi KE(A \underset{c}{\overset{d}{\rightleftharpoons}} B)$ are possible. Channels with security property 0 can be removed, except for $(A \xrightarrow{AC} U \xleftarrow{AC} B)$ which is impossible.*

Proof: Let us prove that UAKE$(A \xrightarrow{AC} U \xleftarrow{AC} B)$ is impossible. In that configuration, Alice and Bob can not receive anything from any secure channel. By removing any interaction with U, we obtain a $\varphi KE(A \underset{0}{\overset{0}{\rightleftharpoons}} B)$ protocol which contradicts Theorem 3 and Fact 1. Other impossible cases follow from Fact 8.

Let us now show that UAKE$(A \underset{b}{\overset{a}{\rightleftharpoons}} U \underset{c}{\overset{d}{\rightleftharpoons}} B)$ is possible for all combinations of φKE limit cases: $\varphi KE(A \underset{A}{\overset{A}{\rightleftharpoons}} B)$, $\varphi KE(A \underset{C}{\overset{C}{\rightleftharpoons}} B)$, $\varphi KE(A \xrightarrow{AC} B)$ and $\varphi KE(A \xleftarrow{AC} B)$. By using symmetries, we restrict to the following limit cases:

- Type 1: $(A \xrightarrow{AC} U \xrightarrow{AC} B)$, $(A \underset{C}{\overset{C}{\rightleftharpoons}} U \underset{C}{\overset{C}{\rightleftharpoons}} B)$, $(A \underset{A}{\overset{A}{\rightleftharpoons}} U \underset{A}{\overset{A}{\rightleftharpoons}} B)$.

- Type 2: $(A \xleftarrow{AC} U \xrightarrow{AC} B)$, $(A \underset{C}{\overset{C}{\rightleftharpoons}} U \xrightarrow{AC} B)$, $(A \underset{C}{\overset{C}{\rightleftharpoons}} U \xleftarrow{AC} B)$.

- Type 3: $(A \underset{A}{\overset{A}{\rightleftharpoons}} U \underset{C}{\overset{C}{\rightleftharpoons}} B)$, $(A \underset{A}{\overset{A}{\rightleftharpoons}} U \xrightarrow{AC} B)$, $(A \underset{A}{\overset{A}{\rightleftharpoons}} U \xleftarrow{AC} B)$, $(A \underset{0}{\overset{AC}{\rightleftharpoons}} U \xleftarrow{AC} B)$.

Fact 9 addresses limit cases of type 1. Theorem 11 and 12 below provide a solution for limit cases of type 2 and 3 respectively. □

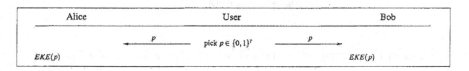

Alice	User	Bob
	$\xleftarrow{\quad p \quad}$ pick $p \in \{0,1\}^r$ $\xrightarrow{\quad p \quad}$	
$EKE(p)$		$EKE(p)$

Figure 4.   UAKE$(A \xleftarrow{AC} U \xrightarrow{AC} B)$

THEOREM 11 *UAKE$(A \xleftarrow{AC} U \xrightarrow{AC} B)$, UAKE$(A \underset{C}{\overset{C}{\rightleftharpoons}} U \xrightarrow{AC} B)$ and UAKE$(A \underset{C}{\overset{C}{\rightleftharpoons}} U \xleftarrow{AC} B)$ are possible by using the protocols from Figures 4, 5 and 6 respectively. The advantage of an adversary which is limited to q oracles queries is at most the best advantage of an adversary to the EKE protocol with the same parameter q plus 2^{-t}.*

Proof: The Figure 4 case is trivial: we assume we can set up a password in a secure way prior to EKE. For the cases of Figures 5 and 6, we note that if the

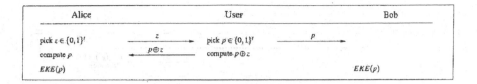

Figure 5. UAKE($A \underset{C}{\overset{C}{\rightleftharpoons}} U \xrightarrow{AC} B$)

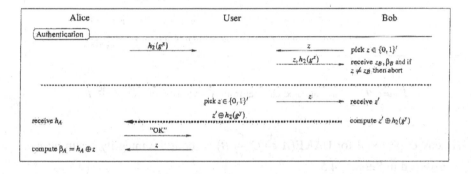

Figure 6. UAKE($A \underset{C}{\overset{C}{\rightleftharpoons}} U \overset{AC}{\longleftarrow} B$)

Alice	User	Bob
Authentication		
	$h_2(g^x)$ →	← z pick $z \in \{0,1\}^t$
		$z, h_2(g^x)$ → receive z_B, β_B and if $z \neq z_B$ then abort
	pick $z \in \{0,1\}^t$	
receive h_A	$z' \oplus h_2(g^y)$ z' → receive z'	compute $z' \oplus h_2(g^y)$
	← "OK"	
compute $\beta_A = h_A \oplus z$	← z	

Figure 7. Authentication step in Figure 3 for UAKE($A \underset{A}{\overset{A}{\rightleftharpoons}} U \underset{C}{\overset{IC}{\rightleftharpoons}} B$)

adversary impersonates User to Alice, since she has no clue about (Bob's) p, Alice will receive an incorrect p with probability $1 - 2^{-t}$ and EKE will fail. □

THEOREM 12 *With the same hypotheses as in Theorem 5, UAKE($A \underset{A}{\overset{A}{\rightleftharpoons}} U \underset{C}{\overset{IC}{\rightleftharpoons}}$ B), UAKE($A \underset{A}{\overset{A}{\rightleftharpoons}} U \xrightarrow{AC} B$), UAKE($A \underset{A}{\overset{A}{\rightleftharpoons}} U \overset{AC}{\longleftarrow} B$), and UAKE($A \underset{0}{\overset{AC}{\rightleftharpoons}} U \overset{AC}{\longleftarrow} B$) are possible by using the sub-protocols from Figures 7, 8, 9, and 10 respectively in the protocol of Figure 3. The advantage of an adversary which is limited to q oracle queries is $O(q.2^{-t}) + O(2^{-s})$. The first part of the protocol on Figure 7 further assumes integrity in the $U \to B$ channel.*

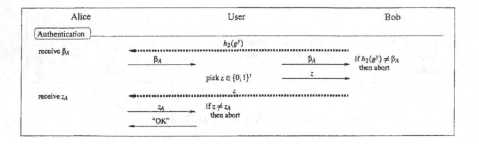

Figure 8. Authenticated channel from Bob to Alice in $A \underset{A}{\overset{A}{\rightleftharpoons}} U \xrightarrow{AC} B$

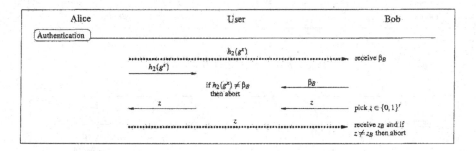

Figure 9. Authenticated channel from Alice to Bob in $A \underset{A}{\overset{A}{\rightleftharpoons}} U \xleftarrow{AC} B$

A heavier protocol for UAKE($A \underset{A}{\overset{A}{\rightleftharpoons}} U \underset{C}{\overset{C}{\rightleftharpoons}} B$) without the integrity assumption is provided in Section 4.5.

Proof (sketch): In Figure 8 (resp. Figure 9), if the adversary impersonates Bob to Alice (resp. Alice to Bob), the random z will never be released, so the protocol cannot succeed but with a probability of 2^{-t}. Figure 10 is similar.

In Figure 7 second part, the adversary has no clue about $h_2(g^y)$ and z until User discloses z. So, if she impersonates Bob to Alice, she can not predict which $h_2(g^y)$ Alice will obtain. Consistency check with the commitment phase in Hoepman's protocol will thus reject with a probability of $1 - 2^{-t}$ (Note that this works because $h_2(g^y)$ is unknown prior to the protocol).□

4. Discussions

4.1 Removing the XORs

We can see that the user has to compute the XOR of two values in protocols from Figures 5 and 6. Those cases have a common particularity: we have a

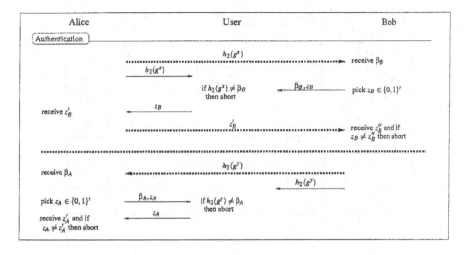

Figure 10. Authentication step in Figure 3 for UAKE$(A \overset{AC}{\underset{0}{\rightleftharpoons}} U \overset{AC}{\leftarrow} B)$

confidential non-authenticated channel between the user and Alice or Bob. In pairing situations those cases may not be relevant: such a channel would e.g. mean for example that the user types some digits on one device in a private way but the device is not sure that the typed digits actually come from the user! Nevertheless, those XORs can be removed by assuming integrity in addition to confidentiality. (Virtual confidential channels achieving this can typically be implemented by using encryption with strong security properties, e.g. IND-CCA2. Using less secure encryption, e.g. CBC encryption requires extra care.) In that case we replace the XORs by the concatenation.

4.2 User operations and bandwidth

One of the crucial points of our protocols are the ease of use for the user, we can thus analyze the number of operations computed by the user on the devices. In Figure 11 is shown the user operations according to the different protocols. According to the previous section, we can remove the XORs; in our table that would mean adding two copied values for each XOR.

4.3 Applications

A typical application of the UAKE problem would be the Bluetooth authentication scheme. The standard Bluetooth pairing assumes the same configuration as the protocol shown on Figure 4: the user types a password on both devices in a private and authenticate way. But according to our analysis of the UAKE

UAKE problems	pick	compare	copy	XOR	receive	send
$A \xrightarrow{AC} U \xrightarrow{AC} B$			t		t	t
$A \overset{C}{\underset{C}{\rightleftharpoons}} U \overset{C}{\underset{C}{\rightleftharpoons}} B$			2t		2t	2t
$A \overset{A}{\underset{A}{\rightleftharpoons}} U \overset{A}{\underset{A}{\rightleftharpoons}} B$			2t		2t	2t
$A \xleftarrow{AC} U \xrightarrow{AC} B$	t		t			2t
$A \overset{C}{\underset{C}{\rightleftharpoons}} U \xrightarrow{AC} B$	t			t	t	2t
$A \overset{C}{\underset{C}{\rightleftharpoons}} U \xleftarrow{AC} B$				t	2t	t
$A \overset{A}{\underset{A}{\rightleftharpoons}} U \overset{IC}{\underset{C}{\rightleftharpoons}} B$	t		3t		2t + 1	4t
$A \overset{A}{\underset{A}{\rightleftharpoons}} U \xrightarrow{AC} B$	t	t	t		3t	3t + 1
$A \overset{A}{\underset{A}{\rightleftharpoons}} U \xleftarrow{AC} B$		t	t		4t	2t
$A \overset{AC}{\underset{0}{\rightleftharpoons}} U \xleftarrow{AC} B$		2t	2t		6t	2t

Figure 11.   User operations in UAKE protocols (in bits)

problem, we can consider many other cases. For example, the user can read a password on device Alice and copy it on the device Bob. Moreover, we can imagine as explained on Figure 9, that the device Bob has only a private and authenticate screen but no keyboard and that the user can read and type data on device Alice in an authenticated way but not in a private way.

Another application could be the establishment of a secure SSL or SSH session without certificates. In the $A \overset{A}{\underset{A}{\rightleftharpoons}} U \overset{A}{\underset{A}{\rightleftharpoons}} B$ case, the user could indeed be two human operators talking (in an authenticated way) over the telephone, i.e. a $A \overset{A}{\underset{A}{\rightleftharpoons}} U_A \overset{A}{\underset{A}{\rightleftharpoons}} U_B \overset{A}{\underset{A}{\rightleftharpoons}} B$ scenario.

Note that problems arise if we do not consider mutual belief in the key as shown by Lowe in Low96. The UAKE protocols should similarly be followed by an acknowledgment protocol.

4.4   Manufacturer aided key exchange

We can consider that a password p_M has been written in the non-volatile memory of one device by the manufacturer, for example for a low-cost device without any keyboard. That would mean a fourth node M in our pairing scheme representing the manufacturer. AC channels from M to Bob and to User can be considered. Note that those channels can only be used once in the first setup. That new assumption would change protocols shown on Figures 4, 5 and 6: we use now p_M for the *EKE* protocol. This works in a $A \xleftarrow{C} U \xleftarrow{AC} M \xrightarrow{AC} B$ setting. Note that obviously this scheme leads to weaker versions of our protocols since the password used for each instance remains always the same.

We can also easily adapt the $\varphi\text{KE}(A \underset{A}{\overset{A}{\rightleftharpoons}} B)$ protocol in Figure 3 to solve the pairing problem in a $A \underset{A}{\overset{A}{\rightleftharpoons}} U \xrightarrow{A} B$ or $A \underset{A}{\overset{A}{\rightleftharpoons}} U \xleftarrow{A} B$ configuration with a prior $U \xleftarrow{AC} M \xrightarrow{AC} B$ setup. We can even restrict one of the two $A \underset{A}{\overset{A}{\rightleftharpoons}} U$ channels to a single bit.

4.5   UAKE($A \underset{A}{\overset{A}{\rightleftharpoons}} U \underset{C}{\overset{C}{\rightleftharpoons}} B$)

We now consider the protocol on Figure 12 as a replacement for the authentication phase in the protocol of Figure 3 using $\text{GF}(2^t)$ arithmetics.

In the first part of the protocol, we consider an optimal adversary who tries to make Bob accept a β of his choice for $X = h_2(g^x)$. (This follows a commitment phase in Hoepman's protocol, so an attack which makes Bob accept a random value is thwarted by the consistency check when opening the commitment.) Note that the right value of X is unknown to the adversary prior to the protocol. Without loss of generality, the adversary replaces (u, v) by $(u', v') = f(u, v)$, the returned (u', v') by $(u'', v'') = g(u', v')$, X by β, and $w = u' + v'\beta$ by $w' = h_X(w)$ for some chosen functions f, g, and h_X.

Let S_w be the set of all (u, v) such that $g(u', v') = (u, v)$ for $(u', v') = f(u, v)$, and $u' + v'\beta = w$. Note that $\#S_w \leq 2^t$. The attack is successful if and only if (X, u, v) is such that there exists w such that $u + vX = h_X(w)$ and $(u, v) \in S_w$. Hence the probability of success p is

$$p = \frac{1}{2^{2t}(2^t - 1)} \sum_w \sum_X \#\{(u, v) \in S_w; u + vX = h_X(w)\}.$$

Given w, let now n_i be the number of X's such that $\{(u, v) \in S_w; u + vX = h_X(w)\}$ has cardinality i. We can view the (u, v) pairs as straight lines. Given a set of i straight lines such that $u + vX = h_X(w)$ for one fixed X and w, we have $i(i-1)/2$ pairs of straight lines intersecting on the same point. If we sum all pairs over all X's, we obtain an overall number of intersecting pairs of at most $\#S_w \times (\#S_w - 1)/2$. Hence

$$\sum_i n_i \times \frac{i(i-1)}{2} \leq \frac{\#S_w \times (\#S_w - 1)}{2} \leq \frac{2^t(2^t - 1)}{2}.$$

We have

$$\sum_X \#\{(u, v) \in S_w; u + vX = h_X(w)\} = \sum_{i=1}^{2^t} i.n_i$$

with the constraint $\sum_i n_i \leq 2^t$. By linear programming results we obtain that

$$\sum_X \#\{(u,v) \in S_w; u + vX = h_X(w)\} \leq O\left(2^{3t/2}\right)$$

hence $p \leq O\left(2^{-t/2}\right)$. This big O is thus a new term to add in Theorem 12 for our protocol without the integrity assumption.

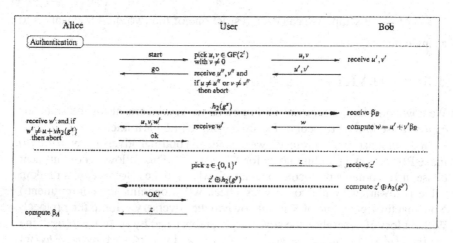

Figure 12. Authenticated step in Figure 3 for $\mathrm{UAKE}(A \underset{A}{\overset{A}{\rightleftharpoons}} U \underset{C}{\overset{C}{\rightleftharpoons}} B)$

5.   Conclusion

We have extended Hoepman's ephemeral pairing problem by introducing the User Aided Key Exchange problem. We studied the minimal assumptions and provided pairing protocols in all cases.

Acknowledgment.   This work was supported in part by the National Competence Center in Research on Mobile Information and Communication Systems (NCCR-MICS), a center supported by the Swiss National Science Foundation under grant number 5005-67322.

References

Steven M. Bellovin and Michael Merritt, *Encrypted key exchange : Password-based protocols secure against dictionnary attacks*, IEEE symposium on Research in Security and Privacy, IEEE Computer Society Press, 1992, pp. 72–84.

Colin Boyd and Anish Mathuria, *Protocols for authentication and key establishment*, Information Security and Cryptography, Springer-Verlag, 2003.

Victor Boyko, Phillip MacKenzie, and Sarvar Patel, *Provably secure password-authenticated key exchange using Diffie-Hellman*, Advances in Cryptology (Eurocrypt'00), Lecture notes in computer science, vol. 1807, Springer-Verlag, 2000, pp. 156–171.

Mihir Bellare, David Pointcheval, and Phillip Rogaway, *Authenticated key exchange secure against dictionary attacks*, Advances in Cryptology (Eurocrypt'00), Lecture notes in computer science, vol. 1807, Springer-Verlag, 2000, pp. 139–155.

Mihir Bellare and Phillip Rogaway, *Entity authentication and key distribution*, Advances in Cryptology (Crypto'93), Lecture notes in computer science, vol. 773, Springer-Verlag, 1994, p. 232.

Whitfield Diffie and Martin E. Hellman, *New directions in cryptography*, IEEE Transactions on Information Theory **IT-22(6)** (1976), 644–654.

Christian Gehrmann and Kaisa Nyberg, *Enhancements to Bluetooth baseband security*, Proceedings of Nordsec 2001, 2001, Copenhagen, Denmark.

———, *Security in personal area networks*, Security for Mobility (2004), 191–230, IEE, London.

Jaap-Henk Hoepman, *The ephemeral pairing problem*, Eighth International Conference on Financial Cryptography, Lecture notes in computer science, vol. 3110, Springer-Verlag, 2004, Key West, FL, USA, pp. 212–226.

———, *Ephemeral pairing on anonymous networks*, To appear in the Proceedings of Second IEEE International Workshop on Pervasive Computing and Communication Security, Lecture notes in computer science, Springer-Verlag, 2005.

Markus Jakobsson, *Method and apparatus for immunizing against offline dictionary attacks*, U.S. Patent Application 60/283,996. Filed on 16th April 2001, 2001.

Barry Jaspan, *Dual-workfactor encrypted key exchange : Efficiently preventing password chaining and dictionnary attacks*, 6th USENIX Security Symposium, San Jose, California, 1996, pp. 43–50.

Gavin Lowe, *Some new attacks upon security protocols*, 9th IEEE Computer Security Foundations Workshop, IEEE Computer Society Press, 1996, pp. 162–169.

Philip MacKenzie, *The PAK suite : Protocols for password-authenticated key exchange*, Tech. Report 2002-46, DIMACS, 2002.

NETWORK SMART CARD
A New Paradigm of Secure Online Transactions

Asad Ali, Karen Lu, and Michael Montgomery
Axalto, Smart Cards Research, 8311 North FM 620 Road, Austin, TX 78726, USA

Abstract: This paper describes the functionality and practical uses of a network smart card: a smart card that can connect to the Internet as a secure and autonomous peer. The network smart card does not require any special middleware on the host device. It uses standard networking protocols PPP and TCP/IP to achieve network connectivity. Network security is accomplished by an optimized SSL/TLS stack on the smart card. The combination of TCP/IP and SSL/TLS stacks on the smart card enables the smart card to establish a secure end-to-end network connection with any standard (unmodified) client or server on the Internet. This opens the door to seamless, secure and novel applications of smart cards in the most ubiquitous network: the Internet. Some of these applications that use the network smart card in confidential online transactions are explained.

Key words: Internet; smart card; network; TCP/IP; SSL/TLS; secure online transaction.

1. INTRODUCTION

Smart cards have been in use for more than two decades. However, their use has so far been on the fringes of mainstream computing. Smart card advantages such as security, portability, wallet-compatible form factor, and tamper resistance make them increasingly useful in many applications. Smart cards are used as agents in off-line transactions, as tokens for controlling access to physical resources, as SIM units in GSM phones, and as secure tokens for storing confidential and personal information. Current applications of smart cards, though useful in their own right, are hindered by the mismatch between smart card communication standards and the communication standards for mainstream computing and networking. When smart cards are connected to computers, host applications cannot communicate with them using standard mainstream network interfaces. Specific hardware and software in the form of smart card reader device drivers and middleware applications are needed to access the smart card services.

The network smart card solves this mismatch of communication standards by implementing standard mainstream networking protocols on the smart card. It supports PPP [1] and TCP/IP [2,3,4,5] as the underlying communication layer. It also supports the SSL/TLS [6,7,8] protocol that adds application level network security. Using this combination of TCP/IP and SSL/TLS protocols, which are ubiquitous in the Internet, a variety of mainstream application frameworks can be supported by the smart card. Currently, a working prototype of the network smart card supports a Telnet server and a secure web server using HTTPS. To the host device or a remote application, the smart card appears as another PC on the network. This seamless integration of network smart card with existing PC applications could pave the way for an unprecedented growth in the use of smart cards by enhancing the security of online web transactions.

2.   MOTIVATION

Smart cards are extremely secure hardware tokens with a programmable microprocessor chip. Although they are useful in a variety of security critical applications, there are two main impediments to widespread acceptance of smart cards for use in online web transactions. The first is the necessity to install middleware on the host PC where the smart card is physically connected. Without this middleware, smart card services cannot be accessed. The second is the lack of end-to-end security between the smart card and any remote application: for example, an online merchant's web server. A key motivation for developing the network smart card was to address both these drawbacks, and thereby achieve the vision of widespread smart card acceptance.

The root cause of both these drawbacks is the communication protocol mismatch between smart card and the host PC. Current smart cards use APDUs and smart card-specific ISO 7816 standards for communication. Host PCs, on the other hand use standard mainstream network standards like PPP and TCP/IP. Middleware software is installed on the host PC to act as a bridge between these two sets of diverse protocols. This software is not only cumbersome to develop, install and maintain across multiple host platforms, it also breaks the end-to-end security model when smart cards are accessed from remote applications. Figure 1 illustrates this break in security as a conventional smart card is accessed from a remote PC.

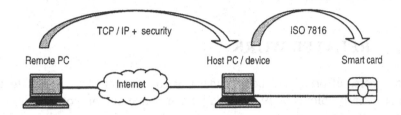

Figure 1. Network connectivity of conventional smart cards.

The remote PC cannot directly communicate with the smart card. It first connects to the host PC using standard mainstream network and security protocols. The host decrypts the information and then passes it to the smart card using ISO 7816 protocols. Due to this protocol conversion, the host PC needs to be trusted. However, PCs are known for their vulnerability to hardware as well as software attacks. As such, even if the ISO 7816 communication is encrypted, the host PC becomes a weak link in the end-to-end security between the remote PC and the smart card.

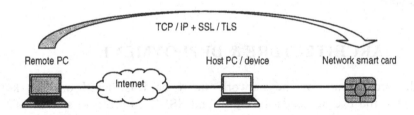

Figure 2. Network connectivity of the Network Smart Card.

Figure 2 describes the same scenario of accessing a smart card from a remote PC using the network smart card. Since the network smart card supports the same mainstream communication protocols as the remote PC, there is no need for any protocol conversion at the Host PC. There is a direct secure connection using the SSL/TLS protocol from the remote PC to the network smart card. The host PC merely acts as a pass-through router and need not be trusted. The OS on the host PC provides services that allow it to be configured as a pass-through router. This eliminates the need for installing any smart card specific middleware. The use of network smart card in an end-to-end secure connection with a remote client creates a much more secure model of conducting online transactions than is possible with current approaches that either do not use smart cards, or use them as auxiliary tokens connected to the host PC.

3.   RELATED WORK

For years, efforts have been underway to connect smart cards to the Internet. Early pioneers [9,10,11,12] have made major contributions to enhancing the connectivity paradigm of smart cards. However, they did not make the smart card a truly independent node on the Internet. These developments relied on some form of host middleware or software service to be installed on the host machine. In addition, they did not support any mainstream network security protocol on the smart card. The network smart card addresses these shortcomings in the earlier efforts by implementing a standard network protocol stack on the smart card. With this approach the smart card can establish an end-to-end secure connection with standard unmodified remote applications. In addition, the use of smart card does not require any smart card specific middleware to be installed on the host PC. The freedom to use the smart card on any PC without requiring smart card specific software on the PC is one of the major contributions of the network smart card.

4.   ARCHITECTURE & DEPLOYMENT

The architecture of the network smart card is based on two key principles: the use of standard TCP/IP and SSL/TLS protocol stacks inside the smart card; and utilizing standard interfaces and drivers built into most operating systems, thereby eliminating the need to install smart card specific software. This allows the network smart card to seamlessly integrate in the existing mainstream computing infrastructure. The deployment of smart cards is no longer encumbered by the middleware. Figure 3 shows some of the ways a network smart card can be connected to the network where unmodified clients can seamlessly access it. The card can be used in the standard credit card size form factor when connected to a host computer or a smart card hub. It can also be used in the SIM form factor when placed in GSM phones and wireless PDAs.

Figure 4 shows details of the network smart card protocol stack. The network smart card is connected via the "direct connection to a host computer" deployment option. It contains a complete network stack consisting of PPP, TCP/IP, and SSL/TLS, and various network applications.

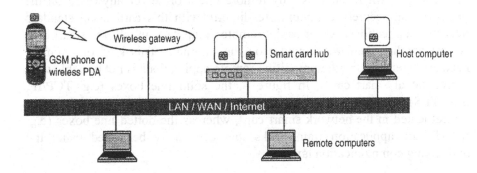

Figure 3. Deployment options for network smart card in various form factors.

The host computer can be any platform that is configured to permit network access from a serial or USB port. This includes most workstation, desktop, and laptop platforms including Windows, Mac, Linux, and Unix platforms, as well as some mobile devices. In the case of Windows platforms, configuration is a simple task requiring a few simple steps via the New Connection Wizard (a standard utility that comes with all Windows operating systems). This establishes a direct connection to another computer. The host is unaware that the computer being connected is a smart card: it treats the smart card as any other computer requesting a connection. Specific details of the network smart card architecture and various

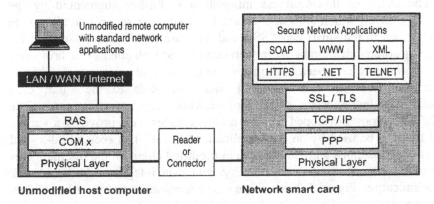

Figure 4. Network protocol stack and connection through a host computer.

connectivity models are described in a separate paper [13].

The host computer functions simply as a router to connect the smart card to the network, where other remote computers may access it. Since the smart card has its own IP address, any remote client or server anywhere on the network can securely communicate directly with this card using standard network applications. As far as the remote computer can tell, the smart card is just another standard computer on the Internet. This kind of seamless network interaction with unmodified remote applications is not possible with conventional smart cards. In figure 4, the solid line boxes (e.g. TCP/IP, SSL/TLS) represent components and services that have already been implemented in the network smart card, whereas the dotted line boxes (e.g. SOAP) are application frameworks that can easily be added using the underlying communication layers.

5.   APPLICATION FRAMEWORKS

A variety of standard application frameworks can be supported on top of the underlying implementation of TCP/IP in the network smart card. Applications running on the smart card do not communicate directly with the TCP/IP layer. Instead they go through a socket interface layer. Since this is how mainstream applications on a PC connect to the network, applications on the smart card can seamlessly integrate with other remote applications on the network. Standard unmodified client applications can connect to the smart card services as if they were connecting to another computer on the network.

The utility of this seamless integration is further augmented by the SSL/TLS layer, which adds application level network security to all remote connections to the smart card. SSL and TLS are the de-facto standards for securing communication between web servers and web clients (the browsers). These protocols have been in use for several years and no critical flaws have been discovered in them. Adding these protocols to the smart card substantially increases the security of network access to smart cards. The SSL/TLS library developed for the network smart card provides a simple API that can be used by on-card applications to establish secure end-to-end network connections with any remote unmodified client on the Internet. This provides authentication, confidentiality, and data integrity for all network communication. Figure 5 shows some of the application frameworks that can be supported on top of socket layer and the SSL/TLS layer. Solid boxes represent components that have already been implemented and demonstrated. Other boxes with dotted outlines can easily be added.

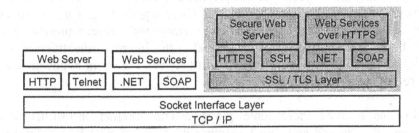

Figure 5. Application frameworks that can be supported in network smart card.

The current version of the network smart card supports a Telnet server and a secure web server that can serve static as well as dynamic HTML content. The static HTML content is served by reading the requested file from the smart card file system and returning its contents. The Dynamic content is supported by a more elaborate mechanism. The web server invokes the requested application, or library through a CGI interface and then redirects the resulting output to the browser. This allows the web sever to generate web pages on the fly, and implement various forms of access control for data stored on the smart card. No smart card specific application is needed to access the network smart card services. The user can simply open a standard Internet browser (IE, Netscape, Mozilla Firefox, etc.) and connect to the smart card in a secure manner over an HTTPS connection. From an administrator's point of view, programming the card can be as simple as creating a new HTML file and uploading it to the smart card using the same browser.

The network smart card also supports a Linux-like shell interface that can be accessed either through Telnet or through a secure HTTPS connection. This provides a very powerful way of interacting with the card, both as a user and as an administrator. New application frameworks such as .NET, SOAP, and Web Services can be easily added.

6. SECURE ONLINE TRANSACTIONS

Because the network smart card supports mainstream networking protocols, it can be seamlessly integrated into any network application. One example of such applications is the web-based online transaction. PC users have become accustomed to conducting online transactions as part of their normal daily activity. However, a closer look at these transactions reveals that they are not as secure as people believe them to be. There are three players in any online transaction: the client application, e.g. a standard Internet browser running on local PC; the remote web server representing the merchant we do business with; and finally, the Internet infrastructure over which our data is carried back and forth between the client application and the remote web server.

Let us analyze these three players. The Internet infrastructure is considered open and any data traversing it can be viewed by anyone who cares to examine the packets. However, security protocols such as SSL/TLS that are used in online transactions encrypt the data in a way that data integrity and confidentiality are not compromised while the data traverses the Internet. The remote web server is under the control of the trusted merchant with which we are conducting our online transactions. We trust the merchant, otherwise, why would we conduct business with them? This leaves us with our local PC on which we run the client application, the Internet browser. PCs are extremely vulnerable to both hardware and software attacks, and therefore form the weak link in any online transaction. Any information that is entered through them can be compromised even before the SSL/TLS layer encrypts it. Similarly, an attacker can capture any data that may have originated on a different device but simply flows through the PC in the clear.

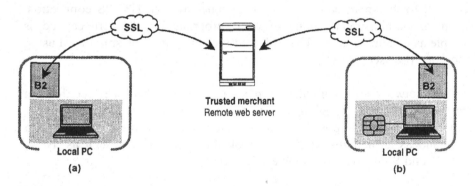

Figure 6. Online transactions: a) without a smart card, b) with a conventional smart card.

Figure 6-a shows this typical scenario of conducting online transactions without any smart card support. Using an Internet browser, B2, the user establishes a secure SSL connection with a trusted merchant. Any confidential data entered by the user is secure while in transit, but is open to attack while it resides on the Local PC. This is because information can be captured before the SSL layer can encrypt it. Current use of smart cards to augment the security of PCs is described in Figure 6-b. It is a step in the right direction but does not completely solve the security problem. The confidential data now resides on the smart card where it is secure. However, in the process of being SSL-encrypted for transmission to the trusted merchant, the data has to flow through the PC in the clear. Since PCs are open to attack, so is any information that flows through them in the clear. Due to their vulnerabilities, PCs are the weak links in any online transaction.

The use of the network smart card in an online transaction solves the problem introduced by the vulnerable nature of PCs. Instead of passing information through the PC, the network smart card establishes a secure connection directly with the remote server of the merchant. All confidential information is passed from smart card directly to the merchant. The PC is used only as an initial means of connecting the smart card to the Internet. Once that connection is established, the PC merely acts as a router that physically passes encrypted information back and forth between the network smart card and the remote merchant. Logically, the smart card has a direct connection with the remote merchant that completely bypasses the PC.

Figure 7 shows this use of network smart card in conducting a truly secure online transaction. The local PC, the trusted merchant, and the network smart card are three independent nodes on the Internet that are capable of establishing secure connections using SSL/TLS. Any two nodes can communicate with each other without disclosing the contents of the communication to the third node, or any other party that happens to be listening. The fact that network smart card is one of these nodes, is a critical distinction that elevates the security of online transactions to a level not

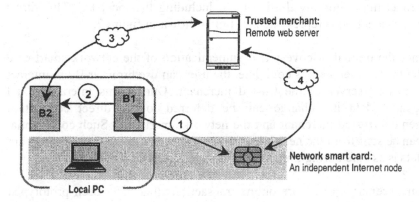

Figure 7. Network smart card in a secure on-line transaction.

possible with conventional smart cards.

Various interactions outlined in Figure 7 are described below. All the bi-directional arrows in this figure represent secure HTTPS connections using the SSL/TLS protocol. The steps listed here represent a simplified set of interactions when using network smart card in an online transaction. More details are covered in a separate paper [14] that talks about the use of network smart card in preventing online identity theft. The sequence of steps in chronological order is as follows:

1. The user opens a web browser on the local PC. This instance of the web browser is referred to as B1. From B1 the user connects to the secure web server running on the network smart card and authenticates himself through some form of card holder verification: a PIN, biometrics, etc. This connection is established over a secure HTTPS link. Once authenticated, he is presented with a list of trusted merchants. The user picks a trusted merchant and asks the network smart card to establish a secure connection with this service provider.

2. The user clicks on a link in B1 to start a new browser. This instance of the web browser is referred to as B2.

3. When browser B2 is launched, it automatically connects to the remote web server of the trusted merchant that was selected by user in step 1. Since the browser was launched from B1, the network smart card address can be passed to the trusted merchant. Alternatively, the smart card can initiate a connection to the trusted merchant and pass its address, as well as other login credentials to it.

4. Regardless of the method used for mutual discovery, the trusted merchant web server and the network smart card can now communicate directly without involving any third party – including the local PC. This direct SSL connection is represented by link 4 shown in figure 7.

Once the mutual discovery and authentication of the network smart card and the trusted merchant is complete, the user can use browser B2 to interact with the web server of the trusted merchant. During this interaction, all confidential data interchange can be deferred to the direct connection between the trusted merchant and the network smart card. Such confidential data can be sent from the network smart card directly to the trusted merchant. The data is as safe during transit as it was when stored on the smart card.

Some scenarios of secure online transactions that have been prototyped with the network smart card are listed in the following sections. These applications have been demonstrated at various smart card conferences [15,16,17]. See Figure 8 for a sample screen-shot from these demonstrations.

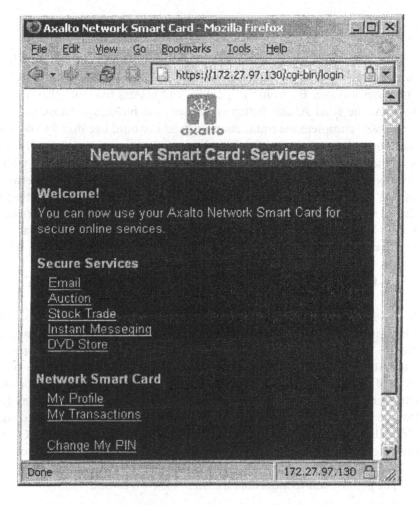

Figure 8. Screen-shot of secure web server running on network smart card.

Figure 8 shows the list of secure services (trusted merchants) that is presented to the user once he has logged on to the network smart card. All information displayed in this browser window is coming directly from the web server running on the network smart card using an HTTPS connection. As evident from the URL, the network smart card has been assigned its own IP address: in this case 172.27.97.130. Standard unmodified Internet client applications can communicate with the smart card using this IP address. The

Mozilla Firefox browser shown in Figure 8 corresponds to the browser B1 in the scenario outlined in Figure 7.

6.1   Online Stock Trade

In this scenario, the trusted merchant is a stock brokerage house where the user wants to sell some stock from his portfolio. As shown in figure 7, the user uses browser B2 from the local PC and selects the stock that need to be sold. As the final submit button is pressed, the brokerage house will not automatically complete the transaction. Instead it would use link 4 to contact the network smart card directly. This is done for two reasons: Firstly, the remote merchant wants to make sure that the network smart card, which represents the user in this online transaction, is still connected to the network. Secondly, it allows the user to manually approve any information that is disclosed by the network smart card. This secure close-back loop is a critical step that is missing in all current on-line transaction scenarios that use conventional smart cards, or are conducted without any smart card support.

6.2   Online Shopping

In the online shopping scenario the network smart card provides a secure way of transferring one's credit card information without having to type it manually. Such confidential details like credit card number and expiration date are stored on the network smart card. Storing this information on a smart card provides a greater level of security than storing it either on the remote merchant server, or on the local PC. Once the user is ready to pay for online shopping, the user instructs the network smart card to send the credit card details to the selected merchant. This secure transfer of credit card information between the smart card and the merchant using an end-to-end SSL connection solves one of the most critical issues with online shopping today. The path of confidential data transfer is shown as link 4 in figure 7.

6.3   Online Auction

The online auction scenario is quite similar to the online stock trade scenario. Instead of trading stocks, the user participates in an online auction by placing bids. However, before the merchant accepts any bids, the user's identity is confirmed by a secure direct connection between the merchant web server and the network smart card. In conventional online auctions, attackers can impersonate the user by compromising the user's username and password. The use of network smart card eliminates such fraudulent bids.

7. CONCLUSION

The network smart card presented in this paper enables smart cards to participate as autonomous nodes on the Internet with no host or remote application changes required. Smart cards may, finally, be accommodated within the existing mainstream computing infrastructure. This innovative technology combines the best aspects of two worlds of computing: the world of PCs, and the world of smart cards. In the world of PCs we have ubiquitous infrastructure access and very strong network security protocols for remote access. However, a PC is a very weak hardware device that is vulnerable to security attacks. Smart cards on the other hand are very secure hardware tokens, but suffer from lack of seamless access to the mainstream communication networks. The network smart card combines the hardware security of smart cards with the ubiquitous access and network security of PC. The result is a new paradigm of portability, enhanced security, and tamper resistance. We foresee this new paradigm triggering an unprecedented growth in the deployment of smart card applications for the Internet and other network environments. The network smart card can bring a level of security to online transactions that has so far not been possible.

REFERENCES

1. Simpson, W. "The Point-to-Point Protocol (PPP)", RFC 1661, July 1994.
2. Postel, J. "Internet Protocol," RFC 791, September 1981.
3. Postel, J. "Transmission Control Protocol," RFC 793, September 1981.
4. Socolofsky, T. "A TCP/IP Tutorial," RFC 1180, January 1991.
5. Almquist, P. "Type of Service in the Internet Protocol Suite," RFC 1349, July 1992.
6. Freier, Alan O., et al. "The SSL Protocol, Version 3.0," Internet Draft, November 18, 1996. Also see the following Netscape URL: http://wp.netscape.com/eng/ssl3/.
7. Dierks, T., Allen, C., "The TLS Protocol, Version 1.0," IETF Network Working Group. RFC 2246. See http://www.ietf.org/rfc/rfc2246.txt.
8. Elgamal, et al. August 12, 1997, "Secure socket layer application program apparatus and method." United States Patent 5,657,390.
9. Rees, J., and Honeyman, P. "Webcard: a Java Card web server," Proc. IFIP CARDIS 2000, Bristol, UK, September 2000.
10. Urien, P. "Internet Card, a smart card as a true Internet node," Computer Communication, volume 23, issue 17, October 2000.
11. Guthery, S., Kehr, R., and Posegga, J. "How to turn a GSM SIM into a web server," Proc. IFIP CARDIS 2000, Bristol, UK, September 2000.
12. Muller, C. and Deschamps, E. "Smart cards as first-class network citizens," 4th Gemplus Developer Conference, Singapore, November 2002.
13. Montgomery, M., Ali, A., and Lu, K. "Implementation of a Standard Network Stack in a Smart Card", CARDIS 2004, Toulouse, France, August 2004.

14. Lu, K., and Ali, A. "Prevent Online Identity Theft - Using Network Smart Cards for Secure Online Transactions," 7th Information Security Conference, Palo Alto, CA, September 2004.

15. Montgomery, M., et al., "Web Identity Card", Axalto booth, CARTES & IT Security 2003, Paris, France, November 2003.

16. Ali, A., et al., "Web Identity Card", Axalto booth, CTST 2004, 14th Annual Conference and Exhibition, Washington D.C., April 2004.

17. Montgomery, M., et al., "Web Identity Card", Axalto booth, CARTES & IT Security 2004, Paris, France, November 2004.

PROTECTION AGAINST SPAM USING PRE-CHALLENGES

Rodrigo Roman[1], Jianying Zhou[1], and Javier Lopez[2]

[1]*Institute for Infocomm Research, 21 Heng Mui Keng Terrace, Singapore 119613;*
[2]*E.T.S. Ingenieria Informatica, University of Malaga, 29071, Malaga, Spain.*

Abstract: *Spam* turns out to be an increasingly serious problem to email users. A number of anti-spam schemes have been proposed and deployed, but the problem has yet been well addressed. One of those schemes is challenge-response, in which a challenge is imposed on an email sender. However, such a scheme introduces new problems for the users, e.g., delay of service and denial of service attacks. In this paper, we introduce a *pre-challenge* scheme that avoids those problems. It assumes each user has a challenge that is defined by the user himself/herself and associated with his/her email address, in such a way that an email sender can simultaneously retrieve a new receiver's email address and challenge before sending an email in the first contact. Some new mechanisms are employed to reach a good balance between security against spam and convenience to email users.

Keywords: electronic mail; anti-spam; internet security.

1.   INTRODUCTION

Email is one of the most valuable tools for Internet users, with which people at anywhere can communicate instantaneously regardless of the distance. However, this tool can be used for bad purposes too, and there is no doubt that the worst use of email is spam.

Spam, or unsolicited commercial email, can be defined as advertising messages (mostly for fraudulent products) neither expected nor desired by the intended receivers. Since it is very easy to flood users' mailboxes with little investment, spam is a big threat to email systems, resulting in the loss of time and money to email users.

A lot of research in the area of anti-spamming has been done in the past years, trying to seek effective solutions to the spam problem. One of those solutions is *challenge-response*. When a sender sends an email, he/she is first given a challenge from the receiver that must be solved before the email reaches the receiver's mailbox. However, such a scheme introduces new problems, for example, *delay of service* (when a sender waits for arrival of a challenge from a receiver), and *denial of service* (when challenges are redirected to a victim's address that is spoofed by spammers as the sender).

Our Contribution. In this paper, we propose a *pre-challenge* scheme, which is based on challenge-response mechanism, preserving its benefits while avoiding its drawbacks. It assumes that each user has a particular challenge associated with his/her email address, in such a way that an email sender can simultaneously retrieve a new receiver's email address and challenge before sending an email in the first contact. Our scheme enables management of mailing list and error messages. Our scheme is easy to be integrated into existing email systems as it is a standalone solution, without changing the other party's software and configuration.

The rest of the paper is organized as follows. In section 2, we summarize the existing solutions against spam and analyze their limitations and/or problems. After that, we present a new solution in section 3, and further discuss it in section 4. Finally, we conclude the paper in section 5.

2.   PREVIOUS WORK

The original SMTP protocol [1] was introduced in 1982, with only minor modifications [2] in the past 20 years. The main problem in SMTP is the lack of authentication. When an email is received, it is not possible to know whether the source of the email is who claims to be. This is precisely the flaw that spammers make use of. However, as the SMTP protocol has been standardized and widely deployed, most of the research focuses on avoiding spam while maintaining the actual SMTP protocol and email infrastructure in order to ensure compatibility. This implies that anti-spamming solutions must be based on the operation with email headers and contents or on specific implementations at the application level.

One of the headers that can provide information regarding an eventual spam of the incoming email is the *"Received:"* headers, which give information about the client MTA. There are some projects [3] that try to identify misconfigured email MTAs or major sources of spam. However, it does not work effectively against individual spammers, and innocent client MTAs might be blocked.

Another header that can be used against spamming is the receiver's address, with policy or password-like extensions. In policy-based systems [4], policies are encoded inside the address and an email is discarded at its destination if the policy is not fulfilled. In password-based systems [5-7], the receiver's address is extended with a sequence of characters that act like a password, which can be obtained with a proof of computational task [10]. These solutions work well in some scenarios (e.g., using mail addresses in computer-based systems like web forums). However, as the email addresses created in such schemes are very hard to remember, they may cause problems when used by humans.

There are several works dealing with email content analysis based on artificial intelligence (AI) and statistical techniques [8,9]. They try to distinguish whether an email comes from a legitimate user or from a spammer by assigning a "spam score" to any incoming message. This approach may lead to false positives, and spammers may try to bypass the classifier algorithms.

Other implementation approaches against spam include *micropayment*, *challenge-response*, and *obfuscation* schemes. Micropayment schemes [10-13] are applied to email systems in order to prevent spammers sending millions of emails. They require the user or client MTA to compute a moderately hard function in order to gain access to the server MTA. As a result, a spammer will not be able to send a large number of emails to a certain server MTA. Such an approach is difficult to be applied to those client devices with very weak computing capability (e.g., mobile phones).

In challenge-response schemes [14,15], whenever an email from an unknown user is received, a challenge is sent back to that user. The solution to that challenge can be simple (e.g., just reply), complicated (e.g., solve a CAPTCHA [17]), or time consuming. Only when the correct response is received, the emails from that user are allowed to enter into the receiver's mailbox. These schemes do not work when a human user is not involved in sending emails (e.g., in the case of mailing lists). Moreover, these schemes may introduce new problems such as delay of service and denial of service.

In the obfuscation scheme, email addresses are displayed in an obfuscated format (e.g., John HIPHEN Smith AT yahoo DOT com), from which senders can reconstruct the real email addresses. It does not require any software from the user side or from the server side. However, the problem with this scheme is the constraints that the human users face when constructing the obfuscated addresses. As the combinations are limited, it allows AI-based harvest programs to easily retrieve real addresses. Moreover, once the email is captured by the spammer, there is no protection against spam (unless other solutions are utilized).

3.   A PRE-CHALLENGE SCHEME

3.1   Overview

As stated, our pre-challenge scheme is based on challenge-response mechanism in the sense that both of them impose a challenge that must be solved by a potential sender. However, in the pre-challenge scheme, the sender retrieves the receiver's email address together with his/her challenge simultaneously (see Fig. 1). Once the challenge is solved, the answer will be included inside the email.

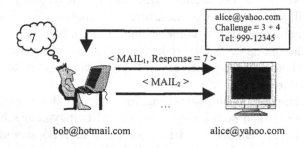

alice@yahoo.com
Challenge = 3 + 4
Tel: 999-12345

< MAIL₁, Response = 7 >

< MAIL₂ >

...

bob@hotmail.com   alice@yahoo.com

Figure 1. Basics of the pre-challenge scheme

When a mail from an unknown sender arrives, the receiver's system tests whether that mail contains an answer to the challenge. If the test turns out positive, the sender is *white-listed*. That means future mails from this sender will get into the receiver's mailbox without being checked again.

The goal of our scheme is to check whether there is really a human behind a sender's computer. The reason is that spammers use automatic programs to send their propaganda, and they feed these systems with email addresses obtained by searching web sites and mail servers. However, it is a bit hard for these programs to retrieve a challenge that matches an email address and even harder to answer each of these challenges. Therefore, whenever a spam arrives to destination, it will be automatically discarded if no correct answer to the challenge is attached.

In comparison with a challenge-response scheme, our pre-challenge scheme preserves its benefits while avoiding its drawbacks, as we explain in the following:

- In a challenge-response scheme, there is a delay in obtaining the receiver's challenge. On the contrary, in our pre-challenge scheme, because the receiver's challenge is available in advance, the sender can directly solve the challenge and send the email to the receiver [13].

* With a challenge-response scheme, if spammers forge a sender's address in their mails, the challenges will be sent to that address, launching a possible DDoS attack [16]. This attack will not take place in the pre-challenge scheme because a receiver need not reply an unknown sender's request for a challenge.
* A challenge-response scheme can work with mailing lists only if some rules are manually introduced, and it cannot handle mail error messages properly. As we will show in section 4.1 and section 4.4, the pre-challenge scheme manages mailing list systems and processes mail error messages without any problem.

Another benefit of the pre-challenge scheme is the continuous protection against email harvesting. When a correct email address is retrieved by a spammer, he/she needs to get the solution of the current pre-challenge at the same time to make the address usable, but the user can change the pre-challenge at any time (see section 3.2), making the combination <email,solution> useless.

3.2 Challenge Retrieval and Update

A challenge is defined by an individual human user. Each user has one challenge at a time to be used by all incoming emails, and the challenge can be updated at any time at his/her own discretion. The challenge can range from a simple question or mathematical operation to a hard-AI problem that only a human can solve [17].

Normally a user's challenge is published next to this user's email address. Since any potential sender must retrieve the email address of the receiver before contacting him/her, challenge and email address can be accessed at the same time. However, in certain cases, a challenge may not be accessed directly. Instead, a URL may be provided to retrieve the challenge.

Since the challenge is not restrained to obfuscate a valid email address, which has a fixed structure (name, domain), the user has more freedom to produce it. When stored inside a website, the challenge can take advantage of its form and content – personal information, the theme, or visual appearance of the website, etc. Challenges may also be retrieved using a majordomo style service [22]. To prevent spammers from using this service as

---

[13] The frequency of challenge update is a security parameter decided by the receiver, based on his/her own experience, to control the risk of replay attacks from spammers.

a collector of valid email addresses, the service must return a false challenge for every non-existent user.

3.3   Data Structures

The pre-challenge scheme requires certain data structures to accomplish its tasks. The two most important structures are the actual challenge (or a URL), and the solution to the challenge. By using these structures, it is possible to advertise the actual challenge and to check whether an incoming mail has solved the challenge. Additionally, the solutions to old challenges must be stored, as discussed later.

Other data structures needed by the scheme are the *white-list* and the *reply-list* (both used by some challenge-response schemes), and the *warning-list*, that is a structure specifically created for our new scheme. Each of those structures contains a list of email addresses and, optionally, a timestamp which indicates the time an email can be in the list.

White-List. The *white-list* contains email addresses in such a way that emails coming from those addresses are accepted without being checked. Some email senders may even be *white-listed* by a receiver at the set-up phase if they are already known. Those senders are marked in order to send a confirmation when receiving their first message (see section 3.5). This list could be manually modified by a human user.

Reply-List. The *reply-list* contains email addresses of those users to which the local user has sent email to, and has not replied yet. The use of this list is justified because the local user is the one who initiated the communication with those users; hence, there is no need to check any challenge when replies are received. This list will be managed automatically by the local user's system.

Warning-List. The *warning-list* contains email addresses of users that have sent an email containing the answer of an old challenge. The existence of this list is justified because an email message with an old response will cause a reply from the receiver indicating the new challenge. With this list, the local user does not need not send that reply more than once. This list will be reset every time when the challenge is updated, and will be managed automatically by the local user's system.

3.4   Security Levels

The pre-challenge scheme can be configured to work at two security levels, *high security* and *low security*. The main difference between these two levels is how the *reply-list* is queried.

The scheme starts working at the high security level of protection. High security means that all queries in the *reply-list* are done by looking for a <user, domain> match, and the matched entry will be erased from the *reply-list*. On the other hand, low security means that all queries in the *reply-list* are done by looking for a <*, domain> match.

The reason why the pre-challenge scheme needs these two levels of security is that some email accounts have different addresses for receiving and for sending email. This usually happens with mailing lists, and this issue will be discussed in section 4.1.

3.5   Architecture

Now we explain the design of our pre-challenge scheme. Suppose user B wants to send an email to user A. To simplify the explanation, we assume that user A is using the pre-challenge scheme while user B is not.

1. A's system checks if B's address is listed in the *white-list*. If this is the case, the email reaches A's mailbox. Additionally, if that mail is the first message A received from B, A sends a confirmation email to B.
2. Otherwise, if B is listed in the *reply-list*, the email reaches A's mailbox and B is added to the *white-list*. We should point out that the query to the *reply-list* is different according to the level of security being applied, as seen in section 3.4. In case of using a high security level, B is erased from the *reply-list* because A received the reply expected from B.
3. Otherwise, A's system checks whether the challenge of the email has been solved. If it is solved, the mail reaches A's mailbox and B is added to the *white-list*. Additionally, B receives a confirmation email.
4. Otherwise, if the email has a solution to an old challenge, A's system checks if B is listed in the *warning-list*[14]. If that is the case, the mail is discarded. If it is not listed, B's address is added to the *warning-list* and B gets a reply containing information about the new challenge.

---

[14] Note, the *warning-list* will be reset whenever the challenge is updated.

5. Otherwise, the email is discarded without any reply to *B* indicating this fact. The problem of accidental discard of a legitimate email will be addressed in section 4.3.

It should be noted, however, that discarding the email does not mean the user cannot read it. The scheme can be configured for labeling the message with a "spam score" and placing it in a special fold of the mailbox if the owner of that mailbox desires so.

3.6   Spam Scenarios

When a spammer wants to send his/her advertisements to a final user that operates the pre-challenge scheme, he/she basically faces two scenarios.

Scenario 1. The spammer only retrieves the email address of a target, but not his/her challenge. When the spam is sent to the target, it will be silently discarded because no solution to a (current or old) challenge is included.

Scenario 2. The spammer only retrieves the email address of a target, and impersonates as a sender that happens to be in the receiver's *white-list*, due to the lack of authentication in the email infrastructure. All schemes that use a *white-list* share this problem, but this is not a serious issue because spammers must find the white-listed senders for all the addresses he/she want to spam. And for millions of addresses to spam, this is unprofitable.

It could seem that a spammer, using little investment (solving one challenge), can send many pieces of spam to a given email address (a replay attack). It could also seem that a group of spammers interchange their solved challenges of the corresponding users in order to lessen each spammer's effort on accessing the victims' mailboxes. However, what spammers want is to send millions of messages. And since the challenges are different for every user and a challenge can be solved only by a human, the task of repeatedly solving or sniffing a new challenge per user, or hiring cheap labor in order to send spam, becomes unprofitable.

4.   FURTHER DISCUSSION

Here we further discuss how our scheme works for users in a mailing list, and whether our scheme can make a challenge easily available to users and make users to be sure on the delivery status of an email. We also discuss how to manage mail error messages.

4.1   Mailing Lists

Mailing lists [18,19] share a common behaviour: upon registration, they send a challenge to the user in order to prove that the user is a real person. As a result, it seems not possible to use challenge-response schemes with mailing lists.

Fortunately, there is a solution to this problem in the pre-challenge scheme. Since all mails from the same mailing list come from the same domain, a user can switch to the low security level (see section 3.4) whenever he/she wants to subscribe to a mailing list. At the low security level, all the incoming mails from the mailing list domain (including all the challenges and all the messages from the mailing list) that have a match in the *reply-list* are accepted into the user's mailbox and their senders are *white-listed*. When the user finally receives the first mail of the mailing list, he/she switches to the high security level (see Fig. 2).

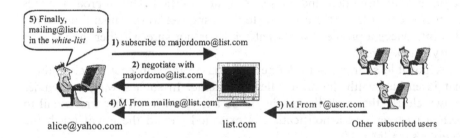

Figure 2. Process of subscription to a mailing list

The risk of inserting a spammer inside the user's *white-list* while the user is at the low security level is very low, because the spammer's email address must have the same domain as the people in the user's *reply-list*, and because a user normally only subscribes to a few mailing lists in a year.

Also, the user can set up the system not for adding the incoming mails to the *white-list* when running at the low security level, but for adding to a temporary *white-list* instead. He/She will decide later whether to add (manually) them into the final *white-list*.

4.2   Availability

It is clear that some availability problems exist when the challenge is not published along with the email address. If a sender cannot obtain the challenge of a new receiver and solve it, his/her email may not be able to reach the receiver's mailbox.

It might be good to provide both the challenge and a URL that points to the challenge for better availability. In case the URL does not work, the challenge (even if outdated) can still be used by an email sender to get in touch with a new receiver. (The receiver will reply with the latest challenge on receiving the answer of an old challenge.)

Finally, there is an availability problem that is common for both pre-challenge and challenge-response schemes: A challenge easy for a normal user might be impossible to solve for a disabled user. For example, a blind user will find impossible to solve a challenge based on images without help.

4.3   Accessibility

One of the main issues in the pre-challenge scheme is that an incoming email from a new sender without the answer of the receiver's challenge is automatically discarded, and the sender is not notified. This approach avoids the increment of Internet traffic due to the responses to spammers' mails, but also introduces a problem: a normal sender is not sure whether the receiver really got the email.

A possible solution is to define a standard prefix in each email address that is enabled with the pre-challenge scheme. In such a way, the sender knows clearly that a challenge should be answered in his/her first email to such a receiver and a notification is expected should the email reach the receiver's mailbox.

There is an alternative solution if the pre-challenge scheme is implemented at the MTA level. In this solution, the sender is warned of the invalid answer of challenge using the error reporting mechanism of the SMTP delivery negotiation protocol. This protocol works as follows:

1. The client MTA sends the contents of the email to the server MTA. After that, the server MTA checks if the email must be accepted or rejected by searching the answer to the pre-challenge.
2. If the negotiation fails, the client MTA creates an email that includes the cause of the error and the undelivered email. That email is sent to the original sender, if the client MTA does not manage his/her emails.

By using this solution, the final user will receive an error message if he/she sends an email with an invalid answer of a challenge, without increasing the Internet bandwidth in most cases. We have more discussions on managing error messages in section 4.4.

4.4 Managing Mail Error Messages

During the SMTP delivery negotiation between two MTAs, if an email cannot be delivered to its recipient, the client MTA has to send the original sender an email containing an error message. Errors can range from an invalid recipient to over-quota mailboxes, or (as seen in the previous section) pre-challenge errors.

A problem arises when the error message is not created by the MTA of the client that implements the pre-challenge scheme. An example is shown in Fig. 3. In the example, the error happens at MTA lvl 2, thus MTA lvl 1 creates and sends an error message back to the original sender. But MTA is a computer and will not include any answer of a challenge inside the error message. Therefore, it will not reach the client's protected mailbox – a problem of availability.

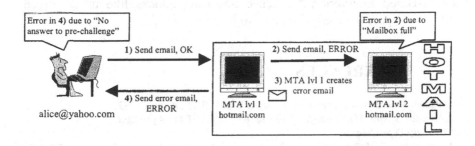

Figure 3. Problems while dealing with error messages

This problem can be solved based on two premises. First, error messages can be identified with the "message/delivery-status" header, and have attached the email that caused the problem. Second, all emails have a unique ID issued by the original client MTA, stored in the "Message-ID" header.

When an error message arrives, the pre-challenge scheme accepts the email if both address of the recipient and ID of the original message are inside the *reply-list*. Thus, it is necessary to add the ID of outgoing emails to the *reply-list*.

A spammer may try to bypass this scheme by forging both the unique ID and the recipient of the original message. This requires the spammer to wiretap the communication channel, which is unprofitable for massive spamming.

5. SUMMARY

In this paper, we presented a pre-challenge scheme for spam controlling, based on challenge-response mechanism but avoiding its drawbacks. Our scheme is a standalone solution, since there is no need to install software or change the configuration in the sender's side. Our scheme allows email senders to have no delay in reaching the receiver's mailbox, and prevents the denial of service attack if the origin of the email is forged. It also manages mailing list messages and error messages properly. Finally, our scheme offers protection against email harvesting.

This scheme can be used jointly with other major anti-spam solutions, because the type of protection that the pre-challenge scheme provides is centered in the protection of email against harvesting, thus leaving the door open to other solutions such as content analysis. Moreover, the scheme could also be integrated with authentication solutions like DomainKeys [20] or Identity-Based Encryption [21], hence thwarting attacks like using forged senders to bypass the *white-list* checking.

REFERENCES

1. J. Postel. *Simple Mail Transfer Protocol.* RFC 821, IETF, August 1982.
2. J. Klensin. *Simple Mail Transfer Protocol.* RFC 2821, IETF, April 2001.
3. SBL. http://spamhaus.org/.
4. J. Ioannidis. *Fighting Spam by Encapsulating Policy in Email Addresses.* NDSS'03, February 2003.
5. E. Gabber, M. Jakobsson, Y. Matias, and A. Mayer. *Curbing Junk E-Mail via Secure Classification.* 1998 Financial Cryptography, pages 198-213, February 1998.
6. R. J. Hall. *How to Avoid Unwanted Email.* Communications of the ACM, 41(3):88-95, March 1998.
7. L. F. Cranor and B. A. LaMacchia. *Spam!* Communications of the ACM, 41(8):74--83, August 1998.
8. M. Sahami, S. Dumais, D. Heckerman, and E. Horvitz. *A Bayesian Approach to Filtering Junk Email.* AAAI'98 Workshop on Learning for Text Categorization, July 1998.
9. P. Cunningham, N. Nowlan, S. J. Delany, and M. Haahr. *A Case-Based Approach to Spam Filtering that Can Track Concept Drift.* ICCBR'03 Workshop on Long-Lived CBR Systems, June 2003.
10. C. Dwork and M. Naor. *Pricing via Processing or Combatting Junk Mail.* Crypto'92, pages 139-147, August 1992.
11. C. Dwork, A. Goldberg, and M. Naor. *On Memory-Bound Functions for Fighting Spam.* Crypto'03, pages 426-444, August 2003.
12. M. Abadi, A. Birrell, M. Burrows, F. Dabek, and T. Wobber. *Bankable Postage for Network Services.* 8th Asian Computing Science Conference, December 2003.
13. Microsoft Penny Black Project. http://research.microsoft.com/research/sv/PennyBlack/.
14. SpamArrest. http://spamarrest.com/.
15. SpamCap. http://www.toyz.org/cgi-bin/wiki.cgi?SpamCap.

16. J. Mirkovic, J. Martin, and P. Reiher. *A Taxonomy of DDoS Attacks and DDoS Defense Mechanisms*. Technical Report #020018, Dept. of Computer Science. Univ. of California.

17. L. von Ahn, M. Blum, N. J. Hopper, and J. Langford. *CAPTCHA: Using Hard AI Problems for Security*. Eurocrypt'03, pages 294-311, May 2003.

18. Ezmlm Mailing List. http://www.ezmlm.org/.

19. Mailman Mailing List. http://www.list.org/.

20. Yahoo DomainKeys. http://antispam.yahoo.com/domainkeys/.

21. D. Boneh and M. Franklin. *Identity Based Encryption from the Weil Pairing*. Crypto'01, pages 213-229, August 2001.

22. Majordomo Mailing List. http://www.greatcircle.com/majordomo/.

AUTOMATICALLY HARDENING WEB APPLICATIONS USING PRECISE TAINTING

Anh Nguyen-Tuong, Salvatore Guarnieri, Doug Greene, Jeff Shirley, David Evans
Department of Computer Science, University of Virginia, 151 Engineer's Way, Charlottesville, VA 22904-4740, USA [1]

Abstract: Most web applications contain security vulnerabilities. The simple and natural ways of creating a web application are prone to SQL injection attacks and cross-site scripting attacks as well as other less common vulnerabilities. In response, many tools have been developed for detecting or mitigating common web application vulnerabilities. Existing techniques either require effort from the site developer or are prone to false positives. This paper presents a fully automated approach to securely hardening web applications. It is based on precisely tracking taintedness of data and checking specifically for dangerous content only in parts of commands and output that came from untrustworthy sources. Unlike previous work in which everything that is derived from tainted input is tainted, our approach precisely tracks taintedness within data values.

Key words: web security; web vulnerabilities; SQL injection; PHP; cross-site scripting attacks; precise tainting; information flow

1. INTRODUCTION

Nearly all web applications are security critical, but only a small fraction of deployed web applications can afford a detailed security review. Even when such a review is possible, it is tedious and can overlook subtle security

[1] This work was funded in part by DARPA (SRS FA8750-04-2-0246) and the National Science Foundation (NSF CAREER CCR-0092945, SCI-0426972).

vulnerabilities. Serious security vulnerabilities are regularly found in the most prominent commercial web applications including Gmail[1], eBay[2], Yahoo[3], Hotmail[3] and Microsoft Passport[4]. Section 2 provides background on common web application vulnerabilities.

Several tools have been developed to partially automate aspects of a security review, including static analysis tools that scan code for possible vulnerabilities[5] and automated testing tools that test web sites with inputs designed to expose vulnerabilities[5-7]. Taint analysis identifies inputs that come from untrustworthy sources (including user input) and tracks all data that is affected by those input values. An error is reported if tainted data is passed as a security-critical parameter, such as the command passed to an exec command. Taint analysis can be done statically or dynamically. Section 3 describes previous work on securing web applications, including taint analysis.

For an approach to be effective for the vast majority of web applications, it needs to be fully automated. Many people build websites that accept user input without any understanding of security issues. For example, *PHP & MySQL for Dummies*[8] provides inexperienced programmers with the knowledge they need to set up a database-backed web application. Although the book does include some warnings about security (for example, p. 213 warns readers about malicious input and advises them to check correct format, and p. 261 warns about <script> tags in user input), many of the examples in the book that accept user input contain security vulnerabilities (e.g., Listings 11-3 and 12-2 allow SQL injection, and Listing 12-4 allows cross-site scripting). This is typical of most introductory books on web site development.

In Section 4 we propose a completely automated mechanism for preventing two important classes of web application security vulnerabilities: command injection (including script and SQL injection) and cross-site scripting (XSS). Our solution involves replacing the standard PHP interpreter with a modified interpreter that precisely tracks taintedness and checks for dangerous content in uses of tainted data. All that is required to benefit from our approach is that the hosting server uses our modified version of PHP.

The main contribution of our work is the development of precise tainting in which taint information is maintained at a fine level of granularity and checked in a context-sensitive way. This enables us to design and implement fully-automated defense mechanisms against both command injection attacks, including SQL injection, and cross-site scripting attacks. Next, we describe common web application vulnerabilities. Section 3 reviews prior work on securing web applications. Section 4 describes our design and implementation, and explains how we prevent exploits of web application vulnerabilities.

2.   WEB APPLICATION VULNERABILITIES

Figure 1 depicts a typical web application. For clarity, we focus on web applications implemented using PHP, which is currently one of the most popular language for implementing web applications (PHP is used at approximately 1.3M IP addresses, 18M domains, and is installed on 50% of Apache servers[9]). Most issues and architectural properties are similar for other web application languages.

A client sends input to the web server in the form of an HTTP request (step 1 in Figure 1). GET and POST are the most common requests. The request encodes data created by the user in HTTP header fields including file names and parameters included in the requested URI. If the URI is a PHP file, the HTTP server will load the requested file from the file system (step 2) and execute the requested file in the PHP interpreter (step 3). The parameters are visible to the PHP code through predefined global variable arrays (including $_GET and $_POST).

The PHP code may use these values to construct commands that are sent to PHP functions such as a SQL query that is sent to the database (steps 4 and 5), or to make calls to PHP API functions that call system APIs to manipulate system state (steps 6 and 7). The PHP code produces an output web page based on the returned results and returns it to the client (step 8).

We assume a client can interact with the web server only by sending HTTP requests to the HTTP server. In particular, the only way an attacker can interact with system resources, including the database and file system, is by constructing appropriate web requests. We divide attacks into two general classes of attacks: *injection attacks* attempt to construct requests to the web server that corrupt its state or reveal confidential information; *output attacks* (e.g., cross-site scripting) attempt to send requests to the web server that cause it to generate responses that produce malicious behavior on clients.

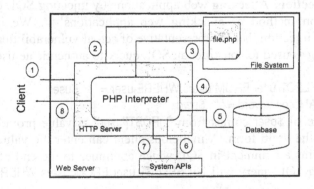

Figure 3. Typical web application architecture

2.1   Command injection attacks

In a command injection attack an attacker attempts to access confidential information or corrupt the application state by constructing an input that allows the attacker to inject malicious control logic into the web application. With the system architecture shown in Figure 1, an attack could attempt to inject PHP code that will be executed by the PHP interpreter, SQL commands that will be executed by the database, or native machine code that will be executed by the web server host directly. We consider only the first two cases. Web application vulnerabilities are far more common than vulnerabilities in the underlying server or operating system since there are far more different web applications than there are servers and operating systems, and developers of web applications tend to be far less sophisticated from a security perspective than developers of operating systems and web servers.

PHP injection. In a PHP injection attack, the attacker attempts to inject PHP code that will be interpreter by the server. If an attacker can inject arbitrary code, the attacker can do everything PHP can and has effectively complete control over the server. Here is a simple example of a PHP injection in phpGedView, an online viewing system for genealogy information[10]. The attack URL is of the form:
    http://[target]/[...]/editconfig_gedcom.php?gedcom_config=../../../../../../etc/passwd

The vulnerable PHP code uses the gedcom_config value as a filename: require($gedcom_config);. The semantics of require is to load the file and either interpret it as PHP code (if the PHP tags are found) or display the content. Thus this code leaks the content of the password file. Abuse of require and its related functions is a commonly reported occurrence[11,12], despite the fact that, properly configured, PHP is impervious to this basic attack. However, additional defenses are needed for more sophisticated injection attacks such as the recently released Santy Worm[13] and the phpMyAdmin attack[14].

SQL injection. Attacking web applications by injecting SQL commands is a common method of attacking web applications[15,16]. We illustrate a simple SQL injection that is representative of actual vulnerabilities. Suppose the following is used to construct an SQL query to authenticate users against a database:
    $cmd="SELECT user FROM users WHERE user = ' " . $user
        . " ' " AND password = ' " . $pwd . " ' ";
The value of $user comes from $_POST['user'], a value provided by the client using the login form. A malicious client can enter the value: ' OR 1 = 1 ; --' (-- begins a comment in SQL which continues to the end of the line). The resulting SQL query will be: SELECT user FROM users WHERE user = ' '

OR 1 = 1 ; -- ' AND password = 'x'. The injected command closes the quote and comments out the AND part of the query. Hence, it will always succeed regardless of the entered password.

The main problem here is that the single quote provided by the attacker closes the open quote, and the remainder of the user-provided string is passed to the database as part of the SQL command. This attack would be thwarted by PHP installations that use the default magic quotes option. When enabled, magic quotes automatically sanitize input data by adding a backslash to all strings submitted via web forms or cookies. However, magic quotes do not suffice for attacks that do not use quotes[17].

One solution to prevent SQL injections is to use prepared statements[18]. A prepared statement is a query string with placeholders for variables that are subsequently bound to the statement and type-checked. However, this depends on programmers changing development practices and replacing legacy code. Dynamic generation of queries using regular queries will continue to be prevalent for the foreseeable future.

2.2   Output attacks

Output attacks send a request to a web application that causes it to produce an output page designed by the attacker to achieve some malicious goal. The most dangerous kind of output attack is a *cross-site scripting* attack, in which the web server produces an output page containing script code generated by the attacker. The script code can steal the victim's cookies or capture data the victim unsuspectingly enters into the web site. This is especially effective in phishing attacks in which the attacker sends potential victims emails convincing them victim to visit a URL. The URL may be a trusted domain, but because of a cross-site scripting vulnerability the attacker can construct parameters to the URL that cause the trusted site to create a page containing a form that sends data back to the attacker. For example, the attacker constructs a link like this:

<a href='http://bad.com/go.php?val=<script src="http://bad.com/attack.js"></script>'>

If the implementation of go.php uses the val parameter in the generated web page output (for example, by doing print "Results for: " . $_GET['val'];), the malicious script will appear on the resulting page. A clever attacker can use character encodings to make the malicious script appear nonsensical to a victim who inspects the URL before opening it.

Five years ago, CERT Advisory 2000-02 described the problem of cross-site scripting and advised users to disable scripting languages and web site developers to validate web page output[19]. Nevertheless, cross-site scripting problems remain a serious problem today. Far too much functionality of the web depends on scripting languages, so most users are unwilling to disable them. Even security-conscious web developers frequently produce websites that are vulnerable to cross-site scripting attacks[1,4,20-22]. As with SQL injection, ad hoc fixes often fail to solve discovered problems correctly—the initial filters develop to fix the Hotmail vulnerability could be circumvented by using alternate character encodings[4]. Hence, we focus on fully automated solutions.

3.   RELATED WORK

Several approaches have been developed for securing web applications including filtering input and output that appears dangerous, automated testing and diversity defenses. The approaches most similar to our proposed approach involve analyzing information flow.

Input and Output Filtering. Scott and Sharp developed a system for providing an application-level firewall to prevent malicious input from reaching vulnerable web servers[23]. Their approach required a specification of constraints on different inputs, and compiled those constraints into a checking program. This requires a programmer to provide a correct security policy specific to their application, so is ill-suited to protecting typical web developers. Several commercial web application firewalls provide input and output filtering to detect possible attacks[24,25]. However, these tools are prone to both false positives and negatives[26].

Automated Testing. There are several web application security testing tools designed specifically to find vulnerabilities[5,27,28]. The problem with these tools is that they have to guess the exploit data in order to expose the vulnerability. For well-known generic classes of vulnerabilities, such as SQL injection, this may be possible. But for novel or complex vulnerabilities, it is unlikely the scanner will guess the right inputs to expose the vulnerability.

Diversity Defenses. Instruction-Set Randomization is a form of diversity in which defenders modify the instruction set used to run applications[29]. Thus, code-injection attacks that rely on knowledge of the original language are detected and thwarted easily. This approach has been advocated for general scripting languages[29] and for protection against SQL injections[30]. There are two main problems with ISR: (1) it is effective only against code injection attacks and incomplete by itself (it does not handle cross-site scripting attacks), and (2), the deployment of ISR is not transparent to developers and requires the transformation of application code.

Information Flow. All of the web vulnerabilities described in Section 2 stem from insecure information flow: data from untrusted sources is used in a trusted way. The security community has studied information flow extensively[31]. The earliest work focused on confidentiality, in particular in preventing flows from trusted to untrusted sources[32]. In our case, we are primarily concerned with integrity. Biba showed that information flow can also be used to provide integrity by considering flows from untrusted to trusted sources[33].

Information flow policies can be enforced statically, dynamically or by a combination of static and dynamic techniques. Static taint analysis has been used to detect security vulnerabilities in C programs[34,35]. Static approaches have the advantage of increased precision, no run-time overhead and the ability to detect and correct errors before deployment. However, they require substantial effort from the programmer. Since we are focused on solutions that will be practically deployed in typical web development scenarios, we focus on dynamic techniques.

Huang et. al developed WebSSARI, a hybrid approach to securing web applications[36]. The WebSSARI tool uses a static analysis based on type-based information flow to identify possible vulnerabilities in PHP web applications. Their type-based approach operates at a coarse-grain: any data derived from tainted input is considered fully tainted. WebSSARI can insert calls to sanitization routines that filter potentially dangerous content from tainted values before they are passed to security-critical functions. Because we propose techniques for tracking taintedness at a much finer granularity, our system can be more automated than WebSSARI: all we require is that the server uses our modified interpreter PHP to protect all web applications running on the server.

4.   AUTOMATIC WEB HARDENING

Our design is based on maintaining precise information about what data is tainted through the processing of a request, and checking that user input sent to an external command or output to a web page contains only safe content. Our automated solution prevents a large class of common security vulnerabilities without any direct effort required from web developers.

The only change from the standard web architecture in Figure 1 is that we replace the standard PHP interpreter with a modified interpreter that identifies which data comes from untrusted sources and precisely tracks how that data propagates through PHP code interpretation (Section 4.1), checks that parameters to commands do not contain dangerous content derived from user input (Section 4.2), and ensures that generated web pages do not contain scripting code created from untrusted input (Section 4.3).

4.1   Keeping track of precise taint information

We mark an input from untrusted sources including data provided by client requests as tainted. We modified the PHP interpreter's implementation of the string datatype to include tainting information for string values at the granularity of individual characters. We then propagate taint information across function calls, assignments and composition at the granularity of a single character, hence *precise tainting*. The application of precise tainting enables the prevention of injection attacks and the ability to easily filter output for XSS attacks. If a function uses a tainted variable in a dangerous way, we can reject the call to the function (as is done with SQL queries or PHP system functions) or sanitize the variable values (as is done for preventing cross-site scripting attacks).

Web application developers often remember to sanitize inputs from GET and POSTs, but will omit to check other variables that can be manipulated by clients. Our approach ensures that *all* such external variables, e.g. hidden form variables, cookies and HTTP header information, are marked as tainted. We also keep track of taint information for session variables and database results.

4.1.1   Taint strings

For each PHP string, we track tainting information for individual characters. Consider the following code fragment where part of the string $x comes from a web form and the other from a cookie:

$x = "Hello " . $_GET['name1'] . ". I am " . $_COOKIE['name2'];

The values of $_GET['name1'] and $_COOKIE['name2'] are fully tainted (we assume they are Alice and Bob). After the concatenation, the values of $x and its taint markings (underlined) are: Hello Alice. I am Bob.

4.1.2   Functions

We keep track of taint information across function calls, in particular functions that manipulate and return strings. The general algorithm is to mark strings returned from function as tainted if any of the input arguments are tainted. Whenever feasible, we exploit the semantics of functions and keep track of taintedness precisely. For example, consider the substring function in which taint markings for the result of the substr call depend on the part of the string they select: substr("precise taint me", 2, 10); // ecise tai

4.1.3   Database values and session variables

Databases provide another potential venue for attackers to insert malicious values. We treat strings that are returned from database queries as untrusted and mark them as tainted. While this approach may appear overly restrictive, in the sense that legitimate uses may be prevented, we show in Section 4.3 how precise tainting and our approach to checking for cross-site scripting mitigates this potential problem. Further, if the database is compromised by some other means, the attacker is still unable to use the compromised database to construct a cross-site scripting attack.

The stateless nature of HTTP requires developers to keep track of application state across client requests. However, exposing session variables to clients would allow attackers to manipulate applications. Well-designed web applications keep session variables on the server only and use a session id to communicate with clients. We modified PHP to store taint information with session variables.

4.2   Preventing command injection

The tainting information is used to determine whether or not calls to security-critical functions are safe. To prevent command injection attacks, we check that the tainted information passed to a command is safe. The actual checking depends on the command, and is designed to be precise enough to prevent all command injection attacks from succeeding while allowing typical web applications to function normally when they are not under attack.

4.2.1   PHP injection

To prevent PHP injection attacks we disallow calls to potentially dangerous functions if any one of their arguments is tainted. The list of functions checked is similar to those disallowed by Perl and Ruby's taint mode[37,38] and consists of functions that treat input strings as PHP code or manipulate the system state such as system calls, I/O functions, and calls that are directly evaluated.

4.2.2   SQL injection

Preventing SQL injections requires taking advantage of precise taint information. Before sending commands to the database, e.g. mysql_query, we run the following algorithm to check for injections:

1. TOKENIZE THE QUERY STRING; PRESERVE TAINT MARKINGS WITH TOKENS.
2. SCAN EACH TOKEN FOR IDENTIFIERS AND OPERATOR SYMBOLS (IGNORE LITERALS, I.E., STRINGS, NUMBERS, BOOLEAN VALUES).
3. DETECT AN INJECTION IF AN OPERATOR SYMBOL IS MARKED AS TAINTED. OPERATOR SYMBOLS ARE ,()[].;:+-*/\%^<>=~!?@#&|`
4. DETECT AN INJECTION IF AN IDENTIFIER IS TAINTED AND A KEYWORD. EXAMPLE KEYWORDS INCLUDE UNION, DROP, WHERE, OR, AND.

Using the example from Section 2.1:
$cmd="SELECT user FROM users WHERE user = ' " . $user
. "' AND password = ' " . $password . " ' ";

The resulting query string (with $user set to ' OR 1 = 1 ; -- ') is tainted as follows: SELECT user FROM users WHERE user = ' <u>' OR 1 = 1 ; --</u> ' AND password = '<u>x</u>'. We detect an injection since OR is both tainted and a keyword.

4.3   Preventing cross-site scripting

Our approach to preventing cross-site scripting relies on checking generated output. Any potentially dangerous content in generated HTML pages must contain only untainted data. We modify the PHP output functions (print, echo, printf and other printing functions) with functions that check for tainted output containing dangerous content. The replacement functions output untainted text normally, but keep track of the state of the output stream as necessary for checking. For a contrived example, consider an application that opens a script and then prints tainted output: print "<script>document.write ($user)</script>";

An attacker can inject JavaScript code by setting the value of $user to a value that closes the parenthesis and executes arbitrary code: " me");alert("yo". Note that the opening script tag could be divided across multiple print commands. Hence, our modified output functions need to keep track of open and partially open tags in the output. We do not need to parse the output HTML completely (and it would be unadvisable to do so, since many web applications generate ungrammatical HTML).

Checking output instead of input avoids many of the common problems with ad hoc filtering approaches. Since we are looking at the generated output any tricks involving separating attacks into multiple input variables or using character encodings can be handled systematically. Our checking involves whitelisting safe content whereas blacklisting attempts to prevent cross-site scripting attacks by identifying known dangerous tags, such as <script> and <object>. The latter fails to prevent script injection involving other tags. For example, a script can be injected into the apparently harmless <b> (bold) tag using parameters such as onmouseover.

Our defense takes advantage of precise tainting information to identify web page output generated from untrusted sources. Any tainted text that could be dangerous is either removed from the output or altered to prevent it being interpreted (for example, replacing < in unknown tags with <). Our conservative assumptions mean that some safe content may be inadvertently suppressed; however, because of the precise tainting information, this is limited to content that is generated from untrusted sources.

5.   CONCLUSION

We have described a fully automated, end-to-end approach for hardening web applications. By exploiting precise tainting in a way that takes advantage of program language semantics and performing context-dependent checking, we are able to prevent a large class of web application exploits without requiring any effort from the web developer. Initial measurements indicate that the performance overhead incurred by using our modified intepreter is less than 10%.

Effective solutions for protecting web applications need to balance the need for precision with the limited time and effort most web developers will spend on security. Fully automated solutions, such as the one described in this paper, provide an important point in this design space.

REFERENCES

1. N. Weidenfeld, *Security Hole Found in Gmail*, (27 October 2004); http://net.nana.co.il/Article/?ArticleID=155025&sid=10.
2. *Report of Ebay Cross-Site Scripting Attack*; http://securityfocus.com/archive/82/246275.
3. *Remotely Exploitable Cross-Site Scripting in Hotmail and Yahoo*, (March 2004); http://www.greymagic.com/security/advisories/gm005-mc/.
4. EyeonSecurity, Microsoft Passport Account Hijack Attack: Hacking Hotmail and More, *Hacker's Digest*.
5. Y.-W. Huang *et al.*, Web Application Security Assessment by Fault Injection and Behavior Monitoring, *Proc. of the World Wide Web Conference (WWW 2003)*, (May 2003).
6. M. Benedikt *et al.*, Veriweb: Automatically Testing Dynamic Web Sites, *Proc. of the World Wide Web Conference*, (May 2002).

7. F. Ricca, and P. Tonella, Analysis and Testing of Web Applications, *Proc. of the IEEE International Conference on Software Engineering*, (May 2001).

8. J. Valade, *Php & Mysql for Dummies*, (Wiley Publishing, 2002).

9. *Netcraft Survey*, (January 2005); http://news.netcraft.com/.

10. JeiAr, *Phpgedview Php Injection*, (Jan 2004); http://xforce.iss.net/xforce/xfdb/14205.

11. Gentoo, *Gallery Php Injection*, (February 2004); http://www.linuxsecurity.com/advisories/gentoo_advisory-4015.html.

12. K. Więsek, Gonicus System Administrator Php Injection, (February 2003).

13. *Santy Worm Used Google to Spread*, (23 December 2004); http://newsfromrussia.com/world/2004/12/23/57537.html.

14. N. Symbolon, *Phpmyadmin Critical Bug*; http://xforce.iss.net/xforce/xfdb/16542.

15. D. Litchfield, *Sql Server Security*, (McGraw-Hill Osborne Media, 2003).

16. K. Spett, "Sql Injection: Are Your Web Applications Vulnerable?" (SPI Labs White Paper, 2002).

17. L. Armstrong, *Phpnuke Sql Injection*, (20 February 2003).

18. *Improved Mysql Extensions*; http://www.php.net/manual/en/ref.mysqli.php.

19. *Malicious Html Tags Embedded in Client Web Requests*, (February 2, 2000); http://www.cert.org/advisories/CA-2000-02.html.

20. G. Hoglund, and G. McGraw, *Exploiting Software: How to Break Code*, (Addison-Wesley, 2004).

21. R. Ivgi, *Cross-Site-Scripting Vulnerability in Microsoft.Com*, (4 October 2004).

22. J. Ley, *Simple Google Cross Site Scripting Exploit*, (17 October 2004).

23. D. Scott, and R. Sharp, Abstraction Application-Level Web Security, *Proc. of the WWW*, (May 2002).

24. *Interdo Web Application Firewall*; http://www.kavado.com/products/interdo.asp.

25. Teros, Inc., *Teros-100 Application Protection System*, (2004); http://www.teros.com/products/aps100/aps.shtml.

26. T. Dyck, *Review: Appshield and Review: Teros-100 Aps 2.1.1*, (May 2003); http://www.eweek.com/article2/0,3959,1110435,00.asp.

27. Tenable Network Security, *Nessus Open Source Vulnerability Scanner Project*, (2005); http://www.nessus.org.

28. J. Offutt *et al.*, Bypass Testing of Web Applications., *Proc. of the IEEE International Symposium on Software Reliability Engineering*, (November 2004).

29. G. S. Kc *et al.*, Countering Code-Injection Attacks with Instruction-Set Randomization., *Proc. of the ACM Computer and Communication Security (CCS)*, (October 2003).

30. S. W. Boyd, and A. D. Keromytis, Sqlrand: Preventing Sql Injection Attacks, *Proc. of the 2nd Applied Cryptography and Network Security (ACNS) Conference*, (June 2004).

31. A. Sabelfeld, and A. C. Myers, Language-Based Information-Flow Security, *IEEE Journal on Selected Areas in Communications* (January 2003).

32. D. E. Bell, and L. J. LaPadula, *Secure Computer Systems: Mathematical Foundations* Mtr-2547, (MITRE Corporation, 1973).

33. K. J. Biba, *Integrity Considerations for Secure Computer Systems* Esd-Tr-76-372, (USAF Electronic Systems Division, 1977).

34. U. Shankar *et al.*, Detecting Format-String Vulnerabilities with Type Qualifiers, *Proc. of the USENIX Security Symposium*.

35. D. Evans, and D. Larochelle, Improving Security Using Extensible Lightweight Static Analysis, *IEEE Software* (January/February 2002).
36. Y.-W. Huang *et al.*, Securing Web Application Code by Static Analysis and Runtime Protection, *Proc. of the World Wide Web Conference*, (May 2004).
37. *Perl 5.6 Documentation: Perl Security*; http://www.perldoc.com/perl5.6/pod/perlsec.html.
38. D. Thomas *et al.*, *Programming Ruby: The Pragmatic Programmer's Guide*, (Pragmatic Programmers, ed. Second, 2004).

TRAFFIC REDIRECTION ATTACK PROTECTION SYSTEM (TRAPS)

Vrizlynn L. L. Thing[1,2], Henry C. J. Lee[2] and Morris Sloman[1]
[1]Department of Computing, Imperial College London, 180 Queen's Gate, London SW7 2AZ, UK, [2]Institute for Infocomm Research, 21 Heng Mui Keng Terrace, Singapore 119613

Abstract: Distributed Denial of Service (DDoS) attackers typically use spoofed IP addresses to prevent exposing their identities and easy filtering of attack traffic. This paper introduces a novel mitigation scheme, TRAPS, whereby the victim verifies source address authenticity by performing reconfiguration for traffic redirection and informing high ongoing-traffic correspondents. The spoofed sources are not informed and will continue to use the old configuration to send packets, which can then be easily filtered off. Adaptive rate-limiting can be used on the remaining traffic, which may be attack packets with randomly-generated spoofed IP addresses. We compare our various approaches for achieving TRAPS functionality. The end-host approach is based on standard Mobile IP protocol and does not require any new protocols, changes to Internet routers, nor prior traffic flow characterizations. It supports adaptive, real-time and automatic responses to DDoS attacks. Experiments are conducted to provide proof of concept.

Key words: Distributed Denial of Service; Attack Response System; Adaptive Security.

1. INTRODUCTION

In Denial of Service (DoS) [1] or Distributed DoS (DDoS) attacks, a large number of malicious packets are sent from single or multiple machines respectively, with the aim of exhausting the target's computational and networking resources. The DDoS attacks that shut down some high-profile Web sites (e.g. Yahoo, Amazon) in February 2000 [2], demonstrated their severe consequences and the importance of efficient defense mechanisms.

Measurements collected in [3] shows the prevalence of DoS attacks in the Internet, whereby more than 12,000 attacks against over 5,000 distinct targets were observed in a 3-week data collection period.

In DDoS attacks, the attack packets are often sent with spoofed IP addresses to hide the attackers' identity. Traceback mechanisms [4-12] have been proposed to trace the true source of the attackers to institute accountability. In [13,14], authenticity of IP packet addresses are verified to eliminate spoofing. Rate limiting [15] can be used to decrease malicious traffic as a response technique when the probability of false positive is high. When the data stream is reliably detected as malicious, filtering mechanisms [16,17] can be used to drop the attack traffic. Reconfiguration mechanisms [18,19] change the topology of the victim's network to isolate the attacks or add more network resources. Detailed discussions continue in Section 6.

This paper proposes a novel comprehensive adaptive DDoS mitigation scheme named "Traffic Redirection Attack Protection System" (TRAPS). It consists of traffic congestion and overloading detection, DDoS alleviation by performing good traffic redirection, bad traffic filtering and suspicious traffic adaptive rate-limiting. This scheme does not require prior traffic flow characterizations compared to most existing DDoS defense systems, and allows for a quick real-time response even when attacks constitute flooding of the victim with legitimate service requests. We examined the various approaches of achieving the TRAPS reconfiguration for redirection functionality, and concluded that the end-host approach is comparatively more efficient and requires the least deployment effort. We used Mobile IP (MIP) [20,21] protocol to implement the end-host approach, to avoid the need for new Internet protocol. Although MIP is used, TRAPS is applicable regardless of whether the victim is a wired or wireless node, at home or in a foreign network, and operating in static or mobile mode.

Section 2 of the paper specifies the design objectives and key assumptions. Section 3 describes TRAPS. The experimentation to prove the concepts is presented in Section 4. Section 5 considers the security issues of the protocol and possible attack scenarios, followed by comparisons with existing techniques in Section 6. Conclusions follow in Section 7.

2.   DESIGN OBJECTIVES AND KEY ASSUMPTIONS

In this section, we present the design objectives and discuss the key assumptions on which the TRAPS' design is based.

Design Objectives:
i)   Should not require any changes to the Internet infrastructure as it would raise conformance issues
ii) Minimal processing and overhead requirements so as not to overload the host or network under attack
iii) Simple and fast algorithms as time and processing power are critical factor and resource during DDoS attacks
iv) Should achieve zero false positive to filter off packets to prevent self-inflicted DoS
v)   Should guarantee QoS for high-bandwidth legitimate users
vi) Should guarantee communication of signals required for mitigation purpose to ensure that victim's "call for help" is not overwhelmed.

Key Assumptions:
i)   If the packets' contents match an attack signature, it could be easily detected and filtered off by an Intrusion Detection Systems (IDS). Co-operation of IDS with TRAPS would allow faster detection of attacks with known signatures and reduce false positive. In this paper, we only focus on attack traffic with seemingly legitimate packet contents and proceed to differentiate them into the good, bad and suspicious types. Prior knowledge of attack signatures and characterization based on packet contents are thus not required. We define 4 classes of DDoS attacks as follows:
Class A: High-bandwidth traffic with legitimate source addresses
Class B: High-bandwidth traffic with randomly generated spoofed source addresses
Class C: Low-bandwidth traffic with legitimate source addresses
Class D: Low-bandwidth traffic with randomly generated spoofed source addresses

   Class A attacks are similar to legitimate user traffic flows and cannot be classified as attacks as they would have the same rights as legitimate users. They are using their own source addresses and transferring legitimate traffic and so will be informed of any host/network reconfiguration by TRAPS. TRAPS should not attempt and would not be able to prevent such attacks. TRAPS assumes these are high-bandwidth users (even if they are zombies), who have negotiated a QoS agreement, and so aims to preserve their QoS. Therefore, this should be handled by mechanisms such as resource allocation at the protected network and we do not consider this form of "attack" here. However, Class A attackers might try to obtain a protected host/network's latest configuration information to support attacks in the other three classes. We discuss this further in Section 5 and propose a solution.

Class B attackers will be sent TRAPS notification of the latest reconfiguration information, but they will not receive them as the addresses are spoofed. Thus, they are not able to send subsequent packets based on the latest protected host/network's configuration, and so the subsequent traffic can be easily identified and filtered off.

Class C and Class D attack traffic are not notified as they constitute a vast distributed set of distinct addresses, so sending individual notifications is not practical. This attack traffic would be treated as suspicious traffic along with any new incoming legitimate requests, and be subjected to lenient treatment (i.e. rate limiting).

ii) We assume that legitimate correspondents are willing to co-operate upon receiving notifications generated by TRAPS. As they would like to have access to the services provided by the protected host/network, they would be motivated to co-operate so they would not block off notifications or refuse to act upon receiving notifications. As such, authentication of notifications becomes an important consideration and we would discuss this further in Section 5.

iii) We assume that the protected network is one under an administrative domain (e.g. enterprise network) and there exists the ability to reconfigure gateways (e.g. for rate limiting) or routers within the network to support TRAPS.

3.   DESIGN OF TRAPS

When severe traffic congestion or overloading is detected at the victim, all the gateways and the victim's access router (AR) are informed to drop packets for it, to maintain resource utilization at a "safe" level. The gateways, with a specified probability, discard packets destined for the victim from external sources. The AR ensures that aggregate traffic destined for the victim does not exceed the "safe" level and performs additional rate limiting if required (to take care of possibility of internal attackers, whereby implications will be discussed later). The above-mentioned step will ease the congestion to prevent the victim and network from being overwhelmed by the flood. This is very important during an attack to allow nodes within the protected network to be able to achieve communication for activating TRAPS mitigation support – *satisfies Objective (vi)*. At the same time, reconfiguration of the victim/network will be performed to support traffic redirection and the victim will determine recent correspondents with high on-going traffic. TRAPS will inform these correspondents to send future traffic based on the new configuration information. Some notifications may fail due to spoofed addresses. When the acknowledgements are received from the correspondents (after allowing time for retries), the victim informs the gateways and its AR to drop all subsequent packets which do not contain the latest configuration information. Legitimate on-going high bandwidth traffic will have received the redirection information and so will be passed through to reach the victim – *satisfies Objective (v)*. Bad on-going high bandwidth traffic using spoofed source addresses, will be filtered off. This traffic detected as attacks, are without doubt from illegitimate users and thus zero false positive is achieved – *satisfies Objective (iv)*.

In DDoS attacks, multiple small-volume bad traffic flows are directed at the victim with randomly spoofed source addresses, and traffic redirection is not feasible. This remaining (Class D) traffic is instead rate-limited (i.e. a more lenient approach) as it might include newly initiated connection requests or small streams of traffic from legitimate sources. In a DDoS attack, a high percentage of the remaining traffic belongs in the category of attacks as compared to the small volume of legitimate traffic and therefore, rate-limiting improves the probability of letting the legitimate requests get through. Next, we propose the various approaches of achieving the TRAPS reconfiguration for traffic redirection.

3.1   En-route Routers Nomination

We propose the network based approach as follows. The victim or a central node nominates routers (e.g. randomly) within the network. These nominated routers are assigned as en-route routers in newly constructed path/s (different set of routers could be nominated for different (set of) correspondents). These new alternative path/s are assigned to the high-bandwidth traffic correspondents, through TRAPS notifications, to allow them to reach the victim. The gateways are then informed of the {correspondent/s' address, victim's address, designated path/s} matching data sets. They will check the incoming packets and if they do not contain any valid designated path information in the packets when checked with the matching data sets, these packets will be dropped. Another set of routers, the Guard Routers, are also randomly chosen within the network and they too, are informed of the matching data sets; Though gateways will be responsible for dropping off attack packets from external attackers, attack packets from internal attackers will bypass gateways and therefore, there exists a need for these Guard Routers to check packets in transit based on the matching data sets and make decisions whether to forward or drop the packets. Guard Routers would filter packets based on both the information in the packets and whether the packets are supposed to visit it. The AR is also informed of the matching data sets to provide a final line of defense. It will perform final checks before forwarding packets to the victim

The disadvantages of this approach are that mechanisms such as source routing have to be used to ensure that the packets follow the designated paths and a new signaling protocol is required to notify the correspondents of the path/s they are assigned. The default route is cut off and the alternate path/s might not be the optimal ones, and high overhead will be incurred as the packets need to encapsulate the en-route routers' addresses. The advantages are that the packets must follow the designated paths or be filtered off and as it would be difficult for the attackers to guess what are the nominated en-route routers (security strength is dependent on the no. of routers selected and no. of address bits (minus away no. of bits for network prefix)), and to derive the exact routers sequence in the alternate path/s. Another advantage would be that the approach could be applied to reduce the load on the gateways by having them perform only random checks and leaving the mandatory verifications to the Guard Routers. Therefore, work distribution across the protected network could be achieved.

3.2   Passcode Approach

Instead of assigning alternate path/s to the correspondents, passcodes could be generated for assignments instead. Packets with matching source address, destination address and valid passcode are allowed to be forwarded. The advantages of this approach are that mechanisms such as source routing is not required (e.g. passcode could be placed in an optional header in the packet), lesser overhead is incurred as passcode is shorter than the entire path information and the default route, which is normally the optimal path, is not "cut off". A disadvantage is that a new signaling protocol for TRAPS notification is still required as in the En-route Router Nomination approach. Attackers having knowledge of this scheme will have a success rate $\alpha\ 1/2^{n+32}$ or $1/2^{n+128}$ of breaking it, for IPv4 or IPv6 networks respectively (i.e. guessing matching correspondent's address and passcode of n bits for each victim it's targetting).

3.3   Virtual Relocation Approach

The following describes the Virtual Relocation Approach, which is end-host based. The victim performs a virtual relocation by requesting a new IP address (different addresses could be used for different correspondents or set of correspondents), while still maintaining its old one for use with correspondents not chosen for notifications. It informs high-bandwidth traffic correspondents of the new IP address, and all gateways and a selected set of Guard Routers in the protected network of the {correspondent/s' address, victim's new address} matching data sets. The AR is also informed of the matching sets to provide a final line of defense. The required forwarding or dropping of packets are performed by the gateways, Guard Routers and AR, based on the matching data sets.

This approach has the least overhead as it does not require additional data in the packets (i.e. just replace the destination field). The gateways, Guard Routers and victim's AR will drop packets with source address = notified correspondents and destination address = victim's old address. The default route need not be "cut off" and multiple paths could still exist between correspondents and victim. The attackers having knowledge of the scheme could guess the new address (having network prefix of m bits) and matching correspondent's address with a success rate α $1/2^{64-m}$ for IPv4 and $1/2^{256-m}$ for IPv6 networks.

Although the possibility of success of attackers breaking the scheme is not very high (e.g. $1.39e^{-17}$ and $2.43e^{-63}$ for Virtual Relocation Approach in IPv4 (assuming 8-bit network prefix) or IPv6 networks (having known 48-bit public topology IDs respectively), security strength could be further increased, by performing dynamic reconfigurations more frequently. However, this increases the signaling overhead.

Comparing the methods proposed, we could see that Virtual Relocation has the least overhead (no additional fields in data packets and minimal signaling within protected network). The processing overhead is low due to its simplicity as it does not require a hashing algorithm or network support in alternate path/s construction – *satisfies Objectives (ii) and (iii)*. The most important factor here is that it requires the least deployment effort – *satisfies Objective (i) (although all the methods do not require modifications to the Internet infrastructure)*. The network based approaches require a signaling protocol for communications with the correspondents and customized TRAPS activation software at all potentially legitimate correspondents. However, with the Virtual Relocation Approach, we could make use of MIP, and thus no special software is needed at the correspondents. As long as the correspondents comply to the MIP standards, they have the necessary mechanisms to support communications and react to relocation of their correspondents. This approach could be used even if their correspondents are not actually mobile.

The following sub-sections describe the details on the components of TRAPS, namely high-bandwidth traffic selection, traffic congestion and overloading detection, rate-limiting, and flooding subsidence.

3.4   High-bandwidth Traffic Selection

A Correspondent Database (CD) is maintained by the victim to record information about the traffic it receives and contains the following fields.
- Source address (S_k - unique key field)
- Amount of traffic (e.g. in bytes), M_k, received from this source

where k (from 0 to K-1) is the sequence number of the entries in the CD, and K is the total number of entries in CD. CD is refreshed every T_u secs to keep the data set updated for monitoring the latest on-going traffic of the last T_p secs ($T_p > T_u$).

When congestion or overloading is detected, the victim looks up its CD to select those correspondents with high-bandwidth ongoing traffics, to be notified about the reconfiguration. This also applies to Class B attacks.

In the event of Class C and D attacks, most of the source addresses in the CD will be widely distributed and short-lived (in the case of Class D). With the record interval, T_p, there will be very little recorded traffic for each unique spoofed source address. However, setting an absolute threshold of traffic received for TRAPS activation would require monitoring normal traffic flow and attack traffic to derive how much traffic are considered heavy good traffic or low unique bad traffic with widely distributed range of source addresses. Therefore, we propose (1) as the first condition for choosing the correspondents to perform notifications. In this case, only entries in the CD with traffic equal or greater than the average traffic received will be chosen. The second condition is that the selected traffic must also be high enough (> threshold, M_T) to justify selection for notifications. This is to prevent massive activation in the event that there are many sources of low bandwidth attack traffic while there is no ongoing high-bandwidth legitimate traffic – this is likely for DDoS.

$$M_k \geq \frac{\sum_{k=0}^{K-1} M_k}{K} \tag{1}$$

3.5   Traffic Congestion and Overloading Detection

The traffic and resource monitoring system on the victim detects flooding and severe resource consumption. A simple method is to observe the resource utilization (i.e. bandwidth and computing resources) at the victim and activate TRAPS when a threshold is reached. Another way would be through monitoring gradual depletion of resources at the victim. For example, in traffic monitoring, the aggregate incoming traffic will be observed for checking bandwidth utilization. Traffic growth rate is then computed, so as to detect seemingly abnormal traffic behavior. As for the computing resource monitoring, parameters such as CPU load or memory consumption would be observed and consumption growth rate could then be computed to detect any signs of attack directed at the victim. The following describes the detection method in details.

1) Let x_n (bandwidth or other resources' utilization in percentage) be the alerting points whereby resource consumption growth rate monitoring has to be started, with $n > 0$ and $x_n > x_{n-1} > \dots > x_2 > x_1$.
2) Let g_n (consumption growth in percentage) correspond to each x_n whereby an alarm has to be triggered and traffic redirection activated. Detection sensitivity has to be increased as the resource utilization gets larger. Therefore, allowable consumption growth rate should be set smaller for increasing monitoring stages.
3) Let t_n be the sampling rate of each stage (in seconds, $n = 0$ for sampling rate before first alerting point and $n > 0$ for sampling rate during alerting stages). Similar to the consumption growth, the detection sensitivity should be increased as the alerting point is advanced. This could be set through the sampling rate by allowing more frequent sampling at later/crucial monitoring stages.
4) Let y be the final alert point or the alarm point, whereby an alarm is immediately generated as soon as the resource utilization reaches or exceeds this point.

3.6   Rate Limiting at Gateways and Victim's AR

After TRAPS is activated, resource consumption at the victim is constantly monitored to adjust the rate-limiting parameters at the gateways and victim's AR in the protected network. An allowable stable resource consumption level, R_c, is configured at the victim. We define the probability of rate-limiting, p, as the probability of dropping the incoming traffic. The initial value of p, p_0, is derived from R_c when alarm is triggered for TRAPS activation. For example, if R_c is 85% of bandwidth and aggregate incoming traffic at the victim is utilizing 95% of it's bandwidth, p_0 will be (95-85)/95, which is approximately 0.105. This value will be sent to the gateways to perform rate limiting for this particular victim (i.e. destination of packets = victim). Resource consumption, which is constantly monitored at the sampling rate, t_n, as in Section 3.4, will be used for adjusting the probability setting. To provide a last line of defense (e.g. in case of internal attackers), victim's AR will be asked to perform further rate-limiting to maintain victim's resource consumption within a "safe" level (e.g. limit victim's aggregate incoming traffic bandwidth at 100kbps).

3.7   Flooding subsidence

To prevent frequent toggling between activation and deactivation of TRAPS resulting in high overhead, three parameters would be used to determine if the DDoS attack has subsided. Therefore, TRAPS will only be deactivated if possible resource consumption without TRAPS is maintained within an acceptable level ($R_a < x_1$, where x_1 is defined in Section 3.4), for at least T_a seconds with a low probability (P_a) of rate limiting at the gateways. Possible resource consumption without TRAPS is measured by totaling resource consumption at the victim, resource conservation due to filtering and rate limiting at the gateways and victim's AR. The choice of the three parameters (R_a, P_a, and T_a) would affect the frequency of toggling as in the following equation.

$$\text{Frequency of toggling } \alpha \ (R_a \times P_a)/T_a \qquad (2)$$

4.   PROOF OF CONCEPT (INCORPORATION WITH MIP)

We used MIP for performing the signaling as it is well-suited for carrying out the required virtual traffic redirection. It is virtual in the sense that traffic is not really redirected to another route but rather to the victim's new address, and the same default or optimal route might still be used. Another reason is that since MIPv4 and MIPv6 are IETF standards, widespread implementations of the protocols are in place (e.g. versions in Windows, Linux, BSD are available). No change will be required in the rest of the Internet infrastructure and the correspondents. In MIP, Home Agents (HAs) are responsible for proxying and intercepting the packets on behalf of Mobile Nodes (MNs, i.e. the victims here), therefore the tasks of filtering and forwarding of the packets destined to MNs can be performed by HAs instead of the gateways. In this case, the gateways are relieved from having to handle all the hosts, which might be activating TRAPS, in the network. In this way, more effective workload distribution and thus higher scalability is achieved.

We developed the TRAPS prototype by implementing the necessary modifications on the MIPL MIPv6 code [22] and additional supporting modules for deployment in a testbed. The systems were running Linux kernel 2.4.22. The supporting modules implemented on the Gateway are the Rate Limiting daemon, which listens for signals from MN and provides rate limiting based on the received parameters, and the Router Bandwidth Monitoring application, which monitors all incoming traffic and records bandwidth utilization for previous interval. The Filtering daemon on the HA listens for signals from MN and filters packets with old correspondent-victim address pair. MN runs the Host Bandwidth Monitoring and TRAPS activation application, which monitors all incoming traffic, computes the bandwidth utilization, monitors the alert stages, sends TRAPS activation signal to the MIP code to trigger TRAPS, notifies gateway regarding rate limiting activation and parameter updates, and notifies HA of filtering updates, and the Test Server, which listens for data transfer from CN before, during and after TRAPS activation to test that there's no cutting off of messages. The Attack module on the Attacker system is an UDP packet generator with adjustable attack rate and configurable spoofed address. The Test Client on CN sends continuous data to MN before, during and after TRAPS activation to test that there's no cutting off of messages

Experiments were performed by setting 3 stages of resource monitoring (2 alert stages at 50 and 60kbps respectively, and the alarm stage at 80kbps) at MN. Test Server module at MN and Test Client module at CN were started to continuously carry out data transfer. The Attacker's spoofed address was set to be CN's IP address. When the attack traffic was gradually increased through each stage corresponding to those set at MN, the alert events and finally the alarm event were triggered. MN then sent rate limiting signal to the gateway and BU to the CN regarding its new IP address. The gateway started rate limiting traffic destined to MN. When CN received MN's BU, it sent a BAck to MN. After that, MN sent the filtering signal to HA to activate filtering on the CN's address, MN's HoA pair. After which, the attack traffic from the Attacker was intercepted by HA and filtered off. On the other hand, the data transfer between the Test Server and Test Client was able to continue.

5. DISCUSSIONS

Security Considerations of Protocol

Traffic redirections as used in TRAPS can pose a major security problem in the Internet if the protocol messages are not properly authenticated. Therefore, we will now consider the MIP related security issues, which are of concern to TRAPS.

In MIPv4, it is specified that each MN, FA, and HA must be able to support a mobility security association for mobile entities, indexed by their security parameter index (SPI) and IP address. Registration messages between MN and its HA must also be authenticated with an authorization-enabling extension. This prevents a malicious node from impersonating MN to redirect away its traffic or HA to intercept MNs' packets.

The MIPv4 Route Optimization Authentication extension [23] is used to authenticate the protocol messages with an SPI corresponding to the source IP address of the message and it must be used in any binding update message sent by the HA or MN to the CNs. The calculation of the authentication data is specified to be the same as in the base MIPv4. This is HMAC-MD5 [24]. A security association must be present between CN, which could be any node in the Internet, and MN/HA. It is suggested in [20] that the mobility security association at a CN could be used for all MNs served by a particular

HA. The effort of establishing such an association with a relevant HA is more easily justified than the effort of doing so with each MN.

In MIPv6, binding updates are protected by the use of IPSec extension headers [25] or the Binding Authorization Data option, which employs a binding management key established through the return routability procedure [21]. It is specified that MN and HA must use an IPSec security association to protect the integrity and authenticity of the binding management messages.

The protection of binding updates to CNs does not require the configuration of security associations or the existence of an authentication infrastructure between the MN and CNs. The return routability procedure is used to prove the authenticity of the MN by testing whether packets addressed to the two claimed addresses (i.e. HoA and CoA) are routed to the MN. MN can only pass the test if it is able to supply proof that it received the keygen tokens which CN sends to those addresses. The return routability procedure also protects CN against memory exhaustion DoS attacks as CN does not need to retain any state about individual MNs until an authentic binding update arrives.

If the gateways are not implemented with the HA functionalities to perform filtering, security associations must be set up between the MN and the gateways, which are responsible for rate-limiting. Finally, it is important to note that TRAPS presents no additional security vulnerability to the MIP protocols.

Random Hit

We mentioned that Class B attacks are singled out by TRAPS for notification of the latest reconfiguration information. As they could not be "reached", they could be easily identified as attack traffic flows and would then be filtered off. However, what if there happens to be a random hit (e.g. randomly generated spoofed addresses by attackers within an address range resulting in an address belonging to one of the attackers)? In this case, that particular attacker would be notified of the latest information and continue attack on the victim using randomly spoofed addresses. However, in this second round of attack, the traffic volume will be lower and distributed across the spoofed address range (and will be rate-limited instead), since the other attackers were "stopped" in the first round. A solution to strengthen the scheme and lower the chances of this happening (recommended in Section 3), is by performing regular dynamic reconfigurations and updates.

Spying by Class A attackers

It was mentioned in Section 2 that there's a possibility that Class A attackers might be used as spies to obtain protected host/network's latest configuration information to support attacks in the other 3 classes. However, even with this information, the other forms of attacks would not be successful as prevention from filtering not only acts on knowledge of this information but also matching correspondent's address. In any case, a solution could be in place to catch the spy. The victim could have multiple sets of configuration information (e.g. multiple addresses in the Virtual Relocation Approach) and provide each set of correspondents with different configuration information. If exploitation of a particular set of configuration information is detected, we would know that a spy is within this set of correspondents. We could narrow down to the exact correspondent by performing iterations of this procedure.

6.   COMPARISONS WITH RELATED WORK

Traceback mechanisms [4-12] have been proposed to trace the true source of the DDoS attackers, as attack packets are often sent with spoofed IP addresses. In traceback, the attack path or graph is constructed to provide information on the route/s the attack packets have taken to arrive at the victim. It is an attacker identification tool which requires further deployment of a detection and mitigation tool to counter DDoS attacks.

Pushback [15] is a rate limiting mechanism which imposes a rate limit on data streams characterized as "malicious". It involves a local mechanism for detecting and controlling high bandwidth aggregate traffic at a single router by rate limiting the incoming traffic, and a co-operative pushback mechanism in which the router can ask upstream routers to control the aggregate. However, all high bandwidth traffic, whether good or bad, will be subjected to this rate limiting. Filtering mechanisms [16,17] on the other hand, filter out attack stream completely. This is used when the data stream is reliably detected as malicious; else, it may run the risk of accidentally denying service to legitimate traffic.

Mechanisms such as traceback, rate limiting, and filtering need to be triggered by a third-party detection tool. The way the detection tools detect an attack is therefore very important to determine how reliable it is and which of the above-mentioned mechanisms is to be used. Detections are classified in two main categories, which are "Anomaly Detection" and "Misuse Detection" [26]. Anomaly detection techniques assume that a "normal activity profile" could be established for a system. Activities not matching the profile would be considered as intrusions. However, an action which is not intrusive but not recorded formerly in the profile would then be treated as an attack, resulting in false positive. Filtering would then result in DoS by the defense system itself. In situations whereby intrusive activities, which are not anomalous, occur, it would result in attacks not detected and therefore false negatives. Such scenarios are possible if DDoS attacks are launched by flooding the victim with legitimate service requests. In misuse detection schemes, the attacks are represented in the form of a pattern or signature so that even variations of the same attack can be detected. However, they can only detect known attacks. For new attacks whereby the characteristics of the attack packets and pattern are unknown, they would of little use. They are also unable to detect attacks that are launched by flooding of legitimate packets. The advantage of TRAPS over these mechanisms is that it does not require prior traffic characterizations.

A preventive measure to DDoS attacks is to ensure the authenticity of packets by eliminating source address spoofing. Ingress filtering [13] filters packets with spoofed source addresses at the first router encountered on entering the Internet. This router typically has information about valid source addresses that are allowed to pass through it. However, enforcement on supporting ingress filtering on all outbound routers to the Internet is difficult. Source Address Validity Enforcement (SAVE) [14] messages propagate valid source address information from the source to all destinations, for en-route routers to build an incoming table that associates each incoming interface of the router with a set of valid source address blocks. Packets with invalid source addresses are identified as attack packets. Widespread deployment is required for this scheme to be effective.

Reconfiguration mechanisms change the topology of the victim or the intermediate network to add resources or isolate attack machines. The Secure Overlay Services (SOS) [18] architecture is constructed using a combination of secure overlay tunneling, routing via consistent hashing, and filtering. The overlay network's entry points perform authentication verification and allow only legitimate traffic. The route taken by the traffic is computed to be designated beacons and then servlets, both of which are kept secret from the correspondents. Potential targets are protected by filtering which only allow traffic forwarded by the chosen secret servlets. Randomness and anonymity is in this way introduced into the architecture, making it difficult for an attacker to target nodes along the path to a specific SOS-protected destination. The XenoService [19] is a distributed network of web hosts that respond to an attack on a web site by replicating it rapidly and widely. It can then quickly acquire more network connectivity to absorb a packet flood and continue providing services.

TRAPS belongs to the category of reconfiguration mechanisms by changing the routes to the victim under attack. However, unlike SOS, an overlay network and complex algorithms (e.g. Chord routing algorithm, consistent hashing) need not be implemented. In SOS, only certain destinations are chosen for protection. These destinations are protected by filtering to only allow traffic forwarded by selected servlets. However, beacons and servlets could be subjected to attacks instead. It is recommended in [18] to have a large number of beacons and servlets to provide redundancy. Nodes overwhelmed by the attacks would then be "removed" and their jobs will be handled by the remaining active ones. In TRAPS, any node running the MN module would be able to bring itself under protection in the event of attacks. Redundancy by providing additional resources is also not required in TRAPS, unlike XenoService.

7. CONCLUSIONS

This paper proposes TRAPS, an adaptive real-time DDoS mitigation scheme. In TRAPS, the victim under attack verifies the authenticity of the source by performing adaptive reconfigurations, either host or network based, and requesting senders of high-bandwidth traffic streams to send subsequent data based on the victim's latest configuration. If the source is illegitimate, it would not be updated with this information. This traffic can be easily identified as attacks, with absolute confidence and be dropped. Suspicious traffic for the victim will be rate limited as most good traffic will have been redirected, leaving mainly attack packets with randomly generated IP addresses.

The basic mechanisms of TRAPS, and various approaches (i.e. En-route Routers Nomination, Passcode and Virtual Relocation) of achieving the TRAPS reconfiguration for redirection were explained in detail. We discussed and evaluated the various approaches, and concluded that the end-host based approach, Virtual Relocation, is comparatively more efficient (e.g. requires least processing at gateways/routers/victim and overhead), and requires the least deployment effort among the proposed approaches. We suggested incorporating this approach with the MIP protocol to avoid proposing new protocols for Internet-wide deployment. Implementation of TRAPS was carried out and deployed in a testbed environment. It was observed that the operations of each module were functioning correctly and TRAPS was able to successfully mitigate an attack launched with spoofed source IP address. The security considerations with regards to MIP are discussed and we showed that TRAPS does not introduce any additional security vulnerability. Other possible scenarios of random hit and spying were also discussed with possible solutions proposed.

Related work on the existing DDoS detection, tracking and mitigation techniques is presented. Comparison of some of their important features with TRAPS is carried out. Advantages of TRAPS over existing DDoS mechanisms are: it does not require prior traffic flow characterizations and allows for a quick real-time response even in the event whereby DDoS attacks constitute brute-force flooding of victim with legitimate service requests; no need for additional resource allocation for providing redundancy; QoS is maintained for good high bandwidth traffic; very suitable for both high-end powerful systems and embedded systems as it is simple to implement and does not require sophisticated algorithms.

ACKNOWLEDGEMENTS

We gratefully acknowledge the support from the Institute for Infocomm Research and the EU funded Diadem Distributed Firewall FP6 IST-2002-002154. We would also like to thank Dr. Robert Deng for the valuable suggestions on the paper.

REFERENCES

1. K. J. Houle, G. M. Weaver, "Trends in Denial of Service Attack Technology", CERT Coordination Center, Oct. 2001
2. L. Garber, "Denial-of-Service attacks rip the Internet", IEEE Computer, Vol. 33, No. 4, pp. 12-17, Apr. 2000
3. David Moore, Geoffrey M. Voelker, Stefan Savage, "Inferring Internet Denial-of-Service Activity", Usenix Security Symposium, Aug. 2001
4. Alex C. Snoeren et al, "Hash-Based IP Traceback", ACM Sigcomm 2001, Aug. 2001
5. Stefan Savage et al, "Practical network support for IP traceback", ACM Sigcomm 2000
6. Dawn Song, Adrian Perrig, "Advanced and authenticated marking scheme for IP traceback", IEEE Infocom 2001
7. K. Park, H. Lee, "On the Effectiveness of Probabilistic Packet Marking for IP Traceback under Denial of Service Attack", IEEE Infocom 2001
8. Steve Bellovin et al, "ICMP Traceback Messages", IETF Internet Draft, Version 4, Feb. 2003 (Work in progress)
9. Allison Mankin et al, "On Design and Evaluation of "Intention-Driven" ICMP Traceback", IEEE International Conference on Computer Communication and Networks, Oct. 2001
10. Henry C. J. Lee, Vrizlynn L. L. Thing, Yi Xu, Miao Ma, "ICMP Traceback with Cumulative Path, An Efficient Solution for IP Traceback, International Conference on Information and Communications Security, Oct. 2003
11. Abraham Yaar, Adrian Perrig, Dawn Song, "Pi: A Path Identification Mechanism to Defend against DDoS Attacks", IEEE Symposium on Security and Privacy, May 2003
12. D. Dean, M. Franklin, A. Stubblefield, "An algebraic approach to IP Traceback", Network and Distributed System Security Symposium, Feb. 2001
13. P. Ferguson, D. Senie, "Network Ingress Filtering: Defeating Denial of Service Attacks which employ IP Source Address Spoofing", BCP 38, RFC 2827, May 2000
14. Jun Li et al, "SAVE: Source address validity enforcement protocol", IEEE Infocom 2002
15. Ratul Mahajan et al, "Controlling High Bandwidth Aggregates in the Network", ACM Sigcomm 2002
16. T. Darmohray, R. Oliver, "Hot spares for DDoS attacks", http://www.usenix.org/publications/login/2000-7/apropos.html
17. Mazu Enforcer, http://www.mazunetworks.com

18. A. D. Keromytis, V. Misra, D. Rubenstein, "SOS: Secure Overlay Services", ACM Sigcomm 2002
19. J. Yan, S. Early, R. Anderson, "The XenoService - A Distributed Defeat for Distributed Denial of Service", Information Survivability Workshop 2000, Oct. 2000
20. C. Perkins, "IP Mobility Support for IPv4", IETF RFC 3344, Aug. 2002
21. D. Johnson, C. Perkins, J. Arkko, "Mobility Support in IPv6", IETF RFC 3775, June 2004
22. MIPL Mobile IPv6 for Linux, http://www.mipl.mediapoli.com
23. C. Perkins, D. B. Johnson, "Route Optimization in Mobile IP", IETF Internet Draft, Version 9, Feb. 2000 (Work in progress)
24. H. Krawczyk, M. Bellare, R. Canetti, "HMAC: Keyed-Hashing for Message Authentication", RFC 2104, Feb. 1997
25. S. Kent, R. Atkinson, "Security Architecture for the Internet Protocol", RFC 2401, Nov. 1998
26. Aurobindo Sundaram, "An Introduction to Intrusion Detection", ACM Crossroads, Vol. 2, Issue 4, pp. 3-7, Apr. 1996

STATISTICAL SIGNATURES FOR EARLY DETECTION OF FLOODING DENIAL-OF-SERVICE ATTACKS

John Haggerty[1], Qi Shi[1] and Madjid Merabti[1]
[1]Liverpool John Moores University, School of Computing & Mathematical Sciences, Byrom Street, Liverpool, L3 3AF. E-mail: {J.Haggerty, Q.Shi, M.Merabti}@livjm.ac.uk

Abstract: A major threat to the information economy is denial-of-service attacks. Despite the widespread deployment of perimeter model countermeasures these attacks are highly prevalent. Therefore a new approach is posited; early detection. This paper posits an approach that utilises statistical signatures at the router to provide early detection of flooding denial-of-service attacks. The advantages of the approach presented in this paper are threefold: analysing fewer packets reduces computational load on the defence mechanism; no state information is required about the systems under protection; and alerts may span many attack packets. Thus, the defence mechanism may be placed within the routing infrastructure to prevent malicious packets from reaching their intended victim in the first place. This paper presents an overview of the early detection-enabled router algorithm and case study results.

Keywords: network attacks; denial of service; statistical signatures; early detection.

1.   INTRODUCTION

The flow of information is the most valuable commodity for organisations and users alike. Information is traded within the networked world and we are becoming ever more reliant on access to data and resources as technologies develop to facilitate this flow. Our reliance on such network technologies has ensured that financially unquantifiable assets, such as people, reputation, and business relations, are amongst the most important to business [1].

A major threat posed to this information economy paradigm is that of denial-of-service attacks. These attacks present a very real threat as they disrupt or interrupt the flow of data that organisations rely on. Such attacks can be launched in a number of ways, from malicious use of common applications such as e-mail, to subverting Internet protocols. The subversion of Internet protocols leads to flooding attacks, whereby large volumes of data are sent to the victim. Denial-of-service attacks may also be a side-effect of other types of attack, such as Internet worms. Irrespective of the *modus operandi*, denial-of-service attacks are prevalent because the tools required are freely available on the Internet, simple to launch, effective, and difficult to prevent. Thus, large numbers of attacks are continuously being launched [2]. In addition, businesses that rely on their connectivity, such as on-line services, can be blackmailed with the threat of a denial-of-service attack if they were not to pay [3].

Yet, despite the prevalence of these attacks, a cost-effective and efficient countermeasure has yet to be proposed. Current defences rely on the perimeter model of network security, where a boundary is established around the nodes under protection. Inside the perimeter is trusted space, whilst outside is viewed as untrustworthy. Denial-of-service attacks remain a significant problem due to the unsuitability of perimeter devices for two reasons. First, perimeter devices are located on the victim system and are therefore under attack at the point of detection. Second, these devices inspect each and every packet in an attack, which in the case of denial of service places a large computational load on the defence mechanism in addition to the large network load caused by the attack.

This paper demonstrates that the use of statistical signatures for early detection of denial-of-service attacks can greatly reduce the volume of packets that are inspected to determine malicious packets from legitimate. The approach employed by this paper has three novel contributions. First, the computational load is reduced on the defence mechanism as fewer packets are analysed. Second, no state information about the networks under protection needs to be held, again reducing computational load. Third, reports of attacks may relate to several packets rather than 'one packet, one alert' techniques employed by traditional countermeasures utilising non-statistical signatures. The reduction of volume enables detection devices to be placed beyond the perimeter and within the routing infrastructure thus enabling attacks to be thwarted prior to their reaching their intended target. As demonstrated in section 4, the approach posited in this paper remains highly efficient despite a significantly reduced number of packets inspected.

This paper is organised as follows. Section 2 discusses related work. Section 3 presents an overview of statistical signatures and the effects that a denial-of-service attack has on a network that is used for early detection. Section 4 presents a case study and results. Finally, section 5 presents conclusions and further work.

2.   RELATED WORK

Flooding denial-of-service attacks are distinct from other attacks, for example, those that execute malicious code on their victim, in that they require a large volume of traffic, and it is this continuing stream of data that prevents the victim from providing services to legitimate users. It is the mass of all packets directed at the victim that poses the threat, rather than the contents of the packets themselves.

Flooding denial-of-service attacks are problematic due to their subversion of normal network protocols. As such, it is these attacks that pose the greatest problem in today's network infrastructures. Subverting the use of protocols, such as the Transmission Control Protocol (TCP) or User Datagram Protocol (UDP), enables the attacker to disrupt on line services by generating a traffic overload to block links or cause routers near the victim to crash [4]. Because they subvert existing protocols the packets involved in these attacks are high-volume without being conspicuous or easily traceable. For example, TCP SYN flooding specifically targets weaknesses in the TCP protocol to achieve its aim. This attack method, which accounts for 94 per cent of denial-of-service attacks [2], is based on exploiting the three-way handshake in TCP.

A number of approaches have been posited to counter the denial-of-service problem. For example, [5] proposes stronger authentication between communicating parties across a network. Alternatively, [6] suggests that network resources should be divided into classes of service, where higher prices would attract less traffic and ensure that an attacker could not afford to launch an attack. Alternatively, [7] suggests that the routing infrastructure should be more robust by securing servers in the first place. These approaches are not without their problems. For example, authentication, whilst attempting to prevent denial of service, paradoxically leaves itself open to such an attack due to the computational load required for the defence. Payment approaches assume that the consumer is willing to pay for different levels of service, which they are usually not. Finally, it is often poor software development practices due to the pressure of getting a product to market that lead to the release of server applications that are subvertable.

These problems have led to the rise of traffic monitoring approaches and fall into two categories; *statistical monitoring* and *adaptation of congestion algorithms*. Statistical monitoring of networks, such as [8, 9], observes a network and detect upsurges in traffic of a particular type or for system compromise. An advantage of this approach is that one alert may cover a number of attack packets, thus reducing network load caused by the reporting of events. In addition, a large upsurge of traffic is indicative of a flooding attack, irrespective of the protocol used by the attacker. Alternatively, congestion algorithms are adapted for detection of denial-of-service attacks. Approaches such as [10, 11], use existing congestion techniques, where routers deal with upsurges in traffic to ensure quality of service, to detect denial of service. These approaches have the benefit of being able to detect the attack in the routing infrastructure, thus being able to halt the attack before it reaches its intended victim.

However, even these more sophisticated approaches are not without their problems. Statistical approaches require human intervention to monitor the networks for upsurges, so is both labour intensive and inefficient. The congestion adaptation approaches may only apply simplistic signatures so as to not impede on the throughput of traffic. In addition, approaches such as [10, 11] require that state information is held on the router. This information is computationally too exhaustive to be effective within the routing infrastructure.

Therefore, a new approach is required that provides early detection of denial-of-service attacks; one that can combine and make use of the advantages of both the statistical and adaptation of congestion algorithm approaches. In this way, the benefits as above are achieved.

3.   STATISTICAL SIGNATURES

Traditional stateful signature analysis applies statistical methods to collected data within the system over a period of time. This data is then analysed to generate some system-specific values: for example, traffic thresholds or user profiles to define normal or abnormal behaviour [12]. By allowing a system to keep state information of the system, detection signatures can be designed to match a complex series of events which fall outside that normal behaviour. A number of techniques are employed in this area and include:

- *Collection of events*. In any system, a number of events may be observed in conjunction to indicate that an attack is under way.
- *Threshold enforcement*. A certain threshold of acceptable events is determined for the system based on prior experience or threats. Once events in the system surpass this threshold, an alert is generated to indicate that an attack is under way.
- *Frequency threshold*. This is a variation on threshold enforcement and is widely used in authentication. If one or more events are

observed, then an alert is raised or services halted until a time limit is reached.

Other approaches that fall into this category include analysis of mean and standard deviation information, the multivariate model, Markov process model, and clustering analysis [12].

These approaches are widely used in anomaly intrusion detection where misuse against known but ill-defined variables is being matched. Despite the requirement for state information to be held by these approaches, statistical monitoring is effective in detecting large volumes of traffic being directed at a victim host. The way in which this is achieved statelessly is presented in section 4.

These approaches require state information to be held about the systems under protection but this is too computationally exhaustive to be used in the routing infrastructure.

The effects on network dynamics of a denial-of-service attack and the applicability of a statistical-based approach can be clearly demonstrated through the use of a probability plot. A probability plot assesses whether a particular distribution fits the given data. The plot points are calculated using a non-parametric method. The fitted line provides a graphical representation of the percentiles. The fitted line is created by calculating the percentiles for the various percents based on the chosen distribution. The associated probabilities are transformed and used as the y variables. The percentiles may be transformed and then used as the x variables. A goodness of fit measure, such as the Anderson-Darling statistic [13], is then applied to the data. This is a measure of how far the plot points fall from the fitted line in the probability plot. The statistic is a weighted squared distance from the plot points to the fitted line with larger weights in the tails of the distribution. A smaller Anderson-Darling "Goodness of Fit" indicates that the distribution fits the data better.

To demonstrate the effect of a denial-of-service attack on the data, attacks are calculated and plotted according to this technique. Figures 1 to 4 compare the results for control traffic, *nuke* which utilises only a small number of packets during an attack, a SYN flood attack, and a UDP flood.

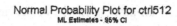
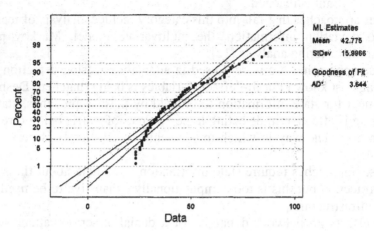

Figure 1. Probability plot for control traffic.

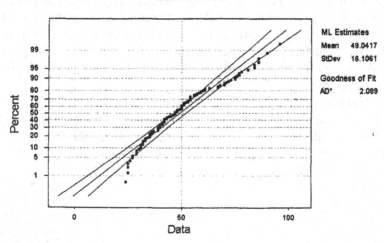

Figure 2. Probability plot for nuke attack traffic.

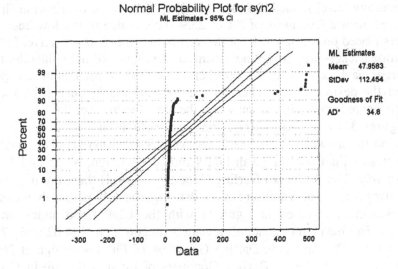

Figure 3. Probability plot for TCP SYN flood attack traffic.

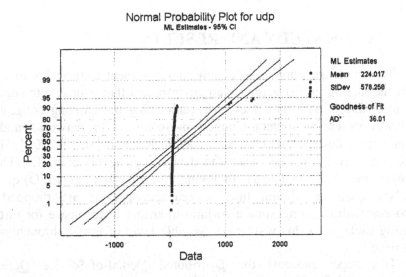

Figure 4. Probability plot for UDP flood attack traffic.

The Anderson-Darling Goodness-of-Fit statistics provides the maximum likelihood and least squares estimation. The control traffic in figure 1 has a low Anderson-Darling Goodness of Fit of 3.64, suggesting that the distribution fits the data well, i.e. an attack is not underway. In vulnerability attacks, such as nuke in figure 2, again the distribution fits the data well with a Goodness of Fit of 2.09. This is due to the few additional packets placed on the network for this attack to achieve its objective. In total, only nine packets are sent to the victim to exploit the kernel vulnerability on the target machine. A further six packets are sent to ensure that the attack has caused the desired effect. Therefore, during an attack, only 15 malicious packets are required. This accounts for the low Goodness of Fit.

Figures 3 and 4 demonstrate that high-volume attacks have a severe effect on the network. In figure 3, the TCP SYN flood attack generates a much higher than control mean and standard deviation, 47.95 and 112.45 respectively. The Anderson-Darling Goodness of Fit is therefore high, 34.8, suggesting that the distribution fits the data significantly poorer than control. This pattern is repeated in figure 4 with the UDP flood attack on the network. The mean and standard deviation remain high, 224.02 and 578.26 respectively. The Anderson-Darling Goodness of Fit is also high at 36. As we can see, the Anderson-Darling Goodness of Fit statistics are ineffective for the detection of attacks that utilise only a small number of packets, such as nuke, but are effective for flooding denial-of-service attacks such TCP SYN flood or UDP flood.

4.   CASE STUDY AND RESULTS

The challenge remains as to how to implement statistical signatures within the routing infrastructure in an environment that is unable to support state information. This is achieved by enhancing the existing congestion algorithms present on routers. Congestion occurs within networks, and so routers employ congestion algorithms, such as RED [14] or CHOKe [15], to ensure that they do not fail when faced with high levels of traffic. These algorithms may be as simple as employing a *first in, first out* (FIFO) queue. Once the queue maximum limit is reached, packets are dropped in accordance with the congestion algorithm to ensure queue space for further incoming packets. In this way, an acceptable level of traffic throughput is maintained.

This paper presents the Distributed Denial-of-Service Defence Mechanism (DiDDeM) architecture [16] for early detection of denial-of-service attacks. DiDDeM is a domain-based system that adapts congestion algorithms within the routing infrastructure. The DiDDeM system comprises a server liaising with a number of DiDDeM-enhanced routers that pre-filter attack traffic outside the traditional perimeter.

Congestion algorithms are adapted for statistical signature matching by detecting large traffic volumes associated with a flooding denial-of-service attack. Rather than purely dropping packets when the router threshold is met, packets to be dropped from the queue are inspected. This enables inference of stateful information about traffic flows and whether these unusual flows are intended for a particular destination thereby suggesting an attack. It is the random inspection that allows the state inference. If two (or more) sampled dropped packets are heading to one destination, they are checked against other (stateless) signatures to confirm an attack.

To demonstrate the way in which this is achieved in the DiDDeM architecture, a *first-in, first-out* (FIFO) queue is used within a *ns2* prototype. The available space within the DiDDeM-enhanced FIFO queue is divided into two sub-queues to allow comparison of packets. If due to bandwidth restrictions the packet cannot be immediately forwarded to the next router, an incoming packet to the router is placed in a queue. These packets are placed in either the first or second sub-queue at the router based on a first-come, first-served basis.

Packets placed in the queue, and its sub-queues, are dequeued and forwarded to their destination. If the threshold of the total queue limit is exceeded the router begins to drop packets to ensure that packets already in the queue are forwarded and that new incoming packets can be placed in the queue. In this way, no stateful information is held about the queue apart from whether the queue limit has been exceeded, thereby reducing the computational overhead placed on the router.

At periods where congestion occurs, packets are dropped. By meeting the threshold of the particular router which invokes packet dropping, an upsurge in traffic can be inferred. However, this may or may not be due to large amounts of traffic, such as would occur during a flooding denial-of-service attack. Therefore, prior to packets being dropped, the IP header is accessed and the destination address obtained. This IP destination address is compared to the previous packet's IP destination address. If they are the same, then the IP destination address is stored for comparison with the next packet and the packet is passed for stateless signature analysis. Stateless signature analysis verifies attacks by applying techniques used in misuse detection whereby fixed byte sequences within the packet are inspected. If the destination addresses are not the same, the destination IP address is still stored for comparison with the next packet, but the packet is dropped. The algorithm is illustrated in figure 5.

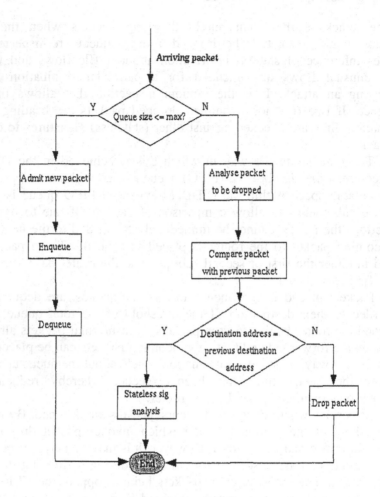

Figure 5. Routing algorithm using a FIFO queue.

A simulation was conducted in which approximately 19,500 attack packets were directed at the victim node by two attacking nodes. This represented an attack consisting of approximately 1,000 packets per second. Once the congestion algorithm was invoked by the router, 798 attack and legitimate packets were to be dropped. Of this number, 742 packets were actual attack packets whilst the remainder were legitimate traffic. Therefore, out of a total of 19,500 attack packets, only 4 per cent of this volume was inspected. DiDDeM detected 697 packets using the algorithm in figure 5. The 697 packets detected out of the 742 inspected by the DiDDeM-enhanced router ensure a 94 per cent detection rate. This system is therefore extremely efficient.

The prototype is tested on two systems to measure the impact of DiDDeM detection on the router. The methodology used is to measure the impact of the simulation on the processor by using the *top* program. This program measures the load by applications on the processor in the UNIX/Linux operating system. To provide a comparison with current network standards, the memory and processor usage were tested for DiDDeM and two routing algorithms; DropTail, and RED. This comparison allows us to see the impact on router efficiency in implementing DiDDeM and is presented in table 1.

Table 1. Impact on memory and processor of DiDDeM versus RED and DropTail algorithms.

Algorithm	PII 400 MHz 128 Mb RAM		AMD 2 GHz 256 Mb RAM	
	Memory usage	Processor load	Memory usage	Processor load
DiDDeM	*4.70%*	*95.50%*	*2.30%*	*41.10%*
RED	*4.70%*	*96.70%*	*2.30%*	*60.00%*
DropTail	*4.60%*	*96.20%*	*2.20%*	*51.80%*

As demonstrated by table 1, the impact of the simulation routing algorithm affected the memory usage of both computers. The DropTail routing algorithm required less memory usage, an improvement of 2.13% (PII processor) and 4.34% (AMD Athlon) compared to both DiDDeM and RED. However, DiDDeM was actively detecting denial-of-service attacks whilst the RED algorithm merely detected congestion at the router. In terms of processor load, DiDDeM proved to be more efficient.

One key measure for DiDDeM is its performance within the network environment. In particular, the DiDDeM algorithm should not have an adverse affect on the network, which would require a trade off between usability and effectiveness. In order to measure the performance of the DiDDeM in the network it is compared to the routing algorithms above; RED and DropTail. Unlike RED, DiDDeM and DropTail do not require any information about the state of the queue. However, to test the impact of DiDDeM on the queue and the network, the number of datagrams and packets passed from a router within the attack domain to the second router is measured. This impact is illustrated in figures 6 to 8 showing DiDDeM, DropTail and RED.

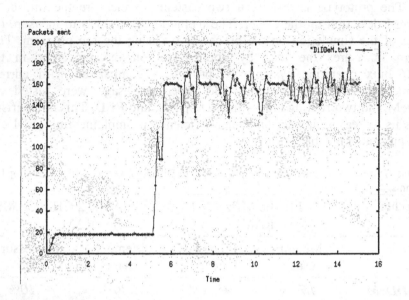

Figure 6. Packets sent by the DiDDeM-enabled router.

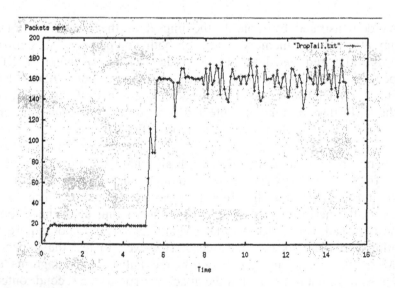

Figure 7. Packets sent by the DropTail-enabled router.

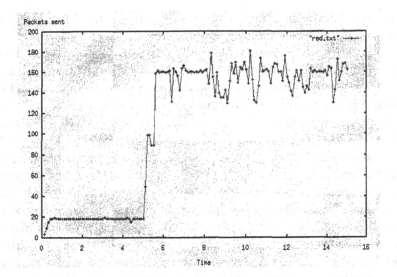

Figure 8. Packets sent by the RED-enabled router.

Figures 6 to 8 show the number of packets sent by router 1 at 0.1 second intervals. The attack is launched just before 5 seconds, as indicated by the sharp increase of traffic. Prior to the attack, all three approaches steadily send the packets comparably, although one slight dip in the number of packets sent is seen in RED at just after 4 seconds. Although all three approaches maintain a queue, they are also able to comparably send packets. In fact, there is very little difference in the number of packets sent between the two routers used in the case study. However, whilst in DropTail and RED, those packets not sent are dropped, DiDDeM is comparing these packets for early detection of denial-of-service attacks. As these figures demonstrate, DiDDeM has little effect on the number of packets sent, and therefore network performance.

5.   CONCLUSIONS AND FURTHER WORK

The flow of information is the most valuable commodity for organisations and users alike, and denial-of-service attacks pose a great threat to this flow. These attacks are highly prevalent despite the widespread deployment of network security countermeasures such as firewalls and intrusion detection systems. These countermeasures find denial of service extremely problematic, therefore a number of other approaches have previously been proposed to counter the problem. However, these approaches are not without their own problems, thus a new approach of early detection of denial-of-service attacks is required.

The paper demonstrates that statistical measures may be used to provide signatures of denial-of-service attacks. However, traditional uses of these techniques are labour intensive and require state information, which is computationally exhaustive within the routing infrastructure. Therefore, this paper has demonstrated the way in which a routing congestion algorithm may be adapted to provide statistical signatures statelessly. The results from our case study in *ns2* demonstrate the applicability of this approach in that it is highly effective in the detection of attacks without impeding on network traffic. This approach provides three benefits: computational load is reduced on the defence mechanism, even during an attack; no state information is required, again reducing computational load; and alerts to attacks may span several packets, thus reducing network load during an attack. In this way countermeasures can be placed beyond the perimeter and within the routing infrastructure to prevent attacks from reaching their intended victim.

Future work concentrates on the application of the system to other attacks that require or generate a high volume of traffic such as worms and malicious mobile code.

REFERENCES

1. Department of Trade and Industry / Price Waterhouse Coopers, "Information Security Breaches Survey 2004," Technical Report, http://www.dti.gov/industries/information-security (2003).
2. Moore, D., Voelker, G.M. & Savage, S., "Inferring Internet Denial-of-Service Activity," in *Proceedings of 10th Usenix Security Symposium*, Washington, DC (2001).
3. Anonymous, "Russian blackmailers arrested: Peace at last for online bookmakers," Computer Fraud and Security, vol. 2004, no. 8, p.1 (2003).
4. Gil, T.M. & Poletto, M., "MULTOPS: a data-structure for bandwidth attack detection," in *Proceedings of USENIX Security Symposium*, Washington, DC, USA (2001).
5. Meadows, C., "A cost-based framework for analysis of denial of service in networks," *Journal of Computer Security*, vol. 9, pp. 143-164 (2001).
6. Brustoloni, J.C., "Protecting Electronic Commerce from Distributed Denial-of-Service Attacks," in *Proceedings of WWW2002*, Honolulu, Hawaii, USA (2001).
7. Papadimitratos, P. & Haas, Z.J., "Securing the Internet Routing Infrastructure," *IEEE Communications Magazine*, vol. 40, pp. 76-82 (2002).
8. Shan, Z., Chen, P., Xu, Y. & Xu, K., "A Network State Based Intrusion Detection Model," in *Proceedings of Computer Networks and Mobile Computing 2001 (ICCNMC)*, Beijing, China (2001).

9. Sterne, D., Djahandari, K., Balupari, R., La Cholter, W., Babson, B., Wilson, B., Narasimhan, P. & Purtell, A., "Active Network Based DDoS Defense," in *Proceedings of DARPA Active Networks Conference and Exposition (DANCE '02)*, San Fransisco, CA, USA (2002).

10. Ioannidis, J. & Bellovin, S.M., "Implementing Pushback: Router-based Defense Against DDoS Attacks," in *Proceedings of Network and Distributed Systems Security Symposium*, San Diego, CA, USA (2002).

11. Kuzmanovic, A. & Knightly, E.W., "Low-Rate TCP-Targeted Denial of Service Attacks," in *Proceedings of Symposium on Communications Architecture and Protocols (SIGCOMM)*, Karlesruhe, Germany (2003).

12. Verwoerd, T. & Hunt, R., "Intrusion detection techniques and approaches," *Computer Communications*, vol. 25, pp. 1356-1365 (2002).

13. Stephens, M.A., "EDF Statistics for Goodness of Fit and Some Comparisons" in the *Journal of American Statistical Association*, vol. 69, pp. 730–737 (1974).

14. Floyd, S. & Jacobson, V., "Random Early Detection Gateways for Congestion Avoidance," *IEEE/ACM Transactions on Networking*, vol. 3, pp. 365-386 (1993).

15. Pan, R., Prabhakar, B. & Psounis, K., "CHOKe: A stateless active queue management scheme for approximating fair bandwidth allocation," in *Proceedings of IEEE INFOCOMM 2000*, Tel Aviv, Israel (2000).

16. Haggerty, J., Shi, Q. & Merabti, M., "DiDDeM: A System for Early Detection of TCP SYN Flood Attacks", in *Proceedings of Globecom 04*, Dallas, TX, USA (2004).

DESIGN, IMPLEMENTATION, AND EVALUATION OF FRITRACE

Wayne Huang, J.L. Cong, Chien-Long Wu, Fan Zhao, S. Felix Wu
University of California, Davis, California 95616

Abstract: Denial-of-Service attacks are more prevalent than ever, despite the loss of media attention after the infamous attacks that shut down major Internet portals such as Yahoo, eBay and E*Trade in February 2000. In these flood-style denial-of-service attacks, attackers send specially formatted IP packets with forged or true IP source addresses at potentially high packet rates in order to overload/waste the resources of the routers or servers they are attacking. Determining the true sources of an attack stream, depending on the DDoS attack types, is a difficult problem given the nature of the IP protocol however it is often beneficial for the victims to acquire this information in a timely manner in order to stop the attack from further denying service to legitimate users. FRiTrace (Free ICMP Traceback) is an IP traceback implementation that can provide victims of flood-style denial-of-service attacks with sufficient information to determine the true sources of an attack, despite forged IP headers and varying attack architectures. In this paper, we present our design, implementation, and evaluation about FRiTrace.

Key words: Attack source tracing; DDoS.

1. INTRODUCTION

Despite being out of the media spotlight, denial-of-service attacks are more prevalent than ever. One study, aimed at analyzing the byproduct of denial-of-service attacks, inferred using their backscatter analysis technique that 12,805 attacks occurred against over 5,000 hosts in just a three-week period (Moore et al., 2001). Due to the lack of spoofed denial-of-service counter measures, the problem of identifying the true source of an attack in the presence of IP spoofing has become even more relevant. If the victims of an attack can discover the true sources in a timely manner, this will allow administrators to coordinate their efforts in stopping the attack by filtering it directly upstream of the source and may provide sufficient information during the investigative phase to prosecute the perpetrators. This provides two benefits. First, the sooner the sources can be identified, the sooner they can be filtered or shutdown to prevent further loss of or degraded service to legitimate users. Second, accountability provides a psychological deterrent to launching further denial-of-service attacks.

In this paper, we present FriTrace, an IP traceback system using the ICMP Traceback (Bellovin, 2000) and Intention ICMP Traceback (Wu et al., 2001) mechanisms as described in their respective IETF drafts. FriTrace, through the use of specially formatted out-of-band messages, can provide the victims of an attack with sufficient information to construct an attack graph consisting of all administrative domains supporting iTrace or Intention iTrace through which attack packets traversed thus allowing the discovery of the true victims or administrative domains containing the victims. In addition, this paper will discuss various design issues with respect to FriTrace and present the results of our deployment of FriTrace in a test-bed environment.

2.   FLOODING ATTACKS

Besides all the existing DDoS attacks, we are considering a new type of statistical attack aimed at IP traceback mechanisms that probabilistically select packets for marking or the generation of an out-of-band message, such as iTrace. With probabilistic schemes, all packets seen at a given router or host are considered equally for marking or generating an iTrace message. Therefore the probability of selecting a packet that is part of an attack stream is based on the percentage of the total packet flow through the router or host that comprises attack packets. Therefore, if it is possible to artificially inflate the background "legitimate" traffic at the router or host, then its possible to decrease the probability of selecting an attack packet, thus reducing the overall effect of the IP traceback mechanism. With a slight modification to the distributed denial-of-service architecture, a statistical attack can be launched. The slaves are split into two logical groups where each slave still attacks the same victim or victims. However, in addition to the attack traffic, each slave in each group sends cover traffic to each slave in the other group.

If there are sufficient slaves, the cover traffic packet rate can be relatively small and thus less detectable. In doing this, depending on the number of slaves, the attackers can generate artificial background traffic thereby reducing the likelihood of attack packets being selected by routers closest to the slaves. In addition, this attack architecture has all of the benefits of the distributed attack including greater anonymity and the ability to generate packet rates of extremely large magnitude.

3.   IP SOURCE ACCOUNTABILITY MECHANISMS

The stateless nature of the IP protocol and lack of appropriate filtering by routers makes it an extremely difficult problem to provide source accountability during spoofed denial-of-service attacks. However, there are currently a number of different techniques that provide varying degrees of source accountability. First, source accountability can be achieved through the use of link testing or other flood-based approaches whereby the victim will flood its upstream links to determine which link is passing the bulk of the attack traffic by measuring the drop in the attack traffic rate when the link is flooded. The second class of mechanisms is logging-based, such as Hash-Based IP Traceback (Snoeren, 2001). These mechanisms require routers to store, in some compact format, information about all the packets it has forwarded for a given interval of time. The last class of source accountability mechanisms select packets probabilistically in which to mark the packet or to generate an out-of-band message describing the links from which the packet arrived at the router and which link the packet was forwarded through. These mechanisms rely on the fact that flooding denial-of-service attacks often generate large packet streams for long durations and thus cannot be used to identify the source of a single packet as with the previously mentioned logging mechanisms. Some examples of these mechanisms include the advanced and authenticated marking scheme (Song et al., 2001) based on the original probabilistic packet marking scheme (Savage et al., 2000). Two other probabilistic schemes, and the schemes implemented in FriTrace are ICMP Traceback (Bellovin, 2000) and Intention-Driven ICMP Traceback (Wu et al., 2001).

4.   INTENTION-DRIVEN ICMP TRACEBACK

ICMP Traceback (iTrace) is a mechanism where packets are selected randomly according to a probability p for generating iTrace messages. Once a packet is selected, a specially formatted out-of-band ICMP message is sent towards the destination of the selected packet. This message contains useful information such as the IP address and MAC address pair of both the upstream and downstream links of the router that selected the packet. In addition, a timestamp and a portion of the selected packet's payload are included for use by the receiver to correlate messages to received packets. Lastly, the iTrace message contains optional message authentication information to prevent a malicious user from forging iTrace messages. In the IP packet containing the iTrace message, the TTL field is always initialized to 64.

When the victim of a denial-of-service attack receives iTrace messages, they can be ordered to generate a graph from the victim to the attacker's domain including all intermediary routers that support iTrace and generated iTrace messages towards the victim. This ordering can be achieved in two ways. First, two adjacent routers will share a link, where one router's downstream link is the other router's upstream link or vice versa. By analyzing the link information in the iTrace packets, it can be inferred that two routers are adjacent if they share a common link. However, this assumes that all routers in the path support iTrace. To allow for incremental deployment of iTrace, ordering can also be deduced based on the value of TTL field in the IP header containing the iTrace message. All routers correctly following the IP protocol, regardless of whether they support iTrace, will decrement the TTL field before forwarding a packet to the next hop. After generating a graph showing all intermediary routers up to or near the attacking domain, the job of locating the actual source, usually done manually, can be expedited by eliminating the need to contact all intermediary administrative domains to determine the next upstream administrative domain passing the attack traffic.

The original ICMP traceback scheme provided sufficient information for the victims of flooding denial-of-service attacks to determine the nearest administrative domain supporting iTrace to the actual attackers, regardless of potentially spoofed IP addresses in the attack packets. In addition, authentication codes can be used to prevent malicious users from forging iTrace messages to obstruct or mislead the construction of the attack graph.

However, the original iTrace scheme is susceptible to the statistical denial-of-service attack described earlier. Specifically, iTrace will select packets randomly with probability p at any given router supporting iTrace. Therefore, the probability of selecting attack packets is simply $p(R_a)$, where R_a is the ratio of attack traffic to total traffic. By artificially increasing the total traffic without affecting the attack traffic, a malicious user can reduce R_a, and thus reduce the overall probability of iTrace selecting attack packets.

Intention-Driven ICMP Traceback (Intention iTrace) is an enhancement to iTrace where packet selection factors in an administrative domain's desire to receive iTrace messages. For each administrative domain, this desire is propagated through the use of an intention bit associated with each network prefix the administrative domain, or autonomous system (AS) advertises through BGP. Using a network intrusion detection system or even manually with an administrator, the intention bit can be dynamically set if the AS detects it is under a denial-of-service attack. BGP will automatically propagate this information to all BGP-speaking routers on the Internet for use with Intention iTrace.

The Intention iTrace consists of two different packet selection algorithms, however both algorithms achieve an average packet selection probability of p while allowing the administrator to specify the percentage p_i of selected packets to subject to the Intention iTrace criteria that an iTrace message only be generated for destinations wishing to receive iTrace messages. For all other $1 - p_i$ selected packets not subject to the Intention iTrace criteria, normal iTrace messages are generated as the normal iTrace.

The first algorithm separates traffic into two classes by analyzing all of the intention bits that are known to a particular router's routing information base (RIB). Packets with destination prefixes having an intention bit of 1 in the RIB will be considered *intention traffic* and all other packets will be considered *normal traffic* with R_{int} representing the ratio of intention traffic to total traffic. Both traffic classes will consider packets equally for normal iTrace, or a probability of $p(1 - p_i)$, however the intention traffic must also be considered for Intention iTrace. Intuitively, this probability should be $p(p_i)$, however this probability would be static. In the ideal case, if R_{int} is extremely small then the probability for generating an Intention iTrace message should be greater than $p(p_i)$, otherwise it is likely that packets from the intention traffic will not be considered. Similarly, as R_{int} approaches 1, the probability for selecting a packet to consider with Intention iTrace should be as close to $p(p_i)$ as possible or packets will no longer be considered for normal iTrace. To strike a balance between packets being considered for normal iTrace and Intention iTrace, the probability for considering a packet for Intention iTrace from the intention traffic is $p(p_i/R_{int})$.

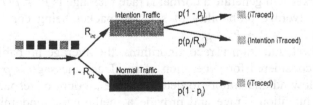

Figure 1. Intention iTrace Scheme 1 (IIS#1) Decision Process

Therefore, the total probability for an intention traffic packet to be considered for iTrace message generation is $P_i = p(1 - p_i) + p(p_i/R_{int})$ and $P_n = p(1 - p_i)$ for packets in the normal traffic class. The total overall probability for generating an iTrace message can then be expressed as $P_{avg} = R_{int} * P_i + (1 - R_{int}) * P_n = p$.

The second algorithm does not segregate traffic into traffic classes and instead, first selects a packet for either iTrace or Intention iTrace consideration with probability p as in the normal iTrace scheme. The selected packet is then sent to the appropriate decision module based on the probability pi, where with probability $p(p_i)$ the packet will be considered by the Intention iTrace decision module and with probability $p(1 - p_i)$ the packet will be considered by the normal iTrace decision module. However, within the normal iTrace decision module, the packet's destination prefix is looked up in the router's RIB and if the intention bit is set, the packet will still be considered under the Intention iTrace criteria. If the destination prefix in the RIB does not have its intention bit set, then the packet generates a normal iTrace message. Based on the ratio R_{int} of intention traffic to normal traffic, it is given that approximately R_{int} of packets sent to the normal iTrace decision module will be considered for Intention iTrace and $1 - R_{int}$ of packets will generate normal iTrace messages.

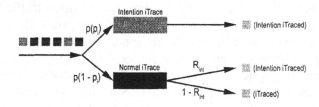

Figure 2. Intention iTrace Scheme 2 (IIS#2), Decision Process

The total probability that a selected packet will be considered using the Intention iTrace criteria is $P_i = p(p_i) + p(1 - p_i) * R_{int}$ and the probability that a selected packet will generate a normal iTrace message is $P_n = p(1 - p_i) * (1 - R_{int})$. The average probability of selected packet being considered for iTrace is thus $P_{avg} = P_i + P_n = p$.

Thus for both Intention iTrace algorithms, the average probability for a packet to be considered for generation of an iTrace message is p, but both algorithms allow administrators to determine the amount of resources to be dedicated to Intention iTrace and provide a method for randomly mixing both normal iTrace and Intention iTrace schemes. Using Intention iTrace will still allow the victim of a statistical denial-of-service attack to receive useful iTrace messages despite the artificial increase in background traffic.

In addition, Intention iTrace will still generate normal iTrace messages to destinations whom may be under a denial-of-service attack but that have not detected the attack and thus have not specified their desire to receive iTrace messages.

5.   FRITRACE DESIGN AND IMPLEMENTATION

FriTrace is our open-source implementation of an IP traceback suite that supports both iTrace and Intention iTrace as well as authentication to prevent spoofed iTrace messages, source IP address spoof detection and source iTrace. The iTrace and Intention iTrace mechanisms were originally designed to work in router platforms where knowledge of upstream and downstream links would be used in the generation of an iTrace message. As such, we developed FRiTrace as a Linux 2.4 Loadable Kernel Module (LKM) to be run on Linux-based routers to provide iTrace and Intention iTrace support. However, most administrative domains use appliance-based routers such as those made by Cisco and Juniper Networks, and therefore the FriTrace LKM would not be as useful in those domains without replacing the appliance-based routers with Linux-based routers. In order to allow such domains to support iTrace and Intention iTrace, a passive implementation of FriTrace was developed that would listen on a network in order to select packets to be iTraced instead of selecting packets while forwarding them, such as in the LKM implementation. In this paper, only the passive approach is discussed.

5.1   FriTrace Probabilistic Packet Selection

Regardless of whether normal iTrace or Intention iTrace is activated in FriTrace, both modes require the ability to probabilistically select packets given a probability p. The naïve approach to the selection method would be to maintain a counter and to select a packet every time the counter modulo $1/p$ was equal to $1/p - 1$. While this maintains an average probability of p to select a packet, it is not secure to specially timed attacks. By selecting packets statically using this method, a crafty attacker can setup a flooding denial-of-service attack where the attack packets are sent periodically with the same frequency. If timed correctly, this attack can exploit the static packet selection technique to avoid selection of packets from the attack stream.

To counter this timed attack, packet selection must occur randomly with probability p. This can be implemented by computing a random number for each packet and selecting the packet if the generated random number is less than p. Assuming a uniformly distributed pseudo random number generator (PRNG), attack packets will be selected at random with a probability p. However this approach suffers in that a random number must be generated for each packet, thus causing per-packet overhead, which should be avoided in order not to affect packet-processing rates.

In FriTrace, random packet selection is achieved by pre-computing a series of n random numbers using a uniformly distributed PRNG for integer values between 0 and $n(1/p) - 1$, inclusive. The window of n random numbers is then sorted. For each packet that is read, a counter is incremented. When the counter value equals a value from the random number window, a packet is selected. When the counter value reaches $n(1/p)$ it is reset to zero and a new window of random numbers is computed. This approach allows packets to be selected at random with a probability p, without incurring the overhead of computing a random number for each packet.

5.2   Source iTrace

IP traceback mechanisms have primarily focused on allowing the victims of denial-of-service attacks to discover the true sources of their attackers. However, depending on the specific attack used, there may be other administrative domains adversely affected. Specifically, the use of most spoofed denial-of-service attacks causes the victim to generate responses towards the IP addresses that were spoofed. Therefore, if iTrace messages are also generated with an independent probability p towards the source of a packet, this would allow administrative domains to discover that their address space is being used by a malicious user in a spoofed denial-of-service attack.

Another distributed denial-of-service attack architecture involves the use of *reflectors*, or innocent hosts on the Internet used to reflect attack packets towards a victim. Instead of the slaves generating the attack traffic directly towards the victim, they instead reflect their attacks off of random hosts on the Internet by using the functionality of the TCP and UDP transport layer protocols. For example, if an attacker sends a SYN packet to a reflector with a forged source address of the victim, any response generated by the reflector will be sent to the victim. With the use of iTrace or Intention iTrace the victim would simply trace back to the reflectors used in the attack, which is not especially useful to the victim. However, with the use of Source iTrace, the packet stream from the slaves to the reflectors will be iTraced with the iTrace messages being sent towards the victim. This information can then be used to discover the slaves in a reflector-based denial-of-service attack.

FriTrace supports the Source iTrace feature and maintains an independent probability to generate Source iTrace messages.

5.3   The Passive Approach and its Implementation

The passive implementation of FriTrace is run entirely in user space as a single process. There are three main components within the passive FriTrace implementation.

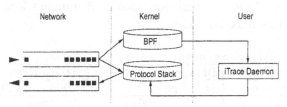

Figure 3. FriTrace, Passive Architecture

The first is the network stack interface using *libpcap* that allows FriTrace passive access to all packets on a network segment using the BSD Packet Filter (BPF) (McCanne et al., 1993) or packet socket functionality. The *libpcap* interface then calls the iTrace decision module for each packet that is observed on the network, similar to the functionality of Netfilter in the active implementation. The iTrace decision module then applies the appropriate packet selection algorithm and if a packet is selected to be iTraced, it will be sent to the iTrace generation component prior to returning control to the *libpcap* interface. The second component, or iTrace generation component, performs the logging of packet information to syslog and generates iTrace messages. Lastly, the third component validates and responds to authentication requests.

It is important to note that because packets are read from the network passively, there is no accompanying information that passive FriTrace can use to truly identify either an upstream or downstream IP link. However, packets intercepted through *libpcap* retain their link layer information, which provides passive FriTrace with information that can be used to generate a MAC address upstream link. Since passive FriTrace will be operating on a network segment, the receiver of an iTrace message can determine the administrative domain from which the message was sent based on the IP address of the host running passive FriTrace. Once the offending administrative domain is located, the MAC address pair can be used to locate individual machines from within that domain.

6.   FRITRACE EVALUATION

The experimental setup consisted of four hosts, a 100 Mbps hub and a router. The *Attacker, Background* and *FRiTrace* hosts were Pentium III 700 MHz servers with 512 MB of memory and a 100 Mbps network interface card, running Red Hat Linux release 7.0 with the Linux 2.2.16 kernel. All three hosts are connected to a 100 Mbps hub, which is in turn connected to a router. The *Victim* host is a dual Pentium III 1000 MHz with 512 MB of memory and a 100 Mbps network interface card, running Red Hat Linux release 7.2 with the Linux 2.4.7 kernel.

Table.1 – Experiment Setup, Experiment Parameters

#	Mode	p_i	Background	Attack Rates
1	Normal	n/a	16,000 pps	1,000 – 16,000 pps
2	Normal	n/a	2,000 – 16,000 pps	2,000 pps
3	IIS#1	1/10	16,000 pps	1,000 – 16,000 pps
4	IIS#1	1/10	2,000 – 16,000 pps	2,000 pps
5	IIS#1	1/4	16,000 pps	1,000 – 16,000 pps
6	IIS#1	1/4	2,000 – 16,000 pps	2,000 pps
7	IIS#1	1/2	16,000 pps	1,000 – 16,000 pps
8	IIS#1	1/2	2,000 – 16,000 pps	2,000 pps
9	IIS#2	1/10	16,000 pps	1,000 – 16,000 pps
10	IIS#2	1/10	2,000 – 16,000 pps	2,000 pps
11	IIS#2	1/4	16,000 pps	1,000 – 16,000 pps
12	IIS#2	1/4	2,000 – 16,000 pps	2,000 pps
13	IIS#2	1/2	16,000 pps	1,000 – 16,000 pps
14	IIS#2	1/2	2,000 – 16,000 pps	2,000 pps

As shown above, separate experiments with different parameters were conducted using FRiTrace on this experimental topology. In the first experiment, the background traffic was fixed at 16,000 pps with the attack rate doubling with each run from 1,000 pps to 16,000 pps. In the second experiment, the attack rate was fixed at 2,000 pps with the background traffic rate doubling with each run from 2,000 pps to 16,000 pps. These experiments were conducted with normal iTrace and both Intention iTrace schemes with $p = 1/20,000$. For the Intention iTrace schemes, p_i was set to 1/10, 1/4 and 1/2. Each experiment was conducted a total of ten times.

Using a modified FRiTrace daemon, information about each iTrace message generated was stored to a log file. This information contained the iTrace packet size and the destination IP address. Once all of the experiments were conducted, the log files, one generated for each run, were post-processed to obtain the total number of iTrace messages generated, the total number of useful iTrace messages, the total packet overhead of FRiTrace and the total data overhead of FRiTrace.

One method of determining the effectiveness of FRiTrace as an IP traceback suite is to measure how many useful iTrace messages were generated with the different schemes, where a useful iTrace message is defined as one generated from a packet belonging to a denial of service attack stream. The higher the number of useful iTrace messages received by the victim of an attack, the more confidence the victim can place on the path generated by post-processing the iTrace messages. Once the data was compiled, all three schemes were compared to each other to show the relative effectiveness of one scheme versus another and to also show how the p_i parameters relates to the overall effectiveness of the Intention iTrace schemes.

Useful Messages (16,000 pps Background Traffic)

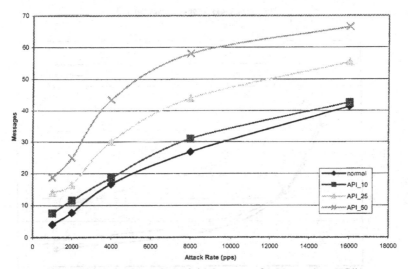

Figure 4. Number of Useful Messages for Normal vs IIS#1

The first comparison was between normal iTrace and Intention iTrace scheme 1 with a fixed background rate and shows how the number of useful messages is affected by changes in the attack rate. As can be seen from this comparison (Figure 4), overall the Intention iTrace scheme 1 showed a higher number of useful iTrace messages. Based on the functionality of Intention iTrace this behavior was expected. In addition, as p_i was increased from 1/10 to 1/2, the number of useful messages also increased. In scheme 1, the Intention iTrace decision module classifies packets into two groups based on the packets' destination addresses. As a result, when the ratio of Intention traffic to total traffic, R_i, is small, the relative increase in the number of useful messages is also small when comparing normal iTrace to Intention iTrace and also between the Intention iTrace runs with varying p_i. However, as R_i increases, the increase in the number of useful messages becomes more realized. Lastly, with Intention iTrace scheme 1, there is roughly a linear relationship between the number of useful messages and p_i at high attack rates. For example, the number of useful messages roughly doubled when p_i was increased from 1/4 to 1/2.

We have also compared the normal iTrace scheme and IIS#2. This scheme does not classify packets into traffic groups, but instead generates a trigger with probability p, after which p_i decides whether the packet should be treated as an Intention packet or a normal packet. Because of this, the change in the number of useful messages as the attack rate increases is not affected by the ratio of Intention traffic to total traffic, R_i.

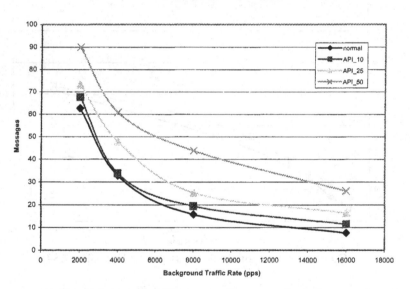

Figure 5. Normal vs. IIS#1, with Variable Background Traffic

Figure 5 shows how the number of useful messages is affected by an increase in the background traffic rate in all the schemes. This experiment is important for determining how effective a statistical denial of service attack is on hiding the true attack packets. In general, as the background traffic rate increases, there is a decrease in the number of useful messages in all runs. However, the rate of decrease as the background traffic rate increases varies for each run. For example, using normal iTrace, as the background traffic rate doubles, the number of useful messages decreases by half. But for Intention iTrace, the rate of decrease decreases as p_i is increased despite increasing the background traffic rate. This can be seen from the scheme 1 run with $p_i = 1/2$. An initial doubling of the background traffic rate drops the number of useful messages from about 90 to 60, a decrease of about 1/2. The next doubling of the background traffic rate yields a decrease of only 1/3, and so on. In general, as p_i is increased towards 1.0, the rate of decrease approaches 0. This is consistent with the functionality of Intention iTrace, which only generates iTrace messages for prefixes with an intention bit of 1. We have also performed the same experiments with IIS#2, and the results are similar.

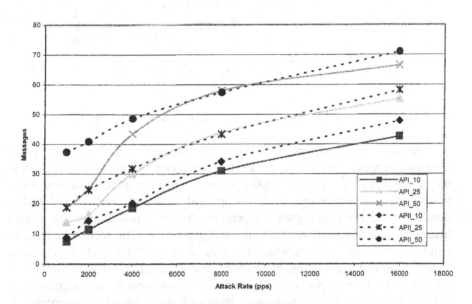

Figure 6. Number of Useful Messages for IIS#1 versus IIS#2.

From these comparisons, the most important result is that both Intention iTrace schemes are resilient to statistical denial of service attacks because as the background traffic rate increases, the rate of decrease of the number of useful messages also decreases. In addition, as p_i is increased, the amount of decrease increases until $p_i = 1$ where a decrease in the number of useful messages no longer occurs.

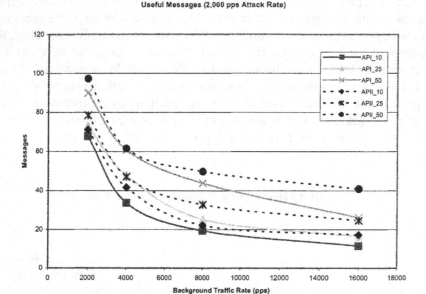

Figure 7. Scheme 1 vs Scheme 2, with Variable Background Traffic

The next comparisons show how the two different Intention iTrace schemes compare in terms of useful messages when there is a fixed background traffic rate of 16,000 pps and a variable attack rate and when there is a fixed attack rate of 2,000 pps and a variable background traffic rate.The solid lines in Figure 6 represent scheme 1 results and the dashed lines represent scheme 2 results with varying p_i. As can be seen, scheme 2, in general generated more useful messages than scheme 1 under identical testing conditions. From this comparison, the effect of classifying traffic into groups and selecting packets independently is apparent by the significantly lower number of useful messages produced by scheme 1 compared to scheme 2 when the ratio of Intention traffic to total traffic R_i is low. For higher values of p_i, it appears that scheme 1 achieves its maximum usefulness around $R_i = 0.5$ when compared to scheme 2. Up to and following $R_i = 0.5$, the number of useful messages decreases compared to scheme 2.

With variable background traffic, as in Figure 8, scheme 2 achieves a higher number of useful messages at all levels of p_i and for all values of R_i. However, more important to note is the fact that scheme 2 is much more resilient to statistical denial of service attacks, which artificially inflate the background traffic rates. Not only do the scheme 2 results generate more useful iTrace messages, but the rate of decrease of the number of useful messages as the background traffic rate increases is much smaller in scheme 2 than in scheme 1. The quicker stabilization of scheme 2 even for small R_i indicates that much more of the background traffic is ignored by scheme 2 and that further increases of the background traffic rate past 16,000 pps would yield only slight decreases in the number of useful messages.

7. SUMMARY

To show the overall effectiveness of FRiTrace, several experiments were conducted to measure the number of useful iTrace messages and the total traffic overhead generated by the various FRiTrace modes. For the Intention iTrace schemes, p_i was also varied to observe its effects on the number of useful messages and total traffic overhead. Our experiments showed that even when the attack traffic represents a small percentage of the overall background traffic that FRiTrace is still able to generate a substantial amount of useful iTrace messages. While it only takes a single message from any of the routers along the path from the victim to the true attacker, the more messages received from the routers along the path the greater confidence the victim can place in the final traced path to the attacker or attackers. By increasing p_i throughout the first set of experiments, it was shown that not only does the Intention iTrace scheme generate more useful messages, but there is also a near-linear relationship between p_i and the relative amount of useful messages, regardless of the attack rate or background traffic rate. Lastly, the first set of experiments showed the true benefits of the Intention iTrace schemes. With a fixed attack rate, as the background rate was increased, the number of useful messages did not drop linearly as in the normal iTrace scheme. Instead, the decrease in the number of useful messages decreased as either p_i or R_i increased and eventually reached a stable value where additional increases in the background traffic would still generate approximately the same number of useful messages. This behavior was not expected or observed in normal iTrace, where there was a linear relationship between the decrease in useful iTrace messages and the increase in the background traffic rate.

Our experiments also showed that even in the worse case, FRiTrace only generated approximately 0.0064% total network traffic overhead. This equates to an increase of approximately 6.5 kbps for every 100 Mbps of traffic analyzed by FRiTrace, a very small amount of overhead. In addition, the amount of overhead is a function of the probability to generate an iTrace message p. A decrease in p would result in a decrease of the network traffic overhead. This parameter is intended to be set by the network administrators of the domain being analyzed based on traffic patterns observed on the network. Aside from the overall low overhead introduced by FRiTrace, several logical trends were observed. As p_i increased for both Intention iTrace schemes, the amount of total traffic overhead also dropped. As R_i increased, total traffic overhead stabilized regardless the increase in either the background traffic rate or attack rate. Lastly, it was observed that scheme 2 produced slightly higher overhead than scheme 1.However, this is a result of the higher efficiency of scheme 2 and is also shown in the first set of experiments where scheme 2 generated more useful iTrace messages than scheme 1 regardless of the parameters to the experiments.

These initial results show that FRiTrace can be an effective tool for allowing victims of denial of service attacks to discover the true attackers despite IP spoofing. While these experiments were conducted on a simple test network representing only a single FRiTrace host, the results would simply scale up to the number of hosts or routers along the attack path that supported iTrace because the probability p to generate an iTrace message is independent of all other hosts or routers along the path. Further, since iTrace messages are sent out of band, there is no interference from downstream routers on previously generated iTrace messages as in the probabilistic packet marking schemes described in an earlier chapter. Because of this, as more routers along the attack path support iTrace, more iTrace messages will be sent to the victim allowing faster path reconstruction and correlation.

REFERENCES

Bellovin, S. M., ICMP Traceback Messages, *IETF Internet Draft*: draft-ietf-itrace-01.txt, October 2000.

McCanne, S, and Jacobson, V., The BSD packet filter: a new architecture for user-level packet capture, in *Proceedings of the Winter 1993 USENIX Conference*, January 1993.

Moore, D, Voelker, G. M., and Savage, S. Inferring Internet Denial-of-Service Activity, in *Proceedings of the 10th USENIX Security Symposium*, August 2001.

Savage, S., Wetherall, D., Karlin, A., and Anderson, T., Practical Network Support for IP Traceback, in *Proceedings of ACM SIGCOMM,* August 2000.

Snoeren, A.C., Partridge, C., Sanchez, L.A., Jones, C.E., Tchakountio, F., Kent, S.T., and Strayer, W.T., Hash-Based IP Traceback, in *Proceedings of SIGCOMM,* August 2001.

Song, D.X., and Perrig, A., Advanced and Authenticated Marking Schemes for IP Traceback, in *Proceedings of IEEE Infocom*, April 2001.

Wu, S. F., Zhang, L., Massey, D., and Mankin, A., Intention-Driven ICMP Traceback, *IETF Internet Draft*: draft-ietf-itrace-intention-00.txt, November 2001.

DESIGN AND IMPLEMENTATION OF A HIGH-PERFORMANCE NETWORK INTRUSION PREVENTION SYSTEM

Konstantinos Xinidis[1], Kostas G. Anagnostakis[2], Evangelos P. Markatos[1]

[1]*Institute of Computer Science, Foundation for Research and Technology Hellas, P.O Box 1385 Heraklio, GR-711-10 Greece {xinidis, markatos}@ics.forth.gr ;* [2]*Distributed Systems Laboratory, CIS Department, Univ. of Pennsylvania, 200 S. 33rd Street, Philadelphia, PA 19104 anagnost@dsl.cis.upenn.edu*

Abstract: Network intrusion prevention systems provide proactive defense against security threats by detecting and blocking attack-related traffic. This task can be highly complex, and therefore, software-based network intrusion prevention systems have difficulty in handling high speed links. This paper describes the design and implementation of a high-performance network intrusion prevention system that combines the use of software-based network intrusion prevention sensors and a network processor board. The network processor acts as a customized load balancing splitter that cooperates with a set of modified content-based network intrusion detection sensors in processing network traffic. We show that the components of such a system, if co-designed, can achieve high performance, while minimizing redundant processing and communication. We have implemented the system using low-cost, off-the-shelf technology: an *IXP1200* network processor evaluation board and commodity PCs. Our evaluation shows that our enhancements can reduce the processing load of the sensors by at least 45% resulting in a system that can handle a fully-loaded Gigabit Ethernet link using at most four commodity PCs.

Key words: Network Intrusion Detection Systems, Network Intrusion Prevention Systems, network processors, load balancing

1.   INTRODUCTION

The increasing importance of network infrastructure and services along with the high cost and difficulty of designing and enforcing end-system security policies has resulted in growing interest in complementary, network-level security mechanisms, as provided by firewalls and network intrusion detection and prevention systems.

High-performance firewalls are rather easy to scale up to current edge-network speeds because their operation involves relatively simple operations such as matching a set of Access Control List-type policy rules against fixed-size packet headers. Unlike firewalls, network intrusion prevention systems (NIPSes) are significantly more complex and, as a result, are lagging behind routers and firewalls in the technology curve. The complexity stems mainly from the need to analyze not just headers but also packet content and higher-level protocols. Moreover, the function of NIPSes needs to be updated with new detection components and heuristics, due to the continuously evolving nature of network attacks.

Both complexity and the need for flexibility make it hard to design high-performance NIPSes. Application-Specific Integrated Circuits (ASICs) lack the needed flexibility while software-based systems are inherently limited in terms of performance. One design that offers both flexibility and performance is the use of multiple software-based systems behind a hardware-based load balancer. Although such a design can scale up to edge-network speeds, it still requires significant resources, in terms of the number of sensors, required rack-space, etc. It is therefore important to consider ways of improving the performance of such systems.

This paper explores the role that high-speed network processors (NPs) can play in scaling up network intrusion prevention systems. We focus on ways for exploiting the performance and programmability of NPs for boosting in-line network intrusion detection. We describe the architecture of a NIPS using commodity Personal Computers (PCs) as network intrusion detection sensors, fed by an *IXP1200*[8] network processor. We present the allocation of operations to components and the trade-offs we faced during designing and prototyping the system. For further details please refer to [20].

The rest of this paper is organized as follows. In Section 2 we describe the architecture and implementation of our system, called *Digenis*[15]. In Section 3 we examine the performance benefits of using NP-based load balancing and acceleration. We discuss work that is related to high-performance intrusion prevention in Section 4. Finally, we summarize and comment on future research directions in Section 5.

---

[15] Digenis Akritas, the ideal medieval Greek hero, is a bold warrior of the Euphrates frontier. He was a proficient warrior by the age of three and spent the rest of his life defending the Byzantine Empire from frontier invaders.

2.   DESIGN AND IMPLEMENTATION

We faced a number of design challenges in constructing *Digenis* with respect to performance, flexibility and scalability:

Performance: The primary metric of interest in the design of a NIPS is throughput. That is, to be able to operate at network speeds of at least 1 Gbit/s without packet losses, so as to detect any attempted attack. Therefore, the system must be capable of analyzing all the incoming traffic under the most stringent conditions. Network intrusion detection systems (NIDSes) based on commodity PCs are able to monitor at speeds much lower than 1Gbit/s[2,5]. This necessitates the use of a distributed design with several intrusion detection sensors operating in parallel and supported by a load balancing traffic splitter[4,11,19]. At the same time, we want to minimize cost and use as few resources as possible. The use of an NP implementing the splitter appears reasonable, since it is likely to be cheaper than a custom ASIC, while load balancing operations seem to be well within the processing capacity of modern NPs. We also want to minimize the number of sensors needed. A key focus of our work is therefore on how to exploit the processing capacity on the NP to reduce the load of the sensors. A second important performance goal is minimizing the latency induced by the NIPS. There is a direct relationship between latency introduced by a networking device and the maximum throughput of TCP flows[16].

If the NIPS will be used at the boundary between an enterprise network and the Internet, latencies in the order of a few milliseconds may be tolerable. If the NIPS is deployed internally, and the network needs to support high-bandwidth local services (such as file sharing, etc.) the latency requirements are even more stringent. Particularly, there is a critical value for the round trip time (RTT) of a packet in each network. If the latency is below this critical value, TCP throughput is unaffected -- it is the line speed of the underlying network which becomes the bottleneck -- above this critical value, however, TCP throughput is negatively impacted. The critical value for RTT in a network supporting Gigabit speeds is 0.5 milliseconds. Thus, if we want the throughput of TCP to be unaffected, we must ensure that the latency imposed by our NIPS is less than 0.5 milliseconds.

However, Gigabit Ethernet links will rarely carry only a single TCP connection. Rather, a Gigabit Ethernet link supports hundreds, if not thousands of TCP connections, and this multiplexing mitigates the impact of latency on the overall throughput of the link[9]. In other words, it is possible

---

[16] Recall that TCP Throughput=Window/RTT where *Window* is the maximum TCP window size (default value is 64 Kbytes) and *RTT* is the round trip time in the network.

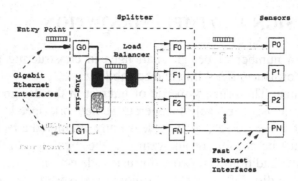

Figure 1. Architecture of *Digenis.*

to impose latency greater than 0.5 milliseconds without affecting the throughput of a link due to the high number of TCP connections.

Flexibility and Scalability: A NIPS needs to be flexible and scalable, both for scaling up to higher link speeds and more expensive detection functions, as well as for updating the detection heuristics. If the protection of a faster link or a more fine-grained detection is required, it would be desirable to reuse as much as possible of the existing hardware. Clearly, this property does not hold for ASIC-based NIPSes. However, it is remarkable that almost all NIPS providers ignore this dimension[8,13,19]. Furthermore, a prerequisite of flexibility is simplicity as extending a complex system may be hard and error-prone. It is therefore desirable for the hard-to-program elements of our system to be as generic as possible.

2.1   Architecture

Digenis is composed of a customized load balancing splitter and a number of content-based network intrusion detection sensors connected with the splitter (Figure 1). The splitter is the entry and exit point of the traffic that runs through the system. The basic task of the splitter is to evenly distribute the traffic across the sensors and to transmit the non-attack packets back to their destination. The sensors are responsible for the heavy task of inspecting the traffic for intrusion attempts. They maintain the required information for recognizing all the malicious traffic and deciding whether to forward or drop the packet. For every input packet, the splitter computes which sensor will be responsible to analyze this packet. Then, it forwards the packet to this sensor for inspection. The sensor searches for known attack patterns contained in the packet. If a pattern is found, then the packet is blocked, otherwise the packet is forwarded back to the splitter. The splitter receives the analyzed packet and transmits it to its destination.

Additionally, *Digenis* supports plug-ins that implement operations necessary to improve the performance of the system. A plug-in has two parts, one running on the splitter and one running on the sensors. These two parts cooperate in order to accomplish their task. In the context of this work we have designed a plug-in for *Digenis* that attempts to minimize the cost of sending a packet from a sensor to the splitter.

Splitter: The functionality of the splitter can be divided into the basic operations and the plug-ins that provide adequate operations to boost performance. The basic part of the splitter integrates the functionality of a load balancer -- it is responsible for distributing the incoming traffic across the output interfaces (ports). However, it differs from a common load balancer in that it must be *flow-preserving*, that is, all the packets belonging to the same flow[17] must be forwarded to the same output interface.

Regarding load balancing, there are two possible approaches that we could use: stateful load balancing that requires from the system to hold state and hash-based load balancing[3,10,16] that experiences greater load imbalances. For the purposes of this paper, we assume that load imbalances are tolerable and use the simpler hash-based method. The input of the hash function is composed of the source and destination IP addresses of the packet.

Sensor: A sensor is a commodity PC that runs a modified popular NIDS and is connected with the splitter (through an Ethernet connection). A sensor receives traffic from the splitter and analyzes it for possible known attacks. In case that an attack is found, it notifies the splitter to block the offending packet(s), otherwise it informs the splitter that the packet(s) should be forwarded. A sensor maintains state about the traffic it analyzes in order to operate correctly. The maintained state includes the active TCP connections it has captured in the near past, TCP connections tagged as offending, fragmented packets and statistics about the connections per second to TCP/UDP destination ports.

2.1.1   Reducing Redundant Packet Transmission

We have designed a plug-in for *Digenis* that is responsible for reducing redundant packet transmission on the system. The idea behind this plug-in is the following: Suppose that the splitter stores temporarily (for a few

---

[17] In case of TCP/UDP traffic, we define a flow to consist of all the traffic of a TCP or UDP connection. Otherwise, a flow consists of all the traffic originating from a particular IP address and destined to a particular IP address.

milliseconds) the packets that it forwards to the sensors for analysis. Then there is no need for the sensors to send back to the splitter the analyzed packet, but only a unique identifier of that packet (PID). Because the splitter has previously stored the packet with this PID, it can infer the referenced packet and forward it to the appropriate destination. The only extra work for the splitter is to tag each packet with a PID, which is a trivial task. Although the additional processing cost to the splitter from this plug-in is minimal, the reduction to the load of the sensors is remarkable. However, this technique requires from the splitter to be equipped with additional memory for the buffering of the packets. As we will present in Section 3, the memory requirements are easily satisfied by modern NPs. Subsequently, we discuss how a sensor communicates the packet information back to the splitter.

Communication between Splitter-Sensor: The splitter communicates with the sensors in order to decide the action that should be performed, that is, forward or drop the packet. This is done with acknowledgments (ACKs) from the sensors to the splitter. An ACK is an ordinary Ethernet packet. It consists of an Ethernet header, followed by two bytes denoting the number of packets acknowledged (ACK factor) and followed by a set of four-byte integers representing the PIDs. There are other possible formats requiring less bytes and supporting higher ACK factors for this configuration. However, this approach is more scalable. There are several options regarding the information that these packets should contain. The sensors may send back to the splitter the following responses:

1. **Positive ACKs:** an ACK for every packet not related to any intrusion attempt.
2. **Positive cumulative ACKs:** an ACK for a set of packets not related to any intrusion attempt.
3. **Negative ACKs:** an ACK for every packet that belongs to an offending session.
4. **Negative cumulative ACKs:** an ACK for a set of packets that belong to an attack session.
5. **The packet received**.

Each of these solutions has its pros and cons. The packet received (PR) scheme, although it has the advantage that it does not require the splitter to temporary hold the packet in memory, it suffers from low performance. In Section 3, we evaluate some of these approaches, with regard to performance. Among positive and negative cumulative ACKs (CACKs) we have chosen the former ones. Negative CACKs have two major drawbacks: First, in order to be able to distinguish when a packet must be forwarded, we have to use a timeout value. Recall that, our NIPS must not drop any packet or an attack might be missed. As a result, we would be forced to choose a timeout for the worst case scenario. The side-effect is that packets will experience a high latency. Second, it is impossible for the splitter to differentiate the case where the analyzed packet contained no intrusion from the case where the packet was dropped due to an error condition. We have chosen positive CACKs (P-CACKs) because they supersede positive ACKs.

2.2   Implementation

We have implemented *Digenis* using low-cost, off-the-shelf technology: an Intel *IXP1200* Ethernet evaluation board and commodity PCs.

Splitter: We have implemented the splitter using an *IXP1200* network processor. The *IXP1200* chip contains six micro-engines with four hardware threads (contexts) each. Also, this chip has a general-purpose StrongARM processor core, a FIFO Bus Interface (FBI) unit and buses for off-chip memories (SRAM and SDRAM). The maximum addressable SRAM and SDRAM memory are 8 Mbytes and 256 Mbytes respectively. The FBI unit interfaces the *IXP1200* chip with the media access control (MAC) units through the IX bus. The FBI also contains a hash unit that can take 48-bit or 64-bit data and produce a 48- or 64-bit hash index. In our evaluation board, an IXF440 MAC unit (with eight Fast Ethernet interfaces) and an IXF1002 MAC unit (with two Gigabit Ethernet interfaces) are connected to the IX bus.

We have developed the application using micro-engine assembly language. The assignment of threads to tasks is done as follows: we assign eight threads for the receive part of the Gigabit Ethernet interface, one thread for the receive part of each of the eight Fast Ethernet interfaces, four threads for the transmit part of the eight Fast Ethernet interfaces, and four threads for the transmit part of the Gigabit Ethernet interface.

For the implementation of hash-based load balancing, we use the hash unit of the *IXP1200*. Also, for the temporary storage of the incoming packets until they are acknowledged we use a circular buffer which resides in SDRAM memory. This circular buffer must be large enough to prevent overwriting packets before their matching ACK is received.

Sensor: The functionality of the sensor has been implemented by modifying the popular NIDS *Snort* version 2.0.2[15]. The functionality of the sensor can be divided into three different phases: (1) the protocol decoding phase, (2) the detection phase, and (3) the prevention phase. In the first phase, the raw packet stream is separated into connections representing end-to-end activity of hosts. A connection, in case of IP traffic, can be identified by the source and destination IP addresses, transport protocol and UDP/TCP ports. Then, a number of protocol-based operations are applied to these connections. The protocol handling ranges from network layer to application layer protocols. Some of the operations applied by the protocol-based handling are IP defragmentation, TCP stream reconstruction and identification of the URI in HTTP requests. The second phase consists of the actual detection. Here, the packet (or an equivalent higher-level protocol data unit) is checked against a database of detection heuristics representing attack patterns. Then follows the prevention phase. The action of this phase depends on the result of the previous one. If no attack is found, the sensor informs the splitter to forward the packets. If malicious activity is observed, then the prevention engine blocks the suspicious traffic by informing the splitter to not forward the packets belonging to the offending connection(s).

Extra Implementations: In addition to our splitter, for comparison purposes, we have implemented the following three configurations on the *IXP1200*:

- A forwarder (FWD) that transmits the traffic arriving at an input Gigabit Ethernet interface to an output Gigabit Ethernet interface.
- A load balancer (LB) that implements a flow-preserving load balancer with the same load-balancing characteristics as our splitter. The *IXP1200* receives traffic from a Gigabit Ethernet interface and transmits the traffic to eight Fast Ethernet interfaces.
- The last configuration (LB + FWD) implements the basic functionality of our splitter (without optimizations).

3. EVALUATION

In this Section we examine the performance of our architecture. We focus on the impact of our enhancements to sensor-splitter communication. In particular, we compare the performance of P-CACK vs. the PR scheme. We also show that such techniques can be efficiently supported by current network processors and that they do not significantly impair forwarding latency.

3.1 Experimental Environment

Splitter: The performance of the configurations running on the *IXP1200* is measured using the *IXP1200* Developer Workbench (version 2.01a)[7]. Specifically, we use the *transactor* provided by Intel. The *transactor* is a cycle-accurate architectural model of the *IXP1200* hardware. We simulate the configurations as they would run on a real *IXP1200* chip. We assume a clock frequency of 232 MHz and a 64-bit IX bus with a clock frequency of 104 MHz.

Sensor: We use a 2.66 GHz Pentium IV Xeon processor with hyper-threading disabled. The PC has 512 Mbytes of DDR DRAM memory at 266 MHz. The PCI bus is 64-bit wide clocked at 66 MHz. The host operating system is Linux (kernel version 2.4.20, Red-Hat 9.0). The Gigabit Ethernet network interface is an Intel PRO/1000 MT Dual Port Server Adapter[6]. The sensor software is a modified *Snort* version 2.0.2, compiled with gcc version 3.2.2. We turn off all preprocessing in *Snort*. Unless noted otherwise, *Snort* is configured with the default rule-set.

Packet Traces: For the evaluation of *Digenis* we use three packet traces. The **FORTH.WEB** trace was captured at **ICS-FORTH** and only contains HTTP traffic. The **FORTH.LAN** trace was also captured at **ICS-FORTH** and contains traffic from an internal Local Area Network (LAN). Both traces contain the real payload of the packets. The **IDEVAL** traces are taken from MIT Lincoln Laboratory and were used in the 1999 DARPA Intrusion Detection Evaluation[12].

3.2   Results

3.2.1   Performance of the Splitter

All the *IXP1200* configurations described in Section 2 (LB, FWD, our splitter, and LB+FWD) handle at most the IP and UDP/TCP header of the incoming packets. Thus, we argue that the most demanding traffic for these configurations is the traffic consisting of a high percentage of small packets, namely 64-byte packets[18]. We simulate the above configurations and the results show that all the configurations are capable of sustaining line speed even with traffic consisting of only 64-byte packets[19]. This is expected as the theoretical forwarding capacity of the *IXP1200* chip is greater than 1600 Mbit/s[7].

While all the configurations sustain line speeds, we use as a metric for comparison the utilization of the micro-engines and the utilization of SRAM

---

[18] This is the smallest possible packet in an Ethernet link including the 4-byte Ethernet CRC.
[19] The splitter uses P-CACK scheme with a factor of eight.

368 *Konstantinos Xinidis, Kostas G. Anagnostakis...*

and SDRAM memories[20]. These are some of the resources that may become the bottleneck, considering that the *IXP1200* specification states that the maximum IX bus throughput is 6 Gbit/s. In Figure 2 we present the average utilization of the micro-engines and the utilization of the SRAM and SDRAM memories for the described configurations. We observe that our approach is efficient and does not consume all the resources of the *IXP1200*, leaving headroom for even more offloading of the sensors. Particularly, the results suggest that the extra cost of the splitter compared to the load balancer is affordable[21].

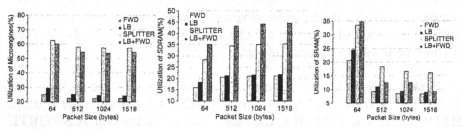

Figure 2. Utilization of the IXP1200 micro-engines, SDRAM and SRAM memories for different packet sizes. It is obvious that the splitter configuration does not consume all the resources of the *IXP1200*.

3.2.2 Performance of the Sensor

We first measure the processing cost of a sensor for different coordination schemes using the default rule-set. In this experiment *Snort* simply reads traffic from a packet trace[22], performs all the necessary NIPS functionality, and then transmits the coordination messages to a hypothetical splitter through a Gigabit Ethernet interface. Figure 3, shows the time that *Snort* spends to process all the packets for the **FORTH.WEB** trace including user and system time breakdown. The results show that the higher the P-CACK factor, the less the total running time for *Snort*. The running time is practically the same with the unmodified *Snort* for P-CACK with factor equal to 128. Also, *Snort* finished 45% faster for P-CACK with factor equal

---

[20] More accurately, we measure the utilization of the buses of SRAM and SDRAM memories.
[21] We have to mention that the increased utilization of the micro-engines in the case of the splitter configuration is caused by the instrumentation code we add to measure the performance of the splitter. While in the other configurations we do not add code for evaluation purposes, we are obliged to do so in the case of the splitter.
[22] We confirm that the hard disk is not the bottleneck by measuring the throughput of the hard disk and the transmit rate of *Snort*. As expected, the transmit rate of *Snort* is smaller than the throughput of the disk.

to 128 compared to the PR scheme. Moreover, we observe that the system time is lower than the user time. This confirms the fact that *Snort* spends most of its processing time in header and content matching which is counted in user time.

We also observe (Figures 3 and 4) that the improvement of the P-CACK scheme compared to the PR scheme depends very much on the trace used: the P-CACK scheme is from 45% to 3.8 times more efficient than the PR scheme. The reason is that the improvement depends on the detection load of the sensor. The smaller the detection load, the bigger the relative improvement. This becomes clearer if we determine where the improvement is coming from. The improvement stems from the fact that the P-CACK scheme reduces the overhead required for sending a packet to the network (*system time* in Figures 3 and 4). If the detection engine of a sensor is overloaded, then this overhead is a small fraction of the total workload of the sensor and reducing it does not lead to much improvement. In contrast, if the detection engine of a sensor is lightly loaded, this overhead consumes a significant fraction of the total workload of the sensor and reducing it results in a more notable improvement. For example, if the traffic is ruleset-intensive, then the detection load of the sensor increases and the relative improvement is small. On the other hand, for traffic that requires fewer rules to be checked for every packet, the detection load of the sensor will be minimal and the improvement will be greater.

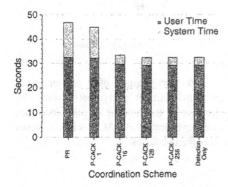 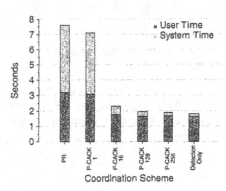

Figure 3. Processing cost of a sensor (time to process all packets in a trace), with user and system time breakdown (**FORTH.WEB** trace). We observe that the P-CACK scheme with factor 256 is 45% more efficient than the PR scheme.

Figure 4. Processing cost of a sensor (time to process all packets in a trace), with user and system time breakdown (**IDEVAL** trace). We observe that the P-CACK scheme with factor 256 is 3.8 times more efficient than the PR scheme.

We also repeat the experiment on a PC with a slower Pentium III processor at 1.13 GHz and the same PCI bus characteristics and Ethernet network interfaces. The results (Figure 5) show that the improvement is smaller compared to the faster machine. When we examine more carefully the results, we observe that while *user time* doubles, the *system time* increases only by 30%. This happens because *user time* is mainly the time spent for content search and header matching, which are processor intensive tasks. On the contrary, *system time* is dominated by the time spent for copying the packet from main memory, over the PCI bus, to the output network interface, handling interrupts and control registers of the Ethernet device. As the speed of processors increases faster than the speed of PCI buses and DRAM memories, we can argue that, as technology evolves, the effect of our enhancements will be even more pronounced – common processors are already running at 3.8 GHz, so the previously reported improvement is in fact a conservative result.

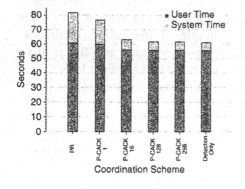 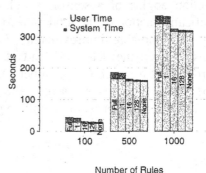

Figure 5. Processing cost of a slower sensor (FORTH.WEB trace). We can see that the improvement is smaller compared to the faster sensor.

Figure 6. Performance of a sensor using incremental number of synthetic rules. We notice that as the number of rules increases the improvement of P-CACK scheme versus PR scheme decreases.

Table 1. Synthetic rule example.

alert tcp any any → any any (ack: 1; flags: S; content: "RPC overflow";)

All the above experiments are performed using the default rule-set of *Snort*. To further understand the correlation between the detection load of a sensor and the P-CACK scheme improvement we also experiment with variable, synthetic rule-sets. An example rule is shown in Table 1. Similarly to the previous experiment, *Snort* reads traffic from a packet trace and sends packets over a Gigabit Ethernet interface. The results are shown in Figure 6. We observe that as the number of rules increases the improvement of P-CACK scheme versus PR scheme decreases. In other words, as detection load increases, improvement decreases.

Another interesting point is that the maximum relative improvement of P-CACK over PR is for small packets of 64 bytes. Small packets require less time for content matching *user time* and communication *system time* is the dominant cost factor. In addition, in the case of 64-byte packets, the bottleneck is not the processor, as in the case of larger packets, but the PCI bus. This is clearly shown in the experiments involving the **IDEVAL** traces (Figure 4), which contain many small packets for emulating certain types of attacks such as SYN flooding. For this trace, the P-CACK scheme is 3 times more efficient compared to the PR scheme. This is also a nice side effect of the P-CACK scheme, in that it makes the NIPS more robust against TCP SYN flood attacks, given that such attacks consist of a big fraction of small packets.

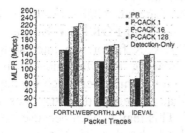

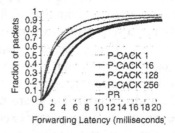

Figure 7. Maximum Loss Free Rate (MLFR) of a sensor using default rule-set.

Figure 8. CDF for latency of a sensor. We notice that latency increases with the P-CACK factor.

3.2.3   Forwarding Latency of the Sensor

The highest portion of the latency imposed by our NIPS is due to content matching on the sensors. This happens due to the fact that content matching is the single most expensive operation in every NIPS. To measure forwarding latency, we use two hosts *A* and *B* with two Gigabit Ethernet network interfaces each, *eth0* and *eth1*. We connect the two interfaces of host *A* with the two interfaces of host *B* back-to-back. Everything that host *A* sends to network interface *eth0/eth1* is received by host *B* on network interface *eth0/eth1*, and vice versa. Host *A* reads a trace from a file and sends traffic to host *B* (using *tcpreplay*[1]). Host *B* runs *Snort*, which receives packets from interface *eth0* and sends replies to interface *eth1*. Host *A* matches the packet transmission time with the arrival time of the reply and computes the latency.

Initially, we estimate the maximum loss free rate (MLFR) of a sensor by replaying a packet trace and measuring the rate at which the sensor started dropping packets (Figure 7). In this experiment we set the input packet buffer size to 16 Mbytes. We use MLFR to compute the latency that a sensor imposes to analyzed packets when reaching its processing capacity.

In this experiment, host *A* replays **FORTH.WEB** trace at the maximum loss free rate **of each communication scheme**. We observe that there are packets that experience very high latency. To understand this phenomenon, we measure the time that *Snort* spends in content and header matching using the *rdtsc*[17] instruction of the Pentium IV processor. The results show that the peaks in time spent for content and header matching coincide with the peaks in latency. This means that, when the required per packet operations increase, so does the latency. A consequence of this property is that packets that require a significant amount of processing slow down other packets that do not. This is a form of head of line (HOL) blocking.

Figure 8 shows the cumulative distribution function (CDF) for all ACK schemes when a sensor receives traffic at the MLFR of **FORTH.WEB** trace. We notice that latency increases with the P-CACK factor. An interesting observation is that the graph is heavy tailed, meaning that while most of the packets experience low latency, 5% of the packets exhibit very high latency (above 20 milliseconds). These are packets that are received from a sensor while the sensor has a temporary excess load. This may happen because, for example, some packets require too many rules to be checked. If too many such packets are received back-to-back, the system reaches (or exceeds) its capacity and the latency increases considerably.

3.2.4   Forwarding Latency of the Splitter

We argue that the overall latency that a packet experiences by our NIPS is due to the processing of the sensors and not the forwarding of the splitter. Also, the cycles spent by the splitter to forward a packet from the input interface to an output interface depend only on the packet length. This means that practically all packets of the same length experience almost the same latency.

3.2.5 Memory requirements

There is a direct relationship between latency imported by the sensors and required memory on the splitter. The splitter needs memory to save incoming packets until they are acknowledged by the sensors. The amount of memory the splitter needs depends on the highest possible latency that our NIPS will tolerate. If we set this value in a reasonable value, for example, 200 milliseconds then according to the fact that our NIPS analyzes traffic at 800 Mbit/s, the required memory is approximately 20 Mbytes. This means that the circular buffer of the *IXP1200* must be at least 20 Mbytes. This is a reasonable requirement considering that the maximum addressable SDRAM memory of the *IXP1200* is 256 Mbytes.

4. SUMMARY AND CONCLUDING REMARKS

We have presented the design of *Digenis*, a high-performance Network Intrusion Prevention System (NIPS). The system consists of a customized load-balancing component built using the *IXP1200* Network Processor, and a number of sensors implemented on commodity PCs. In contrast to off-the-shelf load balancers used in NIPS products, our design exploits the programmability of NPs to move part of the intrusion prevention functionality from the sensors to the NP. We have focused on one method for boosting system performance by optimizing the coordination between the load balancer and the sensors. The result is a 45% improvement in performance, allowing the system to reach speeds of at least 1 Gbit/s.

There are several directions that we are currently pursuing. First, we are re-examining the structure of the sensor software. In particular, we consider the possibility of using a more fine-grained protocol processing model such as the one demonstrated by *Bro*[14], and we try to move part of the protocol processing functionality to the NP. Second, we are looking at ways for building a 10 Gbit/s NIPS using third-generation NPs.

ACKNOWLEDGEMENTS

This work was supported in part by the IST project SCAMPI (IST-2001-32404) funded by the European Union, the GSRT project EAR (GSRT code: USA-022), and by ESTIA, a PAVET-NE project funded by the Greek General Secretariat of Research and Technology (PAVET-NE code: 04BEN8). Kostas Anagnostakis is also supported in part by ONR under Grant N00014-01-1-0795. Konstantinos Xinidis and E. P. Markatos are also with University of Crete. The work of Kostas Anagnostakis was done while at ICS-FORTH.

REFERENCES

1. Aaron Turner and Matt Bing. tcpreplay Tool. http://tcpreplay.sourceforge.net.
2. S. Antonatos, K. G. Anagnostakis, and E. P. Markatos. Generating realistic workloads for intrusion detection systems. In Proceedings of the 4th ACM SIGSOFT/SIGMETRICS Workshop on Software and Performance (WOSP 2004), January 2004.
3. Z. Cao, Z.Wang, and E.W. Zegura. Performance of hashing based schemes for internet load balancing. In Proceedings of IEEE Infocom, pp. 323-341, 2000.
4. Y. Charitakis, K. G. Anagnostakis, and E. Markatos. An active splitter architecture for intrusion detection (short paper). In Proceedings of the Tenth IEEE/ACM Symposium on Modeling, Analysis, and Simulation of Computer and Telecommunications Systems (MASCOTS 2003), October 2003.
5. Y. Charitakis, D. Pnevmatikatos, E. P. Markatos, and K. G. Anagnostakis. Code generation for packet header intrusion analysis on the IXP1200 network processor. In Proceedings of the 7th International Workshop on Software and Compilers for Embedded Systems (SCOPES 2003), September 2003.
6. Intel Corporation. Intel PRO/1000 MT Dual Port Server Adapter. http://www.intel.com.
7. Intel Corporation. Intel IXP1200 Network Processor (white paper), 2000. http://developer.intel.com.
8. Internet Security Systems Inc. http://www.iss.net.
9. Intrusion Prevention Systems Group Test - Edition 1, NSS Group Ltd. http://www.nss.co.uk/acatalog/.
10. L. Kencl and J. Y. L. Boudec. Adaptive load sharing for network processors. In Proceedings of IEEE Infocom, June 2002.
11. C. Kruegel, F. Valeur, G. Vigna, and R. Kemmerer. Stateful intrusion detection for high-speed networks. In Proceedings of the IEEE Symposium on Security and Privacy, pp. 285-294, May 2002.
12. R. Lippmann, J.W. Haines, D. J. Fried, J. Korba, and K. Das. The 1999 DARPA off-line intrusion detection evaluation. Computer Networks, 34(4):579-595, October 2000.
13. Network Associates, Inc. http://www.networkassociates.com.
14. V. Paxson. Bro: A system for detecting network intruders in real-time. In Proceedings of the 7th USENIX Security Symposium, January 1998.
15. M. Roesch. Snort: Lightweight intrusion detection for networks. In Proc. of the second USENIX Symposium on Internet Technologies and Systems, November 1999. (Software available from http://www.snort.org).
16. R. Russo, L. Kencl, B. Metzler, and P. Droz. Scalable and adaptive load balancing on IBM Power NP. Technical report, Research Report - IBM Zurich, August 2002.
17. Time-Stamp Counter. http://www.intel.com/design/Xeon/applnots/24161825.pdf.
18. TippingPoint Technologies Inc. http://www.tippingpoint.com.
19. Top Layer Networks. http://www.toplayer.com.
20. K. Xinidis, K. G. Anagnostakis, and E. P. Markatos. Design and Implementation of a High-Performance Network Intrusion Prevention System. ICS-FORTH Technical Report 333, March 2004.

STRIDE: POLYMORPHIC SLED DETECTION THROUGH INSTRUCTION SEQUENCE ANALYSIS

P. Akritidis[1], E. P. Markatos[1], M. Polychronakis[1], and K. Anagnostakis[2]

[1]*Institute of Computer Science Foundation for Research and Technology Hellas, P.O. Box 1385 Heraklio, GR-711-10 Greece, {akritid, markatos, mikepo}@ics.forth.gr*
[2]*Distributed Systems Laboratory, CIS Department, Univ. of Pennsylvania, 200 S. 33rd Street, Phila, PA 19104, anagnost@dsl.cis.upenn.edu*

Abstract: Despite considerable effort, buffer overflow attacks remain a major security threat today, especially when coupled with self-propagation mechanisms as in worms and viruses. This paper considers the problem of designing network-level mechanisms for detecting polymorphic instances of such attacks. The starting point for our work is the observation that many buffer overflow attacks require a "sled" component to transfer control of the system to the exploit code. While previous work has shown that it is possible to detect certain types of sleds, including obfuscated instances, this paper demonstrates that the proposed detection heuristics can be thwarted by more elaborate sled obfuscation techniques. To address this problem, we have designed a new sled detection heuristic, called STRIDE, that offers three main improvements over previous work: it detects several types of sleds that other techniques are blind to, has a lower rate of false positives, and is significantly more computationally efficient, and hence more suitable for use at the network-level.

Keywords: security; intrusion detection; buffer overflow detection.

1. INTRODUCTION

Buffer overflow attacks, popularized in 1996 by Aleph One[1], have been a major security concern ever since, because exploiting a buffer overflow vulnerability allows an attacker located anywhere on the Internet to execute arbitrary code on the compromised system. The highly interconnected

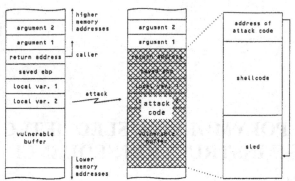

environment of the Internet currently creates tremendous exploitation

Figure 1. Anatomy of a stack-based buffer overflow attack. By overflowing the buffer, the return address can be overwritten with a value pointing somewhere within the sled. The flow of control will be transferred to the start of the shellcode from any location in the sled.

opportunities. In fact, such vulnerabilities in networked services are currently the main means used for propagation of Internet worms.

A tutorial on buffer overflow attacks was provided by Aleph One in 1996[1]. A buffer overflow attack takes advantage of insufficient bounds checking on a buffer located on the stack to overflow the buffer and overwrite the return address of the currently executing function. Figure 1 shows the typical layout of the stack both *before* (left) as well as *after* the buffer overflow attack (middle and right). The attack involves injecting data into the buffer which resides in the lower addresses of the stack. The amount of data injected is larger than the buffer size and the resulting overflow overwrites at least the `local` variables, the `saved ebp`, and the `return address`, resulting in the stack shown in Figure 1 (middle). The return address is hijacked to point to malicious code that is injected by the attacker, usually within the same overflowed buffer, illustrated as the shaded area in Figure 1 (middle). Besides *stack-based* buffer overflow attacks, it is also possible (although, perhaps more difficult) to engineer similar attacks on heap-based and statically-allocated buffers.

Despite considerable efforts in end-system preventive measures[2-8], adoption of these techniques is proceeding at an alarmingly slow pace. Dealing with this threat therefore requires additional perimeter defense mechanisms, as provided by firewalls and network intrusion detection systems. However, obfuscation techniques make it hard to apply simple rule-based detection techniques as currently used in network intrusion detection.

Buffer overflows often have features that can be seen as a weakness that can be exploited by detection heuristics. A key observation is that, although the location of the injected code relative to the start of the buffer is known to the attacker, the *absolute address*, which represents the start of the injected code, is only *approximately* known because the location of the start of the buffer relative to the start of the stack varies between systems, even for the same executable program.

To overcome the lack of exact knowledge on where to divert control, the attacker needs to append the malicious code fragment to a sequence of NOP instructions, typically around a few hundred bytes long. The overwritten return address always transfers control somewhere inside this sequence, and thus, after sliding through the NOP instructions, control will eventually reach the worm code. Due to the sliding metaphor, this sequence is usually called a *sled*. The *exact* location within the sled where execution will start does not matter: as long as the return address causes the system to jump anywhere within the sled, it will always reach the core of the exploit.

To detect buffer overflows, Network Intrusion Detection Systems (NIDSes) rely on signatures, characteristic strings, and regular expressions, such as code sequences included in an attack's shellcode that can identify the attack. However, obfuscation similar to polymorphism[9,10], used in viruses since the early 90s, renders signature-based techniques for zero-day worm detection obsolete[11,12]. Polymorphism usually encrypts the shellcode with a different random key each time and prepends it with a decryption routine. When the malicious program starts executing, the decryption routine will execute first, which in turn will decrypt the shellcode, which will then start executing. Since the decryption routine itself cannot be encrypted, some systems base their zero-day worm detection on detecting the decryption routine itself. Unfortunately, decryption routines are usually obfuscated using *metamorphism*. Metamorphism substitutes (sequences of) instructions with equivalent (sequences of) instructions, making the decryption routine difficult to fingerprint. Metamorphism is also used to obfuscate sleds, by, for example, substituting NOP instructions, with other equivalent instruction sequences.

Many existing detection mechanisms have also focused on detecting the sled component in order to detect buffer overflow attacks. For example, signatures to match simple sleds have been included in the shellcode rule set of the Snort NIDS[13]. In addition, Snort has been extended with the Fnord plugin[14] that searches for obfuscated sleds. Finally, Toth and Kruegel proposed the Abstract Payload Execution (APE) method[15] which further improves the sensitivity of obfuscated sled detection.

The rest of the paper is organized as follows. First, in Section 2 we present a classification of obfuscated sleds, and in Section 3 we discuss some existing detection techniques. Then, in Section 4 we propose STRIDE, a novel detection mechanism. In Section 5 we evaluate the detection mechanisms using generated attacks and real network traces, and in Section

6 we discuss limitations of sled detection in general and of STRIDE in particular. Finally, we conclude in Section 7.

2.   CLASSIFICATION OF SLEDS

The sled is a sequence of instructions responsible for directing the flow of control towards the core code of a buffer overflow attack. Although execution of the sled can start at any position, it always ends up "sliding" inside the core code of the attack. There are many different ways for a sled to achieve its functionality. In this section, we present several types of sleds in order of increasing (perceived) difficulty to detect.

2.1   Simple NOP Sled

The simplest sled consists of a series of NOP (no-operation) instructions. A NOP instruction has no effect on program behavior: it simply advances the program counter. Execution of the sled may start at any position, and the NOPs are used to transfer control, step by step, to the shellcode right after the sled. This simple sled has been demonstrated in the buffer overflow examples of[1] and has been used in many other attacks.

2.2   One-byte NOP-equivalents Sled

A NOP sled can be easily obfuscated by replacing literal NOP instructions with one-byte instructions which have no significant effect, and, for the purposes of the attacker, are practically equivalent to NOPs. For example, instructions that increase or decrease a register which is not used by the attacker, instructions that set or clear a flag, and instructions that push or pop a register, can all be used in a sled instead of NOPs. Current polymorphic buffer overflow attack generators use such sleds to avoid detection. The ADMmutate[16] engine uses this technique with a list of of 55 one-byte NOP-equivalent instructions. The Metasploit Framework[17] extends the ADMmutate engine with 3 additional single-byte NOP replacements. We have enumerated 66 such instructions in the Intel IA-32 architecture[18]. Although not yet seen in the wild, obfuscated sleds are readily available to attackers.

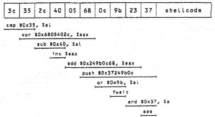

Figure 2. An example of a small sled, executable at every byte offset, which is constructed by interleaving one-byte and multi-byte NOP-equivalent instructions.

2.3   Multi-byte NOP-equivalents Sled

A straightforward extension to one-byte NOP-equivalent sleds is to use multi-byte NOP-equivalent instructions, which, like their one-byte counterparts, simply advance the program counter in order to reach the core of the exploit. However, it is not possible to use *any* multi-byte NOP equivalent instruction available in the instruction set, because a sled must be executable at *every* offset. Therefore, a straightforward way to generate multi-byte NOP-equivalents sleds is to restrict the operands of multi-byte instructions to correspond only to the opcodes of one-byte NOP-equivalent instructions, or to the opcodes of multi-byte NOP-equivalents. Consider for example the multi-byte NOP-equivalents sled shown in Figure 2. If control is transferred to the leftmost byte, it will execute instructions cmp $0x35,%al, sub $0x40,%al, add $0x249b0c68,%eax, etc. Note that the first argument of the first instruction cmp $0x35,%al, is 0x35, which corresponds to the opcode of instruction xor. Therefore, if control is transferred to the penultimate byte from the left, it will execute instructions xor, or, and, etc. leading to the end of the sled. This is true for all instructions in this type of sleds: their arguments are such that if control is transferred to any byte inside the sled, the execution will eventually lead to the end of the sled.

2.4   Four-byte Aligned Sled

Although traditional NOP sleds had to be executable at each and every byte, stack alignment can relax this restriction by constraining the possible placements of the vulnerable buffer. The default behavior of modern compilers is to align the stack at word (4-byte) boundaries[19]. Reference[11] discusses the possibility of exploiting stack alignment to construct sleds that have to be executable every 4 bytes. Pairs of non-destructive 2-byte instructions can be used as NOP-equivalents, but it is also possible to use

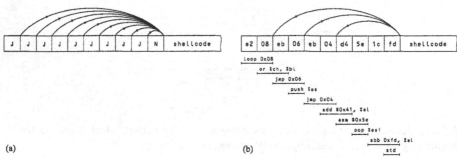

(a)   (b)

Figure 3. (a) The ideal trampoline-sled: flow of control is directed to the shellcode in a single step from any position in the sled. (b) An example of a small trampoline-sled that is executable at every byte offset. Control transfer instructions are placed at every second byte and their relative address operand is chosen so that it is a valid NOP-equivalent opcode.

longer instructions with techniques similar to the multi-byte instruction sled discussed earlier. Code sequences starting at non-word-aligned offsets may contain any kind of instruction, including instructions with destructive side-effects or even illegal ones, which can hinder detection.

2.5   Trampoline Sled

Although typical sleds transfer control to the shellcode by sliding it along their body—hence the name sled—, the same functionality can be achieved by jumping directly to the shellcode, as illustrated in Figure 3(a). The body of such a sled consists of control transfer instructions with relative addresses, all pointing directly to the shellcode. Thus, the flow of control will reach the shellcode in a *single* step from any point it may have entered the sled.

Trampoline-sleds can be directly implemented, relying on four-byte alignment, by cramming a jump instruction together with its operands into every four-byte-long slot of the sled. Even if a trampoline-sled has to be executable at every offset, an attacker can carefully choose the operands of the jump instructions to be valid NOP-equivalent opcodes, as explained in Section 2.3. An example of a small trampoline-sled that is executable at every byte offset is illustrated in Figure 3(b).

The shortest control transfer instructions available are two bytes long. For example, instructions such as jmp and loop take a one-byte operand that specifies the relative address of the jump target. The use of two-byte control transfer instructions places an additional restriction on the maximum jump displacement that can be used for sleds executable at each byte. Generally, the operand of these instructions is encoded as a signed 8-bit immediate value, which allows for a maximum forward relative offset of 127 bytes. Additionally, since the operand must at the same time act as a one-byte NOP-equivalent instruction, the maximum jump displacement is further reduced to the NOP-equivalent opcode with the greater signed integer value that is less than 128. The two NOP replacements with the largest such opcodes that we have come across are push imm8 and push imm32, which result to an offset of 106 and 104 bytes, respectively. Trampoline sleds are still feasible, though, by solely using jumps with relatively large positive displacements, which result to forward execution "bounces". Thus, the flow of control "jumps" and "strides" towards the shellcode.

2.6   Obfuscated Trampoline-sled

Since the number of control transfer instructions that can be used for the construction of trampoline-sleds is limited, one could argue that such sleds can be detected by searching for the specific opcodes of these instructions, much in the same way that Fnord does for NOP-equivalents (cf. Section 3.2).

The entropy of the basic trampoline-sled can be increased in order to evade detection, by interleaving NOP-equivalent instructions along with the jump instructions. In this way, the shellcode is not reached in a single step, but in a number of steps which can be tuned by the attacker. This will result to a sparse distribution of the control transfer instructions, which renders simple detection methods ineffective.

2.7   Static Analysis Resistant Sleds

Sleds of this type attempt to evade detection by making it difficult for detection heuristics to statically infer the outcome of the execution of the sled. When the sled is actually executed, its behavior is that intended by the attacker, correctly leading to the shellcode. This can be achieved by either using branches whose target cannot be determined statically or by using self-modifying code. Static analysis cannot follow branches that cannot be determined statically, such as register or memory indirect jumps, because the contents of the registers or memory are not known during the analysis. Therefore, it cannot continue with the inspection of the corresponding code paths and cannot determine their outcome. Such jumps, however, must specify the target as an absolute address.

Also, a sled could modify itself so that invalid instructions, appearing under static analysis to terminate a code path, are overwritten during execution by previous instructions and are actually executed normally. However, the sled must rely on stack alignment to avoid the execution of illegal instructions before they are fixed-up. Again, like indirect branches, write operations require an absolute address.

To overcome the absolute address problem, present in both indirect branches and self-modifying instructions, the `esp` register, which holds the stack frame's absolute address, can be used to find the buffer and sled addresses. However, the use of the `esp` register could hint for static analysis resistant sleds, but, in fact, the absolute address of the sled can be found even without using this register: knowing the injected return address and maintaining a counter while sliding through the sled provides knowledge of the absolute address of the current sled position. This seems to be relatively hard to implement, especially considering the need for 4-byte alignment.

Table 1. Comparative effectiveness of various sled detection schemes.

Sled Type	Scheme			
	Snort	Fnord	APE	STRIDE
1. NOP instructions	Yes	Yes	Yes	Yes
2. One-byte NOP-equivalents	No	Yes	Yes	Yes
3. Multi-byte NOP-equivalents	No	No	Yes	Yes
4. Four-byte Aligned	No	No	Yes	Yes
5. Trampoline-sled	No	No	No	Yes
6. Obfuscated Trampoline-sled	No	No	No	Yes
7. Static Analysis Resistant	No	No	No	After extension[*]

3.   SLED DETECTION MECHANISMS

In this section we briefly present three techniques which have been proposed for sled detection: NIDS signatures, the Fnord mutated sled detection plugin, and APE. Table 1 summarizes the effectiveness of each technique, along with our proposed detection mechanism, for each sled type.

3.1   NIDS Signatures

Detecting simple NOP sleds such as those described in section 2.1 is relatively straightforward. On the Intel IA-32 architecture, `nop` is a single-byte instruction with opcode 0x90. Thus, to detect a simple sled consisting only of `nop` instructions, a pattern matching rule searching for a sufficiently

---

[*] STRIDE can detect sleds that use indirect jumps, and we discuss how it may be extended to detect self modifying sleds in Section 6.

long sequence of bytes with value 0x90 is enough. Indeed, such rules exist for popular NIDSes, such as Snort[13].

3.2   Fnord

The Fnord[14] mutated sled detection plugin for Snort detects sleds by searching network traffic for long series of one-byte NOP-equivalent instructions. It is, therefore, capable to detect type-2 sleds, such as those described in Section 2.2. It may be the case that its list of NOP-equivalents could be extended with the opcodes of multi-byte NOP-equivalents, making it capable to detect type-3 sleds such as those described in Section 2.3, but we use the standard version here. However, Fnord definitely fails to detect type-4 sleds and above, that exploit the alignment of stack variables.

There also exist various other tools that offer similar sled detection capabilities with Fnord[20, 21]. Since these tools, along with Fnord, all rely on the NOP-equivalents list contained in ADMmutate in order to detect mutated sleds, it is sufficient to consider just one of them.

3.3   Abstract Payload Execution

APE[15] is a detection mechanism that enables the detection of sleds by looking for sufficiently long series of valid instructions: instructions which decode correctly and whose memory operands are within the address space of the process being protected against attacks. To reduce its runtime execution overhead, APE uses sampling to pick a small number of positions in the data from which it will start abstract execution. The number of successfully executed instructions from each position is called the Maximum Executable Length (MEL). When APE encounters a conditional branch, it follows both branches and considers the longest one as the MEL. If the destination of the branch can not be determined statically, APE terminates execution and uses the MEL value computed so far. A sled is detected if a sequence has a MEL value greater than 35. Although APE can be used to detect sleds of type-1 through type-4, it fails, however, to detect sleds of type-5 (trampoline sleds), type-6 (obfuscated trampoline), and type-7 (static-analysis-resistant sleds).

Indeed, although the purpose of type-5, and type-6 sleds is to transfer program control to the shellcode in as few steps as possible using jump instructions, the mechanism that is used by the APE scheme is based on the detection of a sufficiently long execution sequence of instructions, and thus, trampoline-sleds evade detection by having a short sequence of executed

```
stride(input, input_size, sled_length) {
    for (i=0; i < input_size-sled_length; i++) {
      if (find_sled(input+i, sled_length))
        return TRUE;
    }
    return FALSE;
}

find_sled(data, len) {
    for (j = 0; j < 4; j++)
      for (i = j; i < len; i+=4)
        if (!valid_sequence(data+i, len-i))
          return FALSE ;
    return TRUE;
}

is_valid_sequence(data,len) {
    /* decode "len" instructions in buffer "data" */
    res = decode(data, len);
    if (res == VALID_DECODE) return TRUE;
    if (res == ENDS_IN_JMP) return TRUE;
    return FALSE;
}
```

Figure 4. Pseudo-code for STRIDE algorithm

instructions. Static analysis resistant sleds also confuse APE, because it errs on the unsafe side when it cannot decide about a code sequence.

4.   THE STRIDE DETECTION ALGORITHM

In this section we describe STRIDE, our new sled detection mechanism which, compared to previous approaches, is able to detect more types of sleds with less false positives.

STRIDE is given some input data, such as a URL, and searches each and every position of the data to find a sled. If a sled is found, the input data are considered part of an attack. To detect a sled spanning over at least n bytes and starting at position i of the input data, STRIDE searches for all sequences of instructions of length $n - j$ bytes starting at offset $i + j$ of the input data, for all $j \in \{0...n-1\}$. If STRIDE finds all n sequences of instructions to be *valid sequences*, it then concludes that a sled of length n starts at position i.

We call a code sequence, starting at a certain point i in the input data, a "valid sequence of instructions of length n at position i," if it either (1) decodes correctly for n bytes without encountering privileged instructions, or if (2) a jump instruction is encountered along the way. Informally, a valid sequence of instructions is a sequence of instructions which can be used to construct a sled. Such a sequence may only contain valid instructions, and may not contain privileged instructions, i.e., instructions which can be invoked only by the operating system kernel.

Figure 4 gives the pseudo-code for STRIDE. The main routine, `stride`, consists of a loop which tries to find a sled of length `sled_length` at each and every position of input data `input`. Routine `find_sled(data, len)` finds a sled by attempting to valid all valid sequences of length `len`−i which start at position `data+i`, for al values of i. Aligned sleds are accounted-for by checking for valid sequences at every four bytes instead of at every byte but the check is applied for all four possible displacements.

STRIDE is related to previous approaches in several ways. Like Snort[13] and Fnord[14], STRIDE is able to find long sequences of NOP(-equivalent) instructions. Like APE[15], STRIDE is able to decode the input data and identify sequences of instructions that may be part of a sled. However, STRIDE has two major differences from APE:

Table 2. Detection rate of the various detection schemes for traces containing 10,000 different generated sleds of a single type.

Sled Type in Trace	Scheme			
	Snort	Fnord	APE	STRIDE
NOP instructions	100%	100%	100%	100%
One-byte NOP-equivalents	0%	55.4%	100%	100%
Multi-byte NOP-equivalents	0%	0%	100%	100%
Four-byte Aligned	0%	0%	100%	100%
Trampoline-sled	0%	0%	0%	100%
Obfuscated Trampoline-sled	0%	0%	Fig. 5	100%

- APE detects a sled when it finds a sufficiently long execution sequence of instructions. Therefore, although it is able to detect NOP-based sleds, APE can not detect trampoline sleds, because they jump directly to their destination code and, therefore, do not exhibit long execution sequences. In contrast, STRIDE may consider even short execution sequences (such as a single jump instruction) to be part of a valid sled.
- STRIDE verifies that each and every byte of a sled (apart from the cases of word-aligned sleds) is the start of a valid sequence of instructions. On the contrary, for APE it is enough to find only *one* sufficiently long execution sequence to consider it a valid sled.

5.   EXPERIMENTAL EVALUATION

We evaluate the accuracy of the detection rate of our proposed algorithm STRIDE, Snort's shellcode signatures[13], Fnord mutated sled detection plugin for Snort[14], and APE, by generating 10,000 different sleds of each type using the Metasploit Framework v2.2[17], modified to generate sleds ranging from type-1 (simple NOP sleds) up to type-6 (obfuscated trampoline).

We also evaluate the false positives rate of the four methods as in[15], by applying them to HTTP URIs. The URIs were captured from our institution's LAN, which contains about 150 hosts. Sled detection methods which are based on instruction decoding, employ the decoder used in[22].

5.1   Detection Rate

The results of applying all four detection methods on the generated sleds are shown in Table 2. We observe that Snort's shellcode signatures detect simple NOP sleds with 100% success, but fail to detect more elaborate sleds. Fnord is able to detect simple NOP sleds with 100% success too, and in addition is able to detect sleds with one-byte NOP-equivalent instructions with a 55.4% rate. Although it could have achieved a 100% rate for one-byte NOP-equivalent sleds, it achieves a lower-rate due to an incomplete NOP-equivalent instruction list. Fnord also fails to detect sleds with multi-byte NOP-equivalent instructions, but it should be possible to update its list of NOP-equivalents to include them as well. However, this is as far as Fnord can get. Indeed, Table 2 shows that Fnord fails to detect sophisticated sleds, such as 4-byte aligned and trampoline sleds.

Table 2 suggests that the APE method is able to detect simple NOP sleds, sleds with one-byte and multi-byte NOP-equivalent instructions, as well as four-byte aligned sleds with a 100% success rate. However, APE cannot detect trampoline sleds. This was expected, because trampoline sleds reach the core attack code by executing only a small number of jump instructions, while APE bases its detection method on the sequential execution of a long sequence of instructions.

It is interesting, however, to point out that although APE can not detect trampoline sleds, it is able to detect some of the more difficult obfuscated trampoline sleds. Indeed, as Figure 5 shows, APE is able to detect as many as 6% of the obfuscated trampoline sleds for small MEL. This is because the NOP-equivalent instructions that are used for the obfuscation cause an increase of the overall execution steps of the sled, which can now reach a

Finally, STRIDE is able to detect simple NOP sleds, sleds with one-byte or multi-byte NOP-equivalent instructions, as well as four-byte aligned sleds and plain or obfuscated trampoline sleds with 100% success, as expected.

low MEL threshold. Nevertheless, the detection rate of APE is still very low, at 6%, even for the minimum suggested MEL value.

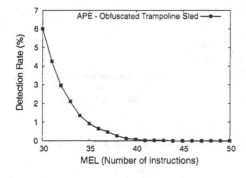

Figure 5. Detection rate for APE when applied to obfuscated sleds as a function of MEL.

Figure 6. Comparative effectiveness of the various detection schemes. The results for APE are for a MEL value of 35 with 100 samples per kilobyte and for STRIDE for sled length 130 bytes.

5.1   False Positives

The results of the false positives rate evaluation for the four methods with real traces are shown in Figure 6. In this experiment STRIDE has a sled length parameter of 130 bytes and APE has a MEL value of 35 instructions with 100 samples per kilobyte. With these parameters APE is sensitive, like STRIDE, to sled lengths of about 130 bytes and above.

The Snort shellcode signatures have zero false positives, because there was no sufficiently long NOP-sequence in our traces. Fnord also has almost 0% false positives, because there were very few sequences of bytes in the traces which corresponded to sequences of NOP-equivalent instructions. Although both Snort and Fnord have an attractive practically 0% of false positives rate, they are severely limited in their ability to detect elaborate sleds, such as trampoline sleds. Figure 6 shows that APE has a false positive rate of 0.006%. Finally, STRIDE has a false positive rate of 0.00027%, close to an order of magnitude smaller than APE. Overall, we see that STRIDE seems to strike a good balance between true positives and false positives. That is, it is able to find more true positives than any other method, while keeping the false positives as low as those of Fnord and Snort.

The interested reader should notice that the exact value of false positives for APE and STRIDE depends heavily on their parameters. To explore the influence of the parameters to the false positive rate of APE and STRIDE,

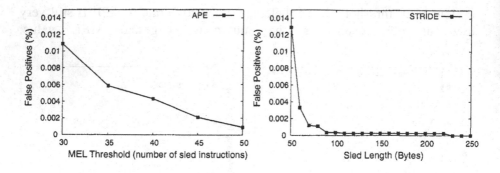

Figure 7. False positives rate for APE and STRIDE with varying parameters.

we investigate the false positives rate for both methods as a function of MEL and sled length, and display the results in Figure 7. We see that as the size of MEL increases, the percentage of false positives for APE decreases. However, we should point out that larger MEL values also decrease the number of detected true positives, as can be seen in Figure 5. Figure 7 also shows that the percentage of false positives for STRIDE decreases with the sled length, and reaches zero for sled length larger than 230 bytes. This is an encouraging result, since typical sleds are usually longer than 250 bytes. Overall, our results suggest that STRIDE is able to have a true positive rate of 100% (Table 2), while having a false positive rate of (close to) 0%.

5.2   PERFORMANCE

Besides being accurate, a worm detection method should also be fast, so as to be able to detect worms in real-time. To evaluate the speed of STRIDE we measured the CPU time consumed by STRIDE with a sled length value of 200 bytes, and compared it to the execution time of APE with a MEL count of 35 on a Pentium 4 machine (2.6GHz clock speed, 512KB cache size) for a trace with 1,093,249 requests. The CPU time for processing this trace was 25 sec for APE (22.9 usec per request) and 4.85 sec for STRIDE (4.4 usec per request). We see that STRIDE outperforms APE by a factor of 5. This is mostly due to the different handling of branch instructions by the two algorithms. Indeed, when APE encounters a branch instruction, whose target can be determined statically it follows both branches, a decision, which may potentially lead to the exploration of an exponential number of execution paths. Unlike APE, when STRIDE encounters a branch instruction, it assumes that it found a valid sequence, without making any attempt to follow the branch. By being conservative, STRIDE avoids the exponential explosion and significantly reduces the associated run-time cost.

6. DISCUSSION

Although our evaluation has shown that STRIDE has some benefits, we should mention that it still has limitations. For example, STRIDE can only be applied to buffer-overflow-based attacks which use sleds. If an attack does not make use of a sled, then it can not be detected by STRIDE. In addition, STRIDE still cannot detect self-modifying sleds. It is, however, possible, to extend it with a decoder that is capable of identifying memory write operations and handle the sequences that contain them the way it currently handles jump instructions: consider them valid, because they cannot be proved invalid with static analysis. Finally, a worm writer could blind STRIDE by adding invalid instruction sequences at suitable locations in the sled. Note that this would most likely lead to a fraction of the infection attempts crashing the remote process and would most certainly slow down the spread of the worm, while also exposing it to other detection components that look for anomalous behavior at the process-level.

7. SUMMARY AND CONCLUDING REMARKS

We have presented STRIDE, a new approach for network-level detection of buffer overflow attacks, which relies on the identification of the sled component that is usually part of such attacks. Because it operates at the network-level, STRIDE can be used for detecting worms that replicate through buffer overflow exploits, even if they involve elaborate obfuscation. Our analysis allows us to make three main observations:

- *STRIDE can detect several classes of sleds that cannot be identified by previous proposals.* As presented in Section 2.5, trampoline sleds can be used by attackers in order to evade current sled-based detection mechanisms. STRIDE detects such sleds, even in their obfuscated variations.
- *STRIDE achieves high detection rates while maintaining low false positive rates.* Snort and Fnord have few false positives but can only detect basic sled types. APE detects more complex sleds, but has an order of magnitude more false positives compared to STRIDE, while also missing two classes of sleds. Our approach can detect all types of sleds presented in this paper, except for Static Analysis Resistant sleds, with a detection rate of 100%, and a false positive rate that reaches 0% for

reasonable algorithm parameters. As suggested in Section 6, it may also be possible to detect Static Analysis Resistant sleds. This question, however, requires additional analysis and is outside the scope of this paper.

- *STRIDE is more efficient in terms of processing cost.* As shown in Table 3, STRIDE has relatively low computational cost, outperforming APE by a factor of 5. This suggests that STRIDE can operate on high-speed links and remain effective even under heavy loads, at a reasonable cost.

The high accuracy, low false positive rate, and low processing cost achieved by STRIDE suggest that it is likely to be highly useful as part of an automated network-level defense mechanism against both targeted attacks and large-scale zero-day worm outbreaks, especially as worms become more aggressive and more sophisticated.

ACKNOWLEDGEMENTS

This work was supported in part by the IST project NoAH (011923) funded by the European Union and the GSRT project EAR (USA-022) funded by the Greek Secretariat for Research and Technology. The work of K. Anagnostakis is also supported by OSD/ONR CIP/SW URI through ONR Grant N00014-04-1-0725. P. Akritidis, E. P. Markatos, and M. Polychronakis are also with the University of Crete. The work of K. Anagnostakis was done while at ICS-FORTH.

REFERENCES

1. Aleph One. Smashing the stack for fun and profit. Phrack, 7(49), Nov. 1996. http://www.phrack.org/phrack/49/P49-14.
2. A. Baratloo, N. Singh, and T. Tsai. Transparent run-time defense against stack smashing attacks. In Proceedings of USENIX Annual Technical Conference, June 2000.
3. C. Cowan, M. Barringer, S. Beattie, and G. Kroah-Hartman. Formatguard: Automatic protection from printf format string vulnerabilities. In Proceedings of the 10th USENIX Security Symposium, August 2001.
4. J. J. C. Cowan, S. Beattie and P. Wagle. PointGuard: Protecting pointers from buffer overflow vulnerabilities. In Proceedings of the 12th USENIX Security Symposium, August 2003.
5. C. Cowan, C. Pu, D. Maier, M. Hinton, J. Walpole, P. Bakke, S. Beattie, A. Grier, P. Wagle, and Q. Zhang. Stackguard: Automatic adaptive detection and prevention of buffer-overflow attacks. In Proc. of the 7th USENIX Security Symposium, January 1998.
6. M. Frantzen and M. Shuey. StackGhost: Hardware facilitated stack protection. In Proceedings of the 10th USENIX Security Symposium, August 2001.

7. I. Goldberg, D. Wagner, R. Thomas, and E. A. Brewer. A Secure Environment for Untrusted Helper Applications: Confining the Wily Hacker. In Proc. of the 5th USENIX Security Symposium, 1996.
8. V. Kiriansky, D. Bruening, and S. Amarasinghe. Secure execution via program sheperding. In Proceedings of the 11th USENIX Security Symposium, 2002.
9. T. Detristan, T. Ulenspiegel, Y. Malcom, and M. Underduk. Polymorphic shellcode engine using spectrum analysis. Phrack, 11(61), Aug. 2003. http://www.phrack.org/phrack/61/p61-0x09_Polymorphic_Shellcode_Engine.txt.
10. C. Nachenberg. Understanding and managing polymorphic viruses. White Paper, July 1996. http://www.symantec.com/avcenter/reference/striker.pdf.
11. O. Kolesnikov, D. Dagon, and W. Lee. Advanced polymorphic worms: Evading IDS by blending in with normal traffic, 2004. http://www.cc.gatech.edu/áok/w/ok_pw.pdf.
12. P. Szor and P. Ferrie. Hunting for metamorphic. White Paper, Sept. 2001. http://www.symantec.com/avcenter/reference/hunting.for.metamorphic.pdf.
13. M. Roesch. Snort: Lightweight intrusion detection for networks. In Proceedings of USENIX LISA 99, Nov. 1999. (software available from http://www.snort.org/).
14. D. Ruiu. Fnord: Multi-architecture mutated NOP sled detector, Feb. 2002. http://www.cansecwest.com/spp_fnord.c.
15. T. Toth and C. Kruegel. Accurate buffer overflow detection via abstract payload execution. In Proceedings of the 5th International Symposium on Recent Advances in Intrusion Detection (RAID), Oct. 2002.
16. K2. ADMmutate. http://www.ktwo.ca/ADMmutate-0.8.4.tar.gz.
17. Metasploit project, 2004. http://www.metasploit.com/.
18. IA-32 Intel Architecture Software Developer's Manual vol. 1-3. http://developer.intel.com/design/ pentium4/manuals/index_new.htm.
19. K. S. Gatlin. Windows data alignment on IPF, x86, and x86-64, Feb. 2003. MSDN Library, http://msdn.microsoft.com/.
20. Prelude IDS. http://www.prelude-ids.org/.
21. NIDSFindShellcode. http://www.ngsec.com/downloads/misc/NIDSfindshellcode.tgz.
22. T. Toth. Apache buffer overflow detector module, Mar. 2002. http://www.infosys.tuwien.ac.at/Staff/tt/abstract_execution/.

PIRANHA: FAST AND MEMORY-EFFICIENT PATTERN MATCHING FOR INTRUSION DETECTION

S. Antonatos[1], M. Polychronakis[1], P. Akritidis[1], K.G. Anagnostakis[2] and E.P. Markatos[1]

[1]*Institute of Computer Science Foundation for Research and Technology Hellas, P.O Box 1385 Heraklio, GR-711-10 Greece {antonat, mikepo, akritid, markatos}@ics.forth.gr;* [2]*Distributed Systems Laboratory, CIS Department, Univ. of Pennsylvania, 200 S. 33rd Street, Philadelphia, PA 19104 anagnost@dsl.cis.upenn.edu*

Abstract: Network Intrusion Detection Systems (NIDS) provide an important security function to help defend against network attacks. As network speeds and detection workloads increase, it is important for NIDSes to be highly efficient. Most NIDSes need to check for thousands of known attack patterns in every packet, making pattern matching the most expensive part of signature-based NIDSes in terms of processing and memory resources. This paper describes Piranha, a new algorithm for pattern matching tailored specifically for intrusion detection. Piranha is based on the observation that if the rarest substring of a pattern does not appear, then the whole pattern will definitely not match. Our experimental results, based on traces that represent typical NIDS workloads, indicate that Piranha can enhance the performance of a NIDS by 11% to 28% in terms of processing time and by 18% to 73% in terms of memory usage compared to existing NIDS pattern matching algorithms.

Key words: network security; intrusion detection; pattern matching; network monitoring; network performance.

1. INTRODUCTION

Network Intrusion Detection Systems (NIDSes) provide a powerful mechanism to defend against well-known attacks on a computer network or detect network abuse. NIDSes are mainly divided into two major categories: *signature-based* and *anomaly detection*. Anomaly-detection NIDS try to spot abnormal behavior on network based on statistics like rate of connections, traffic overload or unusual protocol headers. On the contrary, the detection mechanism of a signature-based NIDS is based on a set of *signatures*, each describing a known attack. As an example, a signature taken from latest Snort, is

alert tcp any any -> HTTP_SERVER 80 (content:"/root.exe"; nocase;)

This signature instructs that if "/root.exe" is found inside the payload of a TCP packet that is originating from any host and any source port and is destined to an HTTP server on port 80, then an attack on the web server is taking place. While this signature requires full packet inspection, there exist simpler signatures that require only header lookups. Pattern matching inflicts a significant cost to the performance of a NIDS. Previous research results suggest that 30% of total processing time is spent on pattern matching[13], while in some cases, like Web-intensive traffic, this percentage raises up to 80%[6]. Apart from processing time, memory demands of a NIDS may reach at high levels due to rule-set growth. Although algorithms with low memory demands have been developed, their performance in comparison with algorithms that consume more memory is still poor. Given the fact that link speed increases every year, pattern matching evolves to a highly demanding process that needs special consideration. Minimizing the demands of pattern matching leaves headroom for further heuristics to be applied for intrusion detection, like anomaly detection or sophisticated preprocessors.

In this paper, we present Piranha, a pattern-matching algorithm designed for and applied to a NIDS. Our experiments with Piranha implemented in *Snort v2.2* indicate that Piranha is faster than existing algorithms by up to 28% in terms of processing time, and requires up to 73% less memory. This improvement relies on the small number of collisions and the compact memory footprint of the algorithm.

The rest of the paper is organized as follows: in Section 2 a description of existing state-of-the-art algorithms is provided, Section 3 depicts the Piranha algorithm, while Section 4 presents the performance of Piranha compared to other algorithms in various traffic scenarios and hardware platforms. Finally, our concluding remarks are discussed in Section 5.

2.   BACKGROUND

In this section we describe how a content matching NIDS operates and summarize the key characteristics of pattern matching algorithms that have been recently used in intrusion detection.

2.1   Basic NIDS model

A NIDS is usually designed as a passive monitoring system that reads packets from a network interface through standard system facilities, such as *libpcap*[10]. After a set of normalization passes (e.g., IP fragment reassembly, TCP stream reconstruction, etc.) each packet is checked against the NIDS rule-set. Some -rather old- NIDS organize their rule-set as a two-dimensional data-structure chain, where each element, often called a *chain header*, tests the input packet against a packet header rule. When a packet header rule is matched, the chain header points to a set of signature tests, including payload signatures that trigger the execution of the pattern matching algorithm. Pattern matching is the single most expensive operation of a NIDS in terms of processing cost. Latest versions of *Snort* (above version 2.0) organize the rules in groups. Rules that check for the same destination port belong to the same group[14]. When a packet arrives, its destination port is used to find the appropriate group. Afterwards, multi-pattern matching is performed on patterns of the group in order to extract a set of rules that possibly match. Each rule of this set is then examined separately.

In order to understand the interaction between pattern matching algorithm, rule-set and experimental workload, we briefly present some of the pattern matching algorithms that are commonly used in intrusion detection systems.

2.2   Pattern matching algorithms

A number of algorithms have been proposed for pattern matching in a NIDS. The performance of each algorithm may vary according to the case in which it is applied. The multi-pattern approach of Boyer-Moore is fast for a few rules, but does not perform well when used for a large set. On the contrary, Wu-Manber behaves perform well when used with large rule-sets. On the contrary, Wu-Manber behaves well on large sets, but its performance starts to degrade when short patterns appear in rules. E^2xB is based on the idea that in most cases we have a mismatch and tries to filter out patterns that do not match. However, E^2xB introduces additional preprocessing cost per packet, which is amortized only after a certain number of rules. In the following subsections a more detailed description for each algorithm is provided.

2.2.1   The Boyer-Moore algorithm

The most well-known algorithm for matching a single pattern against an input was proposed by Boyer and Moore[4]. The Boyer-Moore algorithm compares the search pattern with the input, starting from the rightmost character of the search pattern. This allows the use of two heuristics that may reduce the number of comparisons needed for pattern matching (compared to the naive algorithm). Both heuristics are triggered on a mismatch. The first heuristic, called the *bad character heuristic*, works as follows: if the mismatching character appears in the search pattern, the search pattern is shifted so that the mismatching character is aligned with the rightmost position at which the mismatching character appears in the search pattern. If the mismatching character does not appear in the search pattern, the search pattern is shifted so that the first character of the pattern is one position past the mismatching character in the input. The second heuristic, called the *good suffixes heuristic*, is also triggered on a mismatch. If the mismatch occurs in the middle of the search pattern, then there is a non-empty suffix that matches. The heuristic then shifts the search pattern up to the next occurrence of the suffix in the pattern. Horspool[8] improved the Boyer-Moore algorithm with a simpler and more efficient implementation that uses only the bad-character heuristic. Fisk and Varghese[6] recently developed Set-Wise Boyer-Moore (SWBM), an algorithm based on Boyer-Moore concepts and operating on a set of patterns. SWBM was integrated in *Snort* and tested using a single traffic trace from an enterprise Internet connection.

2.2.2   The E^2xB algorithm

E^2xB is a pattern matching algorithm designed for providing quick negatives when the search pattern does not exist in the packet payload, assuming a relatively small input size (in the order of packet size)[2,9]. As mismatches are by far more common than matches, pattern matching can be enhanced by first testing the input (i.e., the payload of each packet) for *missing* fixed-size sub-strings of the original signature pattern, called *elements*. The collisions induced by E^2xB, i.e., cases with all fixed-size sub-strings of the signature pattern showing up in arbitrary positions within the input, can then be separated from the actual matches using standard pattern matching algorithms, such as Boyer-Moore[4]. The small input assumption ensures that the rate of collisions is reasonably small -experiments have shown collision rates of 10% in the worst case-. In the common case, negative responses can be obtained without resorting to general-purpose pattern matching algorithms. The E^2xB algorithm was evaluated with traffic traces from diverse environments, including traces containing attacks, traces with normal web traffic, and WAN traffic traces from a local ISP.

2.2.3   The Wu-Manber algorithm

The most recent implementation of *Snort* uses a simplified variant of the Wu-Manber multi-pattern matching algorithm[16], as discussed by Snort developers[14]. The "MWM" algorithm is based on a bad character heuristic similar to Boyer-Moore, but uses a one or two-byte bad shift table constructed by pre-processing all the patterns instead of only one. MWM performs a hash on the two-character prefix of the current input to index into a group of patterns, which are then checked starting from the last character, as in Boyer-Moore. The performance of MWM was originally measured using text files and various sets of patterns. The first attempt to measure MWM as the basic algorithm for pattern matching in a NIDS was performed in recent Snort implementation[14]. The results of previous studies[14] show that *Snort* is much faster than previous versions that used Set-Wise Boyer-Moore and Aho-Corasick[1].

3.   IMPLEMENTATION

The Piranha algorithm is based on the idea that if we find the rarest 4-byte substring of a pattern inside the packet payload, then we assume that this pattern matches. Each pattern is now represented by its least popular 4-byte sequence, where popular reflects the number of times that a specific substring exists in all patterns. For all the instances of the rare substring, *Snort* is instructed to check the corresponding rule. Piranha itself can only handle patterns with length greater or equal to 4. For completeness, patterns with length less than 4 are handled separately.

3.1   Preprocessing

Piranha treats every byte-aligned pattern as a set of 32-bit sub-patterns. For example, the pattern *"/admin.exe"* (R1) is considered as the set of its 32-bit byte-aligned sub-patterns, i.e., *"/adm"*, *"admi"*, *"dmin"*, *"min."*, *"in.e"*, *"n.ex"* and *".exe"*. The 32-bit partitioning was chosen as the use of integers results to faster operations. Pattern matching can then be formulated in terms of an AND operation. Every pattern is represented by a gate. The gate has as many inputs as the number of its 32-bit sub-patterns. Each input represents whether the 32-bit sub-pattern has appeared in the payload or not. The gate for pattern R1 can be seen on the top-right part of Figure 1 with all its sub-patterns constituting the inputs of the gate. Initially, all inputs are set to zero, and are being switched on based on the sequences seen on the packet. However, the output must not be regarded as an exact match.

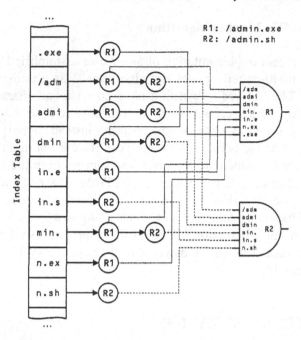

Figure 1. An example of index table and gates for two patterns. When all the inputs of a gate are switched on, then the pattern is possibly matched

For example, if the packet payload is *"/admAAAdmin.exe"*, then, despite the fact that all 4-byte sequences for R1 have appeared, the pattern itself does not match. Each time the output of the gate is switched on, we consider it as collision and *Snort* is instructed for further inspection. In order to find fast which inputs to switch on, an index table is maintained. The index table keeps for all 4-byte sequences a list of all patterns that contain them. For example, if we assume that we only have the patterns *"/admin.exe"* (R1) and *"/admin.sh"* (R2), a view of the table is displayed in Figure 1. Sequences *"/adm"*, *"admi"*, *"dmin"*, and *"min."* appear in both patterns, while *".exe"*, *"in.e"*, and *"n.ex"* exist only in R1, and *"in.s"* and *"n.sh"* only in R2. Each time a node of index table is reached then the appropriate input is switched on. As an example, if the payload is *"min.exe"*, we first access the *"min."* entry of index table and we switch on the *"min."* inputs for R1 and R2, afterwards the *"in.e"* entry and switch on the *"in.e"* input for R1, then we access the *"n.exe"* entry and switch on the input for R1 and finally the *".exe"* entry is traversed. The performance of Piranha for a subset of our packet traces, in terms of running time and collisions per packet, is displayed in Table 1 under the "full gates" column.

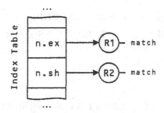

Figure 2. Optimized view of index table

Although gates present a low rate of collisions, their performance is poor as a lot of steps and transitions are needed in order to take a decision whether a pattern matches or not. In a typical case, the index table is firstly accessed, then the appropriate input is switched on and then the whole gate is checked if all inputs are switched on. In our effort to reduce the number of steps, and consequently, memory accesses, an optimization phase takes place. The optimization phase involves the procedure of selecting one input for each gate, a representative sequence. The rarest sequence is chosen as representative. It is defined as the sequence found in the least number of rules and can be found through the index table by counting the number of rules that is contained in. All other inputs are removed from the gate as well as the corresponding nodes from the index table. For example, trying to optimize our previous example we keep sequence "*n.ex*" as representative for pattern R1 and "*n.sh*" for R2. The optimized view of the index table is illustrated in Figure 2. After the optimization phase, every gate has only one input, and thus, it can totally be removed (output is equal to input), as we can use the index table for the searching phase -if a node of the index table is reached then a possible match is triggered-.

The effect of optimization is shown in Table 1, in terms of running time and collisions. The *"full gates"* column represents the unoptimized case of Piranha, and the *"representative sequence"* refers to the optimized case. Although collisions per packet increase as now only one input triggers possible match, the performance increases due to decrease of steps and compactness of memory footprint. Performance is increased by up to 36% even if collisions are two to three times more.

Table 1. Effect of optimizing gate inputs. Collisions increase but running time decreases as less steps and memory are required

	Full gates		Representative sequence	
	Running time	Collisions	Running time	Collisions
forth.web	37.71	0.61	24.06	1.65
forth.tr	36.14	0.29	27.60	0.67
forth.tr2	34.58	0.29	27.04	0.63
ideval2	12.07	0.33	9.58	1.06
ideval3	13.16	0.25	10.68	0.93

With further optimization during the searching phase as it is described in Section 3.2, collisions and running time drop significantly.

3.2   Searching

The searching phase of Piranha is straightforward. For each 4-byte sequence of the packet payload, the index table is consulted in order to find the patterns that contain this sequence. All these patterns are then sent to *Snort* for further inspection. Following our previous example, if the payload is *"/login.sh"*, we have to check sequences *"/log"*, *"logi"*, *"ogin"*, *"gin."*, *"in.s"* and *"n.sh"*. According to the index table, *"n.sh"* is found in pattern R2, so we assume that R2 is matched. The rest of the sequences are not contained in any pattern so no checks are necessary. Such an approach would trigger further inspection multiple times for each packet, as shown in Table 2 (*"No check"* case). We observe that, in the average case, in an unoptimized search we trigger one rule per packet, which is prohibitive in terms of performance. In our effort to reduce collisions, we perform a trivial check before the decision that a pattern is matched. The last two characters of the pattern are checked against the corresponding two characters in the payload, and if the check succeeds then further inspection is triggered. The effect of this optimization is summarized in Table 2. In some cases, up to 75% of triggers are eliminated while the minimum reduction reaches 50%.

4.   EXPERIMENTS

We evaluated the performance of Piranha against E^2xB and MWM algorithms in *Snort 2.2* using a set of packet traces. All Snort preprocessors were disabled.

Table 2. Collisions per packet without and with checking last 2 bytes of pattern against payload

	No check	Check last 2 bytes
forth.web	1.65	0.62
forth.tr	0.67	0.24
ideval2	1.06	0.32

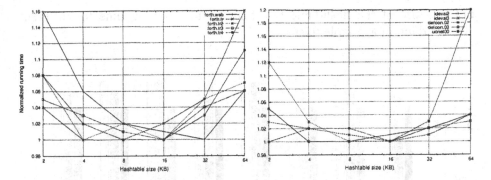

Figure 3. Effect of hash-table size on running time

4.1 Environment

All the experiments were conducted on a machine equipped with a Pentium 4 processor running at 2.80GHz, 8KB of L1 cache, 512KB of L2 cache, and 1GB of main memory. The host operating system was Linux (kernel version 2.4.0, Redhat 9.0). We used five sets of packet traces from diverse environments. The first set consists of a full packet trace containing Web traffic (*forth.web*), generated by concurrently running a number of recursive *wget* requests on popular portal sites from a host within the FORTH network. The second set contains two full packet traces (*forth.tr*) and *forth.tr2*) collected in a local area network at Institute of Computer Science inside FORTH. The third set includes a full-packet trace from the DEFCON "capture the flag" data-set (*defcon.02*). This trace contains numerous intrusion attempts. The fourth set consists of two full-packet traces (*ideval2* and *ideval3*) which were collected during the DARPA evaluation tests at MIT Lincoln Laboratory. Finally, a header-only trace with uniformly random payload (*ucnet00*) collected on the OC3 link connecting the University of Crete campus network (UCNET) to the Greek academic network (GRNET)[5] was used.

4.2 Effect of hash-table size

A complete index table of 32-bit-long patterns would normally contain 2^{32} entries, an outrageous number in terms of memory usage. In order to keep the memory footprint as small as possible, the index table was implemented as a hash-table. Since the memory footprint and locality of accesses is critical to the performance of the algorithm, we determined the optimal size of the hash-table by obtaining the running time for different sizes and for all available traces.

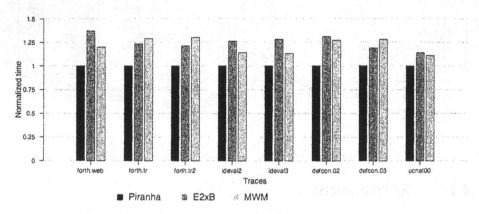

Figure 4. Running time for E2xB, Piranha and MWM for patterns with length greater or equal to 4

Results are summarized in Figure 1. Running times for each set are presented normalized to the lowest value. The time was measured using the *time* facility of the operating system. Small hash-tables suffer from conflicts and consequently longer chains have to be traversed in order to find the correct index. A large hash-table, on the contrary, has fewer conflicts but for every access a performance penalty is paid due to poor cache behavior. We observe that optimal size of the hash table for most of the traces is around 16KB and this is the size we used for all our experiments presented in the paper.

4.3   Comparison against other algorithms

We compared Piranha against MWM[14] and E[2]xB[2,9] on all available traces. In our experiments, we measured running time in user space (kernel time was negligible). Results are presented in Figure 4. Times are presented normalized against the running time of Piranha algorithm.

The performance of Piranha is consistently better compared to other algorithms. Improvement ranges between 10 and 23.50%, with the results remaining the same for the rest of the traces that are not displayed in Figure 4. We also compared our algorithm with AC-Banded[11], an optimized implementation of Aho-Corasick[1], but running time of AC-Banded was two to four times the time of our algorithm. Results in Figure 4 are for patterns with length greater or equal to four, as four is the length that can be natively handled by Piranha. For completeness reasons, the case of small pattern was also implemented. Small patterns impose a performance bottleneck for Piranha and MWM as well as E^2xB. MWM can natively handle patterns with length greater or equal to two while patterns with length one are examined separately. The overhead that small patterns impose in terms of running time can be seen on Table 3. In average case, running time was decreased by 25% for Piranha and 20% for MWM. The effect on E^2xB is smaller as it is not dependent to pattern length but proportional to the number of patterns. In the last two columns of the table we can observe the performance benefit of Piranha against MWM and E^2xB for all pattern lengths. Despite the performance bottleneck, our algorithm still performs better for all available traces, except the case of *defcon.02* trace where improvement is marginal. However, our main contribution is focused on patterns with a fair enough large size as only 3% of patterns have length less than four.

Piranha does not only perform better in terms of processing time but also in terms of memory usage. While MWM requires 45MB of memory to process the full rule-set, AC-Banded 96MB and Aho-Corasick 140MB, Piranha consumes only 37MB. Efforts have been made recently in order to develop algorithms with low memory consumption. Tuck et al.[15]; have developed two modified versions of Aho-Corasick, AC-Bitmap and AC-Path, that reduce memory usage. AC-Bitmap needs 20MB memory while AC-Path only 15MB. However, such algorithms present very high processing time. Comparing Piranha with AC-Bitmap and AC-Path, we observed that they need, in average, three to four times more processing time. *Snort* also comes with *SFKSearch*, an algorithm that requires only 14MB of memory, but its performance compared to others is poor - three to four times more processing time against Piranha -. The tradeoff between memory usage and processing time can be seen on Figure 5. Algorithms with low memory usage need three to four times more processing time, while algorithms with high memory usage present high processing capacity. Although the assumption that low memory means high processing time cannot be generalized, there are strong indications that this tradeoff might hold for other algorithms that are not discussed here.

4.4 Evaluation on different architectures

We evaluated the performance of Piranha on different hardware architectures. Our testing environment, besides the machine described in Section 4.1, consists of a Pentium Xeon 2.4 GHz with 8KB L1 cache, 512KB L2 cache and 512MB main memory, an AMD Athlon MP 1.8GHz with 128KB L1 cache, 256KB L2 cache and 512MB main memory and a Pentium 3 running at 600 MHz with 8KB L1 cache, 256KB L2 cache and *Table 3*. Effect of small-patterns on running time

	Piranha pattern length		MWM pattern length		E2xB pattern length		Piranha vs. MWM	Piranha vs. E^2xB
	>=4	all	>=4	all	>=4	all	%	%
forth.web	21.05	30.17	25.32	33.59	28.86	34.12	10.18	11.57
forth.tr	23.78	30.78	30.80	35.65	29.80	31.18	13.66	1.28
forth.tr2	26.55	30.37	30.23	36.12	29.91	30.46	15.91	0.29
ideval2	8.49	11.36	9.68	12.70	10.84	13.25	10.55	14.26
ideval3	9.88	12.89	11.26	14.58	12.69	15.25	11.59	15.47
defcon.02	7.06	9.91	8.99	9.97	9.42	9.96	0.60	0.50
defcon.03	7.20	8.74	8.59	9.20	8.18	8.99	5.00	2.78
ucnet00	3.11	3.59	3.48	4.21	3.59	3.81	14.72	5.77

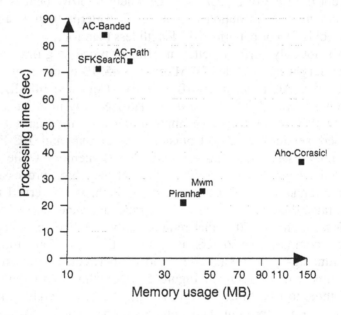

Figure 5. Memory usage against processing time

512MB main memory. Results are presented in Figure 6. Running time is normalized against the time of Piranha running on P4 at 2.8GHz.

Independent of the underlying hardware platform, Piranha performs better for all traces. As processor clock speed decreases, performance of both algorithms decreases as expected. However, the performance gap seems to decrease with the clock speed for specific traces while for others it remains constant. On Pentium Xeon 2.4GHz, improvement waves between 7.8% and 18.8% while on Pentium 3 600MHz between 10.86% and 14.83% (leaving out the *ucnet00* trace where improvement is marginal). Similar results apply to the AMD Athlon architecture, where improvement is ranged between 7.32% and 18.21% (again *ucnet00* trace is omitted).

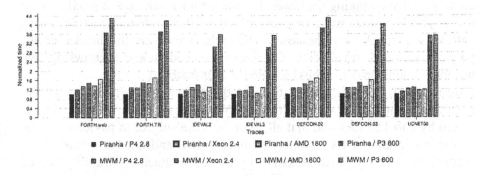

Figure 6. Performance of Piranha and MWM on different architectures

4.5 Performance under attack

Intrusion detection systems are themselves subject to being attacked. Some types of attack try to evade NIDS by exploiting weaknesses in protocol handling, like IP defragmentation or TCP reassembly[7,12]. Other attacks aim at overloading the detection engines by exploiting weaknesses in the internal algorithms used, in our case pattern matching. The attacker sends packets with carefully crafted payload in order to force the pattern matching engine to spend more processing time than it would require for an innocent packet. Most of the traffic is then dropped by the NIDS, including packets containing attack, giving the attacker the chance to evade detection. Our previous work on such attacks has shown that the processing time of *Snort* can be raised by up to 25 times[3]. Although the worst case scenario for each algorithm and the *Snort* itself is extremely difficult to be generated, we provide some hints on how a NIDS can be heavily overloaded. For performance reasons, *Snort* firstly performs the multi-pattern matching and then for all possible matches the whole rule is checked: header processing and exact string matching for all patterns that the rule contains[14]. Examining the groups of rules that are processed during packet inspection, it can be observed that rule

alert tcp any any → any any (ack:0; flags:SFU12; content:"AAAAAAAAAAAAAAAA"; depth:16;)

is found in all groups as it applies to all source and destination ports. That means that for all packets examined, *Snort* will try to locate the pattern ``AAAAAAAAAAAAAAAA" and for all possible matches will check the rest of the rule. In our example, after the pattern matching phase the acknowledgment number and the TCP flags will be verified. We constructed an attack trace by taking the headers of the *forth.web* and placing only "A" in the payload. In that way, in every offset of the payload *Snort* finds that pattern and checks for the rest of the rule. However, the header of the packets is normal (no special TCP flags are turned on and acknowledgment number in non-zero) and thus the rule is never matched.

Forcing *Snort* to generate matches and checks in every offset is very expensive as it can be seen on Table 4. We observe that processing time is raised by 3 to 15 times and that all algorithms are subject to payload attacks, as the way *Snort* performs detection is exploited and not the nature of the algorithms. Such overload factors can provide the attacker the ability to hide his attack among legitimate traffic. Other payloads were also crafted, like payload including only "a". As the packet payload is capitalized, possible matches are also generated and the overloading still takes place. In the case of MWM, running time is increased further as there are some patterns that start with "aa" and trigger more inspections on the internal structures of MWM. The Aho-Corasick-like algorithms try -as an optimization- to verify their match by calling *memcmp()* for pattern against the payload before forcing Snort to check the whole rule. The cost of memory-comparing is increasingly high as in each offset a comparison is performed. However, there are some cases where a specific payload can cause Piranha to generate collisions in most of the payload offsets but Aho-Corasick-like algorithms are not affected. This payload can be made by replacing the last character of *"AAAAAAAAAAAAAAAA"* pattern with another character, like *"B"*. Piranha decides that pattern matches only by seeing the appearance of an *"AAAA"* but the whole pattern is not really matched. Aho-Corasick algorithm detect that the whole pattern cannot be matched so their time remains practically the same. As Table 4 shows, only Piranha and MWM suffer from this payload attack. Focusing on the worst overall performance (the *"worst overall"* column) among all attacks described above, Piranha needs 3 times less running time than other algorithms.

Table 4. Completion time and overhead factor (attack completion time / original completion time) for different attack payloads. "Time" denotes completion time and "factor" denotes overhead factor

		Packet payload						
	Original	AAAAA...		aaaaa...		AAA...B...		Worst overall
	Time	Time	Factor	Time	Factor	Time	Factor	
Piranha	21.94	120.01	5.46	118.50	5.40	91.47	4.16	120.01
MWM	25.91	233.73	9.02	376.72	14.53	204.88	7.90	376.72
AC	35.71	417.72	11.69	361.45	10.12	28.98	0.81	417.72
AC-path	81.59	357.84	4.38	212.62	2.60	78.81	0.96	357.84
AC-Bitmap	72.87	409.74	5.62	241.88	3.31	110.65	1.51	409.74

5.   CONCLUDING REMARKS

We have presented the design of Piranha, a novel pattern matching algorithm for NIDS and evaluated its performance under various network traffic characteristics using a diverse set of packet traces. Our comparison against existing algorithms shows that an improvement of up to 28% can be achieved. The improvement is due to its quick decisions on which patterns may match and to its compact memory footprint which infers good cache behavior. Our results on different architectures indicate that Piranha performs consistently better, with the performance gain increasing along with processor speed. Furthermore, we have concluded to some general remarks for pattern matching on NIDS: small patterns inflict a significant performance overhead that needs to be examined carefully, and cache-conscious programming of a NIDS pattern-matching algorithm is a key element to its performance.

ACKNOWLEDGEMENTS

This work was supported in part by the IST project SCAMPI (IST-2001-32404) funded by the European Union and in part by the i-Guard GSRT Project (02-PRAXE-212) funded by the Greek General Secretariat for Research and Technology through PRAXE A. Work of K.G. Anagnostakis is also supported in part by ONR under Grant N00014-01-1-0795. E. P. Markatos, S. Antonatos, M. Polychronakis and P. Akritidis are also at University of Crete. Work of K. G. Anagnostakis was done while at ICS-FORTH. We would also like to thank Vasilis Siris for providing the UCnet traces.

REFERENCES

1. Aho and M. Corasick, Fast pattern matching: an aid to bibliographic search. Commun. ACM, 18(6):333-340, June 1975.
2. K. G. Anagnostakis, E. P. Markatos, S. Antonatos, and M. Polychronakis, E2xB: A domain-specific string matching algorithm for intrusion detection. In Proceedings of the 18th IFIP International Information Security Conference (SEC2003), May 2003.

3. S. Antonatos, K. G. Anagnostakis, and E. P. Markatos, Generating realistic workloads for network intrusion detection systems. SIGSOFT Softw. Eng. Notes, 29(1):207-215, 2004.
4. R. Boyer and J. Moore, A fast string searching algorithm. Commun. ACM, 20(10):762-772, October 1977.
5. C. Courcoubetis and V. A. Siris, Measurement and analysis of real network traffic. In Proceedings of the 7th Hellenic Conference on Informatics (HCI'99), August 1999.
6. M. Fisk and G. Varghese, An analysis of fast string matching applied to content-based forwarding and intrusion detection, Technical Report CS2001-0670 (updated version), University of California - San Diego, 2002.
7. M. Handley, V. Paxson, and C. Kreibich, Network intrusion detection: Evasion, traffic normalization, and End-to-End protocol semantics, In Proceedings of USENIX Security Symposium, pages 115-134, 2001.
8. R. Horspool, Practical fast searching in strings. Software - Practice and Experience, 10(6):501-506, 1980.
9. E. P. Markatos, S. Antonatos, M. Polychronakis, and K. G. Anagnostakis, ExB: Exclusion-based signature matching for intrusion detection, In Proceedings of CCN'02, November 2002.
10. S. McCanne, C. Leres, and V. Jacobson, libpcap. Lawrence Berkeley Laboratory, Berkeley, CA, available via anonymous ftp to ftp.ee.lbl.gov.
11. M. Norton, Optimizing Pattern Matching for Intrusion Detection, July 2004. http://docs.idsresearch.org/ OptimizingPatternMatchingForIDS.pdf.
12. T. H. Ptacek and T. N. Newsham, Insertion, evasion, and denial of service: Eluding network intrusion detection, Technical report, Secure Networks, Inc., Jan. 1998.
13. M. Roesch, Snort: Lightweight intrusion detection for networks. In Proceedings of the 1999 USENIX LISA Systems Administration Conference, November 1999. http://www.snort.org/.
14. Sourcefire, Snort 2.0 - Detection Revisited. October 2002. http://www.snort.org/ docs/Snort_20_v4.pdf.
15. N. Tuck, T. Sherwood, B. Calder, and G. Varghese, Deterministic memory-efficient string matching algorithms for intrusion detection, In Proceedings of the IEEE Infocom Conference, March 2004.
16. S. Wu and U. Manber, A fast algorithm for multipattern searching, Technical Report TR-94-17, University of Arizona, 1994.

DESIGNATED-VERIFIER PROXY SIGNATURE SCHEMES

Guilin Wang
Instititue for Infocomm Research (I^2R)
21 Heng Mui Keng Terrace, Singapore 119613
glwang@i2r.a-star.edu.sg

Abstract: In a proxy signature scheme, a user delegates his/her signing capability to another user in such a way that the latter can sign messages on behalf of the former. In this paper, we first propose a provably secure proxy signature scheme, which is based on a two-party Schnorr signature scheme. Then, we extend this basic scheme into designated-verifier proxy signatures (DVPS). More specifically, we get two versions of DVPS: weak DVPS and strong DVPS. In both versions, the validity of a proxy signature can be checked only by the designated verifier. In a weak DVPS scheme, however, the designated verifier can further convert such proxy signatures into public verifiable ones, while a strong DVPS scheme does not have the same property even if the designated verifier's secret key is revealed willingly or unwillingly. In addition, we briefly discuss some potential applications for DVPS.

Keywords: proxy signature, digital signature, information security.

1. INTRODUCTION

Proxy Signatures. In a proxy signature scheme, one user Alice, called *original signer*, delegates her signing capability to another user Bob, called *proxy signer*. After that, the proxy signer Bob can sign messages on behalf of the original signer Alice. Upon receiving a proxy signature on some message, a verifier can validate its correctness according to a given verification procedure, and further be convinced of the original signer's agreement on the signed message. Proxy signature schemes have been suggested for use in a number of applications, including electronic commerce, mobile agents, and distributed shared object systems etc [15, 4, 28].

Most existing proxy signature schemes are constructed in the following way. The original signer Alice sends a specific message and the corresponding signature to the proxy signer Bob, who then uses this information to derive a proxy secret key. With this secret key, Bob can produce proxy signatures by employing a specified standard signature scheme. When a proxy signature is given, a verifier first recovers the proxy public key from some public information, and then checks its validity according to the corresponding standard signature verification procedure.

Mambo et al. firstly introduced the concept of proxy signatures and proposed several constructions in [17,18]. After that, a number of new schemes and improvements have been proposed [4, 14-16, 29]; however, most of them do not fully meet the desired security requirements (see Section 2.2). In [14], Kim et al. introduced the concept of partial delegation by warrant, and proposed a threshold proxy signature, in which the original signer's signing ability is shared among a delegated group of n proxy singers such that only t or more of them can generate proxy signatures cooperatively. Lee et al. [15] constructed mobile agents for e-commerce applications from non-designated proxy signature, in which a warrant does not specify the identity of a proxy signer so any possible proxy signer may respond this delegation and become a proxy signer. Furthermore, Lee et al. [16] investigated whether a secure channel for delivery of a signed warrant is necessary in existing schemes. Their results show that if secure channels are not provided, the MUO scheme [17] and the LKK scheme [15] all are insecure. To avoid the usage of secure channels and overcome some other weaknesses, they proposed new improvements. However, Wang et al. [28] showed that all of those schemes and improvements proposed in [15-16] are insecure by demonstrating several kinds of attacks. Boldyreva et al. [4] presented the formal model and security notion for proxy signature, i.e., the existential unforgeablity against adaptive chosen-message attacks [10].

Designated-Verifier Signatures. In 1996, Jakobsson et al. introduced a new primitive called *designated-verifier proofs* [13]. Such proofs enable a prover Alice to convince a designated verifier Bob that a statement is true. However, Bob cannot use such proofs to convince a third party of this fact. The reason is that Bob himself can simulate such proofs. Here is their basic idea. When Alice wants to convince only the designated verifier Bob a statement Θ, she actually proves the statement "Θ *is true* or *I knows Bob's secret key*". Upon receiving such a proof, Bob is convinced that the statement Θ must be true, since he knows that this proof is not generated by himself and that his secret key is not compromised. However, a third party cannot accept the statement Θ from such a proof since this proof may be generated by Bob even if Θ is false. Furthermore, Jakobsson et al. proposed an elegant non-interactive designated-verifier proof for Chaum's zero-knowledge undeniable signature scheme [7] to avoid blackmailing [9, 12] and mafia attacks [8]. In other words, they introduced a *designated-verifier signature scheme* in the sense that only the designated verifier can be convinced that a signature is issued by the claimed signer. However, Wang [27] pointed out this scheme is insecure since a dishonest signer can cheat a designated verifier easily.

Note that in Jakobsson et al.'s scheme any verifier can validate a signature though he does not know whether this signature is produced by the signer or simulated by the designated verifier. In [22], however, Saeednia et al. recently proposed a *strong designated-verifier signature scheme* in the sense that without the knowledge of the designated verifier's secret key, any third party cannot check the validity of such signatures. Compared with Jakobsson et al.'s scheme, their scheme is very efficient in both respects of communications and computation. In addition, Steinfeld et al. introduced a new type of signature scheme called *universal designated-verifier signature* (UDVS) [25, 26]. Such a scheme enables *any* holder of a signature (not necessarily the signer) to designate the signature to a third party as the designated-verifier.

Our Work. As mentioned above, Wang et al. [28] demonstrated several attacks on several DLP-based proxy signature schemes. Those attacks mainly result from the fact that a valid proxy key pair can be forged by an adversary, including the original signer and the proxy signer. However, Wang et al. did not provide improvements to avoid such attacks. In this paper, we first propose a new proxy signature scheme, which is based on the two-party Schnorr signature scheme proposed by Nicolosi et al. [20]. The new scheme is provably secure and as efficient as the schemes in [15-16,4].

Then, by combining the ideas of proxy signatures and designated-verifier signatures, we extend this basic scheme to designated-verifier proxy signatures (DVPS for short). More specifically, we get two versions of DVPS: weak DVPS and strong DVPS. In both versions, the validity of a proxy signature can be checked only by the designated proxy signer. In a weak DVPS scheme, however, the designated verifier can further convert such proxy signatures into public verifiable ones, while a strong DVPS scheme does not have the same property even if the designated verifier's secret key is revealed willingly or unwillingly.

In addition, we briefly discuss some potential applications for DVPS in electronic commerce settings.

Structure. The rest of this paper is organized as follows. Section 2 introduces the computational assumptions, security requirements for proxy signature schemes, and notations. We introduce the new (basic) proxy signature scheme in Section 3, and then extend this scheme into designated-verifier proxy signatures (DVPS) in Sections 4. Finally, Section 5 concludes the paper and points out future work.

2.   PRELIMINARIES

2.1   Assumptions

We review the following computational assumptions that are related to the security of our proxy signature schemes constructed in this paper.

Assumption 1: Discrete Logarithm (DL) assumption. *Let $G_q = <g>$ be a cyclic multiplicative group generated by g of order q. Then, on inputs $(g, g^x) \in G_q^2$ where $x \in Z_q$ is a random (unknown) number, there is no probabilistic polynomial-time (PPT) algorithm that outputs the value of x with non-negligible probability.*

Assumption 2: Computational Diffie-Hellman (CDH) assumption. *Let $G_q = <g>$ be a cyclic multiplicative group generated by g of order q. Then, on inputs $(g, g^x, g^y) \in G_q^3$ where $x, y \in Z_q$ are random (unknown) numbers, there is no PPT algorithm that outputs the value of g^{xy} with non-negligible probability.*

Assumption 3: Decisional Diffie-Hellman (DDH) assumption. *Let $G_q = <g>$ be a cyclic multiplicative group generated by g of order q. Then, on inputs $(g, g^x, g^y, g^z) \in G_q^4$ where $x, y, z \in Z_q$ are random (unknown) numbers, there is no PPT algorithm that distinguishes with non-negligible probability whether g^{xy} and g^z are equal.*

Those computational assumptions are widely believed to be true for many cyclic groups, such as the multiplicative subgroup $G_q = <g>$ of the finite field Z_p, where p is a large prime and q is a prime factor of p-1. In practice, $|p| = 1024$ and $|q| = 160$ are considered to be suitable for most current security applications. More discussions on those assumptions can be found in [5, 2].

2.2   Definitions

Definition 1. A *proxy signature scheme* is usually comprised of the following procedures:

- **Setup**: On input of a security parameter l, this probabilistic algorithm outputs two secret/public key pairs (x_A, y_A) and (x_B, y_B) for the original signer Alice and the proxy signer Bob. Note that those key pairs may be used in a standard signature scheme at the same time.
- **Proxy Key Pair Generation**: The original signer Alice and the proxy signer Bob execute this interactive randomized algorithm to generate a proxy key pair (x_P, y_P) for Bob, such that only Bob knows the value of x_P, while y_P is public or publicly recoverable.
- **Proxy Signature Generation**: The proxy signer Bob runs this (possibly probabilistic) algorithm to generate a proxy signature σ for a message m by using the proxy secret key x_P.
- **Proxy Signature Verification**: A verifier runs this deterministic algorithm to check whether an alleged proxy signature σ for a message m is valid with respect to a specific original signer and a proxy signer.

The security requirements for proxy signature are first specified in [17,18], and later are kept almost the same besides being enhanced in [15], and formalized in [4].

Definition 2. A *secure* proxy signature scheme should satisfy the following requirements:

- **Verifiability**: From the proxy signature, a verifier can be convinced of the original signer's agreement on the signed message.
- **Identifiability**: Anyone can determine the identities of the corresponding original signer and proxy signer from a proxy signature.
- **Unforgeability**: Only the designated proxy signer can create a valid proxy signature on behalf of the original signer. In other words, the original signer and other third parties who are not designated as proxy signers cannot create a valid proxy signature.
- **Undeniability**: Once a proxy signer creates a valid proxy signature on behalf of an original signer, he cannot repudiate the signature creation against anyone else.
- **Prevention of misuse**: The proxy signer cannot use the proxy secret key for purposes other than generating valid proxy signatures. In case of misuse, the responsibility of the proxy signer should be determined explicitly.

2.3   Notations

Throughout this paper, p and q are two large primes such that $q|(p\text{-}1)$ and $G_q=\langle g\rangle$ is a q-order multiplicative subgroup of Z_p^* generated by an element $g\in Z_p^*$. The discrete logarithm problem in G_q is assumed to be difficult. Hereafter, we call three such integers (p, q, g) a *DLP-triple*. Let $h(\ )$ and $h'(\ )$ be two secure cryptographic hash functions. In addition, we suppose that the original signer Alice and the proxy signer Bob possess *certified key pairs* $(x_A, y_A = g^{x_A} \bmod p)$ and $(x_B, y_B = g^{x_B} \bmod p)$, respectively. Here, a certified key pair (x_A, y_A) means that Alice knows the private key x_A and has to prove her knowledge of x_A when she registers her public key certificate with a certificate authority (CA). Actually, this is a recommended practice for issuing public key certificates [1,19], and can be used to prevent rogue-key attacks [4]. In addition, we denote by m_w the *warrant* which specifies the delegation period, what kind of message m is delegated, and the identities of the original signer and the proxy signer, etc.

3.   BASIC PROXY SIGNATURE SCHEME

In this section, we propose a new proxy signature scheme. The basic idea is that the provably secure two-party Schnorr signature scheme proposed in [20] is used to generate a proxy key pair (x_P, y_P) such that

$$g^{x_P} = y_P = (y_A \cdot y_B)^{h(m_w, r_P)} \cdot r_P \bmod p), \qquad (1)$$

where r_P is a public value, and m_w is a warrant which specifies the related information about a proxy delegation. In fact, (r_P, x_P) is exactly a two-party Schnorr signature on message m_w. The point is that (a) *only* Bob knows the value of x_P, and (b) a valid tuple (r_P, x_P) can *only* be generated by Alice and Bob *jointly*. Therefore, x_P can be used as the proxy secret key to generate proxy signatures according to a standard DLP-based signature scheme. At the same time, a verifier can validate such proxy signatures after recovering the public proxy key y_P from Eq. (1).

In the following description of our scheme, it is assumed that Alice and Bob have agreed on a warrant m_w before generating a proxy key pair for Bob. In addition, as pointed in [20], the hash function $h'(\ )$ can be replaced by any secure commitment scheme.

<u>Proxy Key Generation</u>. To generate a proxy key pair (x_P, y_P) for the proxy signer Bob, Alice and Bob execute the following interactive protocol jointly.

(1) Alice picks a random number $k_A \in Z_q^*$, computes $r_A = g^{k_A} \bmod p$ and $c = h'(r_A)$, and then sends c to Bob.

(2) Similarly, Bob first chooses a random number $k_B \in Z_q^*$, then computes $r_B = g^{k_B} \bmod p$ and replies Alice with (c, r_B).

(3) When (c, r_B) is received, Alice checks whether $r_B^q \equiv 1 \bmod p$. If this is true, she computes $r_P = r_A \cdot r_B \bmod p$, $s_A = k_A + x_A \cdot h(m_w, r_P) \bmod q$, and sends the pair (r_A, s_A) to Bob.

(4) Upon receiving (r_A, s_A), Bob computes $r_P = r_A \cdot r_B \bmod p$, and then checks whether $r_A^q \equiv 1 \bmod p$, $c \equiv h'(r_A)$, and $g^{s_A} \equiv y_A^{h(m_w, r_P)} \cdot r_P \bmod p$. If all validations pass, he calculates $s_B = k_B + x_B \cdot h(m_w, r_P) \bmod q$, and finally sets his proxy key pair (x_P, y_P) by

$$x_P = s_A + s_B \bmod q, \quad \text{and} \quad y_P = g^{x_P} \bmod p. \tag{2}$$

It is easy to know that the above defined proxy key pair (x_P, y_P) satisfies Eq. (1), i.e., (r_P, x_P) is a standard Schnorr signature [23] on the warrant m_w with respect to the public key $y_A y_B \bmod p$.

In addition, note that in the above proxy key generation procedure, we do not assume the communication channel between Alice and Bob is secure. Namely, public channel could be used unless delegation privacy is required. The reason is that the exchanged data, i.e., m_w, r_A, s_A, r_B, s_B etc., are useless for other party (to forge proxy key pairs or proxy signatures).

Proxy Signature Generation. To generate a proxy signature on a message m that conforms to the warrant m_w, the proxy signer Bob performs the same operations as in the standard Schnorr signature scheme [23]. That is, he first selects a random number $k \in Z_q^*$, then computes $r = g^k \bmod p$ and $s = k + x_P \cdot h(m, m_w, r) \bmod q$. The resulting proxy signature on message m is $\sigma = (m_w, r_P, r, s)$.

Proxy Signature Verification. To verify the validity of an alleged proxy signature σ for message m, a verifier operates as follows:

(1) Check whether the message m conforms to the warrant m_w. If not, stop. Otherwise, continue.

(2) Check whether Alice and Bob are specified as the original signer and the proxy signer in the warrant m_w, respectively.

(3) Recover the proxy public key y_P from public information by computing $y_P = (y_A \cdot y_B)^{h(m_w, r_P)} \cdot r_P \bmod p$.

(4) Accept the proxy signature σ *if and only if* the following equality holds:

$$g^s = y_P{}^{h(m,m_w,r)} \cdot r \bmod p. \tag{3}$$

In the above proxy scheme, when the proxy singer Bob generates a proxy signature the warrant m_w is embedded in the input of the hash function $h(\ )$. The aim is to use m_w as an identifier of proxy signatures.

We now discuss the security of our above scheme. According to the results in [20] and [21], we have the following Proposition 1 and 2. Then, Proposition 3 holds.

Proposition 1 (Theorems 1 and 2 of [20]). *In the random oracle model, if an adversary, who may compromise the original signer or the proxy signer (but not both), can forge a proxy key pair (x_P, y_P) that satisfies Eq. (1) with respect to a pair (m_w, r_P) in probabilistic polynomial time (PPT) with non-negligible probability, then the discrete log problem in the multiplicative subgroup <g> can be solved in PPT with non-negligible probability.*

Proposition 2 [21]. *Under the assumption that the discrete log problem in the multiplicative subgroup <g> is intractable, the Schnorr signature scheme is secure in the random oracle model.*

Proposition 3. *Under the assumption that the discrete log problem in the multiplicative subgroup <g> is intractable, the proposed proxy signature scheme is secure in the random oracle model.*

Proof (*Sketch*): In our scheme, we use Nicolosi et al.'s *provably secure* two-party Schnorr signature scheme [20] to generate proxy key pair (x_P, y_P). That is, in their scheme a two-party Schnorr signature for a message can *only* be generated by the two related parties jointly. In our scheme, a valid proxy key pair (x_P, y_P) (defined by Eq. (1)) implies that (r_P, x_P) is exactly Alice and Bob's valid two-party Schnorr signature on the warrant m_w in Nicolosi et al.'s scheme. Therefore, anybody (including Alice and Bob) cannot generate a valid proxy key pair independently. Meanwhile, without a valid proxy key pair anybody cannot generate a proxy signature such that Eq. (3) is satisfied. Because the proxy signature generation algorithm is just the Schnorr scheme [23], which is also provably secure [21] in the random oracle model [3]. Therefore, we conclude that our proxy scheme is unforgeable. Other security requirements are also met in our new scheme, since we can provide similar security analysis as done in [15-17].

4.   DESIGNATED-VERIFIER PROXY SIGNATURES

We now present two designated-verifier proxy signature schemes, in which proxy signatures can be only verified by a designated-verifier. Those DVPS schemes are constructed from the basic proxy signature scheme introduced in Section 3. However, note that the Triple Schnorr proxy signature proposed in [4] could also be used as the basic scheme in a similar way.

In the following description, it is supposed that the original signer Alice and the proxy signer Bob have agreed on a warrant m_w before generating a proxy key pair. In addition, we assume Cindy be the designated verifier with certified key pair $(x_C, y_C = g^{x_C} \bmod p)$. Other system parameters are the same as in previous section. Proxy key generation procedure is the same as our basic scheme described in Section 3. That is, the original signer Alice and the proxy signer Bob jointly generate a proxy key pair (x_P, y_P) for Bob such that the proxy public key y_P can be recovered from Eq. (1), and that only Bob knows the value of the proxy secret key x_P.

4.1   Weak Designated-Verifier Proxy Signature Scheme

<u>Proxy Signature Generation</u>. To generate a weak designated-verifier proxy signature on a message m that conforms to the warrant m_w, the proxy signer Bob performs as follows. He first selects a random number $k \in Z_q^*$ at uniform, then computes (r, r', s) by Eq. (4), and sends the proxy signature $\sigma' = (m_w, r_P, r', s)$ to the designated verifier Cindy.

$$r = g^k \bmod p,$$
$$r' = y_C^k \bmod p, \qquad\qquad (4)$$
$$s = k + x_P \cdot h(m, m_w, r) \bmod q.$$

<u>Proxy Signature Verification</u>. To verify the validity of a weak DVPS σ', the designated verifier Cindy operates as follows:

(1) Check whether the message m conforms to the warrant m_w. If not, stop. Otherwise, continue.
(2) Check whether Alice and Bob are specified as the original signer and the proxy signer in the warrant m_w, respectively.
(3) Recover the values of r and the proxy public key y_P by computing $r = (r')^{x_C^{-1}} \bmod p$ and $y_P = (y_A \cdot y_B)^{h(m_w, r_P)} \cdot r_P \bmod p$.

(4) Accept the proxy signature σ' *if and only if* the following equality holds:

$$g^s = y_P^{h(m,m_w,r)} \cdot r \bmod p. \tag{5}$$

The essence of the above scheme is that to restrict the publicly verifiability of a proxy signature, we simply encrypted the value r by releasing $r' = y_C^k \bmod p$. Therefore, only the designated verifier Cindy can recover r from r' by using her secret key x_C, and then check the validity of such an encrypted proxy signature. Note that in our above scheme, the designated verifier Cindy can convince any third party to accept such a proxy signature σ' by simply releasing $\sigma = (m_w, r_P, r, s)$. Weak designated-verifier proxy signature schemes might be suitable in the settings where both the proxy signer and the designated verifier want that without their help, any third party cannot validate proxy signatures. We now state the security of the above scheme as follows.

Proposition 4. *Under the assumption that the Diffie-Hellman problem in the multiplicative subgroup <g> is intractable, the proposed weak designated-verifier signature scheme is secure in the random oracle model.*

Proof: We first prove that except the designated-verifier Cindy (and the proxy signer Bob), any third party (including the original signer Alice) cannot check the validity of a weak DVPS $\sigma' = (m_w, r_P, r', s)$ for message m. First of all, note that without the value of r the third party cannot check the validity of σ'. In other words, to validate σ' the third party has to recover the value of r from public information. Under the assumptions that DL is difficult and that $h(\ )$ can be modeled as a random function [3], neither (m_w, r_P, y_P, y_C) nor s can be used to reveal x_C or recover r. Consequently, the third party can only use y_C and r' to recover the value of r. That is, on input $(g, y_C = g^{x_C} \bmod p, r' = g^{k \cdot x_C} \bmod p)$, the third party wants to output $r = g^k \bmod p$. In [2], it is proved that this problem (called *Divisible Computation Diffie-Hellman Problem*) is as difficult as the CDH problem. Therefore, under the CDH assumption, only the designated verifier Cindy can check the validity of a week designated-verifier proxy signature.

Now, we prove that the unforgeability. According to Proposition 3, any adversary who is not delegated as a proxy signer by Alice cannot forge a valid proxy key pair (x_P, y_P). Furthermore, we claim that given a proxy public y_P even the designated verifier Cindy, who knows her secret key x_C but does not know the proxy secret key x_P corresponding to y_P, cannot forge a valid proxy signature for a new message m that never appears in Cindy's records of known signature-message pairs. If this is not the fact, i.e., Cindy can forge a valid weak DVPS $\sigma' = (m_w, r_P, r', s)$ for a new message m. Then,

Cindy can compute $r = (r')^{x\bar{c}^{-1}} \bmod p$. The latter means that Cindy can forge a Schnorr signature $\bar{\sigma} = (r, s)$ for message (m, m_w) with respect to the public key y_P. This is contrary to the Proposition 2, i.e., the provable security of the Schnorr signature scheme. Therefore, only the delegated proxy signer Bob can generate valid weak designated-verifier proxy signatures.

4.2 Strong Designated-Verifier Proxy Signature Scheme

Based on the basic proxy scheme proposed in Section 3 and Saeednia et al.'s strong designated-verifier signature scheme [22], we now construct a strong designated-verifier proxy signature scheme. In the new scheme, the designated-verifier Cindy can verify that a proxy signature is signed by the proxy signer Bob, but she is unable to convince anyone else of this fact. Because others know that such signatures may be simulated by the designated-verifier Cindy.

Proxy Signature Generation. To generate a strong designated-verifier proxy signature on a message m that conforms to the warrant m_w, Bob performs as follows. He first selects two random numbers $k \in Z_q$ and $t \in Z_q^*$, then computes (r, c, s) by Eq. (6), and sends the proxy signature $\sigma = (m_w, r_P, c, s, t)$ to the designated verifier Cindy.

$$
\begin{aligned}
r &= y_C^k \bmod p, \\
c &= h(m, m_w, r), \\
s &= kt^{-1} - x_P \cdot c \bmod q.
\end{aligned}
\tag{6}
$$

Proxy Signature Verification. To verify the validity of a strong DVPS σ, the designated-verifier Cindy operates as follows:

(1) Check whether the message m conforms to the warrant m_w. If not, stop. Otherwise, continue.

(2) Check whether Alice and Bob are specified as the original signer and the proxy signer in the warrant m_w, respectively.

(3) Recover the proxy public key $y_P = (y_A \cdot y_B)^{h(m_w, r_P)} \cdot r_P \bmod p$.

(4) Accept the proxy signature σ *if and only if* the following equality holds:

$$
c \equiv h(m, m_w, \bar{r}), \quad \text{where } \bar{r} = (g^s y_P^c)^{t \cdot x_c} \bmod p. \tag{7}
$$

Proxy Signature Simulation. To simulate a strong designated-verifier proxy signature σ' for any message m that conforms to the warrant m_w, Cindy picks $s' \in Z_q$ and $r' \in Z_q^*$, at random, and computes the following values:

$$r = g^{s'} y_P^{r'} \bmod p, \quad c = h(m, m_w, r),$$
$$l = r'c^{-1} \bmod q, \quad s = s'l^{-1} \bmod q, \quad t = l x_C^{-1} \bmod q. \tag{8}$$

$\sigma' = (m_w, r_P, c, s, t)$ is the simulated proxy signature for message m with respect to the proxy public key y_P. It is easy to check that σ' is also a valid proxy signature, i.e., it satisfies Eq. (7).

Now we discuss the security of the above scheme. From the results of [22], we obtain Proposition 5, and then Proposition 6 can be proved in a similar way as we did in Proposition 4.

Proposition 5. *If a valid strong designated-verifier proxy signature for Cindy can be generated without the knowledge of the proxy secret key x_P or Cindy's secret key x_C, then the computational Diffie-Hellman problem may be solved in PPT. Furthermore, the transcripts simulated by Cindy are indistinguishable from those generated by the proxy signer Bob.*

Proposition 6. *Under the CDH assumption, our strong designated-verifier proxy signature scheme is secure in the random oracle model.*

4.3   Applications

In this section, we briefly introduce two potential applications for designated-verifier proxy signatures (DVPS). Other applications are also possible. Let us first consider the following scenario. A corporate manager Alice will have vacation for one or two weeks, however, some current businesses need to be processed continuously during this period. Naturally, Alice could delegate her signing capability to several assistants to deal with each business with different customer. For example, assistant Bob, as the representative of Alice, is assigned to negotiate a business contract with customer Cindy in this period. During this procedure, some intermediate documents will be produced with digital signatures for authentication or non-repudiation. However, to protect the confidentiality and authenticity of those documents, it may be highly expected that the corresponding signatures could be validated only by the designated receiver. In this case, DVPS could be used. More specifically, Bob's proxy signatures can only be verified by Cindy. Furthermore, if non-repudiation service is required, weak DVPS could be exploited.

Another example is about on-line shopping. When a customer Cindy buys a digital product m from an Internet vendor Bob, who sells some digital products (e.g. digital music, movies, and books etc.), she needs a digital receipt from Bob to guarantee the quality, authenticity, and legality of m. This is reasonable since Cindy does not completely trust Bob and his goods.

Furthermore, Cindy would expect the receipt is bounded with not only the identity of the vendor Bob but also that of the goods producer, say Alice. With such receipts, Cindy will be convinced that digital product m is produced by Alice and sold by Bob. At the same time, to prevent Cindy from illegally re-selling m to others, Alice and Bob want the validity of Cindy's receipt can only be validated by Cindy herself. In such situations, strong designated-verifier proxy signatures, instead of ordinary digital signatures, can be used as such receipts. That is, Alice delegates her signing capability to Bob so that he can generate strong designated-verifier proxy signatures as digital receipts to all potential customers. Note that this approach cannot prevent Cindy to send a copy of digital product to her friends. To deal with this problem, one could exploit some techniques from digital right managements (DRM), such as watermarking and fingerprinting etc.

5. CONCLUSION AND FUTURE WORK

In this paper, based on the two-party Schnorr signature scheme proposed in [20], we first proposed a provably secure proxy signature scheme. Then, we extended this basic scheme into designated-verifier proxy signature (DVPS) schemes. Actually, we constructed two versions of DVPS: weak DVPS and strong DVPS. In both versions, only the designated verifier can check the validity of a proxy signature. In a weak DVPS scheme, however, the designated verifier can further convert such proxy signatures into public verifiable ones, while a strong DVPS scheme does not meet the same property even if the designated verifier's secret key is revealed willingly or unwillingly. Finally, we introduced some potential applications for DVPS.

Some other variations can be obtained directly, such as blind proxy signatures from the blind Schnorr signature [24], universally designated-verifier proxy signature scheme by using the techniques in [26], and fully distributed proxy signatures by using the techniques in [11] (though this concrete scheme is insecure [28]). Another interesting work is to design forward-secure proxy schemes.

REFERENCES

1. C. Adams and S. Farrell, Internet X.509 public key infrastructure: Certificate management protocols, RFC 2510, March 1999.
2. F. Bao, R.H. Deng, and H. Zhu, Variations of Diffie-Hellman problem, in: Information and Communicatins Security (ICICS'03), LNCS 2836, pp. 301-312, Springer-Verlag, 2003.

3. M. Bellare and P. Rogaway, Random oracles are practical: A paradigm for designing efficient protocols, in: Proc. of 1st ACM Conference on Computer and Communications Security (CCS'93), pp. 62-73, ACM Press, 1993.

4. A. Boldyreva, A. Palacio, and B. Warinschi, Secure proxy signature schemes for delegation of signing rights, Cryptology ePrint archive; http://eprint.iacr.org/2003/096, May 2003.

5. D. Boneh, The decision Diffie-Hellman problem, in: Proc. of the Third Algorithmic Number Theory Symposium (ANTS'98), LNCS 1423, pp. 48-63, Springer-Verlag, 1998.

6. D. Chaum and H. van Antwerpen, Undeniable signatures, in: CRYPTO'89, LNCS 435, pp. 212-216, Springer-Verlag, 1989.

7. D. Chaum, Zero-knowledge undeniable signatures, in: EUROCRYPT'90, LNCS 473, pp. 458-464, Springer-Verlag, 1991.

8. Y. Desmedt, C. Coutier, and S. Bengio, Special uses and abuses of the Fiat-Shamir passport protocol, in: CRYPTO'87, pp. 21-39, Springer-Verlag, 1987.

9. Y. Desmedt and M. Yung, Weakness of undeniable signature schemes, in: EUROCRYPT'91, LNCS 547, pp: 205-220, Springer-Verlag, 1991.

10. S. Goldwasser, S. Micali, and R. Rivest, A digital signature scheme secure against adaptive chosen-message attacks, SIAM Journal of Computing, 17(2): 281-308 (April 1988).

11. J. Herranz and Sáez, Verifiable secret sharing for general access structures, with applications to fully distributed proxy signatures, in: Financial Cryptography (FC'03), LNCS 2742, pp. 286-302, Springer-Verlag, 2003.

12. M. Jakobsson, Blackmailing using undeniable signatures, in: EUROCRYPT'96, LNCS 950, pp.: 425-427, Springer-Verlag, 1994.

13. M. Jakobsson, K. Sako, and R. Impagliazzo, Designated verifier proofs and their applications, in: EUROCRYPT'96, LNCS 1070, pp. 143-154, Springer-Verlag, 1996.

14. S. Kim, S. Park, and D. Won, Proxy signatures, revisited, in: Information and Communications Security (ICICS'97), LNCS 1334, pp. 223-232, Springer-Verlag, 1997.

15. B. Lee, H. Kim, and K. Kim. Secure mobile agent using strong non-designated proxy signature, in: Information Security and Privacy (ACISP'01), LNCS 2119, pp. 474-486, Springer-Verlag, 2001.

16. J.-Y. Lee, J. H. Cheon, and S. Kim, An analysis of proxy signatures: Is a secure channel necessary? in: Topics in Cryptology - CT-RSA 2003, LNCS 2612, pp. 68-79, Springer-Verlag, 2003.

17. M. Mambo, K. Usuda, and E. Okamoto, Proxy signatures: Delegation of the power to sign messages, IEICE Trans. Fundamentals, Vol. E79-A, No. 9, pp. 1338-1353 (Sep. 1996).

18. M. Mambo, K. Usuda, and E. Okamoto, Proxy signatures for delegating signing operation, in: Proc. of 3rd ACM Conference on Computer and Communications Security (CCS'96), pp. 48-57, ACM Press, 1996.

19. M. Meyers, C. Adams, D. Solo, and D. Kemp, Internet X.509 certificate request format, RFC 2511, March 1999.

20. A. Nicolosi, M. Krohn, Y. Dodis, and D. Mazieres, Proactive two-party signatures for user authentication, in: Proc. of 10th Annual Network and Distributed System Security Symposium (NDSS'03); http://www.isoc.org/isoc/conferences/ndss/.

21. D. Pointcheval and J. Stern, Security arguments for digital signatures and blind signatures, Journal of Cryptology, 13(3): 361-369 (2000).

22. S. Saeednia, S. Kremer, and O. Markowitch, An efficient strong designated verifier signature scheme, in: Information Security and Cryptology - ICISC 2003, LNCS 2971, pp. 40-54, Springer-Verlag, 2004.

23. C. Schnorr, Efficient signature generation by smart cards, Journal of Cryptography, 4(3): 161-174 (1991).

24. C. Schnorr, Security of blind discrete log signatures against interactive attacks, in: Information and Communications Security (ICICS'01), LNCS 2229, pp. 1-12, Springer-Verlag, 2001.

25. R. Steinfeld, L. Bull, H. Wang, and J. Pieprzyk, Universal designated-verifier signatures, in: ASIACRYPT'03, LNCS 2894, pp. 523-542, Springer-Verlag, 2003.

26. R. Steinfeld, H. Wang, and J. Pieprzyk, Efficient extension of standard Schnorr/RSA signatures into universal designated-verifier signatures, in: PKC 2004, LNCS 2947, pp. 86-100, Springer-Verlag, 2004.

27. G. Wang. An Attack on not-interactive designated verifier proofs for undeniable signatures, Cryptology ePrint archive; http://eprint.iacr.org/2003/243/, Nov. 2003.

28. G. Wang, F. Bao, J. Zhou, and R. H. Deng, Security analysis of some proxy signatures, in: Information Security and Cryptology - ICISC 2003, LNCS 2971, pp. 305-319, Springer-Verlag, 2004.

29. K. Zhang, Threshold proxy signature schemes, in: Information Security Workshop (ISW'97), LNCS 1396, pp. 282-290, Springer-Verlag, 1997.

TRIPARTITE CONCURRENT SIGNATURES

Willy Susilo and Yi Mu
Centre for Information Security Research
School of Information Technology and Computer Science
University of Wollongong
Wollongong 2522, Australia
Email: {wsusilo, ymu}@uow.edu.au

Abstract: Fair exchange in digital signatures has been considered as a fundamental problem in cryptography. The notion of concurrent signatures was introduced in the seminal paper of Chen, Kudla and Paterson in Eurocrypt 2004 Chen et al., 2004. In this paper, we partially solve an open problem proposed in Chen et al., 2004. We extend the notion of two party concurrent signatures to tripartite concurrent signature schemes. In tripartite concurrent signatures, three parties can exchange their signatures in such a way that their signatures will be binding *concurrently*. We present a model of tripartite concurrent signatures together with a concrete scheme based on bilinear pairings. It was noted in Chen et al., 2004 that extending concurrent signatures to a multi-party scheme, where there are three or more participants, cannot be achieved by trivially modifying their construction in Chen et al., 2004.

Key words: Tripartite Concurrent Signatures, Multi-party Fair Exchange.

1.   INTRODUCTION

Fair exchange in digital signatures has been considered as a fundamental problem in cryptography. Fair exchange is a necessary feature in many applications for electronic commerce. Typical applications include contract signing where two parties need to exchange their signature on a contract.

Two party fair exchange has been studied extensively in the literature. In general, the method can be broadly divided into two types, namely *with* or *without* a trusted party *TTP*. It was believed that fair exchange without a TTP is not practical, since it requires a large number of communication rounds, until the recent work of Chen, Kudla and Paterson in Chen et al., 2004 that shows a *weaker* version of two party fair exchange can be done efficiently *without* any involvement of a TTP. In concurrent signatures, two parties can produce two signatures in such a way that from any third party's point of view, both signatures are ambiguous. However, after additional information, called the *keystone*, is released by one of the parties, both signatures are binding concurrently. It was noted in Chen et al., 2004 that this type of signature scheme falls just short of providing a full solution to the problem of fair exchange of signatures. In the same paper, they questioned the existence of multi party concurrent signatures. They noted that if multi party concurrent signatures can be constructed and modeled correctly, this will move closer to the full solution of multi party fair exchange. They also mentioned that their scheme *cannot* be trivially extended to include multiple matching signers, since the fairness of the scheme will not be achieved.

Our Contribution

In this paper, we present a novel model of tripartite concurrent signatures that allows three parties to exchange their signatures in a fair way. Our model guarantees *fairness* as in the seminal paper of Chen et al., 2004. We also provide a concrete scheme that satisfies our model, based on bilinear pairings. We provide a set of security analysis for our concrete scheme.

1.1   Related Work

In Rivest et al., 2001, the notion of *ring signatures* was formalized and an efficient scheme based on RSA was proposed. This signature can be used to convince any third party that one of the people in the group (who know the trapdoor information) has authenticated the message on behalf of the group.

The authentication provides *signer ambiguity*, in the sense that no one can identify who has actually signed the message.

Designated Verifier Proofs were proposed in Jakobsson et al., 1996. The idea is to allow signatures to convince only the intended recipient, who is assumed to have a public-key. As noted in Rivest et al., 2001, ring signature schemes can be used to provide this mechanism by joining the verifier in the ring. However, it might not be practical in the real life since the verifier might not have any public key setup. In Desmedt, 2003, Desmedt raised the problem of generalizing the designated verifier signature concept to a multi designated verifier scheme. This question was answered affirmatively in Laguillaumie and Vergnaud, 2004, where a construction of multi designated verifiers signature scheme was proposed.

2. PRELIMINARIES

2.1 Basic concepts on Bilinear Pairings

Let G_1, G_2 be cyclic additive groups generated by P_1, P_2, respectively, whose order are a prime q. Let G_M be a cyclic multiplicative group with the same order q. We assume there is an isomorphism $\Psi : G_2 \rightarrow G_1$ such that $\Psi(P_2) = P_1$. Let $\hat{e} : G_1 \times G_2 \rightarrow G_M$ be a bilinear mapping with the following properties:

1. *Bilinearity*: $\hat{e}(aP, bQ) = \hat{e}(P, Q)^{ab}$ for all $P \in G_1, Q \in G_2, a, b \in Z_q$.
2. *Non-degeneracy*: There exists $P \in G_1, Q \in G_2$ such that $\hat{e}(P, Q) \neq 1$.
3. *Computability*: There exists an efficient algorithm to compute $\hat{e}(P, Q)$ for all $P \in G_1, Q \in G_2$.

For simplicity, hereafter, we set $G_1 = G_2$ and $P_1 = P_2$. We note that our scheme can be easily modified for a general case, when $G_1 \neq G_2$.

Bilinear pairing instance generator is defined as a probabilistic polynomial time algorithm $\mathcal{IG}$ that takes as input a security parameter ℓ and returns a uniformly random tuple $param = (p, G_1, G_M, \hat{e}, P)$ of bilinear parameters, including a prime number p of size ℓ, a cyclic additive group G_1 of order q, a multiplicative group G_M of order q, a bilinear map $\hat{e} : G_1 \times G_1 \rightarrow G_M$ and a generator P of G_1. For a group G of prime order, we denote the set $G^* = G \setminus \{\mathcal{O}\}$ where $\mathcal{O}$ is the identity element of the group.

2.2   Complexity Assumptions

Definition 1. Computational Diffie-Hellman (CDH) Problem.

Given two randomly chosen $aP, bP \in G_1$, for unknown $a, b \in Z_q$, compute $Z = abP$.

Definition 2. Computational Diffie-Hellman (CDH) Assumption.

If $\mathcal{IG}$ is a CDH parameter generator, the advantage $\mathrm{Adv}_{\mathcal{IG}}(\mathcal{A})$ that an algorithm $\mathcal{A}$ has in solving the CDH problem is defined to be the probability that the algorithm $\mathcal{A}$ outputs $Z = abP$ on inputs, where $(G_1, G_M, \hat{e})$ is the output of $\mathcal{IG}$ for sufficiently large security parameter ℓ, P is a random generator of G_1 and a,b are random elements of Z_q. The CDH assumption is that $\mathrm{Adv}_{\mathcal{IG}}(\mathcal{A})$ is negligible for all efficient algorithms $\mathcal{A}$.

2.3   Signature of Knowledge

The first signature based on proof of knowledge (SPK) was proposed in Camenisch, 1998 ; Camenisch, 1997. We will use the following definition of SPK from Camenisch, 1998.

Let q be a large prime and $p=2q+1$ be also a prime. Let G be a finite cyclic group of prime order p. Let g be a generator of Z_p^* such that computing discrete logarithms of any group elements (apart from the identity element) with respect to one of the generators is infeasible. Let $H : \{0,1\}^* \rightarrow \{0,1\}^\ell$ denote a strong collision-resistant hash function.

Definition 3. *A pair $(c,s) \in \{0,1\}^\ell \times Z_q$ satisfying $c=H$ $(g\|y\|g^s y^c\|m)$ is a signature based on proof of knowledge of discrete logarithm of a group element y to the base g of the message $m \in \{0,1\}^*$ and is denoted by $SPK\{\alpha : y = g^\alpha\}(m)$.*

A $SPK\{\alpha : y = g^\alpha\}(m)$ can only be computed if the value (secret key) $\alpha = \log_g(y)$ is known. This is also known as a non-interactive proof of the knowledge α.

Definition 4. *A pair (c,s) satisfying $c=H(h\|g\|z\|y\|h^s z^c\|g^s y^c\|m)$ is a signature of equality of the discrete logarithm problem of the group element z with respect to the base h and the discrete logarithm of the group element y with respect to the base g for the message m. It is denoted by $SPKEQ\{\alpha : y = g^\alpha \wedge z = h^\alpha\}(m)$.*

This signature of equality can be seen as two parallel signatures of knowledge $SPK\{\alpha : y = g^\alpha\}(m)$ and $SPK\{\alpha : z = h^\alpha\}(m)$, where the exponent for the commitment, challenge and response are the same. It is straightforward to see that this signature of equaiity can be extended to show the equality of n parallel signatures of knowledge *SPK* using the same technique. This technique can be applied to elliptic curve domain. For completeness, we illustrate the technique as follows.

Definition 5. *A pair (c,s) satisfying $c=H(P\|Q\|sP+cQ\|m)$ is a signature based on proof of knowledge of elliptic curve discrete logarithm of a group element Q to the base P of the message $m \in \{0,1\}^*$ and is denoted by $ECSPK\{\alpha : Q = \alpha P\}(m)$.*

We note that $ECSPK\{\alpha : Q = \alpha P\}(m)$ can only be computed iff the value of a, where $Q=aP$, is known. It can be computed as follows. Firstly, select a random $z \in Z_q^*$ and compute $c=H(P\|Q\|zP\|m)$, and then, compute $s = z - ca \pmod q$. Using the same technique, the following definition can be derived.

Definition 6. *A pair (c,s) satisfying $c=H(U\|P\|S\|Q\|sU+cS\|sP+cQ\|m)$ is a signature of equality of the elliptic curve discrete logarithm problem of the group element S with respect to the base U and the discrete logarithm of the group element Q with respect to the base P for the message m. It is denoted by $ECSPKEQ : \{\alpha : Q = \alpha P \wedge S = \alpha U\}(m)$.*

3. FORMAL DEFINITIONS

3.1 Tripartite Concurrent Signature Algorithms

In this section, we provide a formal definition of a tripartite concurrent signature scheme. In our system, the three participants are polynomially bounded in the security parameter ℓ.

Definition 7. *A tripartite concurrent signature scheme is a digital signature scheme that consists of the following algorithms.*

- SETUP: *A probabilistic algorithm that on input a security parameter ℓ, outputs descriptions of the set of participants $\mathcal{U}$, the message space $\mathcal{M}$, the signature space $\mathcal{M}$, the keystone space $\mathcal{K}$, the keystone fix space $\mathcal{F}$ and a function $KGEN : \mathcal{K} \rightarrow \mathcal{F}$. The algorithm also outputs the public parameters* **param***, together with all public keys of the participants $\{\mathcal{P}_i\}$, where each participant retaining their private key s_i.*

- ASIGN: *A probabilistic algorithm that on inputs $(m, f, \mathcal{P}_i, \mathcal{P}_j, \mathcal{P}_k, s_i)$, where $f \in \mathcal{F}, m \in \mathcal{M}$. $\mathcal{P}_i, \mathcal{P}_j$ and $\mathcal{P}_k$ are the participants' public keys and s_i is the associated secret key for public key $\mathcal{P}_i$, outputs an ambiguous signature $\sigma \in S$ on m.*

- **AVERIFY:** *A deterministic algorithm that on inputs* $(m, f, \sigma, \mathcal{P}_i, \mathcal{P}_j, \mathcal{P}_k)$, *where* $f \in \mathcal{F}, m \in \mathcal{M}$. $\mathcal{P}_i, \mathcal{P}_j$ *and* $\mathcal{P}_k$ *are the participants' public keys and* $\sigma \in S$, *outputs* accept *or* reject.

- **RELEASE:** *A deterministic algorithm that accepts* $f \in \mathcal{F}$ *and a set of valid signatures* $\{\sigma_i, \sigma_j, \sigma_k\}$ *for message* $\{m_i, m_j, m_k\}$ *and outputs the correct* $\hat{k} \in \mathcal{K}$ *used in the* $f = KGEN(\hat{k})$ *function, together with some necessary information,* info, *to confirm the published signatures.*

- **PROOF-VERIFY:** *A deterministic algorithm that accepts a keystone* $\hat{k} \in \mathcal{K}$, *a keystone fix* $f \in \mathcal{F}$, *some required information produced by the* **RELEASE** *algorithm,* info. *This algorithm verifies the correctness of the keystone together with* info. *If they are correct, then output* accept. *Otherwise, output* reject.

- **VERIFY:** *A deterministic algorithm that accepts* $(f, \hat{k}, \sigma)$ *and some necessary information produced by the* **RELEASE** *algorithm,* info, *and executes* **PROOF-VERIFY** $(\hat{k}, f, \text{info})$ *and* **AVERIFY** $(m, f, \sigma_u, \mathcal{P}_i, \mathcal{P}_j, \mathcal{P}_k)$ *algorithm, for* $u \in (i, j, k)$, *to produce* accept *or* reject, *respectively.*

- **DENY:** *A probabilistic algorithm that accepts* $(m, f, \sigma_1, \sigma_2, \mathcal{P}_i, \mathcal{P}_j, \mathcal{P}_k, s_i)$ *where* $m \in \mathcal{M}$, $f \in \mathcal{F}$, s_i *is the associated secret key for* $\mathcal{P}_i$ *and* $\sigma_1, \sigma_2 \in S$ *are signatures on* m, *and tests whether both signatures are valid, i.e. pass* **AVERIFY** *test, and confirm that one of them is a forgery. If forgery happens, then output* accept. *Otherwise, output* reject.

3.2   Tripartite Concurrent Signature Protocol

We will describe a tripartite concurrent signature protocol among three parties, Alice, Bob and Charlie (or A, B and C, respectively). One of the three parties needs to create a keystone and send the first ambiguous signature to the other two parties. We call this party the *initial signer*. Then, another party will respond to this initial signature by creating another ambiguous signature with the same keystone fix. We call the second party as a *first matching signer*. Finally, the third party will respond to the first two signatures by creating his own ambiguous signature. We call this party a *second matching party*. Without losing generality, we assume A to be the initial singer, B the first matching signer and C the second matching signer. From here on, we will use subscripts A, B and C to describe initial signer A, first matching signer B and second matching signer C. The signature works as follows.

A, B and C run SETUP algorithm to determine the public parameters of the scheme. We assume that participants i's secret and public keys are indicated by P_i and s_i, respectively, for $i \in (A, B, C)$. Hence, A's public key is P_A and her secret key is s_A and so forth.

1. A picks random keystone $\hat{k} \in \mathcal{K}$ and computes $f = KGEN(\hat{k})$. A takes her own public key P_A, together with the other parties' public key, P_B, P_C and picks a message $m_A \in \mathcal{M}$ to sign. A then computes her ambiguous signatures as $\sigma_A = \mathsf{ASIGN}(m_A, f, P_A, P_B, P_C, s_A)$ and sends this to B and C.

2. Upon receiving A's ambiguous signature σ_A, B and C verifies the signature by testing whether $\mathsf{AVERIFY}(m_A, f, \sigma_A, P_A, P_B, P_C) = $ accept holds with equality. If not, B and C abort. Otherwise, B picks a message $m_B \in \mathcal{M}$ to sign and computes his ambiguous signature $\sigma_B = \mathsf{ASIGN}(m_B, f, P_B, P_A, P_C, s_B)$ using the same keystone fix $f \in \mathcal{F}$ and sends this to A and C.

3. Upon receiving B's ambiguous signature σ_B, A and C verifies the signature by testing whether $\mathsf{AVERIFY}(m_B, f, \sigma_B, P_B, P_A, P_C) = $ accept holds with equality. If not, A and C abort. Otherwise, C picks a message $m_C \in \mathcal{M}$ to sign and computes his ambiguous signature $\sigma_C = \mathsf{ASIGN}(m_C, f, P_C, P_A, P_B, s_C)$ using the same keystone fix $f \in \mathcal{F}$ and sends this to A and B.

4. Upon receiving C's ambiguous signature σ_C, A and B verifies the signature by testing whether $\mathsf{AVERIFY}(m_C, f, \sigma_C, P_C, P_A, P_B) = $ accept holds with equality. If not, A and B abort. Otherwise, A executes VERIFY algorithm to release the keystone $\hat{k}$ (together with several other confirmation messages, info, whenever necessary) to B and C, and all signatures are binding concurrently.

Any third party can be convinced with the authenticity of the signatures by executing VERIFY algorithm.

3.3 Security Requirements

As the original model of concurrent signatures in Chen et al., 2004, we require a tripartite concurrent signature to satisfy *correctness*, *unforgeability*, *ambiguity* and *fairness*. Intuitively, these notions are described as follows.

- *Correctness*: If a signature σ has been generated *correctly* by invoking ASSIGN algorithm on a message $m \in \mathcal{M}$, then AVERIFY algorithm will return accept with an overwhelming probability, given a signature σ on *m*. Moreover, after the keystone $\hat{k} \in \mathcal{K}$ is released, then the output of VERIFY algorithm will be accept with an overwhelming probability.

- *Unforgeability*: There are two different cases that we need to consider. Case 1) When an adversary $\mathcal{A}$ does not have any knowledge of the respective secret key s_i, then no valid signature that will pass the AVERIFY algorithm can be produced. Otherwise, one of the underlying hard problems can be solved by using this adversary's capability. Case 2) Any party cannot *frame* the other party that he/she has indeed signed message. We require that although both signatures are ambiguous, any party who would like to frame (or cheat) the others will not be able to produce a valid keystone with an overwhelming probability.

- *Ambiguity*: We require that given the two ambiguous signatures, any adversary will not be able to distinguish who was the actual signer of the signatures *before* the keystone is released.

- *Fairness*: We require that any valid ambiguous signatures generated using the same keystone will all become binding *after* the keystone is released. Hence, a matching signer cannot be left in a position where a keystone binds his signature to him whilst the initial signer's signature is not binding to her. Additionally, we also require that *only* the party who generates a keystone can use to create a binding signature. We do not require that the matching signers will definitely receive the necessary keystone.

Definition 8 *A tripartite concurrent signature scheme is secure if it is existentially unforgeable under a chosen message attack, ambiguous and fair.*

4.   **A CONCRETE TRIPARTITE CONCURRENT SIGNATURE SCHEME**

A tripartite concurrent signature scheme is defined by the following algorithms. Our scheme is developed using the technique proposed in

Laguillaumie and Vergnaud, 2004. In the following, we denote the participants by $\mathcal{U}_i$, $i \in \{A,B,C\}$ for convenience and clarify of the presentation.

- SETUP: On input security parameter ℓ, the algorithm selects a uniformly random tuple $param = (p, G_1, G_M, \hat{e}, P)$ of bilinear parameters, including a prime number q of size ℓ, a cyclic additive group G_1 of order q, a multiplicative group G_M of order q, a bilinear map $\hat{e}: G_1 \times G_1 \to G_M$ and a generator P of G_1. The algorithm also selects a secret key $s \in Z_q^*$ and computes the associated public key $P_{pub} = sP$, for a random generator $P \in G_1$. The algorithm also publishes two cryptographic hash function $H_0: \{0,1\}^* \to G_1$ and $H_1: \{0,1\}^\ell \to Z_q^*$. Each user $\mathcal{U}_i \in \mathcal{U}$, $i \in \{A,B,C\}$, selects his/her secret key s_i and publishes his/her public key $P_i = s_i P$. At the end of the algorithm, the parameter $param = (p, G_1, G_M, \hat{e}, P)$ is published, together with the public key P_{pub} and public key of the participants P_A, P_B, P_C. The algorithm also sets $\mathcal{M} = \mathcal{F} = \mathcal{K} = Z_q$. The $KGEN(\cdot)$ function is defined to be $H_1(\cdot)$.

- ASIGN: The algorithm accepts $(m, f, P_i, P_j, P_k, s_i)$ as input, where $P_i = s_i P$, for s_i is $\mathcal{U}_i$'s secret key, P_j and P_k are public keys published by $\mathcal{U}_j$ and $\mathcal{U}_k$, $m \in \mathcal{M}$ and $f \in \mathcal{F}$, and performs the following.

  - Select a random $r \in Z_q^*$.
  - Compute $M = H_0(m \| f)$
  - Compute $Q_1 = s_i^{-1}(M - r(P_j + P_k))$ and $Q_2 = rP$
  - Output $\sigma = (Q_1, Q_2)$ as the signature on m.

- AVERIFY: The algorithm accepts $(m, f, \sigma, P_i, P_j, P_k)$, for $\sigma = (Q_1, Q_2)$, $m \in \mathcal{M}$, $f \in \mathcal{F}$, and verifies whether $\hat{e}(Q_1, P_i)$ $\hat{e}(Q_2, P_j + P_k) = \hat{e}(H_0(m \| f), P)$ holds. If it does not hold, then output **reject**. Otherwise, output **accept**.

- RELEASE: This algorithm accepts a keystone $\hat{k} \in \mathcal{K}$ together with a valid set of signatures $\{(m_A, (Q_1^A, Q_2^A)), (m_B, (Q_1^B, Q_2^B)), (m_C, (Q_1^C, Q_2^C))\}$ and performs the following. Hereafter, we abuse the notation $\sigma_i = (Q_1^i, Q_2^i)$ to indicate a signature that is produced by $\mathcal{U}_i$. Since $Q_1, Q_2 \in G_1$, where G_1 is an additive group, then this notation is clear from its context.

We note that each $(m_i, (Q_1^i, Q_2^i))$ will pass the AVERIFY algorithm.

- Computes $Q_{ij} = s_i Q_2^j$ and $Q_{ik} = s_i Q_2^k$, where Q_2^u denotes Q_2 that was generated by u using ASIGN algorithm (and therefore, it implies that AVERIFY$(m_i, f, \sigma_i, \mathcal{P}_i, \mathcal{P}_j, \mathcal{P}_k)$= accept holds with equality, for $\sigma_i = (Q_1^i, Q_2^i)$).

- Produces the following signatures of knowledge.

$$\Gamma = ECSPKEQ\{\alpha : Q_{ij} = \alpha Q_2^i \wedge Q_{ik} = \alpha Q_2^j\}(\varepsilon)$$

- Outputs $(f, \hat{k}, Q_{ij}, Q_{ik}, \Gamma)$, where $\hat{k} \in \mathcal{K}$ and $H_1(f) = \hat{k}$ holds. Notice that f is only known by the initial signer, and hence, this algorithm can only be performed correctly by the initial signer.

- ● PROOF-VERIFY: In the following description, the initial signer is denoted by $\mathcal{U}_i$. This algorithm accepts and verifies whether $\Gamma = ECSPKEQ\{\alpha : Q_{ij} = \alpha Q_2^i \wedge Q_{ik} = \alpha Q_2^j\}(\varepsilon)$ holds. If it does not hold, then output reject. Then, it verifies whether the following equations

$$\hat{e}(Q_1^j, \mathcal{P}_j)\hat{e}(Q_{ij}, P)\hat{e}(Q_2^j, \mathcal{P}_k) \stackrel{?}{=} \hat{e}(M, P)$$
$$\hat{e}(Q_1^k, \mathcal{P}_k)\hat{e}(Q_{ik}, P)\hat{e}(Q_2^k, \mathcal{P}_j) = \hat{e}(M, P)$$

hold with equality. If it does not hold, then output reject. Finally, verify whether $f = KGEN(\hat{k})$ holds. If not, output reject. Otherwise, output accept.

- ● VERIFY: The algorithm accepts $(m, f, k, \mathcal{P}_i, \mathcal{P}_j, \mathcal{P}_k, Q_{ij}, Q_{ik}, \Gamma)$, for $\sigma = (Q_1, Q_2), m \in \mathcal{M}, f \in \mathcal{F}, \hat{k} \in \mathcal{K}$, and performs the following verification steps.

  - test whether $H_1(f) \stackrel{?}{=} \hat{k}$ holds. If not, then output reject.
  - execute PROOF-VERIFY. If not hold, then output reject.
  - execute AVERIFY with parameter $(m, f, \sigma_u, \mathcal{P}_i, \mathcal{P}_j, \mathcal{P}_k)$ for the three message-signature pairs, $u \in \{i, j, k\}$. The output of VERIFY is the output of AVERIFY algorithm.

- ● DENY: This algorithm accepts $(m, f, \sigma_1, \sigma_2, \mathcal{P}_i, \mathcal{P}_j, \mathcal{P}_k, s_i)$, where $m \in \mathcal{M}, f \in \mathcal{F}$, are signatures on m, $u = \{1, 2\}$, s_i is $\mathcal{U}_i$'s secret key associated with the public key $\mathcal{P}_i$ (which implies $\mathcal{P}_i = s_i P$) and $\mathcal{P}_j, \mathcal{P}_k$ are the public keys of $\mathcal{U}_j, \mathcal{U}_k$, respectively. The algorithm performs the following.

  - Test whether AVERIFY accept $(m, f, \sigma_u, \mathcal{P}_i, \mathcal{P}_j, \mathcal{P}_k)$, for $u = \{1, 2\}$. If it does not hold, then terminate the algorithm and output reject.
  - Compute $\delta_1 = s_i Q_1 - M + P$ for $M = H_0(m \| f)$.

- Perform the following verification
$$(\hat{e}(Q_2, \mathcal{P}_j + \mathcal{P}_k)\hat{e}(\delta_1, P) \overset{?}{=} \hat{e}(P, P))$$
for both signatures σ_1, σ_2.
- If the result of the above verification for either σ_1 or σ_2 is true, return accept with δ_1 as the proof. Otherwise, return reject.

Correctness.

The correctness of the AVERIFY algorithm is justified as follows.
$$\hat{e}(H_0(M \parallel f), P) = \hat{e}(Q_1, \mathcal{P}_i)\hat{e}(Q_2, \mathcal{P}_j + \mathcal{P}_k)$$
$$= \hat{e}(s_i^{-1}(M - r(\mathcal{P}_j + \mathcal{P}_k)), \mathcal{P}_i)\hat{e}(rP, \mathcal{P}_j + \mathcal{P}_k)$$
$$= \hat{e}(M - r(\mathcal{P}_j + \mathcal{P}_k), P)\hat{e}(r(\mathcal{P}_j + \mathcal{P}_k), P)$$
$$= \hat{e}(M - r(\mathcal{P}_j + \mathcal{P}_k) + r(\mathcal{P}_j + \mathcal{P}_k), P)$$
$$= \hat{e}(M, P)$$
$$= \hat{e}(H_0(M \parallel f), P) \qquad \qquad \square$$

4.1 Security Analysis

Lemma 1. *If a participant A has signed a message m to generate σ_A, both B and C will be convinced with the authenticity of the signature, but no other third party will.*

Proof. When a signature $\sigma_A = (Q_1, Q_2)$ is generated, firstly either B and C needs to execute the AVERIFY algorithm. If the signature passes this test, then B and C will believe that his signature was indeed generated by A, because they have not colluded to generate this signature. We note that no other third party can be convinced with the authenticity of this signature, since if B and C collude, they can collaboratively compute $Q_1' = rP, Q_2' = (s_B + s_C)^{-1}(M - r\mathcal{P}_A)$, for a random $r \in Z_q^*$, which is valid and indistinguishable signature from any third party's point of view. Hence, the signature cannot be used to convince any other third party other than B and C. $\qquad \square$

Theorem 1. *The DENY algorithm is correct and sound. This algorithm is used to protect a participant against a collusion of two malicious participants.*

The proof of this theorem is shown in terms of the following lemmas.

Lemma 2. *Any two participants can collude and frame another participant that he has signed a message.*

Proof. To show the correctness of the DENY algorithm, we need to show a successful attack that is launched by a conspiracy of two participants to frame the other participant. Without losing generality, we assume B will conspire with C to frame A, i.e. to accuse that A had signed a message that he has not signed. The attack is as follows.

- B and C collaboratively perform the following.

  • Select a random $\hat{r} \in Z_q^*$.

  • Compute $Q_1' = \hat{r}P$

  $$Q_2' = (s_B + s_C)^{-1}(M - \hat{r}\mathcal{P}_A) \text{ for } M = H_0(m\|f)$$

- Output (Q_1', Q_2') as a signature on m.

One can verify that the signature (Q_1', Q_2') will pass the AVERIFY algorithm, due to the following.

$$\hat{e}(H_0(m\|f), P) \overset{?}{=} \hat{e}(Q_1', \mathcal{P}_A)\hat{e}(Q_2', \mathcal{P}_B + \mathcal{P}_C)$$
$$= \hat{e}(\hat{r}P, \mathcal{P}_A)\hat{e}((s_B + s_C)^{-1}(M - \hat{r}\mathcal{P}_A), \mathcal{P}_B + \mathcal{P}_C)$$
$$= \hat{e}(\hat{r}P, \mathcal{P}_A)\hat{e}(M - \hat{r}\mathcal{P}_A, P)$$
$$= \hat{e}(\hat{r}\mathcal{P}_A, P)\hat{e}(M - \hat{r}\mathcal{P}_A, P)$$
$$= \hat{e}(\hat{r}\mathcal{P}_A + M - \hat{r}\mathcal{P}_A, P)$$
$$= \hat{e}(M, P)$$
$$= \hat{e}(H_0(m\|f), P)$$

Lemma 3. *Any collusion attack can be prevented by performing the* DENY *algorithm.*

Proof. As illustrated in Lemma 2, a valid signature (Q_1', Q_2') can be generated by a collusion of two participants. The signature will be in one of the following forms.

- $(\hat{r}P, (s_B + s_C)^{-1}(M - \hat{r}\mathcal{P}_A))$.
- $(-\hat{r}P, (s_B + s_C)^{-1}(M + \hat{r}\mathcal{P}_A))$.

We denote the above signatures as σ_1 and σ_2. We can easily verify that both signatures will pass the AVERIFY algorithm. Now, we shall demonstrate that A can provide a proof that a forgery has happened, by performing the DENY algorithm. Basically, the algorithm will compute following.

- Compute $\delta_1 = s_A Q_1' - M + P$

- We note that the value of δ_1 will be one of the following.

$$\delta_1 = \hat{r}\mathcal{P}_A - M + P \text{ or } \delta_1 = P - \hat{r}\mathcal{P}_A - M$$

depending on the forged signature above.

When the conspiracy happens, one of the following tests will return true.

$$(\hat{e}(Q_2, P_B + P_C)\hat{e}(\delta_1, P)) \overset{?}{=} \hat{e}(P, P)$$

for σ_1 and $_?\sigma_2$. This is due to

$$\hat{e}(P, P) = (\hat{e}(Q_2, P_B + P_C)\hat{e}(\delta_1, P))$$
$$= (\hat{e}((s_B + s_C)^{-1}(M - \hat{r}P_A), P_B + P_C)\hat{e}(\hat{r}P_A - M + P, P))$$
$$= (\hat{e}(M - \hat{r}P_A, P)\hat{e}(\hat{r}P_A - M + P, P))$$
$$\overset{?}{=} (\hat{e}(P, P))$$

$$\hat{e}(P, P) = (\hat{e}(Q_2, P_B + P_C)\hat{e}(\delta_1, P))$$
$$= (\hat{e}((s_B + s_C)^{-1}(M + \hat{r}P_A), P_B + P_C)\hat{e}(P - \hat{r}P_A - M, P))$$
$$= (\hat{e}(M + \hat{r}P_A, P)\hat{e}(P - \hat{r}P_A - M, P))$$
$$= (\hat{e}(P, P))$$

We note that if the signature is not forged (i.e. generated by A), then the above verification will not return true. Hence, the DENY algorithm will return true iff forgery has happened due to collusion of two participants. □

Theorem 2. (Ambiguity) *Before the keystone is released using the* RELEASE *algorithm, both signature are ambiguous.*

Proof. As shown in the proof of Lemma 2, a collusion of two participants can always produce a set of signatures that will pass the verification AVERIFY algorithm. We will illustrate this attack as follows. Without losing generality, we assume A colludes with C. Firstly, A produces (Q_1^A, Q_2^A) by herself, and then collaboratively with C, they can produce $(Q_1^{B'}, Q_2^{B'})$ and claim that this signature was indeed signed by B. Finally, C can produce (Q_1^C, Q_2^C). We note that all the signatures are produced using a legitimate ASIGN algorithm. From any third party's point of view, these signatures are indistinguishable from a set of signatures that are genuinely created by the three participants. In this scenario, before the keystone is released using the RELEASE algorithm, B can always invoke DENY algorithm at any time to deny that he has not signed the message. Hence, the signatures are ambiguous from any third party's point of view. We also note that A and C cannot collude and frame that B has signed a message that he has not signed, as A cannot invoke the RELEASE algorithm correctly since $(Q_1^{B'}, Q_2^{B'})$ are not in the "correct" form, i.e. $(s_B^{-1}(M - r(P_A + P_C)), rP)$.

Lemma 4. *When the output of* VERIFY *is* accept, *then any third party can be sure who has generated the signature.*

Proof. We note that VERIFY algorithm will test three different components. The first component is to verify whether the keystone $\hat{k}$ is generated correctly, using the *KGEN* function. The second verification is to make sure that the published signatures are in the correct form. This is guaranteed with the following test

$$\hat{e}(Q_1^j, \mathcal{P}_j)\hat{e}(Q_{ij}, P)\hat{e}(Q_2^j, \mathcal{P}_k) \overset{?}{=} \hat{e}(M, P)$$
$$\hat{e}(Q_1^k, \mathcal{P}_k)\hat{e}(Q_{ik}, P)\hat{e}(Q_2^k, \mathcal{P}_j) \overset{?}{=} \hat{e}(M, P)$$

We note that the above tests will be satisfied, iff $Q_{ij} = s_i Q_2^j$ and $Q_{ik} = s_i Q_2^k$ hold. Finally, the last verification will confirm that the signatures were indeed generated correctly by each participant, so that they will pass the verification test. Hence, any third party can be convinced with the authenticity of the signature.

Theorem 3. (Fairness) *Any signature that is generated with the same keystone will be binding concurrently when the keystone is released.*

Proof. Suppose any of the participants tries to cheat by signing more than one signature. By testing the VERIFY algorithm, all the signatures will be binding concurrently. This way, the fairness is guaranteed. Moreover, as illustrated in the proof of Theorem 2, two colluding participants cannot frame another participant by generating a forged signature and later on confirm it as if it was signed by the framed participant. This is due to the inability of the initial signer to execute the RELEASE algorithm correctly.

Theorem 4. (Unforgeability) *The scheme presented in this section is existentially enforceable under a chosen message attack in the random oracle model, assuming the hardness of the Computational Diffie-Hellman problem.*

Proof. We use the notion of existential unforgeability against a chosen message attack from Chen et al., 2004. We consider an **EF-CMA** adversary $\mathcal{A}$ that outputs an existential forgery (M^*, σ^*) with probability $\text{Succ}_{\text{EF-CMA}}^{\mathcal{A}}(k)$ within the time t. We denote the number of queries from the random oracle $\mathcal{H}$ by $q\mathcal{H}$ and from the signing oracle Σ by $q\Sigma$. For simplicity, we show a simulation where $\mathcal{A}$ corrupts one of the receiver B or C (but not both of them) to produce a forgery for the initial signer A. The game between the adversary $\mathcal{A}$ and the challenger $\mathcal{C}$ is defined as follows.

- SETUP: $\mathcal{C}$ runs SETUP to a given security parameter ℓ to obtain descriptions of $\mathcal{U}, \mathcal{M}, \mathcal{S}, \mathcal{K}, \mathcal{F}$ and *KGEN*: $\mathcal{K} \to \mathcal{F}$. In addition, SETUP also generates the public key of each participant $\mathcal{P}_i \to s_i P$, for $i \in \{A, B, C\}$. The associated secret values s_i's are delivered securely to participant $\mathcal{U}_i, i \in \{A, B, C\}$. The public keys $\{\mathcal{P}_A, \mathcal{P}_B, \mathcal{P}_C\}$ are published together with the system parameters. Let $\Upsilon = xP$ and $\Psi = xyP$ be the CDH challenge and kept secret by C at this stage.

-   KGen Queries: $\mathcal{A}$ can request that C selects a keystone $\hat{k} \in \mathcal{K}$ that it used to generate a keystone fix $f \in \mathcal{F}$, by invoking $f = KGEN(\hat{k})$. $\mathcal{A}$ can also select his own keystone $\hat{k} \in \mathcal{K}$ and the compute the keystone fix by himself by running the **$KGEN(k)$** function.

-   KReveal Queries: $\mathcal{A}$ can request the challenger C to reveal the keystone $\hat{k}$ that is used to produce a keystone fix $f \in \mathcal{F}$ in a previous KGen Query. If f was not asked before, then C outputs invalid. Otherwise, C returns $\hat{k} \in \mathcal{K}$.

-   Hash Queries: $\mathcal{A}$ can query the random oracle $\mathcal{H}$ at any time. When (m, f) are requested, firstly C checks his H-list. If it exists, then returns the value from the list. Otherwise, C selects a random $h \in Z_q^*$ at random and computes $M = h\Psi$. The value (m, f, h, M) is recorded in the H-list and M is returned to $\mathcal{A}$.

-   ASIGN Queries: $\mathcal{A}$ can request an ambiguous signature for any input of the form $(m, f, \mathcal{P}_A, \mathcal{P}_B, \mathcal{P}_C)$ for published values $(\mathcal{P}_A, \mathcal{P}_B, \mathcal{P}_C)$. C checks the H-list for the existence of m. If it does not exist, then C calls the ASIGN algorithm to sign the message as usual. However, if m exists, then C selects $a_2, r_1, r_2 \in Z_q^*$ at random and sets $a_1 = r_1 - a_2 r_2$. Finally C sets $Q_1 = a_1 P$ and $Q_2 = a_2 P$, and stores $(m, f, r_1, r_1 \Upsilon)$. Return (Q_1, Q_2) to $\mathcal{A}$ as a valid signature.

-   AVerify and Verify Queries: $\mathcal{A}$ cannot request an answer for these queries since he can compute them for himself using AVERIFY and VERIFY algorithms.

-   Output: Eventually, $\mathcal{A}$ outputs a forgery (m^*, Q_1^*, Q_2^*) and by definition of the existential forgery, there is in the H-list a quadruple (m^*, f^*, h^*, M^*) such that $\Phi = (h^*)^{-1}(Q_1^* + r_2 Q_2^*) = yP$.

We note that the success probability of the attack is defined by

$$\left(\frac{1}{2} \mathrm{Adv}_{\mathrm{EF-CMA}}^{\mathcal{A}} - \frac{q\mathcal{H}q\Sigma + 1}{2^\ell} \right)^2 \leq \mathrm{Succ}_{CDH}(\ell)$$

where ℓ is the security parameter. The time to execute the attack is defined by $t' \leq 2(t + q\mathcal{H} + 2q\Sigma + O(1)T_G + q\Sigma T_{GM})$, where T_G denotes the time complexity to perform a scalar multiplication in G and T_{GM} denotes the time complexity to perform an exponentiation in G_M.

Solving two instances of this problem will lead us to a solution to the CDH problem using the technique in Boneh et al., (2003). Hence, we complete the proof.   □

Theorem 5. *Our tripartite concurrent signature scheme is secure in the random oracle model, assuming the hardness of the discrete logarithm problem.*

Proof. The proof can be derived from Theorems 1, 2, 3 and 4 and Lemma 4.   □

5.   OPEN PROBLEMS

In this paper, we presented a tripartite concurrent signature scheme. Using a similar idea, we can obtain a multi party concurrent signature, but we have not defined the notion of fairness in that scenario. Hence, the open problem left in this area is how to define a formal model for a multi party concurrent signature scheme, where there are n parties involved ($n>3$).

REFERENCES

Abe et al., (2002) Abe, Masayuki, Ohkubo, Miyako, and Suzuki, Koutarou (2002). 1-out-of-n Signatures from a Variety of Keys. *Asiacrypt 2002, LNCS 2501*, pages 415-432.

Boneh et al., (2003) Boneh, Don, Gentry, Craig, Lynn, Ben, and Shacham, Hovav (2003). Aggregate and verifiably encrypted signatures from bilinear maps. *Eurocrypt 2003 LNCS 2656*, pages 416-432.

Camenisch, 1997 Camenisch, Jan (1997). Efficient and generalized group signatures. *Eurocrypt'97, LNCS 1233*, pages 465–479.

Camenisch, 1998 Camenisch, Jan (1998). Group signature schemes and payment systems based on the discrete logarithm problem. *PhD thesis, ETH Zürich.*

Chen et al., 2004 Chen, Liqun, Kudla, Caroline, and Paterson, Kenneth G. (2004). Concurrent signatures. In *Eurocrypt 2004, LNCS 3027*, pages 287–305.

Desmedt, 2003 Desmedt, Yvo (2003). Verifier-Designated Signatures. *Rump Session, Crypto 2003.*

Jakobsson et al., 1996 Jakobsson, Markus, Sako, Kazue, and Impagliazzo, Russell (1996). Designated Verifier Proofs and Their Applications. *Eurocrypt'96, LNCS 1070*, pages 143 – 154.

Laguillaumie and Vergnaud, 2004 Laguillaumie, Fabien and Vergnaud, Damien (2004). Multi-Designated Verifiers Signatures. *Sixth Intl Conf on Inf and Comm Security (ICICS 2004)*.

Rivest et al., 2001 Rivest, Ronald L., Shamir, Adi, and Tauman, Yael (2001). How to Leak a Secret. *Asiacrypt 2001, LNCS 2248*, pages 552 – 565.

SIGNCRYPTION IN HIERARCHICAL IDENTITY BASED CRYPTOSYSTEM
(Extended Abstract)

Sherman S.M. Chow[1], Tsz Hon Yuen[2], Lucas C.K. Hui[1], and S.M. Yiu[1]

[1]Department of Computer Science
University of Hong Kong
Pokfulam, Hong Kong
{smchow, hui, smyiu}@cs.hku.hk

[2]Department of Information Engineering
Chinese University of Hong Kong
Shatin, Hong Kong
thyuen4@ie.cuhk.edu.hk

Abstract: In many situations we want to enjoy confidentiality, authenticity and non-repudiation of message simultaneously. One approach to achieve this objective is to "sign-then-encrypt" the message, or we can employ special cryptographic scheme like signcryption. Two open problems about identity-based (ID-based) signcryption were proposed in [16]. The first one is to devise an efficient forward-secure signcryption scheme with public verifiability and public ciphertext authenticity, which is promptly closed by [10]. Another one which still remains open is to devise a hierarchical ID-based signcryption scheme that allows the user to receive signcrypted messages from sender who is under another sub-tree of the hierarchy. This paper aims at solving this problem by proposing two concrete constructions of hierarchical ID-based signcryption.

Key words: signcryption, hierarchical identity-based cryptosystem, bilinear pairings

1.   INTRODUCTION

In traditional public key infrastructure, certificates leak data and are not easily located. Strict online requirement removes offline capability, and validating policy is time-consuming and difficult to administer. Moreover, traditional PKI may not provide a good solution in many scenarios. For example, in tetherless computing architecture (TCA) [24] where two mobile hosts wanting to communicate might be disconnected from each other and also from the Internet. As exchange of public keys is impossible in this disconnected situation, identity-based (ID-based) cryptosystem fits in very well since the public key can be derived from the identity of another party [23].

In many situations we want to enjoy confidentiality, authenticity and non-repudiation of message simultaneously. A traditional approach to achieve this objective is to "sign-then-encrypt" the message, or we can employ special cryptographic scheme like signcryption which can be more efficient in computation than running encryption and signature separately. A recent direction is to merge the concept of ID-based cryptography [22] and signcryption [26]. Two open problems about ID-based signcryption were proposed in [16]. The first one is to devise an efficient forward-secure signcryption scheme with public verifiability and public ciphertext authenticity, which is promptly closed by [10]. Another one which still remains open is to devise a hierarchical ID-based signcryption scheme that allows the user to receive signcrypted messages from sender who is under another sub-tree of the hierarchy. This paper aims at solving this problem.

1.1   Applications

ID-based cryptography is suitable for the use of commercial organizations. In their settings, the inherent key-escrow of property is indeed beneficial, where the big boss has the power to monitor his/her employees' Internet communications if necessary. Hierarchical structure is common in nowadays' organizations, single trusted authority for generation of private key and authentication of users may be impractical; all these motivated the need of hierarchical ID-based cryptosystem.

Moreover, hierarchical ID-based cryptosystem is also useful in other scenarios, such as in TCA, a computing architecture with the concept of "regions", which can be viewed as a branch of the hierarchy [15,23].

1.2   Related Work

Malone-Lee gave the first ID-based signcryption scheme [18]. This scheme is not semantically secure as the signcrypted text produced is a concatenation of a signature by a variant of Hess's ID-based signature [14] and a ciphertext by a simplified version of Boneh and Franklin's ID-based encryption [4]. In short, the signature of the message is visible in the signcrypted message.

On the other hand, Nalla and Reddy's ID-based signcryption scheme [20] cannot provide public verifiability as well as public ciphertext authenticity since the verification can only be done with the knowledge of recipient's private key. Libert and Quisquater proposed three ID-based signcryption schemes [16]. None of them can satisfy the requirements for public verifiability and forward security at the same time.

Boyen's multipurpose ID-based signcryption scheme [5] is the first scheme that provides public verifiability and forward security and is also provably secure. However, this scheme aims at providing ciphertext unlinkability and anonymity. So, a third party cannot verify the origin of the ciphertext, thus the scheme does not satisfy the requirement of public ciphertext authenticity. We remark that Boyen's scheme is very useful in applications that require unlinkability and anonymity.

The public verifiability of the signcrypted message usually can only be checked with some ephemeral data computed by the intended recipient of the signcrypted message. The notion of verifiable pairing was introduced in [8] to ensure the non-repudiation property of the ID-based signcryption by disallowing the intended recipient to manipulate the ephemeral data.

In 2004, [19] claimed that they were the first one closing the open problem proposed by [16]; however, the open problem was indeed closed by [10] in 2003. Recently, a simple but secure ID-based signcryption scheme was proposed in [7] and an ID-based signcryption scheme with exact security was proposed in [17]. The first blind ID-based signcryption scheme was proposed in [25]. This scheme offers the option to choose between authenticated encryption and ciphertext unlinkability. The generic group and pairing model was also introduced in this paper. Notice that none of the previously mentioned schemes works with hierarchical ID-based cryptosystem.

2.   PRELIMINARIES

Before presenting our results, we give the definition of a hierarchical ID-based signcryption scheme by extending the framework in previous work (e.g. [10,25]). We also review the definitions of groups equipped with a bilinear pairing and the related complexity assumptions.

2.1   Framework of Hierarchical ID-based Signcryption

An ID-based signcryption (IDSC) scheme consists of six algorithms: Setup , Extract , Sign , Encrypt , Decrypt and Verify . Setup and Extract are executed by the private key generators (PKGs henceforth). Based on the security level parameter, Setup is executed to generate the master secret and common public parameters. Extract is used to generate the private key for any given identity. The algorithm Sign is used to produce the signature of a signer on a message, it also outputs some ephemeral data for the use of Encrypt ; Encrypt takes the message, the signature, the ephemeral data produced by Sign and the recipient's identity to produce a signcrypted text. Decrypt takes the input of secret key and decrypt the signcrypted text to give the message and the corresponding signature, finally Verify is used by any party to verify the signature of a message.

In the hierarchical ID-based signcryption (HIDSC henceforth), PKGs are arranged in a tree structure, the identities of users (and PKGs) can be represented as vectors. A vector of dimension ℓ represents an identity at depth ℓ . Each identity ID of depth ℓ is represented as an ID-tuple $ID \mid \ell = \{ID_1, \cdots, ID_\ell\}$. The algorithms of HIDSC have similar functions to those of IDSC except that the Extract algorithm in HIDSC will generate the private key for a given identity which is either a normal user or a lower level PKG. The private key for identity ID of depth ℓ is denoted as $S_{ID \mid \ell}$ (or S_{ID} if the depth of ID does not related to the discussion). The functions of Setup , Extract , Sign , Encrypt , Decrypt and Verify in HIDSC are described as follows.

- Setup : Based on the input of a unary string 1^k where k is a security parameter, it outputs the common public parameters *params* , which include descriptions of a finite message space, a finite signature space and a finite signcrypted text space. It also outputs the master secret s , which is kept secret by the root private key generator (PKG).

- Extract : Based on the input of an arbitrary identity ID of depth j , it makes use of the secret key $S_{ID \mid j-1}$ (if $j = 1$, the input of the algorithm is s , which is the master secret of the root PKGs, instead of $S_{ID \mid j-1}$) to output the private key $S_{ID \mid j}$ for ID .

- Sign : Based on the input (M, S_{ID}) , it outputs a signature σ and some ephemeral data r .

- Encrypt : Based on the input $(M, S_A, ID_B, \sigma, r)$, it outputs a signcrypted message C .

- Decrypt: Based on the input (C, S_B, ID_B), it outputs the message M, the corresponding signature σ and the purported signer ID_A.
- Verify: Based on the input (σ, M, ID), it outputs $\top$ for "true" or $\bot$ for "false", depending on whether σ is a valid signature of message M signed by ID or not.

These algorithms must satisfy the standard consistency constraint of hierarchical ID-based signcryption, i.e. if $\{\sigma, r\} = \text{Sign } (M, S_A)$, $C = \text{Encrypt } (S_A, ID_B, M, \sigma, r)$ and $\{M', ID_{A'}, \sigma'\} = \text{Decrypt } (C, S_B)$, we must have $M = M'$, $ID_A = ID_{A'}$ and $\top = \text{Verify } (\sigma', M, ID_A)$.

2.2 Bilinear Pairing

Let $(G, \cdot)$ and $(G_1, \cdot)$ be two cyclic groups of prime order q and g be a generator of G. The bilinear pairing is given as $\hat{e}: G \times G \to G_1$, which satisfies the following properties:

1. *Bilinearity*: For all $u, v \in G$ and $a, b \in Z$, $\hat{e}(u^a, v^b) = \hat{e}(u, v)^{ab}$.
2. *Non-degeneracy*: $\hat{e}(g, g) \neq 1$.
3. *Computability*: There exists an efficient algorithm to compute $\hat{e}(u, v)$ $\forall u, v \in G$.

2.3 Diffie-Hellman Problems

DEFINITION 1. The computational Diffie-Hellman problem (CDHP) in G is defined as follows: Given a 3-tuple $(g, g^a, g^b) \in G^3$, compute $g^{ab} \in G$. We say that the (t, ε)-CDH assumption holds in G if no t-time algorithm has advantage at least ε in solving the CDHP in G.

DEFINITION 2. The bilinear Diffie-Hellman problem (BDHP) in G is defined as follows: Given a 4-tuple $(g, g^a, g^b, g^c) \in G^4$ and a pairing function $\hat{e}(\cdot, \cdot)$, compute $\hat{e}(g, g)^{abc} \in G_1$. We say that the (t, ε)-BDH assumption holds in G if no t-time algorithm has advantage at least ε in solving the BDHP in G.

DEFINITION 3. The decisional bilinear Diffie-Hellman problem (DBDHP) in G is defined as follows: Given a 5-tuple $(g, g^a, g^b, g^c, T) \in G^4 \times G_1$ and a pairing function $\hat{e}(\cdot, \cdot)$, decides whether $T = \hat{e}(g, g)^{abc}$. We say that the (t, ε)-DBDH assumption holds in G if no t-time algorithm has advantage at least ε in solving the DBDHP in G.

3.   SECURITY MODEL

We present our security model for indistinguishability, existential unforgeability and ciphertext authenticity for HIDSC.

3.1   Indistinguishability

Indistinguishability for HIDSC against adaptive chosen ciphertext attack (IND-CCA2) is defined as in the following IND-CCA2 game.
1. The simulator selects the public parameter and sends the parameter to the adversary.
2. There are three oracles except the random oracles (hash oracles).
   - **Key extraction oracle** KEO: Upon the input of an identity, the key extraction oracle outputs the private key corresponding to this identity.
   - **Signcryption oracle** SO : Upon the input of the message M , the sender ID_A , the recipient ID_B , the signcryption oracle produces a valid signcryption C.
   - **Unsigncryption oracle** UO: Upon the input of the ciphertext C , the sender ID_A and the recipient ID_B , the unsigncryption oracle outputs the decryption result and the verification outcome.

   The adversary is allowed to perform a polynomial number of oracle queries adaptively, but oracle query to KEO with input ID_B is not allowed.
3. The adversary generates M_0, M_1 , ID_A , ID_B , and sends them to the simulator. The simulator randomly chooses $b \in_R \{0,1\}$ and delivers the challenge ciphertext C to the adversary where $\{\sigma, r\} = \text{Sign} (M, S_A)$ and $C = \text{Encrypt} (S_A, ID_B, M_b, \sigma, r)$. M_0 and M_1 should be of equal length, and no oracle query have been made and will be made to SO with input (M_0, ID_A, ID_B) and (M_1, ID_A, ID_B) throughout the game.
4. The adversary can again perform a polynomial number of oracle queries adaptively, but oracle query to UO for the challenge ciphertext (defined later) from the simulator is not allowed.
5. The adversary tries to compute b .

The adversary wins the game if he can guess b correctly. The *advantage* of the adversary is the probability, over half, that he can compute b accurately.

DEFINITION 4. (Indistinguishability)   A   hierarchical   ID-based signcryption scheme is *IND-CCA2* secure if no PPT adversary has a non-negligible advantage in the IND-CCA2 game.

Our security notion above is a strong one. It incorporates previous security notions including *insider-security* in [1] and *indistinguishability* in [18].

Notice that if we set the adversary to send the recipient identity ID_B to the simulator before step 1 (say, in an initialization stage) in the game, the security is reduced to the indistinguishability against *selective identity*, adaptive chosen ciphertext attack (IND-sID-CCA2).

3.2   Existential Unforgeability

Existential unforgeability against adaptive chosen message attack (EU-CMA2) for HIDSC is defined as in the following EU-CMA2 game. The adversary is allowed to query the random oracles, KEO, SO and UO (which are defined above) with the restriction that oracle query to KEO with input ID_A is not allowed.

The game is defined as follows:

1. The simulator selects the public parameter and sends it to the adversary.
2. The adversary is allowed to perform a polynomial number of oracle queries adaptively.
3. The adversary delivers a recipient identity ID_B and a ciphertext C.

The adversary wins the game if he can produce a valid (C, ID_B) such that C can be decrypted, under the private key of ID_B, to a message M, a sender identity ID_A and a signature σ which passes the verification test and no SO request that resulted in a ciphertext C, whose decryption under the private key of ID_B is the claimed forgery (σ, M, ID_A).

DEFINITION 5. (Existential Unforgeability) A hierarchical ID-based signcryption scheme is *EU-CMA2* secure if no PPT adversary has a non-negligible probability in winning the EU-CMA2 game.

The adversary is allowed to get the private key of the recipient in the adversary's answer. This gives us an *insider-security* as defined in [1].

Notice that if we set the adversary to send the sender identity ID_A to the simulator in Step 1 in the game, the security is reduced to the existential unforgeability against *selective identity*, adaptive chosen ciphertext attack (EU-sID-CMA2).

3.3   Ciphertext Authenticity

Ciphertext authenticity against adaptive chosen message attack (AUTH-CMA2) for HIDSC is defined as in the following AUTH-CMA2 game. The adversary is allowed to query the random oracles, KEO, SO and UO, which are defined above. The game is defined as follows:

1. The simulator selects the public parameter and sends the parameter to the adversary.
2. The adversary is allowed to perform a polynomial number of oracle queries adaptively.

3. The adversary delivers a recipient identity ID_B and a ciphertext C.

The adversary wins the game if he can produce a valid (C, ID_B) such that C can be decrypted, under the private key of ID_B, to a message M, sender identity ID_A and a signature σ which passes the verification test.

Oracle query to *KEO* with input ID_A and ID_B is not allowed. The adversary's answer (C, ID_B) should not be computed by *SO* before.

DEFINITION 6. (Ciphertext Authenticity) A hierarchical ID-based signcryption scheme is *AUTH-CMA2* secure if no PPT adversary has a non-negligible probability in winning the AUTH-CMA2 game.

Outsider-security is considered in this model since the adversary is not allowed to get the private key of the recipient in the adversary's answer. This model represents the attack where a signature is re-encrypted by using a public key with unknown secret key.

4.   SCHEME 1

4.1   Construction

Let ℓ be the number of levels of the hierarchy to be supported. Let H_1, H_2 and H_3 be three cryptographic hash functions where $H_1 : \{0,1\}^* \to G$ and $H_2 : \{0,1\}^* \to G$, and $H_3 : G_1 \to \{0,1\}^{k_0 + k_1 + n}$ where k_0 is the number of bits required to represent an element of G, k_1 is the maximum number of bits required to represent an identity (of depth ℓ) and n is the maximum number of bits of a message to be signcrypted. Our first construction of a hierarchical ID-based signcryption scheme is given below. The construction is based on the idea in [13].

Setup: On the input of a security parameter $k \in N$, the root PKG uses the BDH parameter generator [4] to generate G, G_1, q and $\hat{e}(\cdot, \cdot)$, where q is the order of groups G and G_1. Then the root PKG executes the following steps.

1. Select an arbitrary generator P_0 from G.
2. Pick a random s_0 from Z_p, which is the system's master secret key.
3. Compute $Q_0 = P_0^{s_0}$.
4. The public system parameters are
$$\text{params} = <G, G_1, \hat{e}(\cdot, \cdot), q, P_0, Q_0, H_1(\cdot), H_2(\cdot), H_3(\cdot)>.$$

KeyGen: For an entity with $ID | k - 1 = \{ID_1, ID_2, \cdots, ID_{k-1}\}$ of depth $k - 1$ (for root PKG, its depth is defined as 0 and its identity is defined as empty string ε), it uses its secret key $S_{ID|k-1}$ (or the master secret s_0 of the root PKGs, if $k = 1$) to generate the secret key for a user $ID | k$ (where the first $k - 1$ elements of $ID | k$ are those in $ID | k - 1$) as follows.

1. Compute $P_{ID|k} = H_1(ID_1, ID_2, \cdots, ID_{k-1}, ID_k)$.
2. Pick random s_{k-1} from Z_p (this step is not necessary for the root PKG as s_0 is already defined).
3. Set the private key of the user to be $S_{ID|k} = S_{ID|k-1} \cdot P_{ID|k}{}^{s_{k-1}} = \prod_{i=1}^{k} P_{ID|i}{}^{s_{i-1}}$, where $S_{ID|0}$ is defined as the identity element in G.
4. Send the values of $Q_i = P_0{}^{s_i}$ for $1 \leq i \leq k-1$ as "verification points" to the user.

Sign: For a user $A|k = \{A_1, A_2, \cdots, A_k\}$ with secret key $S_{A|k} = \prod_{i=1}^{k} P_{A|i}{}^{s_{i-1}}$ and the points $Q_i = P_0{}^{s_i}$ for $1 \leq i \leq k-1$ to sign on a message M, he/she follows the steps below.
1. Pick a random number r from Z_p^*.
2. Compute $P_M = H_2(M)$.
3. Compute $\sigma = S_{A|k} \cdot P_M{}^{r}$.
4. Return $\{\sigma, Q_1, Q_2, \cdots, Q_{k-1}, Q_M = P_0{}^{r}\}$ as the signature and return r as the ephemeral data for Encrypt.

Encrypt: To signcrypt the message M to user $B|l$, the steps below are used.
1. Compute $P_{B|j} = H_1(B_1, B_2, \cdots, B_j)$ for $1 \leq j \leq l$.
2. Pads the identity A with a chain of zero bits if it is not of depth ℓ.
3. Return ciphertext $C =$
$$\{P_{B|2}{}^{r}, \cdots, P_{B|l}{}^{r}, (M \| \sigma \| A) \oplus H_3(\hat{g}^r), Q_1, Q_2, \cdots, Q_M\}$$
where $\hat{g} = \hat{e}(Q_0, P_{B|1}) \in G_1$ and $\oplus$ represents the bitwise XOR.

Decrypt: For user $B|l$ with secret key $S_{B|l} = \prod_{i=1}^{l} P_{B|i}{}^{s_{i-1}}$ and the points $Q'_i = P_0{}^{s_i}$ for $1 \leq i \leq l$ to decrypt the signcrypted message c, the steps below are used.
1. Let $C = \{U_2, \cdots, U_l, V, Q_1, Q_2, \cdots, Q_{k-1}, Q_M\}$
2. Compute $V \oplus H_3(\hat{e}(Q_M, S_{B|l}) / \prod_{i=2}^{l} \hat{e}(Q'_{i-1}, U_i)) = M \| \sigma \| A$.
   (for $l = 1$, $\prod_{i=2}^{l} \hat{e}(Q'_{i-1}, U_i)$ is defined as the identity element in G_1.)
3. Return $\{M, \sigma, A, Q_1, Q_2, \cdots, Q_{k-1}, Q_M\}$.

Verify: For A's signature $\{\sigma, Q_1, Q_2, \cdots, Q_{k-1}, Q_M\}$, everyone can do the following to verify its validity.
1. Compute $P_M = H_2(M)$.
2. Compute $P_{A|i} = H_1(A_1, A_2, \cdots, A_i)$ for $1 \leq i \leq k$.
3. Return $\top$ if $\hat{e}(P_0, \sigma) / \prod_{i=2}^{k} \hat{e}(Q_{i-1}, P_{A|i}) = \hat{e}(Q_0, P_{A|1})\hat{e}(Q_k, P_M)$.
   (for $k = 1$, $\prod_{i=2}^{k} \hat{e}(Q_{i-1}, P_{A|i})$ is defined as the identity element in G_1.)

4.2   Efficiency Analysis

We first consider the communication efficiency of the scheme. The signcrypted message is shortened by one G_1 element, as compared with using the schemes HIDE and HIDS in [13] together. Moreover, the size of the signcrypted message can be further reduced if the sender and the receiver have a common low-level PKG ancestor. The modification incurred includes using a fixed s_{k-1} instead of a random one for each invocation of KeyGen . For verification side, since the sender and the receiver share some common "verification points", these points can be omitted from the transmission. For encryption side, the ciphertext size can be reduced by using the concept of "Dual-HIDE" in [13], which can be seen as an extension of the concept of non-interactive key sharing in [22]. The basic idea behind non-interactive key sharing is that a same value can be computed either from the sender's private key and the recipient's public key or from the recipient's private key and the sender's public key. The sender is required to get his/her private key before the encryption can be done, but there is no practical difference in the case of signcryption since the sender who are going to sign the message must have his/her private key ready anyway. In our proposed construction, the "non-interactive agreed secret key" created by the sender ID whose the common ancestor with the receipt is at level m is $\hat{e}(S_{ID|m}, P_0) = \hat{e}(P_0, S_{ID}) / \prod_{i=m+1}^{l} \hat{e}(Q_{i-1}, P_{ID|i})$. To utilizing it, simply replace $\hat{g}$ with this agreed secret key.

For the computational efficiency, chosen ciphertext secure HIDE requires the transformation in Section 3.2 of [13], while our scheme does not require such transformation as the integrity checking of the ciphertext is obtained from the signature. Notice that the above modification from the concept of "Dual-HIDE" distributes the computational effort of the sender and that of the recipient in a more even way.

4.3   Security Analysis

THEOREM 1. Suppose that the (t, ε)-BDH assumption holds in G, then the above scheme is $(t', q_S, q_H, q_E, q_R, \varepsilon)$-adaptive chosen ciphertext (IND-CCA2) secure for any $t' < t - o(t)$.

THEOREM 2. Suppose that the (t, ε)-CDH assumption holds in G, then the above scheme is $(t', q_S, q_H, q_E, q_R, \varepsilon)$-adaptive chosen message (EU-CMA2) secure for any $t' < t - o(t)$, $\varepsilon' > \varepsilon/e^2 q_S q_E$.

THEOREM 3. Suppose that the (t, ε)-CDH assumption holds in G, then the above scheme is $(t', q_S, q_H, q_E, q_R, \varepsilon)$-adaptive chosen message (AUTH-CMA2) secure for any $t' < t - o(t)$.

Proofs are omitted due to the length constraint. Please refer to the full version of this paper [11].

5. SCHEME 2

5.1 Construction

Let H be a cryptographic hash function where $H : \{0,1\}^* \to Z_p$. We use $H(\cdot)$ to hash the string representing the identity into an element in Z_p^k, the same hash function will be used in the signing algorithm too. Similar to [3], H is not necessarily a full domain hash function. Notice that the identity string is hashed to Z_p instead G in scheme 1, so we use I_i to denote $H(ID_i)$ for $1 \le i \le \ell$, where ℓ is the number of levels of the hierarchy to be supported. Our second construction of HIDSC, based on the ideas in [9] and [3], is given below.

Setup: On the input of a security parameter $k \in N$, the root PKG uses the BDH parameter [4] to generate G, G_1, q and $\hat{e}(\cdot,\cdot)$, where q is the order of groups G and G_1. Then the root PKG executes the following steps.

1. Select α from Z_p^*, $h_1, h_2, \cdots, h_\ell$ from G and two generators g, g_2 from G^*, where ℓ is the number of levels of the hierarchy to be supported.
2. The public parameters are: $\{g, g_1 = g^\alpha, g_2, h_1, h_2, \cdots, h_\ell, \hat{e}(g_1, g_2)\}$.
3. The master secret key is $d_{ID|0} = g_2^\alpha$.

KeyGen: For a user $ID|k-1 = \{ID_1, ID_2, \cdots, ID_{k-1}\}$ of depth $k-1$, he/she uses his/her secret key $d_{ID|k-1}$ to generate the secret key for a user $ID|k$ (where the first $k-1$ elements of $ID|k$ are those in $ID|k-1$) as follows.

1. Pick random r_k from Z_p.
2. $d_{ID|k} = \{d_0 F_k(I_k)^{r_k}, d_1, \cdots, d_{k-1}, g^{r_k}\}$, where $F_k(x)$ is defined as $g_1^x h_k$.

Sign: For a user $ID|k$ with secret key $\{g_2^\alpha \prod_{j=1}^k F_j(I_j)^{r_j}, g^{r_1}, \cdots, g^{r_k}\}$ to sign on a message M, he/she follows the steps below.

1. Pick a random number s from Z_p^*.
2. Compute $h = H(M, \hat{e}(g_1, g_2)^s)$.
3. Repeat Steps 1-3 in case the unlikely event $s + h = 0$ occurs.
4. For $j = \{1, 2, \cdots, k\}$, compute $y_j = d_j^{s+h}$.
5. Compute $z = d_0^{s+h}$.
6. Return $\{s, y_1, y_2, \cdots, y_k, z\}$ as the signature.

Encrypt: To signcrypt a message $M \in G_1$ to user $ID|l = \{ID_1, ID_2, \cdots, ID_l\}$, the ciphertext to be generated is
$$\{F_1(I_1)^s, F_2(I_2)^s, \cdots, F_l(I_l)^s, \hat{e}(g_1, g_2)^s \cdot M, g^s, y_1, y_2, \cdots, y_k, z\}.$$

Decrypt: For a user $ID'|l$ with secret key $\{d'_0 = g_2^\alpha \prod_{j=1}^l F_j(I'_j)^{r'_j}, d'_1 = g^{r'_1}, \cdots, d'_l = g^{r'_l}\}$ to decrypt the signcrypted text $\{u_1, \cdots, u_l, v, w, y_1, y_2, \cdots, y_k, z\}$, he/she follows the steps below.

1. Compute $\sigma = \hat{e}(g_1, g_2)^s$ by $\hat{e}(w, d'_0) / \prod_{j=1}^{l} \hat{e}(u_j, d'_j)$.
2. Obtain the message M by $v \cdot \sigma^{-1}$.

Verify: For $ID \mid k = \{ID_1, ID_2, \cdots, ID_k\}$'s   signature
$\{\sigma, y_1, y_2, \cdots, y_k, z\}$, everyone can do the following to verify its validity.

1. Compute $h = H(M, \sigma)$.
2. Return $\top$ if $\hat{e}(g, z) = \sigma \cdot \hat{e}(g_1, g_2^h \prod_{j=1}^{k} y_j^{l_j}) \prod_{j=1}^{k} \hat{e}(y_j, h_j)$, $\perp$ otherwise.

5.2   Efficiency Analysis

We first analyze the computational efficiency. For the proposed scheme 1, admissible encoding scheme [4] are required for the hash function H_1 and H_2, which is computationally expensive as such scheme requires $\log_2(q/p)$-bit scalar multiplication in $E(F_q)$ where F_q is the field on which G is based and p is the size of the group G. Using the example from [21], if $\log_2 p = 512$ and the embedding degree of pairing is 6, then $\log_2 q$ should be at least 2560 and hence 2048-bit scalar multiplication is needed. Scheme 2's hash function does not rely on such admissible encoding scheme. Moreover, chosen ciphertext secure HIDE requires the transformation in Section 4 of [6], while our scheme does not require such transformation as the integrity checking of the ciphertext is obtained from the signature.

For the communication efficiency of the scheme, the signcrypted message is shortened by one G_1 element, as compared with using the scheme in [9] and [3] together.

5.3   Security Analysis

THEOREM 4. Suppose that the (t, ε)-Decision BDH assumption holds in G, then the above scheme is $(t', q_S, q_H, q_E, q_R, \varepsilon)$-selective identity, adaptive chosen ciphertext (IND-sID-CCA2) secure for arbitrary q_S, q_H, q_E, q_R, and any $t' < t - o(t)$.

THEOREM 5. Suppose that the (t, ε)-CDH assumption holds in G, then the above scheme is $(t', q_S, q_H, q_E, q_R, \varepsilon')$-selective identity, adaptive chosen message (EU-sID-CMA) secure for any $t' < t - o(t)$, $\varepsilon' > \varepsilon \cdot (1 - q_S(q_H + q_S)/q)$.

THEOREM 6. Suppose that the (t, ε)-CDH assumption holds in G, then the above scheme is $(t', q_S, q_H, q_E, q_R, \varepsilon)$-selective identity, adaptive chosen message (AUTH-sID-CMA2) secure for any $t' < t - o(t)$.

Proofs are omitted due to the length constraint. Please refer to the full version of this paper [11].

6.   CONCLUSION

Two concrete constructions of hierarchical identity based signcryption are proposed, which closed the open problem proposed by [16]. Our schemes are provably secure under the random oracle model [2]. Moreover, our schemes do not require transformation which is necessary for the case of hierarchical identity based encryption as the integrity checking of the ciphertext is obtained from the signature. We believe that hierarchical identity based signcryption schemes are useful in nowadays commercial organization and also in new network architecture such as tetherless computing architecture. Future research directions include further improvement on the efficiency of hierarchical identity based signcryption schemes and achieving other security requirements such as public ciphertext authenticity ([10,16]) or ciphertext anonymity ([5]).

ACKNOWLEDGEMENT

This research is supported in part by the Areas of Excellence Scheme established under the University Grants Committee of the Hong Kong Special Administrative Region (HKSAR), China (Project No. AoE/E-01/99), grants from the Research Grants Council of the HKSAR, China (Project No. HKU/7144/03E and HKU/7136/04E), and grants from the Innovation and Technology Commission of the HKSAR, China (Project No. ITS/170/01 and UIM/145).

REFERENCES

[1]   Jee Hea An, Yevgeniy Dodis, and Tal Rabin. On the Security of Joint Signature and Encryption. In Lars R. Knudsen, editor, Advances in Cryptology - EUROCRYPT 2002, International Conference on the Theory and Applications of Cryptographic Techniques, Amsterdam, The Netherlands, April 28 - May 2, 2002, Proceedings, volume 2332 of Lecture Notes in Computer Science, pages 83–107. Springer-Verlag Heidelberg, 2002.

[2]   Mihir Bellare and Phillip Rogaway. Random Oracles are Practical: A Paradigm for Designing Efficient Protocols. In Proceedings of the 1st ACM Conference on Computer and Communications Security, pages 62–73, 1993.

[3]   Dan Boneh and Xavier Boyen. Efficient Selective-ID Secure Identity-Based Encryption Without Random Oracles. In Christian Cachin and Jan Camenisch, editors, Advances in

Cryptology - EUROCRYPT 2004, International Conference on the Theory and Applications of Cryptographic Techniques, Interlaken, Switzerland, May 2-6, 2004, Proceedings, volume 3027 of Lecture Notes in Computer Science, pages 223–238. Springer, 2004.

[4] Dan Boneh and Matt Franklin. Identity-Based Encryption from the Weil Pairing. In Joe Kilian, editor, Advances in Cryptology - CRYPTO 2001, 21st Annual International Cryptology Conference, Santa Barbara, California, USA, August 19-23, 2001, Proceedings, volume 2139 of Lecture Notes in Computer Science, pages 213–229. Springer-Verlag Heidelberg, 2001.

[5] Xavier Boyen. Multipurpose Identity-Based Signcryption : A Swiss Army Knife for Identity-Based Cryptography. In Dan Boneh, editor, Advances in Cryptology - CRYPTO 2003, 23rd Annual International Cryptology Conference, Santa Barbara, California, USA, August 17-21, 2003, Proceedings, volume 2729 of Lecture Notes in Computer Science, pages 382–398. Springer, 2003.

[6] Ran Canetti, Shai Halevi, and Jonathan Katz. Chosen-Ciphertext Security from Identity-Based Encryption. In Christian Cachin and Jan Camenisch, editors, Advances in Cryptology - EUROCRYPT 2004, International Conference on the Theory and Applications of Cryptographic Techniques, Interlaken, Switzerland, May 2-6, 2004, Proceedings, volume 3027 of Lecture Notes in Computer Science, pages 207–222. Springer, 2004.

[7] Liqun Chen and John Malone-Lee. Improved Identity-Based Signcryption. In Serge Vaudenay, editor, Public Key Cryptography - PKC 2005: 8th International Workshop on Theory and Practice in Public Key Cryptography, Les Diablerets, Switzerland, January 23-26, 2005. Proceedings, volume 3386 of Lecture Notes in Computer Science, pages 362–379. Springer, 2005. Also available at Cryptology ePrint Archive, Report 2004/114.

[8] Sherman S.M. Chow. Verifiable Pairing and Its Applications. In Chae Hoon Lim and Moti Yung, editors, Information Security Applications: 5th International Workshop, WISA 2004, Jeju Island, Korea, August 23-25, Revised Selected Papers, volume 3325 of Lecture Notes in Computer Science, pages 170–187. Springer-Verlag, 2004.

[9] Sherman S.M. Chow, Lucas C.K. Hui, S.M. Yiu, and K.P. Chow. Secure Hierarchical Identity Based Signature and its Application. In Javier Lopez, Sihan Qing, and Eiji Okamoto, editors, Information and Communications Security, 6th International Conference, ICICS 2004, Malaga, Spain, October 27-29, 2004, Proceedings, volume 3269 of Lecture Notes in Computer Science, pages 480–494. Springer-Verlag, 2004.

[10] Sherman S.M. Chow, S.M. Yiu, Lucas C.K. Hui, and K.P. Chow. Efficient Forward and Provably Secure ID-Based Signcryption Scheme with Public Verifiability and Public Ciphertext Authenticity. In Jong In Lim and Dong Hoon Lee, editors, Information Security and Cryptology - ICISC 2003, 6th International Conference Seoul, Korea, November 27-28, 2003, Revised Papers, volume 2971 of Lecture Notes in Computer Science, pages 352–369. Springer, 2003.

[11] Sherman S.M. Chow, Tsz Hon Yuen, Lucas C.K. Hui, and S.M. Yiu. Signcryption in Hierarchical Identity Based Cryptosystem, 2004. Extended abstract appeared in Security and Privacy in the Age of Ubiquitous Computing, IFIP TC11 20th International Conference on Information Security (SEC 2005), May 30 - June 1, 2005, Chiba, Japan. Full version available at Cryptology ePrint Archive, Report 2004/244.

[12] Jean-Sébastien Coron. On the Exact Security of Full Domain Hash. In Mihir Bellare, editor, Advances in Cryptology - CRYPTO 2000, 20th Annual International Cryptology Conference, Santa Barbara, California, USA, August 20-24, 2000, Proceedings, volume 1880 of Lecture Notes in Computer Science, pages 229–235. Springer, 2000.

[13] Craig Gentry and Alice Silverberg. Hierarchical ID-Based Cryptography. In Yuliang Zheng, editor, Advances in Cryptology - ASIACRYPT 2002, 8th International Conference on the Theory and Application of Cryptology and Information Security, Queenstown, New Zealand, December 1-5, 2002, Proceedings, volume 2501 of Lecture Notes in Computer Science, pages 548–566. Springer, 2002. Available at http://eprint.iacr.org.

[14] Florian Hess. Efficient Identity Based Signature Schemes based on Pairings. In Kaisa Nyberg and Howard M. Heys, editors, Selected Areas in Cryptography, 9th Annual International Workshop, SAC 2002, St. John's, Newfoundland, Canada, August 15-16, 2002. Revised Papers, volume 2595 of Lecture Notes in Computer Science, pages 310–324. Springer, 2003.

[15] Berkeley. Intel Research. Identity Based Cryptosystem for Secure Delay Tolerant Networking.

[16] Benoît Libert and Jean-Jacques Quisquater. New Identity Based Signcryption Schemes from Pairings. In IEEE Information Theory Workshop, pages 155–158, 2003. Full Version Available at http://eprint.iacr.org.

[17] Benoît Libert and Jean-Jacques Quisquater. The Exact Security of an Identity Based Signature and its Applications. Cryptology ePrint Archive, Report 2004/102, 2004. Available at http://eprint.iacr.org.

[18] John Malone-Lee. Identity Based Signcryption. Cryptology ePrint Archive, Report 2002/098, 2002. Available at http://eprint.iacr.org.

[19] Noel McCullagh and Paulo S. L. M. Barreto. Efficient and Forward-Secure Identity-Based Signcryption. Cryptology ePrint Archive, Report 2004/117, 2004. Available at http://eprint.iacr.org.

[20] Divya Nalla and K.C. Reddy. Signcryption Scheme for Identity-Based Cryptosystems. Cryptology ePrint Archive, Report 2003/066, 2003. Available at http://eprint.iacr.org.

[21] Dong Jin Park, Kihyun Kim, and Pil Joong Lee. Public Key Encryption with Conjunctive Field Keyword Search. In Chae Hoon Lim and Moti Yung, editors, Information Security Applications: 5th International Workshop, WISA 2004, Jeju Island, Korea, August 23-25, Revised Selected Papers, volume 3325 of Lecture Notes in Computer Science, pages 73–86. Springer-Verlag, 2004.

[22] Ryuichi Sakai, Kiyoshi Ohgishi, and Masao Kasahara. Cryptosystems based on Pairing over Elliptic Curve. In Proceedings of Symposium on Cryptography and Information Security (SCIS 2000) C-20, 2000.

[23] Aaditeshwar Seth. Personal Communication, September 2004.

[24] Aaditeshwar Seth, Patrick Darragh, and Srinivasan Keshav. A Generalized Architecture for Tetherless Computing in Disconnected Networks. Manuscript.

[25] Tsz Hon Yuen and Victor K. Wei. Fast and Proven Secure Blind Identity-Based Signcryption from Pairings. In A. J. Menezes, editor, Topics in Cryptology - CT-RSA 2005, The Cryptographers' Track at the RSA Conference 2005, San Francisco, CA, USA, Febrary 14-18, 2005, Proceedings, volume 3376 of Lecture Notes in Computer Science, San Francisco, CA, USA, February 2005. Springer. To Appear. Also available at Cryptology ePrint Archive, Report 2004/121.

[26] Yuliang Zheng. Digital Signcryption or How to Achieve Cost (Signature & Encryption) << Cost(Signature) + Cost(Encryption). In Burton S. Kaliski Jr., editor, Advances in Cryptology: Proceedings of CRYPTO 1997 5th Annual International Cryptology Conference, Santa Barbara, California, USA, August 17-21, 1997, volume 1294 of Lecture Notes in Computer Science, pages 165–179. Springer-Verlag, 1997.

PROTECTING GROUP DYNAMIC INFORMATION IN LARGE SCALE MULTICAST GROUPS

Yongdong Wu[1], Tieyan Li[1], and Robert H. Deng[2]

[1]Instititue for Infocomm Research(I²R), Singapore, {wydong,litieyan}@i2r.a-star.edu.sg;
[2]School of information Systems, Singapore Management Univerist, robertdeng@smu.edu.sg

Abstract: Existing key management schemes can secure group communication efficiently, but are failed on protecting the Group Dynamic Information (GDI) that may undermine group privacy. Recently, Sun et al.[1] proposed a scheme to hide the GDI with batch updating and phantom members inserting so that an adversary is not able to estimate the number of group members. In this paper, we first point out that their scheme is only applicable in departure-only group communication instead of the common conference groups. Secondly, we introduce our method of estimating the group size at a higher confidence level given a prior departure probability. Further, to enhance GDI protection and extend the application fields, we propose to protect GDI with two new mechanisms: chameleon member identifications and virtual departure events. The proposed scheme is effective to protect both centralized groups and contributory groups. The simulation shows that our scheme is better on protecting the GDI.

1. INTRODUCTION

With the development of network applications such as pay-TV, remote education and videoconference, secure multicast distribution of copyright-protected or confidential material is more and more important. At the heart of the applications, a center broadcasts encrypted data to a large group of receivers so that only a predefined subset is able to decrypt the protected data. Broadcast encryption[2] deals with methods to efficiently broadcast information to a dynamically changing group of end members. It includes three major modules: Registration, Join and Departure[3]. Since the processing for registration and join requests is relatively easy, the main challenge in managing a group is how to exclude some subset of the members efficiently (e.g., Dodis *et al.*[4]). Technically, this *blacklisting Problem* is how to distribute content decryption keys over a shared insecure channel so that only intended receivers can get the key efficiently in terms of key storage[5] and/or communication overhead[6]. Based on the survey[7], the key management methods are categorized into two classes: (1). **Centralized Group Key Management:** A secure multicast scheme allows one or more group controllers (GCs) to send key updating messages securely over a multicast channel to a dynamically changing group of members[8]. For example, in logical-tree-hierarchy (LKH) scheme[9] and LKH+[10] scheme, the leaf nodes represent the member keys, and the non-leaf nodes are key encryption keys (KEKs) which are used to encrypt the control traffic. Specially, the root represents the group key. (2). **Contributory key management:** There is no GC but each member has the same task in managing the group key so as to remove the bottleneck of GC and increase the security. For instance, Rodeh *et al.*[11] divided the group members into sub-groups which have leaders. The leaders exchange keys on behalf of the sub-group members to agree a group key.

The state-of-the-art key management schemes are designed to prevent unauthorized access to the multicast content, but unfortunately provide opportunities for unauthorized parties to obtain group dynamic information (GDI). That is to say, the number of join members, departure members and the size of the group are disclosed. As noted in Sun *et al.* scheme[1] (hereafter called as SL scheme): "some applications such as subscription services, and military, it is highly undesirable to disclose instant detailed dynamic membership information to competitors, who would develop effective competition strategies by analyzing the statistical behavior of the audience."

Sun *et al.*[1] not only described the leakage of GDI in the existing key management schemes, but also improved the design of current key management schemes such that both GDI and the multicast content are protected. Technically, Sun *et al.* proposed a protection method by batch re-keying and phantom members. Their upgraded scheme provided some protection on the GDI, but we think it is not sufficient in terms of applications and security strength.

The present paper focuses on the protection of group dynamics, in particular to the large scale group because the adversary is hard to accurately estimate the GDI of a small group. Since non-tree-based schemes (e.g. Safaeli *et al.*[7]) are merely applicable for small groups, we deal with the binary tree-based scheme only in the following sections. We denote a path as the node sequence from leaf to root. The path length is the number of nodes in the path. $l(x)$ represents the number of nodes in the path from node x to root. Particularly, the depth of the root is 0. Each key has a field SN Sequence Number which indicates the position of a key in the tree. Our main contributions include: (1) address the security flaw of the scheme[1]; (2). propose methods to protect GDI with virtual events and chameleon members. The proposed schemes are effective in protecting both centralized group and contributory group.

The remainder of the present paper is organized as follows. Section 2 refreshes the SL scheme. Section 3 addresses our analysis on SL scheme. Section 4 depicts our proposed GDI protection method and its extension to contributory group key management. Section 5 describes the comparison between SL scheme and our scheme. A conclusion is drawn in Section 6.

2. SL HIDING SCHEME

In many group communications, group dynamic information is confidential and should not be disclosed to either inside group members or outsiders. An important goal in SL scheme[1] is to produce an observed re-keying process that reveals the least amount of information about the GDI. The tools in SL scheme to hide GDI include batch re-keying and phantom member insertion.

Periodical batch re-keying is to postpone the updates of keys such that several members can be added to or removed from the key tree altogether. Compared with updating keys immediately after each member joins or departures, batch re-keying reduces the communication overhead at the expense of allowing the joining/leaving member to access a small amount of information before/after his join/departure. But batch updating provides little contribution for the GDI protection if the attacker can intercept and parse the re-keying messages.

Phantom members inserting is a way to hide the number of the real members. These phantom members, as well as their join and departure behavior, are created by GC. As a result, the combined effects of the phantom members and the real members lead to an artificial GDI which is observed by the attackers.

According to notation in SL scheme, N_k, J_k and L_k are the number of the real members, and the real join event, and real departure event at time k respectively, $N_a(k)$, $J_a(k)$ and $L_a(k)$ are the total number of members, total number of join events, and total number of departure events respectively. $N_a(k)$, $J_a(k)$, and $L_a(k)$ are referred to as the artificial GDI whose target values are N_0, J_0 and L_0 respectively. Thus, the artificial GDI functions are:

$$N_a(k) = \max \{N_k, N_0\} \tag{1}$$

$$J_a(k) = \max \{J_k, L_k, L_0\}$$
$$L_a(k) = N_a(k-1) - N_a(k) + J_a(k)$$

Due to Eq.(1), the number of key tree leaves is that of the real members if $N_k \geq N_0$. Therefore, an attacker can estimate the group size if he is permitted to join the group without limitation. For example, an attacker floods N_0 join requests to GC so as to exclude all phantom members. Thanks to the group size technologies described in Section 3, the attacker can obtain the GDI easily. To defeat this attack, N_k must always be less than N_0 by imposing some limitations on the join requests, for example, restrict the number of join requests by checking the identification of the requestor.

3.   DISCLOSING GDI PROTECTED WITH SL SCHEME

This section describes an approach to estimate GDI protected with methods in SL scheme. To this end, subsection 3.1 firstly introduces how to estimate the group size with the tree depth since a key-tree tends to be balanced in a long term communication. Secondly, Subsection 3.2 describes how to refine the key-tree with the key SNs based on the re-keying messages. Finally, the number of real members is estimated thanks to some prior knowledge on the member departure.

3.1   Step 1: Grossly estimating the group size

Generally, the key-tree is changed from time to time due to the group dynamics. However, the tree will tend to some stationary status with the updating strategy in SL scheme: (1) The number of tree leaves is invariable, i.e., the umber of departure members is equal to that of the join members at any updating time. (2) Every member (i.e., tree leaf) has the same departure probability, but the join member is always located in the shortest path for the sake of lightest communication overhead.

This subsection discloses the key-tree asymptotic property in SL scheme to estimate the number of tree leaves. Denote the initial tree as T_0, the tree is T_k at time k. $l_{max,k}$ is the length of the longest path and $l_{min,k}$ is the length of the shortest path in the tree T_k.

Theorem: If a join member is always mounted in the shortest path, and a member departures from any leaf at an identical probability, an arbitrary binary tree will tend to be balanced given a departure event is always followed by a join event. Formally, let $P(X)$ to be the probability of random variable X, $\lim_{k \to \infty} P(l_{max,k} - l_{min,k} \leq 1) = 1$

Proof: see appendix.

Although the theorem is expressed for two events at each updating time, it holds for any even number of events. Consequentially, the key tree will tend to be a balanced tree if the probability of departure event is identical to that of join event. Thus the size of re-keying messages is almost constant for each departure event[23] in a long term. For example, Fig. 1 illustrates the dynamic property of the key tree. Initially, the tree of depth 10 is very unbalanced ($l_{max,0} - l_{min,0} = 8$), but $l_{max,k} - l_{min,k}$ is smaller and smaller with more and more events, i.e., the tree is more and more balanced. Because $l_{min,k}$ and/or $l_{max,k}$ are publicly available from the re-keying messages, the adversary estimates the group size N_0 as

$$2^{l_{min,k}} < N_0 < 2^{l_{max,k}}$$

for a sufficiently large k. This method grossly estimates the group size without employing the re-keying message content. In the following, the adversary exploits the key SNs from the re-keying messages so as to refine the number of tree leaves.

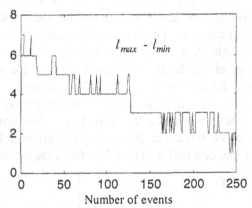

Figure 1. Tree depth difference dynamics. With more and more join/departure events, the difference $l_{max,k} - l_{min,k}$ tends to 1.

3.2   Step 2: Refining the estimation

Assume at time t_0, the attacker starts to monitor the network traffic so as to reconstruct the key tree T_{t_0} at time t_0. To accomplish this, the adversary analyzes the re-keying messages one by one and updates the reconstructing

---

[23] In the Subsection II-C of SL scheme, the departure position ($L_1, L_2, ..., L_W$) is assumed to be *i.i.d.* We think this assumption is not true for a long duration communication. Thus, the attack AII addressed SL scheme is impractical.

key tree with the following heuristic knowledge: for each departure event, if the departure member does not join after t_0, the member must be in T_{t_0} ; for each join event, the new leaf position must be empty in T_{t_0} unless it is blank due to the departure events after t_0.

Since phantom members are added into the group such that the group size is invariable in SL scheme, i.e., $N_a(k)= N_0$, what the adversary obtains is the artificial size N_0 which includes the number of the phantom members. Therefore, the adversary has to filter out the number of phantom members with the following method.

3.3   Step 3: Filtering the phantom members

According to the experiments[12,13] on member departure, the member arrival process can be modelled as Poisson; the membership duration of short sessions is accurately modelled using an exponential distribution; and the membership duration of long sessions is accurately modeled using the *Zipf* distribution (http://www.useit.com/alertbox/zipf.html). Therefore, the adversary is able to exploit the difference between real member and phantom member. The former departures at its own probability which may be available in advance with the domain knowledge, and the departure probability of the latter is decided by GC. Before elaborate the process of estimating the real members, let's play a game.

Game: With respect to Fig.2, there are 2 boxes called G_1 and G_2. The first box G_1 has a black balls and b white balls, another box G_2 has x black balls and y white balls, where a, b, $x+y=N$ are known, but x and y are unknown. At each time, one ball will be taken from the boxes.

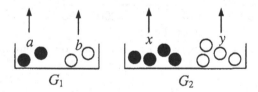

Figure 2. Experiments for simulating group dynamics. The black balls represent real members, while white balls represent the phantom members. The tested members map to the balls in box G_1.

The rules for taking balls with repetition are as follows:
(1) the black balls will be taken with a predefined probability f. Each black ball has the same probability to be taken.
(2) only after no black ball is taken in step (1), the white balls will be selected uniformly. Define the random events
X_1: a black ball taken from the first box G_1

X_2: a white ball taken from the first box G_1
X_3: a black ball taken from the second box G_2
X_4: a white ball taken from the second box G_2
Then the probability of the taken ball is from the first box G_1 as

$$P(X_1 \cup X_2) = P(X_1) + P(X_2) = \frac{a}{a+x} f + \frac{b}{N-a-x}(1-f) \qquad (2)$$

Given that the number of balls in the first box is much smaller than that of the second one ($a+b \ll x+y$), and the observer can tell which box the ball is taken from, but he can not tell the color of the taken ball. After many experiments (return the ball to the original box), the occurrence frequency α of balls taken from the first box is

$$\alpha \approx P(X_1 \cup X_2) \qquad (3)$$

That is to say, the observer can estimate the number of white (or black) balls according to Eq.(2) and Eq.(3). Referencing to Table 1, the adversary selects all the joining members as a tested set of members at a updating time, and other members are regarded as non-tested. With the following ways, an adversary can obtain the parameters to play the game on the GDI.

- In an updating period, the attacker can detect all the requests (join and/or departure) by eavesdropping the network. Equally, the attacker knows a in the above experiment. Meanwhile, the attacker can deduce the number b from a re-keying message in the same period. But the attacker can not distinguish the real members from the phantom members due to batch updating. Thus, the adversary can construct a box G_1.
- The number N of the old members (balls in G_2) is known based on Subsections 3.1 and 3.2.
- Since GC does not remove a real member otherwise the victim will protest. Thus, the real member will quit at her own probability f. According to the experiments[13], the probability f may be available.
- A phantom member can be selected and excluded uniformly if no real member quits.

Careful readers may notice some minor difference between the non-tested set and G_2 in that non-tested set may be changed from time to time, while the balls in G_2 are identical all the time. But we think this difference will not incure big estimation error if $a+b \ll N$. In addition, the game simplifies the group dynamics with only one ball leaving at each time,

However, it can be extended to multiple balls leaving like the case of batch updating.

Strictly speaking, f varies with the number of real members. To increase the estimation of f, we can repeat the above experiment with the estimated number x till the difference between two estimations is smaller than a threshold. As a result, the number x of real members is estimated according to the above game.

Table 1. Mapping the game to GDI. #X: the size of X

	Game	GDI
G_1	tested balls	tested real members
G_2	non-tested balls	non-tested members
a	# black balls in G_1	# tested real members
b	# white balls in G_1	# tested phantom members
x	# black balls in G_2	# non-tested real members
y	# white balls in G_2	# non-tested phantom members
N	# balls in G_2	# non-tested members
f	probability of black ball leaving	probability of real member leaving

4.   PRESENT GDI PROTECTION SCHEME

The present approach employs chameleon ID and virtual member switch for hiding GDI. Chameleon ID means that each user has more than one ID, or multiple leaf positions in the key tree; and virtual member switch is to remove an innocent member and then artificially insert him again. With these technologies, as well as batch updating and phantom members, the phantom member and real member are statistically indistinguishable such that the estimation method in Section 3 is of no use.

4.1   Chameleon Members

In the SL scheme[1], the approach of inserting phantom members provides GDI protection to some degree. But it may be vulnerable in the group communication, in particular to conferences because the phantom members are always silent. If the attacker treats the members who are dumb for a long time as phantom members, the attacker is able to separate the real members from the phantom members. Therefore, the protection methods in SL scheme are merely applicable to the content delivery, i.e., one member acts as a server or speaker, and others are listeners.

In order to hide GDI in all kinds of the group communications, we propose a GDI protection based on the idea of chameleon IDs. Fig.3

illustrates the response process of join event with chameleon IDs. The request message is R_M=(JOIN, M, $E(ID\|n, K)$), where $E(.)$ is an encryption function, ID is a random number for identifying the requestor M, and key K is known to both GC and the requestor and n is the number of chameleon IDs for the requestor M.

With the secret key K, GC obtains ID and n by decrypting R_M and stores them for virtual departure (see Subsection 4.2), then unicasts a response $\{ID_i, E(C_i, K)\}^n$ (i=1,2, ..., n) to the requestor, where ID_i=$H(ID \| i)$, C_i is a set of re-keying messages (SN_j, K'_{SN_j}), SN_j is the key sequence number in the whole key tree. Here, assume that the attacker can intercept all the re-keying messages but can not identify whether the recipients of two messages are identical or not. To be efficient, a weaker assumption is that the attacker can not identify a majority of recipients of the re-keying messages. K'_{SN_j} is derived from the old K_{SN_j} with a one-way function. Hence the requestor joins the group as n members. Because all the ID_i are non-linkable and the number of artificial join requests is unknown to the attacker, the attacker can not distinguish the real join re-keying message from the artificial ones. Furthermore, the chameleon IDs are used at the same probability in the communication sessions such that no dumb members exist.

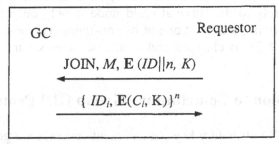

Figure 3. Diagram for member join. After receiving the join request from a new member, GC adds several leaves in the key tree for the new member.

4.2 Virtual Departure

In the attack proposed in Section 3, the adversary exploits the statistics difference between the real member and phantom member. In order to hide *a prior* departure probability of real members, GC will exclude a real member, afterwards let her join. To accomplish this virtual departure, GC will select an innocent member, send the exclusion message as usual. Optionally, not all of her chameleon IDs are removed such that the member can still involve in the group communication and postpone the join process to a later time. When an innocent member is revoked for no reason, she knows that she can obtain a new key from the re-keying message sooner or later. For example, the re-keying messages with $ID_i=H(ID \parallel n+t)$ indicates her new join KEKs, where t is the updating time. Therefore, the number of the members is invariable. Since either real member or phantom member has the same departure probability, the attack addressed in Section 3 is foiled.

4.3   Mixed Departure with Join

This section addresses a batch way to reduce the overhead efficiently. To deal with join and departure events simultaneously, whether the join/departure is virtual or real, all the keys from the removed leaf node to the root must be changed. Based on our observation in Section 3.1, most of leaves are located in $D = \lfloor log_2 N_0 \rfloor$ and D-1 levels. It is natural to insert the new members into the departure positions. The rules for processing the member departure are as follows:

- Prune the sub-trees which include all and exactly the departure members.
- For each remaining leaf node, if its sibling node departures, its new parent KEK is unicast to him securely.
- For each remaining internal node of level $l < D$-1, its new KEK will be multicast securely to the survival child node if only one child node is removed. Otherwise, if at least one of its non-immediate descendant node departures, its KEK is changed and multicast twice securely to its two children nodes.

4.4   Extension to Contributory Group GDI Protection

In the paper[1], "Contributory key management schemes are generally not suitable for the applications with confidential GDI because each group member need to be aware of other group members in order to establish the shared group key in the distributed manner." However, this oracle is not true based on our chameleon IDs.

Technically, in the setup stage of a contributory group key, the members broadcast the messages so that all the members share the same key. For the sake of clarity, we adopt the distributed LKH[11] to explain the protection process for contributory group communication. At the stage of forming a group key, each member has several chameleon IDs and each ID corresponds to a leaf of the key tree. After constructing the tree with Rodeh's scheme[11], the root is the shared key. In the conversation stage, the member talks with different IDs from time to time. In this protected contributory GDI, the height of the key tree is public, but the number of chameleon IDs of each member is secret, thus neither the insider nor the outsider can obtain the number of the real members.

In the Distributed LKH, the key tree depth increases $log_2 N_1 - log_2 N_0$ if the number of members is increased from N_0 to N_1, thus the communication overhead and computational cost increases $log_2 N_1 - log_2 N_0$ on average to achieve GDI protection.

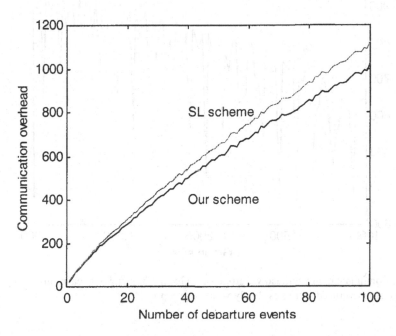

Figure 4. Communication overhead vs. the number of departure events (x-axis). Given a group of 4000 members, the members leave the group uniformly. The overhead is represented with the number of re-keying messages.

5. PERFORMANCE

In the following simulations, suppose that the adversary knows the total number of members, as well as the departure probability of the real members.

5.1 Network overhead

With the proposed approach of batch departure in Subsection 4.3, the communication overhead are shown in Fig.4 and Fig.5. The size of re-keying messages of the proposed batch departure (solid line) is smaller than that in SL scheme (dashed line) since we do not multicast KEKs of pruned sub-trees.

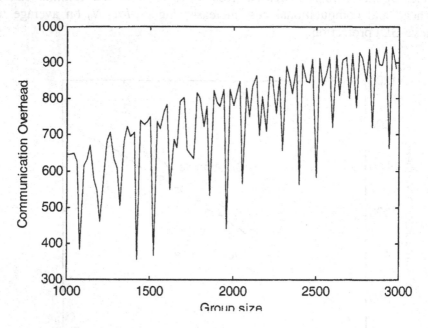

Figure 5. Communication overhead vs. group size. Given 100 members leaves uniformly, the communication overhead increases slowly with the group size.

5.2 GDI Protection Capacity

5.2.1 Estimation on GDI protected with SL scheme

Fig.6 illustrates the estimation result for the real members protected with SL scheme. With reference to Table I, assume G_1 includes a=10 real members and b=10 phantom members, and G_2 has N=4000 members including x=1000 real members and y=3000 phantom members. The experiment is repeated with 10000 times, the departure probability f of the real member is selected as 0.01. Denote $\tilde{N}$ as the estimated number of real members. Define $Z=(\tilde{N}/N-1)$ as the normalized estimation error. In Fig.6, mean error $\mu_0=|\mathbf{Mean}(Z)|=0.15$ and standard deviation $\sigma_0=\mathbf{Var}(Z)=0.1232$. From the experiments, we come to a conclusion that the size of the real members can be estimated to some extent. If the probability f is smaller, the estimation result is better. Thus, the attacker may select the time slice for estimation when few real members departure.

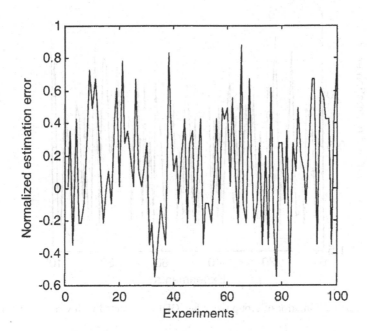

Figure 6. Estimation error of group size when GDI is protected with SL scheme.

5.2.2 Estimation of GDI protected with our protocol

Fig.7 illustrates the estimation result for the real members protected with the virtual departure. Its configuration parameters are the same as those in Subsection 5.2.1. In Fig.8, the mean error $\mu_1 = |\mathbf{Mean}(Z)| = 0.3458 > \mu_0$. And the standard deviation $\sigma_1 = \mathbf{Var}(Z) = 0.587 > \sigma_0$. It means that the virtual departure results in much greater estimation error, and the estimation results are not stable due to high variance.

With the same parameters as Fig. 6, Fig. 8 illustrates the estimation result for the real members protected with chameleon IDs (here the number of

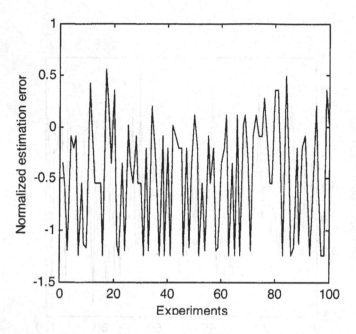

Figure 7. Estimation error of group size when GDI is protected with virtual departure.

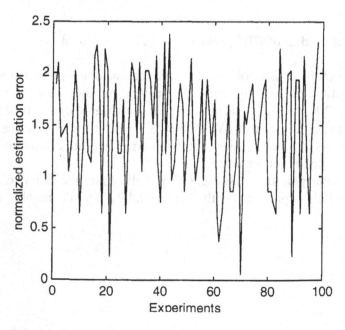

Figure 8. Estimation error of group size when GDI is protected with chameleon IDs.

chameleon ID is equal to the real member). The mean of estimation error $\mu_2=|\mathbf{Mean}(Z)|=0.476 >> \mu_0$, and standard deviation $\sigma_2=\mathbf{Var}(Z)=0.2956 > \sigma_0$. Therefore, the estimated number of real members is far from true value, and the estimation results are not stable due to high variance.

6. CONCLUSION

This paper investigates the protection schemes of hiding GDI and points out the security flaw in SL scheme. We further propose a method to protect GDI, such as the number of group members, from disclosing. In this paper, we consider the membership dynamic in the whole process including set up, join/leave, and communication session. Although the adversary is assumed to have more knowledge (e.g., knowing the key tree) than that in SL scheme, she is unable to even closely approximate the number of real members due to the technologies of chameleon IDs and virtual departure. In addition, our method is applicable to both centralized group and contributory group.

REFERENCE

1. Yan Sun and K.J. Ray Liu, "Securing Dynamic Membership Information in Multicast Communications," IEEE Infocom, 2004
2. A. Fiat and M. Naor, "Broadcast Encryption," Crypto '93, LNCS 773, pp. 480-491, 1993
3. R. Canetti, J. Garay, G. Itkis, D. Miccianancio, M. Naor and B. Pinkas, "Multicast Security: a Taxonomy and Some Efficient Constructions," IEEE Infocom, vol. 2, pp. 708-716, 1999.
4. Yevgeniy Dodis and Nelly Fazio, "Public Key Trace and Revoke Scheme Secure against Adaptive Chosen Ciphertext Attack," Public Key Cryptography, LNCS 2567, pp. 100 - 115, 2003
5. Yuh-Min Tseng, "A Scalable Key-management Scheme with Minimizing Key Storage for Secure Group Communications," International Journal of Network Management, pp.419-425, 2003
6. Jack Snoeyink, Subhash Suri and George Varghese, "A Lower Bound for Multicast Key Distribution," IEEE Infocom, pp.422-431, 2001
7. Sandro Rafaeli and David Hutchison, "A Survey of Key Management for Secure Group Communication," ACM Computing Surveys, 35(3):309-329, 2003
8. Alan T. Sherman and David A. McGrew, "Key Establishment in Large Dynamic Groups Using One-Way Function Trees," IEEE Trans. Software Eng. 29(5):444-458, 2003.
9. D.M. Wallner, E.J. Harder and R.C. Agee, "Key Management for Multicast: Issues and Architectures," Internet Request for Comments 2627, June, 1999. ftp.ietf.org/rfc/rfc2627.txt

10. M. Waldvogel, G. Caronni, D. Sun, N. Weiler and B. Plattner, "The VersaKey framework: Versatile Group Key Management," IEEE Journal on selected areas in communications, 17(9):1614-1631, 1999.
11. O. Rodeh, K. Birman and D. Dolev, "Optimized Group Re-key for Group Communication Systems, " Network and Distributed System Security, pp. 39-48, 2000
12. K. Almeroth and M. Ammar, "Multicast Group Behavior in the Internet's Multicast backbone (MBone), " IEEE Communications Magazine, 35(6):124-129, 1997.
13. S. Acharya, B. Smith and P.Parnes, "Characterizing User Access to Videos on the World Wide Web, " ACM/SPIE Multimedia Computing and Networking, 2000

APPENDIX

Proof of Theorem 1: Let $D= \lfloor log_2 N_0 \rfloor$. The member set $U_k=\{M_i \mid \mathbf{l}(M_i) > D \}$ is defined after the updating time k, but before the updating time $k+1$, where M_i is a node in the tree, $\mathbf{l}(x)$ is the depth of a leaf x.

Define two events as Departure event e_k and Join event E_k at time k.

(1) The leaves in U_k will vanish gradually.

Because $N_a(k)=N_0 \le 2^D$, the join member will be inserted into the levels smaller than $D+1$. Thus, the join event has no impact on U_k. Thus, $U_0 \subseteq U_1 \subseteq ...$, the size of U_k is monotonically deceasing, $|U_0| \ge |U_1| \ge ... \ge 0$

Denote the departure event of member m as

$$e_k = \begin{cases} 1: & m \in U_{k-1} \\ 0: & otherwise \end{cases}$$

If $|U_{k-1}| > 0$, $P(e_k=1) = |U_{k-1}|/N_0 \ge 1/N_0$. Considering e_1, e_2, ... are independent, for any arbitrary positive ε_1, there is some c such that
$$P(E_1 +E_2 + ...+E_c = |U_0|) \ge 1-\varepsilon_1. \text{ i.e., } P(|U_c|=0) \ge 1-\varepsilon_1.$$

(2) The leaves in any level $l< D-1$ will vanish gradually.

Assume the length of the shortest path is $l_{min,0}<D-1$ initially, and S_k is the set of the leaves in level $l_{min,0}$. At time k, denote the departure event of member m as

$$E_k = \begin{cases} 1: & m \notin S_k \\ 0: & otherwise \end{cases}$$

Suppose $n_0= |S_0|$ is the number of leaves in the level $l_{min,0}$ at the initial time. Due to $P(E_k=1)= |S_k|/N_0 \ge 1/N_0$, if $|S_k| \ne 0$, for any arbitrary positive $\varepsilon_2 >0$, there is some r such that
$$P(E_1 +E_2 + ...+E_k = n_0) \ge 1- \varepsilon_2. \text{ i.e., } P(|S_r|=0) \ge 1- \varepsilon_2.$$

According to Eq.(1), the leaves in the shortest path will disappear gradually. That is to say, the length of shortest path will increase. Repeat the above steps, the shortest path will be longer and longer till D-1 at the probability of at least $1-\varepsilon_2$. That is to say, for any level $l = l_{min,0}, \ldots, D$-2, there is some r_l, no leaves exist at all given arbitrary positive ε_2 after r_l join events.

(3) After sufficient number of events, there is no leaf x where either $l(x) > D$ or $l(x) < D$-1 at a high probability. Formally, For any arbitrary positive $\varepsilon = \min(\varepsilon_1, \varepsilon_2)$, after updating θ times, the leaves are in levels D and D-1 at the probability at least $1-\varepsilon$, where

$$\theta = c + r_{l_{min,0}} + r_{l_{min,0}+1} + \ldots + r_{D-2}.$$

SECURE GROUP COMMUNICATION WITH DISTRIBUTED GENERATION OF PRIVATE KEYS FOR AD-HOC NETWORKS

Shrikant Sundaram[1], Peter Bertok[1] and Benjamin Burton[2]
[1]*School of Computer Science and Information Technology;*
[2]*School of Mathematical and Geospatial Science;*
RMIT University, Melbourne, Australia

Abstract: Mobile ad-hoc networks are emerging as important computing platforms, and their users expect the security to be comparable to fixed, infrastructural networks. With the increase of group-oriented applications, secure communication has to be provided between a group of nodes that may join and leave the network in an unplanned manner. Previous solutions have achieved limited success, for example did not provide manageable confidentiality of messages. This paper proposes a fast, scalable group encryption key model that is suitable for ad-hoc networks and is resistant against multiple compromised nodes.

Key words: key-insulated, ad hoc networks, key generation, threshold cryptography

1.   INTRODUCTION

With the increasing accessibility of wireless networks, securing information exchange is becoming more and more important as these networks are easy to penetrate. The increasing popularity of collaborative applications has raised interest in secure group communication, primarily message authentication and confidentiality. In large groups, setting up secure channels for each communicating pair of nodes can be very expensive, so a common communication platform, including keys for the whole group is often required.

Managing group communication can become quite complex as the keys should be available to all group members but not to outsiders. On the other hand, new members have to be provided with the key, possibly without access to earlier massages, and when a member leaves, it should have no access to later messages. Some solutions utilise a group leader approach to calculate and distribute keys[1] while other approaches arrange the nodes in a key tree that connects group members for efficient communication[7].

Reliability of communication can also be a problem in wireless networks. However, some solutions[12] have already been proposed to overcome this problem, so we do not address it here.

In this paper we propose a key management approach, primarily for ad-hoc networks. In our method, there is only one public key for the whole group and each member has a unique private key that operates with the public key and the member's own unique identity (IP or MAC Address). A certain number of current group members collaborate to provide a new member with the private key. The private key's validity is limited in time and has to be updated on a regular basis to prevent the member from unlimited access to group messages. The validity is certified by a *key certificate* that each group member carries. The *key certificate* is signed with a secret key that never appears in its entirety at any node, but is distributed among the group members. With a minor extension described later, our proposed solution also securely operates in networks where messages between group members have to pass through other nodes in the key distribution phase, a common scenario in ad-hoc networks.

We assume that an authentication scheme, for example one based on certificates, is in use in the network.

In the following section we describe some related work in the area of security in ad hoc networks. In Section 3 we describe our proposed solution and in Section 4 we explain the mathematical background. Section 5 describes the results of our implementation of the proposed solution, followed by discussion and future work in Section 6. Finally, we provide our conclusions in Section 7.

2.   PREVIOUS WORK

2.1   Existing Communication Management Solutions

Secure communication in ad hoc networks has been widely discussed recently, and it is very challenging due to the dynamic nature of the network and the roaming devices. There have been a number of solutions proposed, for authentication and key generation/distribution that range from leader based key management schemes[1] to self-organized schemes[2]. Threshold cryptography[8-10, 13] can greatly improve the key generation and distribution process and can tolerate a certain number of node failures. We assume that the reader is familiar with threshold cryptography; in the next section we explain the key insulated scheme that forms the basis of our solution.

2.2 Key-Insulated Cryptography and Signature Scheme

Dodis et al.[3, 4] describe *Key-insulated cryptography and signature schemes* to enable small insecure devices to periodically refresh their private keys at discrete time intervals, without changing the public keys.

The insecure devices update their private keys with the help of another device that we call a base station. The base station is considered to be physically secure. An example scenario can be a mobile phone or PDA that updates its private key with the help of the personal computer of the user.

A *[t, N]* key insulated scheme is one where an adversary who compromises a physically insecure device and obtains private keys for up to *t* time periods is unable to decrypt or sign messages for all other *(N - t)* time periods.

The device gets its initial private key and the public key from a key generation algorithm. The key generation algorithm also generates a master private key that is stored at the base station. All encryption/decryption and signing is done on the insecure device with the private keys and the time period information. The base station only helps the device to update its private keys.

When a device needs to update its private key, the base station calculates a partial key and sends it to the device. The device calculates a new private key from the partial key and its previous private key. The updating algorithms allow random access key updates given any two time periods. Even though the base station is considered to be secure, if it is compromised and the master key is exposed, the adversary cannot gain knowledge of any private keys. However, the keys need to be updated from a single, secure device that may not be possible in ad hoc networks.

2.3 Our Contribution

Our aim is to facilitate secure group communication in two ways: reduce the burden of encryption key management by using fewer keys, and to assist mobile and ad-hoc type networks by not relying on a single, fixed and well secured base station.

Our approach has common roots with the threshold scheme[8-10, 13]; however, we provide key generation rather than authentication. We derive our solution from key insulated schemes[3, 4], but with two major additions.

Firstly, we use a single public key for group communication, and individual nodes have their own, individual private keys that are refreshed periodically. Secondly, we do not store the master private key anywhere in the system; instead it is destroyed after initialisation. We dispense with the secure base station used there, and the keys are generated and updated in a distributed manner such that no node, except the intended node, can gain the knowledge of the new private key. Minor improvements, such as using the combination of node identifiers and time period information instead of time only, provide further security in the proposed solution. A small extension makes our solution work for communication between non-adjacent nodes; as is frequently the case in wireless networks.

3.   PROPOSED SOLUTION

An ad hoc group consists of several nodes, all which are identified by a certificate and hence all nodes have a public key-private key pair. This key pair is used by the nodes to authenticate themselves to the group. Once they have been authenticated they are issued a public key - private key pair for use within the group. In this key pair the public key is common for the whole group while the private key is known only by that node. The private key is generated by a group of nodes in the group using the principles of threshold cryptography. The private key is based on the identity information of the node and time information. The identity can be the public key of the node or its MAC/IP address and is to be unique for that node. The time information is a time period value that specifies the validity of the private key.

The model supports encryption, authentication and integrity of messages. To send an encrypted message to another member of the group, we encrypt the message with the group public key and the unique identifier of the receiver member. The node with the given identifier decrypts the message with its private key. To sign messages we encrypt with the sender's private key and it can be verified by all with the global public key. These algorithms are the same as in the key insulated schemes [3, 4] and are not discussed further here. Our focus is on distributed key generation, distributed updating and validation of the public and private keys.

The group is a *(t, N)* secure group where an adversary would need to compromise t nodes out of the total N nodes in the group to break the security of the group.

3.1   Initialisation

In the initialisation phase, a trusted server or dealer generates the public key, the master private key and initialises $t + 1$ nodes with their private keys. The dealer calculates the private keys of $t + 1$ nodes, deletes the master key and disappears from the network. This is a common assumption in threshold cryptography based models[8-10, 13].

3.2   Key Generation and Updating

The nodes update their private keys when the time period information changes. The reason to update the private keys is to renew trust. A node receives its private key when it first comes online, but an adversary can compromise it at some later time. Compromises can go undetected to some nodes and it may be difficult for the group as a whole to generate a revocation certificate. In addition, not all nodes in the group may receive the revocation certificate. Hence, the best possible way, in our view, is to renew trust by refreshing the private keys.

A node gets its private keys from a coalition of $t + 1$ nodes that are within one hop distance of that node. We require the authenticating nodes to be within one hop distance, as in ad hoc networks, nodes cannot trust the intermediate nodes to route the authentication information. In section 4.4 we propose extensions whereby the authenticating nodes can be more than one hop away but at the price of additional encryption that affects performance.

The joining node first finds a coalition of $t + 1$ nodes in its neighbourhood. It then generates a list of the identifiers of the $t + 1$ nodes and sends the list, along with a request for a group private key, to all of the $t + 1$ nodes in the list. It can append additional information about itself (like a certificate) to facilitate authentication.

The authenticating $t + 1$ nodes verify the node's identity and check the node's validity by checking their revocation lists. Once they are satisfied with the legitimacy of the node, they will generate partial private keys and send them to the node. Finally, the joining node will construct a full private key from the individual partial private keys.

3.3   Key Certificate Generation

A node v_j uses its private key to sign outgoing messages and decrypt incoming messages sent by other nodes in the network. As many of these communicating nodes may not have been part of the coalition that authenticated and generated the private key for node v_j, nodes need to be able verify that v_j got the keys from legitimate nodes. In addition, our model requires that nodes update their private keys regularly for improved security. Thus all nodes need to know when and how the node got its private key. For this end, each node carries a *key certificate* that contains the node's identity, the coalition that authenticated the node, the time at which the private key was given and the expiration time of the private key.

3.4   Communication Establishment and Termination

Once a node has been authenticated and has been given a private key, the node can communicate with any node in the group with the group public key and its private key pair. This key pair can either directly protect the whole communication, or be used only for negotiating a temporary session key that is symmetric in nature. The manner of communication is left open and can be decided by the group policies and requirements.

4.   ALGORITHMS AND MATHEMATICAL BASICS

4.1   Mathematical Basics

Our solution is based on key insulated cryptography and signature schemes [3, 4]; first we give some background information about the algorithms and terminology in the key-insulated schemes.

In key insulated schemes, a key generation algorithm takes as input a security parameter k that is the bit length of a prime number q. The prime number q is chosen such that $p = 2q + 1$ is also prime. This defines a unique subgroup $G \subset Z_p^*$ of size q where the Decision Diffie-Hellman (DDH) assumption[11] is assumed to hold. Two random elements $g, h \in G$ are selected and two random polynomials $f_x(\tau) = \sum_{j=0}^{t} x_j^* \tau^j$ and $f_y(\tau) = \sum_{j=0}^{t} y_j^* \tau^j$ of degree t are defined over Z_q. The public key consists of g, h and $z_0^* , z_1^* , ... z_t^*$, where $z_0^* = g^{x_0^*} h^{y_0^*} ...... z_t^* = g^{x_t^*} h^{y_t^*}$. The initial private key is (x_0^*, y_0^*), that is the initial coefficient of the polynomial, and the remaining coefficients $(x_1^*, y_1^*, ..., x_t^*, y_t^*)$ are stored in the base station as the master private key. The remaining private keys of the device are the two polynomial evaluations $f_x(i)$ and $f_y(i)$ with i as the time period. These private keys are updated with the master private key and the old and new time period. Fig. 1 sums up the terminology used in the key insulated schemes.

$$PK := (g, h, z_0^*, ..., z_t^*) \text{ where } z_0^* = g^{x_0^*} h^{y_0^*}, ... z_t^* = g^{x_t^*} h^{y_t^*}$$

$$SK^* := (x_1^*, y_1^*, x_2^*, y_2^*, ..., x_t^*, y_t^*);$$

$$SK_0^x := x_0^* \quad SK_0^x := y_0^*$$

$$SK_i^x = x_0^* + x_1^* i + x_2^* i^2 + ... + x_t^* i^t = \sum_{s=0}^{t} x_s^* i^s$$

$$SK_i^y = y_0^* + y_1^* i + y_2^* i^2 + ... + y_t^* i^t = = \sum_{s=0}^{t} y_s^* i^s$$

where $1 \leq i \leq N$ is time period information

Figure 1. Background terminology

Thus, in key insulated schemes, nodes maintain a public and private key pair and change their private keys at discrete time intervals without changing the public key.

In our proposed solution, we make two important changes to the key insulated schemes. Firstly, each node maintains its own unique individual private key (SK_i^x, SK_i^y) based on the parameter i that is now a combination of the time period information and the unique identity of the node. Thus the private keys are generated for all nodes in the group for a single global public key. From now on we refer to i as the unique identifier of a node.

Secondly, we do not maintain the master private key anywhere in the system. The private keys for nodes are generated and updated by a coalition of $(t+1)$ neighbouring nodes, where t is the degree of the two random polynomials described earlier.

4.2   Private Key Generation and Updation

Without loss of generality, let us assume that the new node v_j finds a coalition $\beta = \{v_1, v_2, ..., v_{t+1}\}$. It sends this list to all nodes in β with a request for a private key. Each node v_i in the coalition β would authenticate   the   requesting   node   and   generate   partial   shares

$$Psh_i^x = SK_i^x . \prod_{r=1, r \neq i}^{t+1} \frac{v_j - v_r}{v_i - v_r} \text{ and } Psh_i^y = SK_i^y . \prod_{r=1, r \neq i}^{t+1} \frac{v_j - v_r}{v_i - v_r}$$

Using Lagrange's interpolation, node v_j can calculate its private key by adding the x and y components of the above partial shares respectively.

$$SK_j^x = Psh_1^x + Psh_2^x + ... + Psh_{t+1}^x = \sum_{r=1}^{t+1} Psh_r^x = \sum_{s=0}^{t} x_s^* v_j^s$$

$$SK_j^y = Psh_1^y + Psh_2^y + ... + Psh_{t+1}^y = \sum_{r=1}^{t+1} Psh_r^y = \sum_{s=0}^{t} y_s^* v_j^s$$

4.3   Distributed Generation of Key Certificates

The *key certificate* contains a message M that runs something like, "This is to certify that node v_j has been issued a private key by coalition β at time T and it would expire at time $T + T_{lifetime}$." The value $T_{lifetime}$ is the duration for which the certificate is valid. The certificate is signed with a signing key that is known by none of the nodes in the group, which is the initial private key (SK_0^x, SK_0^y). This key can also be used to generate group messages to other groups in the network. No single node knows this secret signing key as all nodes have non-zero identity values. However, $t+1$ nodes can generate a signature using Lagrange's interpolation, without revealing the key at any time.

The coalition of nodes β generates the *key certificate* of node v_j along with the private key. Since all nodes in the coalition know the time periods T and $T_{lifetime}$ together with the coalition list β, the certificate message M is well known. Thus, each node in the coalition β can construct the certificate message M independently. The certificate-signing algorithm is similar to the signing algorithm in key-insulated signature schemes except that the certificate signature is generated in a distributed manner.

To sign the *key certificate* of node v_j, each node v_i generates two random numbers $r_{i,1}$ and $r_{i,2} \leftarrow \mathbf{Z}_q$. It then calculates $w_i = g^{r_{i,1}} h^{r_{i,2}}$ and sends w_i to all nodes in the coalition ß so that all nodes in the coalition can calculate $w = w_1 \times w_2 \times ..... \times w_{t+1}$.   This   is   equivalent   to $w = g^{r_{1,1} + r_{2,1} + ... + r_{t+1,1}} h^{r_{1,2} + r_{2,2} + ... + r_{t+1,2}} = g^{r_1} h^{r_2}$ where $r_1 = r_{1,1} + r_{2,1} + ... + r_{t+1,1}$ and $r_2 = r_{1,2} + r_{2,2} + ... + r_{t+1,2}$ and $r_1, r_2 \in \mathbf{Z}_q$.

Once all nodes have calculated w, each node generates a hash $\tau = H(0, M, w)$ of the certificate message M and w. Each node v_i further calculates $a_i = r_{i,1} - \tau P_i^x$ and $b_i = r_{i,2} - \tau P_i^y$ where $P_i^x = SK_i^x \cdot \prod_{r=1, r \neq i}^{t+1} (\frac{v_r}{v_r - v_i})$

and $P_i^y = SK_i^y \cdot \prod_{r=1, r \neq i}^{t+1} (\frac{v_r}{v_r - v_i})$.

Similar to the private key generation algorithm, node v_j gets the full signature of the certificate by adding up $a = a_1 + a_2 + ... + a_{t+1}$ and $b = b_1 + b_2 + ... + b_{t+1}$. With Lagrange's interpolation we get $a = r_1 - \tau SK_0^x$ and $b = r_2 - \tau SK_0^y$ which is equivalent to a message signed with the initial private keys in the key insulated schemes. The final generated certificate is of the form $(M, [0, (w, a, b)])$. The certificate can be verified by anyone who has the public key $PK := (g, h, z_0^*, ..., z_t^*)$.

Steps			
1	Let $\beta = \{v_1, v_2, v_3, ... v_{t+1}\}$	6	$w = w_1 \times w_2 \times ... \times w_{t-1}$
2	for each node v_i in $\mathfrak{B}$	7	$\tau = Hash(0, M, w)$
	$r_{i,1}, r_{i,2} \leftarrow \mathbb{Z}_q$	8	$a_i = r_{i,1} - \tau P_i^x, b_i = r_{i,2} - \tau P_i^y$
3	$w_i = g^{r_{i,1}} h^{r_{i,2}}$		where $P_i^x = \sum_{s=0}^{t} x_s^* v_i^s \cdot \prod_{r=1, r \neq i}^{t+1} (\frac{v_r}{v_r - v_i})$
4	send w_i to all nodes in $\mathfrak{B}$		and $P_i^y = \sum_{s=0}^{t} y_s^* v_i^s \cdot \prod_{r=1, r \neq i}^{t+1} (\frac{v_r}{v_r - v_i})$
5	collect all $w_1, ... w_{t+1}$	9	node v_j calculates $a = a_1 + a_2 + ... + a_{t+1}$ and $b = b_1 + b_2 + ... + b_{t+1}$

Figure 2. Key certificate generation algorithm

4.4 Protecting the Partial Key Shares

Whenever a node v_i sends a share Psh_i^x, Psh_i^y, there is a serious risk that an adversary may eavesdrop and extract the private key of the node. If an adversary gets $Psh_i^x = SK_i^x \cdot \prod_{r=1, r \neq i}^{t+1} \frac{v_j - v_r}{v_i - v_r}$ it can calculate the private key of the node v_s, since the second parameter in the equation $\prod_{r=1, r \neq i}^{t+1} \frac{v_j - v_r}{v_i - v_r}$ can be easily calculated if the coalition β is known. This is a serious security concern as even the node that requests the key can try to gain knowledge of the private keys of its authenticators.

There are two ways of protecting the partial shares of the authenticating nodes. One is by mixing the partial shares with another value so that the transferred value does not reveal any information about the actual partial shares. This is known as shuffling. The second way is to encrypt the partial shares with the public key of the receiving node so that only that node can decrypt it.

4.4.1   Shuffling

All nodes in the group securely exchange secret factors with each other. Specifically, two nodes v_a and v_b in the coalition β can securely exchange a factor $d_{a,b}$. The node with the higher id, say a, would then add the factor to the partial share Psh_a while the other node b would subtract the factor from its share Psh_b. These nodes would similarly exchange more such factors with other nodes in the network. When the joining node adds up all the partial shares it receives from the coalition, the factors cancel out and the final result is still the new private key of the node. Luo and Lu[9] first described this process.

4.4.2   Public Key Encryption of Shares

The partial shares can also be encrypted with the public key of the joining node. This public key is not the group public key but a personal public key of the node that could be authenticated in the form of a certificate. When the nodes generate the partial shares, they would encrypt the shares with the public key of the joining node so that only the intended node can decrypt the share with its private key.

This method puts additional burden on the joining node, as it has to decrypt the shares, but it is more robust than the shuffling method as the authenticating nodes can be more than one hop distance away from the joining node.

4.5   Certificate Revocation

We require that nodes continuously update their private keys so that faulty nodes can gracefully leave the network. Another advantage is that compromised nodes cannot operate long in the network. However, we still need a mechanism to evict a compromised node immediately from the network as keys are refreshed after some time period and we may not want a compromised node to be in the network till its time for it to update its keys.

To revoke the certificate of compromised or faulty nodes we propose to use a model similar to those suggested by[6, 8-10]. When a node is found to be faulty, $t+1$ nodes need to sign a special counter certificate that would nullify the presence of the node in the network. The counter certificate is signed by the same key that was used to sign certificates. The counter certificate is sent to all possible nodes in the network by flooding.

A node needs to be accused by at least $t+1$ nodes before being marked as 'faulty' and evicted from the group. Any number of nodes less than the coalition size $t+1$ cannot evict a node from the group.

5.   RESULTS

We implemented the algorithms in Java and used the Jini platform for communication. We used Java's built-in classes for hashing messages and support for large integers (java.math.BigInteger). The Jini platform allowed the program to be independent of the network configuration and allowed a certain degree of interoperability. We can, however, replace Jini with any other communication platform.

5.1   Setup

We tested the authentication process in a laboratory environment with wired and wireless networks, and with different types of machines being the nodes of the ad hoc network. The experiments tested the computational costs of a worst-case scenario and compared it with network latency. There was no other network traffic that could affect the results.

The testing environment had three PCs with Pentium III, 1 GHz processors, 256Mb RAM and running Debian Linux. These PCs were connected to other PC's through a wired Ethernet switch. Two more PC's with the same configuration but running Gentoo Linux were connected via wireless LAN. We name two of the wired Pentium PCs as *pentium1* and *pentium2* and the Wireless PCs as *wireless1* and *wireless2*.

Besides these, there were two Sun Blade 150 workstations, each with a 650MHz processor and 512Mb RAM (*blade1* and *blade2*) and two Sun Netra X1 machines, with 500MHz processors and 1024Mb RAM (*sparc1* and *sparc2*). All Sun machines were running the Sun Solaris 8 operating system.

All these machines were connected via Ethernet (100Mbps) while the wireless machines were connected to the wired network through an access point. The wireless connections were operating at 56Mbps. All machines had Java 1.4.2 and Jini 2 installed locally.

We tested the algorithms for a threshold limit t of 8. A Jini lookup service provider was running on another Sun Netra machine that facilitated the locating of the nodes. Once the remote nodes had been located, a proxy object was downloaded and used to call methods through RMI.

The joining node sent the requests and received the partial shares from the coalition nodes in a serial order. The serial order of requests and replies gave us the worst-case scenario of the working of the algorithms. We could improve the speed by using multithreading and multicasting, but this was not implemented.

5.2   Key and Signature Generation Results

The results of the experiments are given in Fig. 3 and Fig. 4. As expected, the processing time increases as key length increases, and the faster processors perform better than the others. An interesting point to note is that even for a reasonable high key length of 1280 bits, the *process time* is under 10sec for all machines.

The total processing time is the time taken for a node to generate and send the partial key requested by a joining node.

6.   DISCUSSION AND FUTURE WORK

In the previous sections, we presented a design for distributed generation of private keys and signatures and showed the implementation results. We now discuss some of the design issues we considered for the proposed (t, N) secure model.

Initialisation of t nodes: In section 3.1 we discussed that at the time of the creation of the ad hoc group, an initialisation program or dealer would initialise some number of nodes in the group. This is a common assumption in secret sharing[5, 8-10, 13]. One possible future direction is on how to initialise t or more nodes in a distributed fashion. Distributed initialisation can also lead to better pro-active security for the model.

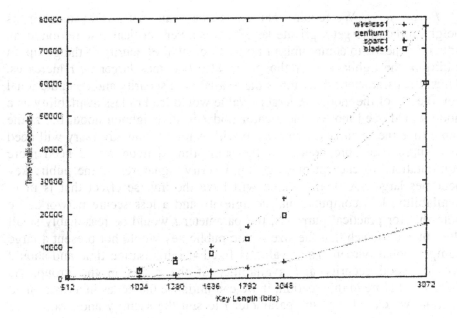

Figure 3. Authentication times for different machines (Part 1)

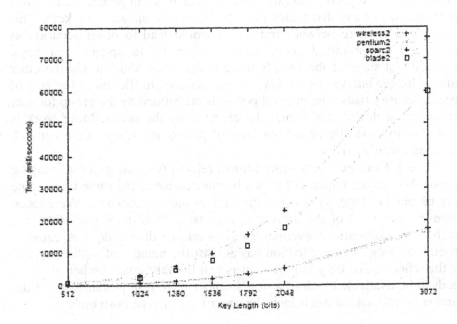

Figure 4. Authentication times for different machines (Part 2)

Parameter t: We require that a node have at least t valid nodes in its neighbourhood to get a private key. This is a very critical assumption as an adversary needs to compromise t nodes to break the security of the group. In addition, the public key of the group also becomes larger as t increases. Thus, the parameter t determines the availability, security and computational complexity of the model. A larger t value would lead to less availability as a node would need more authenticator nodes in its neighbourhood, but at the same time the security of the system will increase as an adversary will need to compromise more nodes. At the same time a node would need more computations to encrypt messages and verify signatures as the public key becomes larger. A small t value will have the inverse effect that is more availability, less computational complexity and a less secure network. We note that for practical purposes, the parameter t would be reasonably small (less than ten) such that the size of the public key would not present a large computational burden. The t value is fixed at initialisation time and should provide good security and availability of the nodes in the group. As discussed before in this section, if we can initialise the nodes in a distributed fashion, we can change the parameter t to suit the security and availability requirements of the group.

Renewal of Private Keys: We require that nodes in the group update their private keys at regular intervals. The interval is an important factor, as a short interval means that nodes need to frequently update their keys. This would lead to more network traffic but would lead to better security, as faulty and compromised nodes would be unable to update their keys, provided that some of the neighbouring nodes know about it. On the other hand, a longer interval would lead to less network traffic but at the cost of greater security risks. The interval period is predefined by the group founder (initialisation dealer) and cannot be changed by the nodes. More work is needed to dynamically adjust the interval period according to the network traffic and security risks.

Network Failures: We assume limited reliability in the group messaging service. We do not require all nodes be contactable at the same time, some may be out of range, as is often the case in mobile networks. We expect, however, that most of the messages, e.g. those about node exclusion will reach their destinations eventually. Network partitions do not cause a problem as long as each partition has at least the number of nodes required by the scheme. The only requirement is that the threshold number of nodes needs to be available to admit a new member or refresh private keys. If the number of available nodes is less, the operation has to be postponed.

7.   CONCLUSIONS

In this paper we proposed an architecture that generates individual private keys for a common public key dynamically, in a distributed manner. We combined the benefits of the group key based approaches and the public key based approaches, while we do not need to maintain lists of public keys or a shared secret key.

Our proposed solution is based on the key insulated cryptography and signature schemes, and on threshold cryptography. We modified the key insulated schemes to suit ad hoc networks, and generate and update keys in a distributed fashion. We do not require any base stations or servers to provide keys, except at the initialisation phase.

We also generate a certificate showing that the node has received the key from t legitimate nodes and that it would expire after a predetermined time. The certificate is signed with a secret signing key that is known by none of the nodes but t nodes can produce a signature without revealing the key.

The proposed model is (t, N) secure as an adversary who compromises less than t nodes in the group cannot gain any more information. In addition, t nodes can generate a private key and a certificate for a node signed with a secret signing key. The network traffic is restricted to the local neighbourhood of the nodes and hence the performance is good. The results of our experiments are promising, as even for large key lengths the distributed key and signature generation algorithms work well.

ACKNOWLEDGEMENTS

This research was performed on equipment partially provided by Sun Microsystems grant no. 7832-030217-AUS.

REFERENCE

1. T. Aura and S. Mäki. Towards a survivable security architecture for ad-hoc networks. In *Security Protocols*, volume 2467, pages 63–73. Springer-Verlag Heidelberg, Lecture Notes in Computer Science, 2002.
2. S. Čapkun, L. Buttyán and J.P. Hubaux. Self-organised public-key management for mobile ad hoc networks, *IEEE Transactions on Mobile Computing*, 2(1):52–64, Jan-Mar 2003.
3. Y. Dodis, J. Katz, S. Xu, and M. Yung. Key-insulated public key cryptosystems. In *Advances in Cryptology- EUROCRYPT 2002*, volume 2332. Springer-Verlag Heidelberg, Lecture Notes in Computer Science, April 2002.
4. Y. Dodis, J. Katz, S. Xu, and M. Yung. Strong key-insulated signature schemes. In *Workshop on Public Key Cryptography (PKC)*, volume 2567. Springer-Verlag Heidelberg, Lecture Notes in Computer Science, January 2003.

5. Y. Frankel, P. Gemmell, P. D. MacKenzie, and M.Yung. Proactive RSA. In *Advances in Cryptology - CRYTPTO 1997*, volume 1294, pages 440–454. Springer-Verlag, Heidelberg, Lecture Notes in Computer Science, 1997

6. S. Kaliaperumal. Securing authentication and privacy in ad hoc partitioned networks. In *Symposium on Applications and the Internet Workshops (SAINT'03 Workshops)*, pages 354–357, Orlando, FL, January 2003.

7. Y. Kim, A. Perrig, and G. Tsudik. Group key agreement efficient in communication. *Computers, IEEE Transactions on*, 53(7):905–921, July 2004.

8. J. Kong, P. Zerfos, H Luo, S Lu, and L. Zhang. Providing robust and ubiquitous security support for mobile ad hoc networks. In *IEEE 9th International Conference on Network Protocols (ICNP '01)*, 2001.

9. H. Luo and S. Lu. Ubiquitous and robust authentication services for ad hoc wireless networks. *UCLA Computer Science Technical Report*, 200030, October 2000.

10. H. Luo, P. Zerfos, J. Kong, S. Lu, and L. Zhang. Self securing ad hoc networks. In *Proceedings Seventh International Symposium on Computers and Communications (ISCC 2002)*, pages 567–574, July 2002.

11. A. Menezes, P. van Oorschot, and S. Vanstone. *Handbook of Applied Cryptography*. CRC Press, 1997.

12. K. Obraczka, K. Viswanath, and G. Tsudik. Flooding for reliable multicast in multi-hop ad hoc networks. *Wireless Networks*, 7(6):527–634, November 2001.

13. Lidong Zhou and Zygmunt J. Haas. Securing ad hoc networks. *Network, IEEE*, 13(6):24–30, Nov-Dec 1999.

ENSURING MEDIA INTEGRITY ON THIRD-PARTY INFRASTRUCTURES

Jana Dittmann[1], Stefan Katzenbeisser[2], Christian Schallhart[2], Helmut Veith[2]
[1]Otto-von-Guericke Universität Magdeburg, Germany, `jana.dittmann@iti.cs.uni-magdeburg.de` *[2] Technische Universität München, Germany* `{katzenbe,schallha,veith}@in.tum.de`

Abstract: In many heterogeneous networked applications the integrity of multimedia data plays an essential role, but is not directly supported by the application. In this paper, we propose a method which enables an individual user to detect tampering with a multimedia file without changing the software application provided by the third party. Our method is based on a combination of cryptographic signatures and fragile watermarks, i.e., watermarks that are destroyed by illegitimate tampering. We show that the proposed system is provably secure under standard cryptographic assumptions.

Key words: Digital Signatures, Media Integrity

1. INTRODUCTION

Commercial transactions as well as private communication are increasingly performed using electronic infrastructures—most importantly the WWW. These applications are typically built using standard software and existing third-party network infrastructures (such as standard Web browsers, standardized network protocols or off-the-shelf video streaming tools). This infrastructure is commonly used to transmit multimedia information (like video or audio files or facsimiles). The integrity of these data can be of crucial importance and may have legal and commercial consequences. Unfortunately, the level of security a user can achieve depends on the services provided, as the users themselves typically cannot change the architecture of the overall system. In this paper, we describe a method for ensuring the integrity and authenticity of multimedia data which overcomes this limitation. We use a combination of cryptographic signatures and digital watermarks to encode authentication information in the media objects themselves, which allows us to completely decouple the authentication mechanism from the rest of the application. In essence, the security information is "tunneled" through the third-party system. No modification of the infrastructure is necessary, which makes our approach feasible for a wide range of applications; the approach is also targeted towards applications built on large portions of legacy code, which can only be changed with considerable effort or cost.

Consider the following example scenarios:

- During an online auction of a precious item, a potential bidder wants to make sure that an image of an item displayed by the auction software adequately displays the item for sale.

- In a video conferencing tool that is used to facilitate online business, both parties may (for documentation purposes) require a proof of authenticity and integrity of the video stream.

- Digital images (or digitally recorded video streams) cannot readily be used as evidence in legal cases, as they can easily be forged or altered. In particular, portions of the images can be blurred or two images can be pasted together to remove suspicious parts.

- Some government agencies store documents (such as drivers licenses, birth certificates, etc.) only in digitized form. Typically, these documents are scanned and the resulting images are retained in an image database. Given the availability of sophisticated image processing software, it is obvious that the integrity of such electronic documents must be protected.

In all cases, both the integrity and authenticity of a multimedia object is of central importance for the security of the overall software system. Integrity and authenticity can be guaranteed by cryptographically signing the multimedia objects in question; in a naive solution, a media object is always stored together with its corresponding signature (Friedman, 1993). Tightly integrating a multimedia authentication mechanism into such applications can be a serious obstacle in practice, as this may require considerable changes to existing software architectures or the adoption of nonstandard application software. From a software-engineering point of view, modular media authentication mechanisms would be desirable that can be added as plug-ins or front-ends to existing software systems and third-party network infrastructures at ease.

In this paper, we give a construction for a modular—but yet provably secure—media authentication mechanism that can readily be plugged into existing software systems. Technically, we use *invertible fragile watermarks* to store cryptographic signatures directly in the media objects themselves. Using this system, the authentication mechanism can be *completely separated from the application software*, as all authentication information is readily encoded (in a format-compliant way) in the media file itself.

In contrast to classical digital watermarks which are designed to resist signal processing attacks, *fragile watermark* encode additional data imperceptibly in some media file *without providing any robustness*; that is, signal processing attacks will destroy the embedded watermark. For an overview of digital watermarking technology see (Katzenbeisser and Petitcolas, 2000). Fragile watermarks can be seen as a kind of alerter that reports whether a media object was tampered. In our approach, we use *invertible fragile watermarks*, which allow to insert a fragile watermark into an object as usual, but facilitate the lossless removal of the watermark from an untampered watermarked object. More precisely, if a watermark is successfully detected, the information contained in the recovered watermark, together with the watermark key, suffices to remove the watermark completely from the object.

Fragile watermarks are natural candidates for assuring the integrity of image files and were proposed by various authors (e.g., Schneider and Chang, 1996; Xie and Arce, 1998). In these approaches, image-dependant patterns are encoded as fragile watermarks in digital images. An image is considered authentic if and only if it is possible to correctly recover these patterns. If a file with such a watermark is modified, then either the watermark cannot be detected any more or the recovered patterns do not match the image. In both cases, the image is considered to be tampered. Unfortunately, this approach has the apparent drawback that it is not possible to formally prove its security in a cryptographically precise way, as properties of the watermark embedder or detector become security-critical.

In this paper we provide the first construction for a *provably secure watermark-based authentication scheme for digital media files*. In contrast to previous schemes, we use watermarks merely as a communication channel for transmitting a cryptographic signature; the scheme therefore draws its security entirely from this signature. In our approach, the signing algorithm produces an authenticated object $\overline{O}$ out of a media file O in such a way that the integrity and authenticity of $\overline{O}$ can readily be asserted using only a public key and $\overline{O}$. As $\overline{O}$ will be perceptually similar to O, $\overline{O}$ can readily be used as a replacement for O in existing application software, thereby adding a layer of security.

The rest of the paper is organized as follows. After reviewing the necessary watermarking technology in Section 2, we introduce our framework for media authentication in Section 3. Finally, we present two provably secure constructions for media authentication schemes in Sections 4 and 5; the second construction can be used for large media files or in streaming applications.

2. INVERTIBLE WATERMARKS

While almost all previous watermarking schemes introduced some small amount of irreversible distortion in the data during the embedding process,

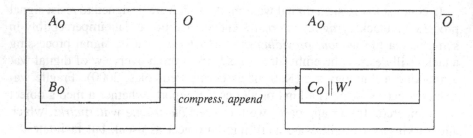

Figure 1.   Invertible watermarking. An object O is divided into two parts A_O and B_O. The watermark consists of the compressed part B_O, denoted by C_O, and the watermark payload W'.

invertible watermarks (Honsinger et al., 1999; Fridrich et al., 2001; Fridrich et al., 2002a; Steinebach and Dittmann, 2003a; Maas et al., 2002; Dittmann and Benedens, 2003) can be removed completely from a watermarked object, thereby recovering the original.

Fridrich et al., 2001 introduced a general framework that allows to construct an invertible fragile watermarking scheme out of a fragile one. The general idea is to divide the object O, dependent on a public key K_W, into two (not necessarily consecutive) parts A_O and B_O. The latter part contains perceptually insignificant portions of the object that can be overwritten by a watermark without lowering the object quality, whereas A_O contains perceptually visible parts that must be preserved. To provide invertibility, the original part B_O is compressed and stored in the watermark; denote the compressed part B_O with C_O. The watermark W consists of the watermark payload W' and C_O, thus $W = C_O \| W'$. W replaces the part B_O in the watermarked object $\overline{O}$. This general framework is depicted in Figure 1.

The distortion of the watermark can easily be removed by separating the marked object $\overline{O}$ into the two parts $A_{\overline{O}}$ and $B_{\overline{O}}$. During the watermark insertion process, only $B_{\overline{O}}$ was modified, so $A_{\overline{O}} = A_O$. Now, $B_{\overline{O}}$ has the form $W = C_O \| W'$; decompressing C_O yields to the part B_O of the original object O. By overwriting $B_{\overline{O}}$ with B_O in the object $\overline{O}$, O can be completely recovered. This procedure works only if $\overline{O}$ was not altered; it is therefore a fragile watermarking scheme.

In the rest of the paper, we denote an invertible watermarking scheme as a tuple of two probabilistic polynomial algorithms $\langle \text{SEPARATE}, \text{JOIN} \rangle$. On input O and K_W, SEPARATE produces the tuple $\langle A_O, B_O \rangle$. JOIN inverts the algorithm SEPARATE, i.e., on input $\langle A_{\overline{O}}, B_{\overline{O}} \rangle$ and K_W it outputs $\overline{O}$. Except with negligible probability, we require that

$$\text{JOIN}(K_W, \text{SEPARATE}(K_W, O)) = O,$$

for all objects O and keys K_W with $\text{SEPARATE}(K_W, O) \neq \text{FAIL}$.

From the previous description it is obvious that it is not possible to embed an invertible watermark in every object. In case the part B_O cannot be sufficiently compressed, there is not enough room to store both the watermark payload and the compressed part C_O. However, typical multimedia files (such as images or audio files) contain enough redundant, compressible information so that the watermarking operation *works for virtually all relevant objects*.

This general construction principle can be instantiated for different types of media objects: for example (Fridrich et al., 2002b) gave two different constructions that operate on JPEG images; (Steinebach and Dittmann, 2003b) showed how audio files can be watermarked in an invertible manner. Other implementations can be found in (Fridrich et al., 2001; Fridrich et al., 2002a; Dittmann et al., 2002).

It is important to note that the construction of media authentication schemes given in this paper is general and *works for any invertible watermarking scheme. We will show that the security of the scheme only depends on the cryptographic primitives involved, while the watermark serves merely as insecure communication channel and thus does not affect the security of the authentication scheme*; a formal proof of this claim will be given in Sections 4 and 5.2. For this reason, we will treat an invertible watermarking scheme as black box in the remaining parts of the paper.

3. MEDIA AUTHENTICATION SCHEMES

Media authentication schemes based on invertible watermarks can be described by a tuple of four probabilistic polynomial time algorithms $\langle \text{GENKEY}, \text{PROTECT}, \text{VERIFY}, \text{RECONSTRUCT} \rangle$. GENKEY denotes the key-generation process; by using a private key, PROTECT authenticates an object O and outputs its signed version $\overline{O}$. Signed objects can be verified by the algorithm VERIFY and a public key; VERIFY either outputs TRUE or FALSE. In the first case, the object is deemed authentic; in the latter case, the object is considered modified. RECONSTRUCT reverses the protection mechanism and losslessly reconstructs O out of $\overline{O}$.

3.1 Definition

Formally, an *invertible media authentication scheme* is defined as follows:

- Algorithm GENKEY generates keys for the application. On input 1^n, GENKEY produces a triple of strings $\langle K_P, K_V, K_R \rangle$ with $|K_P \| K_V \| K_R| = n$; the operation $\|$ denotes string concatenation. The key K_P will be used in the protection step, whereas K_V and K_R are used for verification and recovery. The verification key K_V is a public key, whereas K_P and K_R are private keys.

- Algorithm PROTECT takes K_P, K_R and an object O. The output of the algorithm consists of an authenticated object $\overline{O}$.

- Algorithm VERIFY takes the verification key K_V and an object $\overline{O}$ and outputs a boolean variable.

- Algorithm RECONSTRUCT takes the keys K_R and K_V and an object $\overline{O}$ and restores the original object O.

Note that we have defined all algorithms as probabilistic, which implies that they can fail on certain instances (as noted above it may not be possible to embed a watermark in an invertible manner); in this case, the algorithms output a special symbol FAIL. We require that the media authentication scheme "works" for almost all objects that can be watermarked. In particular, VERIFY(PROTECT$(O, K_P, K_R), K_V)$ = TRUE and RECONSTRUCT(PROTECT$(O, K_P, K_R), K_R, K_V) = O$ must hold except for a negligible fraction of all objects O with PROTECT$(O, K_P, K_R) \neq$ FAIL.

As usual, we will denote a cryptographic *signature scheme* as triple of probabilistic polynomial algorithms $\mathbb{S} = \langle \text{GENSIGN}, \text{SIGN}, \text{SIGVERIFY} \rangle$; GENSIGN denotes the key generation, SIGN the signing and SIGVERIFY the signature verification algorithm. A signature scheme is said to be secure, if it is secure against existential forgery of signatures under a chosen-message attack (Goldwasser et al., 1988); that is, if the attacker is unable—even with access to a signing oracle—to forge a valid pair of a message and a corresponding signature.

3.2   Attacker Model

Sticking to Kerckhoffs' principle, we assume that an attacker possesses complete knowledge of the system. The general goal of an attacker will be to *forge an authentic object* relative to a public key K_V, whose corresponding secret key K_P is unknown to him.

Similar to attacks against cryptographic signature schemes, we can distinguish several types of attacks against media authentication schemes according to the possibilities for an attacker to interfere with the system. It seems natural to assume that an attacker will know several protected media files under one verification key K_V, as such objects might be freely available on the Internet. A more powerful attacker may even launch a *chosen message attack*. In this setup, an attacker is able to obtain protected objects of his own choice. That is, he can obtain an authenticated version $\overline{O}$ of object O chosen during the attack. In imaging applications, such an attack is particularly realistic, as long as the attacker has physical access to the imaging device and can take pictures of his own choice.

For this reason, we adopt the notion of *existential forgery under chosen message attacks* for the present scenario. For a given public key K_V, the attacker attempts to output an object O so that $\text{VERIFY}(O, K_V) = \text{TRUE}$. During his attack, he is free to use an oracle that delivers him arbitrary authenticated objects on request. We say that an attack is successful, if the attacker manages to output (with non-negligible probability) an object $\overline{O}$ together with an alleged original O such that $\text{VERIFY}(\overline{O}, K_V) = \text{TRUE}$, where the original object O was *not* presented to the oracle previously. It is generally agreed that this is the *most general attacker model* one can define for the signature scenario.

DEFINITION 1 *Let* $\langle \text{GENKEY}, \text{PROTECT}, \text{VERIFY}, \text{RECONSTRUCT} \rangle$ *be a media authentication scheme and* QUERY_{K_P} *be an oracle that computes* $\overline{O} \leftarrow \text{PROTECT}(O, K_P)$ *on input* O. *Furthermore, let* $\langle K_P, K_V, K_R \rangle \in [\text{GENKEY}(1^{n_K})]$. *An attack is a probabilistic algorithm* ATTACK *with oracle access to* QUERY_{K_P} *and success probability* $\varepsilon_{\text{ATTACK}}$ *such that*

$$
\text{ATTACK}(1^n, K_V) =
\begin{cases}
\langle O, \overline{O} \rangle & \text{such that } \text{VERIFY}(\overline{O}, K_V) = \text{TRUE}, \\
& |O| = n, \ \overline{O} \in [\text{PROTECT}(O, K_P)] \\
& \text{and } O \neq O_i \text{ for all } 1 \leq i \leq l, \\
& \text{with probability } \varepsilon_{\text{ATTACK}} \\
\text{FAIL} & \text{with probability } 1 - \varepsilon_{\text{ATTACK}},
\end{cases}
$$

where O_i *denotes the input to the i-th oracle query* QUERY_{K_P}. *The probability is taken over all coin tosses of* ATTACK *and all keys* $\langle K_P, K_V, K_R \rangle$.

We say that a media authentication scheme is secure, if the success probability of *every* probabilistic polynomial time attack is negligible:

DEFINITION 2 *A media authentication scheme is secure against existential forgery of authenticated objects, if every probabilistic polynomial time attack* ATTACK *has negligible success probability.*

4. OFFLINE AUTHENTICATION

In this section, we describe a media authentication scheme that needs access to the whole media file at once during the protection algorithm (such schemes will be called offline media authentication schemes). This construction is therefore primarily suitable for image files or short video sequences. A separate construction for streaming media will be given in Section 5.

Let $\mathbb{S} = \langle \text{GENSIGN}, \text{SIGN}, \text{SIGVERIFY} \rangle$ be a cryptographic signature scheme producing signatures of length k, ENCRYPT and DECRYPT be the encryption and decryption function of a symmetric cipher and COMPRESS be the compression algorithm of a lossless compression scheme. Furthermore, we fix an invertible watermarking scheme $\langle \text{SEPARATE}, \text{JOIN} \rangle$ that can embed watermark strings of length k.

Loosely speaking, the media authentication scheme stores a cryptographic signature of the unmodified portion of the object (the part A_O) and the encrypted, compressed part B_O as an invertible watermark. During the verification process, we check whether we can extract a valid cryptographic signature out of the watermark. The construction is as follows:

- GENKEY runs GENSIGN to obtain a key pair $\langle K_{SS}, K_{VS} \rangle$; furthermore, it computes a key K_E for the symmetric cipher and a random string K_W. Let $K_P = K_{SS} \| K_W$, $K_V = K_{VS} \| K_W$ and $K_R = K_E \| K_W$.

- PROTECT, on input O, $K_P = K_{SS} \| K_W$ and $K_R = K_E \| K_W$, separates O, using algorithm SEPARATE and key K_W, into two parts A_O and B_O. The latter part is compressed to obtain C_O. Denote with W' the string $W' = X \| s$, where
$$X \leftarrow \text{ENCRYPT}(K_E, C_O \| H(O)),$$
H is a hash function and
$$s \leftarrow \text{SIGN}(K_{SS}, A_O \| X).$$

  Note that C_O is stored encrypted so that the reconstruction of the original is only possible by knowing the key K_E. PROTECT runs JOIN on K_W and $\langle A_O, W' \rangle$ to obtain the authenticated object $\overline{O}$ or FAIL. If JOIN fails, PROTECT outputs FAIL, otherwise $\overline{O}$.

- VERIFY, on input $\overline{O}$ and $K_V = K_{VS} \| K_W$, runs SEPARATE on K_W and $\overline{O}$ to obtain the two parts $A_{\overline{O}}$ and $B_{\overline{O}}$ of $\overline{O}$. The latter part has the form $B_{\overline{O}} = X \| s$, where X is an arbitrary string and s is a cryptographic signature. VERIFY outputs the Boolean value $\text{SIGVERIFY}(K_{VS}, A_{\overline{O}} \| X, s)$.

- RECONSTRUCT, on input $\overline{O}$, $K_R = K_E \| K_W$ and $K_V = K_{VS} \| K_W$, first runs VERIFY to assure the integrity of $\overline{O}$; in case VERIFY outputs FALSE, RECONSTRUCT exits with FAIL. Otherwise, it separates $\overline{O}$ (using SEPARATE and key K_W) into the two parts $A_{\overline{O}}$ and $B_{\overline{O}}$. The latter part has the form $B_{\overline{O}} = X \| s$. By using K_E, RECONSTRUCT decrypts X to obtain $C_O \| h$, where h denotes a hash; the part C_O is decompressed to obtain B_O. Finally, the part $B_{\overline{O}}$ of $\overline{O}$ is overwritten with B_O to obtain an object O. If $H(O) = h$, RECONSTRUCT outputs O, otherwise FAIL.

Intuitively, the scheme is secure because of the following argument: in case an attacker modified the part $A_{\overline{O}}$ of $\overline{O}$, the embedded cryptographic signature s is matched against a modified media file. On the other hand, if any bit in $B_{\overline{O}}$ is modified, then at least one bit of the embedded fragile watermark (containing either the signature s or the compressed part B_O) is destroyed. In all cases, the tampering will be detected during the signature verification. Formally, we can state this result as a theorem:

THEOREM 3 *If $\mathbb{S}$ is a cryptographic signature scheme secure against existential forgery of messages under a chosen message attack, then the above scheme is a secure media authentication scheme.*

Proof. Suppose, for the sake of contradiction, that there exists an attack ATTACK (with access to the media authentication oracle QUERY$_{K_P}$) against the scheme, which succeeds with non-negligible probability. We show that in this case there exists also an attack FORGE (with access to a signing oracle SIGNQUERY$_{K_{SS}}$) against $\mathbb{S}$, which contradicts the assumption.

We construct the signature forging algorithm FORGE (for the public signature key K_{VS}) in the following manner. On input K_{VS}, FORGE chooses random keys K_E and K_W and simulates ATTACK. Whenever ATTACK makes an oracle query QUERY$_{K_P}(O_i)$, this query is replaced by the following probabilistic algorithm, which utilizes the signing oracle SIGNQUERY$_{K_{SS}}$; here, K_{SS} denotes the corresponding secret signature key:

$\langle A_{O_i}, B_{O_i} \rangle \leftarrow$ SEPARATE(K_W, O_i)
compress B_{O_i} to obtain C_{O_i}
$X_i \leftarrow$ ENCRYPT$(K_E, C_{O_i} \| H(O_i))$
query SIGNQUERY$_{K_{SS}}(A_{O_i} \| X_i)$ for signature s
$W_i' = X_i \| s$
output JOIN$(K_W, \langle A_{O_i}, W_i' \rangle)$

Note that JOIN either outputs FAIL or the object $\overline{O}_i$.

When the simulation of ATTACK is finished, ATTACK either outputs FAIL or obtains a tuple $\langle O, \overline{O} \rangle$. In the first case, FORGE exits with FAIL. Otherwise, FORGE runs SEPARATE on $\overline{O}$ and K_W, resulting in the tuple $\langle A_{\overline{O}}, B_{\overline{O}} \rangle$; $B_{\overline{O}}$ has the form $B_{\overline{O}} = X \| s$. Finally, FORGE outputs the pair $\langle A_{\overline{O}} \| X, s \rangle$. It is easy to see that FORGE perfectly simulates ATTACK so that a valid pair of a message and a signature is produced if and only if ATTACK succeeded.

It remains to show that the message $A_{\overline{O}} \| X$ was not presented to the signature oracle previously. For this, assume the contrary, i.e., that there exists an index i such that $A_{\overline{O}} \| X = A_{O_i} \| X_i$. This can only be the case if $A_O = A_{\overline{O}} = A_{O_i}$ and $X = X_i$, i.e., ENCRYPT$(K_E, C_O \| H(O)) =$ ENCRYPT$(K_E, C_{O_i} \| H(O_i))$. This requires that both O and O_i agree on part A; furthermore, by ENCRYPT being uniquely decipherable, we have $C_O \| H(O) = C_{O_i} \| H(O_i)$. This can only be the case if both O and O_i agree on part C and thus also on part B. We conclude that $O = O_i$, but this contradicts the definition of a successful attack against the media authentication scheme. This completes the proof. $\Box$

5.   ONLINE AUTHENTICATION

The authentication method of the previous section assumes that the full media O is present when the media file is authenticated. However, for many multimedia applications such a solution is unacceptable, e.g., in audio or video streaming. In this section we present an online authentication scheme that operates only on fixed-length chunks of media at a time, but nevertheless allows the full media object to be authenticated. For this purpose, an object O is considered to consist of n chunks of equal length $O_1, \ldots, O_n$; in abuse of notation, we write $O = O_1 \| \cdots \| O_n$.

The online media authentication scheme presented in this paper is targeted towards applications where it must be possible to produce authenticated *excerpts*, i.e., small consecutive portions of the media stream. For example, consider the evidence produced by eavesdropping a telephone, which might be automatically authenticated by a recording device; in a court hearing only a small and relevant part of the overall evidence is presented to the public. In order to prevent tampering, this excerpt should be produced *without* access to the secrets of the eavesdropping system (i.e., the protection key K_P). Nevertheless the integrity and authenticity of the excerpt should be publicly verifiable.

Given an object O, we call an object O' an *excerpt* of O, if O' may be obtained from O by removing some chunks from the beginning and the end of O. Formally, $O' = O'_1 \| \cdots \| O'_m$ is an excerpt of $O = O_1 \| \cdots \| O_n$, written as $O' \preceq O$, if $m \leq n$ and there exists an index $1 \leq i \leq n - m$ so that $O'_1 = O_i, \ldots, O'_m = O_{i+m}$.

Given an original object O, it is possible with the proposed system to generate a signed object $\overline{O}$ such that *each* excerpt of the signed object $\overline{O}' \preceq \overline{O}$ can be checked for its integrity and authenticity without further preprocessing. More precisely, the verification algorithm described below will detect any modifications in an excerpt of $\overline{O}$ and will report the presence of non-consecutive chunks.

Formally, the attacker model we use for online authentication schemes is similar to the one presented in Section 3.2, with the sole exception that the production of excerpts is not considered an attack. Again, an attacker is forced to perform a selective forgery under a chosen message attack. However, the media object obtained at the end of the attack must not be an excerpt of an object submitted to the signing oracle previously.

DEFINITION 4 *Let* $\langle$GENKEY, PROTECT, VERIFY, RECONSTRUCT$\rangle$ *be an online authentication scheme and* QUERY$_{K_P}$ *be an oracle that, on input* O, *computes* $\overline{O} \leftarrow$ PROTECT$'(O, K_P)$. *Furthermore, let* $\langle K_P, K_V, K_R \rangle \in$ [GENKEY$'(1^{n_K})$]. *An attack is a probabilistic algorithm* SATTACK *with oracle access to* QUERY$_{K_P}$ *and success probability* $\varepsilon_{\text{SATTACK}}$ *such that*

$$
\text{SATTACK}(1^n, K_V) = \begin{cases} \langle O, \overline{O} \rangle & \text{such that VERIFY}'(\overline{O}, K_V) = \text{TRUE}, \\ & |O| = n, \ \overline{O} \in [\text{PROTECT}'(O, K_P)] \\ & \text{and } O \not\preceq O^{(i)} \text{ for all } 1 \leq i \leq l, \\ & \text{with probability } \varepsilon_{\text{SATTACK}} \\ \text{FAIL} & \text{with probability } 1 - \varepsilon_{\text{SATTACK}}, \end{cases}
$$

where $O^{(i)}$ denotes the input to the i-th oracle query QUERY$_{K_P}$. *The probability is taken over all coin tosses and all keys* $\langle K_P, K_V, K_R \rangle$.

Again, we say that an online media authentication scheme is secure, if *every* probabilistic polynomial attack has only negligible success probability.

5.1 Construction

In this section, we provide the construction of an online media authentication scheme that operates block by block on the media content. Essentially, we apply the authentication scheme described in the previous section on each chunk O_i, with the exception that the there is some linkage between the chunks, computed by a hash function. Technically, we rely on the concept of hash chains introduced by (Gennaro and Rohatgi, 1997).

Fix any collection of hash functions

$$\mathbb{H} = \left\langle H_h : \{0,1\}^* \to \{0,1\}^{\ell(|h|)} \mid h \in \{0,1\}^* \right\rangle$$

for any super-logarithmically growing function $\ell : \mathbb{N} \mapsto \mathbb{N}$. Denote with k_h an index to $\mathbb{H}$; furthermore, let k be the length of the cryptographic signatures. We assume that both k_h and k are polynomial in the security parameter. For the construction we use an invertible watermarking scheme that is capable of storing $k + \ell(k_h)$ bits. The construction is as follows:

- GENKEY runs GENSIGN to obtain a tuple of keys $\langle K_{SS}, K_{VS} \rangle$; furthermore it computes a key K_E for a symmetric cipher and a random string K_W. GENKEY$'$ outputs the keys $K_P = K_{SS} \| K_W$, $K_V = K_{VS} \| K_W$ and $K_R = K_E \| K_W$.

- PROTECT, on input $O = O_1 \| \cdots \| O_n$, K_P and K_R, performs the following steps:

  > $h_0 \leftarrow$ RANDOM$(\ell(k_h))$
  > **for** $i = 1, \ldots, n$ **do**
  > $\langle A_{O_i}, B_{O_i} \rangle \leftarrow$ SEPARATE(K_W, O_i)
  > compress B_{O_i} to obtain C_{O_i}
  > $X_i \leftarrow$ ENCRYPT$(K_E, C_{O_i} \| H_h(O_i))$
  > $s_i \leftarrow$ SIGN$(K_{SS}, A_{O_i} \| X_i \| h_{i-1})$
  > $h_i \leftarrow H(A_{O_i} \| X_i \| h_{i-1})$
  > let $W_i = X_i \| h_{i-1} \| s_i$
  > $\overline{O}_i \leftarrow$ JOIN$(K_W, \langle A_{O_i}, W_i \rangle)$
  > **if** $\overline{O}_i =$ FAIL, **exit with** FAIL
  > **end for**
  > **output** $\overline{O} = \overline{O}_1 \| \cdots \| \overline{O}_n$

- VERIFY, on input $\overline{O} = \overline{O}_1 \| \cdots \| \overline{O}_n$ and K_V, performs the following steps:

for $i = 1, \ldots, n$ **do**

$\quad \left\langle A_{\overline{O}_i}, B_{\overline{O}_i} \right\rangle \leftarrow \text{SEPARATE}(K_W, \overline{O}_i)$

$\quad B_{\overline{O}_i}$ has the form $X_i \| h_{i-1} \| s_i$

$\quad$ **if** $i > 1$ **and** $h_{i-1} \neq \tilde{h}$ **exit with** FAIL

$\quad$ let $\tilde{h} = H_h(A_{\overline{O}_i} \| X_i \| h_{i-1})$

$\quad b_i \leftarrow \text{SIGVERIFY}(K_{VS}, A_{\overline{O}_i} \| X_i \| h_{i-1}, s_i)$

$\quad$ **if** $b_i = \text{FALSE}$, **exit with** FALSE

end for

exit with TRUE

- RECONSTRUCT applies the reconstruction algorithm of Section 4 on the chunks of $\overline{O}$.

5.2   Security Against Forgeries

In a similar way as in Theorem 3, the security of the above scheme can be established:

THEOREM 5 *If $\mathbb{S}$ is a cryptographic signature scheme secure against existential forgery of messages under a chosen message attack and if $\mathbb{H}$ is a collection of preimage- and collision-resistant hash functions, then the above scheme is a secure online media authentication scheme.*

Proof. Suppose, for the sake of contradiction, that there exists an attack SATTACK against the above scheme, which succeeds with a non-negligible probability. We show that in this case there exists also an attack FORGE against $\mathbb{S}$, which contradicts the assumption.

We construct the signature forging algorithm FORGE (for the public signature verification key K_{VS}) in the following manner. On input K_{VS}, FORGE first chooses random keys K_E and K_W. Finally, FORGE invokes SATTACK. In the rest of the proof, denote with $O^{(i)}$ the input to the i-th query to the oracle QUERY$_{K_P}$, whereas $O_j^{(i)}$ denotes the j-th chunk of $O^{(i)}$; the number of chunks in $O^{(i)}$ is given by n_i.

Whenever SATTACK makes an oracle query QUERY$_{K_P}(O^{(i)})$ in order to obtain a signed stream $\overline{O}^{(i)}$, given $O^{(i)} = O_1^{(i)} \| \ldots \| O_{n_i}^{(i)}$, this query is simulated by the following probabilistic computation that uses a signature oracle SIGNQUERY$_{K_{SS}}$ (essentially, this code is equivalent to that of PROTECT):

$s_{i,0} \leftarrow \text{RANDOM}(\ell(h_k))$

for $j = 1, \ldots, n_i$ **do**

$\quad \left\langle A_{O_j^{(i)}}, B_{O_j^{(i)}} \right\rangle \leftarrow \text{SEPARATE}(K_W, O_j^{(i)})$

$\quad$ compress $B_{O_j^{(i)}}$ to obtain $C_{O_j^{(i)}}$

$\quad X_j^{(i)} \leftarrow \text{ENCRYPT}(K_E, C_{O_j^{(i)}} \| H_h(O_j^{(i)}))$

$$s_j^{(i)} \leftarrow \text{SIGNQUERY}_{K_{ss}}(A_{O_j^{(i)}} \| X_j^{(i)} \| h_{j-1}^{(i)})$$

$$h_j^{(i)} \leftarrow H_h(A_{O_j^{(i)}} \| X_j^{(i)} \| h_{j-1}^{(i)})$$

$$\text{let } W_j^{(i)} = X_j^{(i)} \| h_{j-1}^{(i)} \| s_j^{(i)}$$

$$\overline{O}_j^{(i)} \leftarrow \text{JOIN}\left(K_W, \left\langle A_{O_j^{(i)}}, W_j^{(i)} \right\rangle \right)$$

if $\overline{O}_j^{(i)} = \text{FAIL}$, **exit with** FAIL
end for
output $\overline{O}^{(i)} = \overline{O}_1^{(i)} \| \cdots \| \overline{O}_{n_i}^{(i)}$

Up to here, ATTACK perfectly simulates SATTACK. When the simulation of SATTACK is finished it obtains (with non-negligible probability) a tuple $\langle O, \overline{O} \rangle$, where $\overline{O}$ is a signed media stream with n chunks and $O \npreceq O^{(i)}$ for all $1 \le i \le l$. If SATTACK fails, ATTACK fails as well.

Denote with

$$\mathbf{Q} = \{A_{O_j^{(i)}} \| X_j^{(i)} \| h_{j-1}^{(i)} \mid 1 \le i \le l, 1 \le j \le n_i\}$$

the set of oracle queries. For all $1 \le k \le n$, ATTACK runs SEPARATE on $\overline{O}_k$ and K_W to obtain $A_{\overline{O}_k} = A_{O_k}$ and $B_{\overline{O}_k}$; the latter string has the form $B_{\overline{O}_k} = X_k \| h_{k-1} \| s_k$. Consider two cases:

- Case 1: there exists an index $1 \le k \le n$ such that $A_{O_k} \| X_k \| h_{k-1} \notin \mathbf{Q}$. Then, ATTACK outputs the tuple

$$\langle A_{\overline{O}_k} \| X_k \| h_{k-1}, s_k \rangle$$

  as signature forgery. By assumption, this tuple is a valid forgery.

- Case 2: for all indices $1 \le k \le n$ we have $A_{\overline{O}_k} \| X_k \| h_{k-1} \in \mathbf{Q}$. In this case, ATTACK fails. We argue later that this case can happen only with negligible probability.

ATTACK can distinguish the two cases in polynomial time; furthermore, the success probability of ATTACK equals the success probability of SATTACK, up to a negligible quantity (resulting out of case 2). This contradicts the assumption.

It remains to show that case 2 happens only with negligible probability. Note that, by assumption, O (and thus also $\overline{O}$) contains at least two chunks, as otherwise trivially $O \preceq O^{(i)}$ for some index $1 \le i \le l$. Consider the last chunk $\overline{O}_n$; its decomposition according to SEPARATE is given by $\langle A_{\overline{O}_n}, X_n \| h_{n-1} \| s_n \rangle$. By assumption, there exist indices $1 \le i \le l$ and $1 \le j \le n_i$ such that

$$A_{O_j^{(i)}} \| X_j^{(i)} \| h_{j-1}^{(i)} = A_{O_n} \| X_n \| h_{n-1}.$$

In particular, also $h_{j-1}^{(i)} = h_{n-1}$. Distinguish two cases:

- Case (a): We have $j = 1$. Now, as both $\overline{O}$ and $\overline{O}^{(i)}$ are valid,

$$h_{n-1} = H_h(A_{O_{n-1}} \| X_{n-1} \| h_{n-2}).$$

By assumption, $h_{n-1} = h_0^{(i)}$, showing that $A_{O_{n-1}} \| X_{n-1} \| h_{n-2}$ is a pre-image of the random string $h_0^{(i)}$.

- Case (b): We have $j > 1$. Again, as both $\overline{O}$ and $\overline{O}^{(i)}$ are valid, $h_{n-1} = H_h(A_{O_{n-1}} \| X_{n-1} \| h_{n-2})$ and $h_{j-1}^{(i)} = H_h(A_{O_{j-1}^{(i)}} \| X_{j-1}^{(i)} \| h_{j-2}^{(i)})$. By assumption, $h_{n-1} = h_{j-1}^{(i)}$. If

$$A_{O_{n-1}} \| X_{n-1} \| h_{n-2} \neq A_{O_{j-1}^{(i)}} \| X_{j-1}^{(i)} \| h_{j-2}^{(i)},$$

we have found a collision of H_h. Otherwise, $A_{O_{n-1}} = A_{O_{j-1}^{(i)}}$, $h_{n-2} = h_{j-2}^{(i)}$ and $X_{n-1} = X_{j-1}^{(i)}$. The latter equation implies

$$\underbrace{\textsc{Encrypt}(K_E, C_{O_{n-1}} \| H(O_{n-1}))}_{X_{n-1}} = \underbrace{\textsc{Encrypt}(K_E, C_{O_{j-1}^{(i)}} \| H(O_{j-1}^{(i)}))}_{X_{j-1}^{(i)}}.$$

Since $\textsc{Encrypt}$ is uniquely decipherable, $C_{O_{n-1}} = C_{O_{j-1}^{(i)}}$, implying that $B_{O_{n-1}} = B_{O_{j-1}^{(i)}}$. This shows that now O and $O^{(i)}$ also agree on their second-last chunk. By assumption, O must therefore have at least one more chunk (as otherwise trivially $O \preceq O^{(i)}$). Applying this argument inductively, we either find a collision or have $n > j$. In the latter case, as in case (a), $A_{O_{n-j-1}} \| X_{n-j-1} \| h_{n-j-2}$ is a pre-image of $h_0^{(i)}$.

In summary, if case 2 happens, then we can either find a pre-image of a random string with respect to H_h or a collision of H_h (a formal proof of this claim uses again a reducibility argument). By the assumptions on $\mathbb{H}$, this can happen only with negligible probability. This completes the proof. $\square$

6.   CONCLUSIONS

In this paper, we provided two constructions which solve the data integrity problem for multimedia applications by combining methods from cryptography and watermarking. Technically, we used digital watermarks to encode cryptographic signatures directly in multimedia files. One construction is suitable for images and short media files, whereas the other one is targeted towards streaming applications. Both schemes were shown to be secure under standard cryptographic assumptions and can easily be incorporated into existing software systems or used as front-end programs to applications whose code is not under control of the user.

Acknowledgement. The work described in this paper has been supported in part by the European Commission through the IST Programme under Contract IST-2002-507932 ECRYPT. The information in this document reflects only the author's views, is provided as is and no guarantee or warranty is given that the information is fit for any particular purpose. The user thereof uses the information at its sole risk and liability.

References

Dittmann, J. and Benedens, O. (2003). Invertible authentication for 3d-meshes. In *Proceedings of the SPIE vol. 5020, Security and Watermarking of Multimedia Contents V*, pages 653–664.

Dittmann, J., Steinebach, M., and Ferri, L. (2002). Watermarking protocols for authentication and ownership protection based on timestamps and holograms. In *Proceedings of the SPIE vol. 4675, Security and Watermarking of Multimedia Contents IV*, pages 240–251.

Fridrich, J., Goljan, M., and Du, R. (2001). Invertible authentication. In *Proceedings of the SPIE vol. 3971, Security and Watermarking of Multimedia Contents III*, pages 197–208.

Fridrich, J., Goljan, M., and Du, R. (2002a). Lossless data embedding—new paradigm in digital watermarking. *EURASIP Journal on Applied Signal Processing*, (2):185–196.

Fridrich, J., Goljan, M., and Du, R. (2002b). Lossless data embedding for all image formats. In *Proceedings of the SPIE vol. 4675, Security and Watermarking of Multimedia Contents IV*, pages 572–583.

Friedman, G. L. (1993). The trustworthy digital camera. *IEEE Transactions on Consumer Electronics*, 39(4):905–910.

Gennaro, R. and Rohatgi, P. (1997). How to sign digital streams. In *Advances in Cryptology (CRYPTO'97)*, volume 1294 of *Lecture Notes in Computer Science*, pages 180–197. Springer.

Goldwasser, S., Micali, S., and Rivest, R. (1988). A digital signature scheme secure against adaptive chosen-message attacks. *SIAM Journal on Computing*, 17(2):281–302.

Honsinger, C. W., Jones, P., Rabbani, M., and Stoffel, J. C. (1999). Lossless recovery of an original image containing embedded data. US patent application, Docket No: 77102/E/D.

Katzenbeisser, S. and Petitcolas, F. A. P., editors (2000). *Information Hiding Techniques for Steganography and Digital Watermarking*. Artech House.

Maas, D., Kalker, T., and Willems, F. M. (2002). A code construction for recursive reversible data-hiding. In *Proceedings of the ACM Workshop on Multimedia*, pages 15–18.

Schneider, M. and Chang, S.-F. (1996). A robust content based digital signature for image authentication. In *IEEE International Conference on Image Processing, Proceedings*, Lausanne.

Steinebach, M. and Dittmann, J. (2003a). Watermarking-based digital audio data authentication. *EURASIP Journal on Applied Signal Processing*, (10):1001–1015.

Steinebach, M. and Dittmann, J. (2003b). Watermarking-based digital audio data authentication. *EURASIP Journal on Applied signal processing*, 10:1001–1015.

Xie, L. and Arce, G. R. (1998). A blind wavelet based digital signature for image authentication. In *European Signal Processing Conference, Proceedings*, Rhodes, Greece.

A NEW FRAGILE MESH WATERMARKING ALGORITHM FOR AUTHENTICATION

Hao-Tian Wu and Yiu-Ming Cheung
Department of Computer Science, Hong Kong Baptist University, Hong Kong, China

Abstract: In this paper, we propose a new fragile watermarking algorithm based on the global characteristics of the mesh geometry to authenticate 3D mesh models. In our method, a sequence of data bits is adaptively embedded into the mesh model by properly adjusting the vertex positions, and the bit information can be blindly extracted from the watermarked mesh model using a key. The embedding process is adaptive to the mesh model so that the watermarked mesh is perceptually indistinguishable from the original. We show that the embedded watermark is invariant to affine transformation but sensitive to other operations. Besides, the embedding strength is adjustable and can be controlled to a certain extent that even a trivial tampering with the watermarked mesh would lead the watermark signal to change. Therefore, unauthorized modifications of the mesh models can be detected and estimated.

Key words: 3D models; fragile watermarking; mesh authentication; dither modulation

1.   INTRODUCTION

While the prevalence of network facilitates people's acquirement and distribution of multimedia works, it also challenges the protection of digital works' copyrights. As a potential technique for copyright protection of digital works, digital watermarking for multimedia data (e.g. digital images, video and audio streams) has been proposed and arduously studied in the literature[1,2].

Recently, watermarking 3D objects, such as 3D polygonal meshes and various 3D geometric CAD data, has received much attention in the community and considerable progress has been made. In the literature[3-24], a variety of watermarking algorithms have been proposed to embed watermarks into 3D models, mainly 3D polygonal meshes. For instance, the algorithms[3, 9-16] embed the watermarks by modifying the geometry of the meshes such as vertex coordinates, surface normal distribution, and so forth. They have shown the robustness against some operations to which 3D models are routinely subjected, e.g., affine transformations and mesh simplification. Furthermore, some algorithms[14, 15] modify the topology, i.e. the connectivity, to embed watermarks robust against geometrical operations, but weak to topological modifications. Additionally, some works[16, 24] have used the appearance attributes associated with mesh models to embed the watermarks. In the paper[17], data embedding algorithms for NURBS and other types of parametric curves and faces are also proposed.

To enhance the robustness of 3D watermarking systems, some frequency approaches[18-23] have been recently proposed. In the paper[19, 20], an algorithm that employs multi-resolution wavelet decomposition of polygonal mesh models is presented. Furthermore, the paper[21] proposes an informed watermarking algorithm that constructs a set of scalar basis functions over the mesh vertices, through which the watermark is embedded into the "low frequency" of the polygonal meshes. In the paper[22], Guskov's multi-resolution signal processing method[27] is adopted and a 3D non-uniform relaxation operator is used to construct a Burt-Adelson pyramid[28] for the mesh to embed watermark information into a suitable coarser mesh. Mesh spectral analysis techniques[26] are also employed to transform the original meshes to the frequency domain and watermark information is embedded into the low frequency of the meshes[18, 23].

Nevertheless, few fragile algorithms[4-8] have been proposed to authenticate the integrity of 3D models. Actually, the first fragile watermarking of 3D objects for verification purpose was addressed by Yeo and Yeung[4], as a 3D version of the approach proposed for 2D image watermarking. Because their algorithm heavily relies on the vertex position, a translation operation, which does not affect the integrity of the mesh model, would easily break the authentication mechanism.

In this paper, we shall present a new fragile watermarking algorithm to authenticate 3D mesh models. Compared to the former methods, our approach makes the embedded watermark invariant to integrity-reserved affine transformation, including translation, rotation and uniformly scaling, but sensitive to other geometrical or topological operations. The rest of this paper is organized as follows. In the following section, a new fragile mesh watermarking algorithm is proposed to authenticate the integrity of 3D mesh models. The experiment results using the proposed method are given in Section 3. Finally, we draw a conclusion in Section 4.

2. A NEW FRAGILE MESH WATERMARKING ALGORITHM

We perform the watermarking process on meshes, which are the "lowest common denominator" of surface representations. It is easy to convert other representations of 3D models to meshes. The mesh geometry can be denoted by a tuple (K, V) [25], where K is a simplical complex specifying the connectivity of the mesh simplices (the adjacency of the vertices, edges, and faces), and $V= \{v_1, ..., v_m\}$ is the set of vertex positions defining the shape of the mesh in V^3.

2.1 Extending Dither Modulation to 3D Meshes

We aim to authenticate the integrity of the mesh model, i.e., both the positions and connectivity of vertices need to be verified not having been modified. Our approach extends an implementation of quantization index modulation (QIM) [1] called dither quantization [2] to 3D meshes, and embeds a sequence of data bits by properly adjusting the distances from the faces to the centroid of the mesh geometry.

To extend dither quantization to the mesh model, we choose the quantization step adaptive to the mesh geometry. Suppose $V= \{v_1, ..., v_m\}$ is the set of vertex positions in V^3, the position of the mesh centroid is defined by

$$v_c = \frac{1}{m}\sum_{i=1}^{m} v_i .$$ (1)

The Euclidean distance d_i from a given vertex with the position v_i to the mesh centroid is given by

$$d_i = \sqrt{(v_{ix} - v_{cx})^2 + (v_{iy} - v_{cy})^2 + (v_{iz} - v_{cz})^2} ,$$ (2)

where $\{v_{ix}, v_{iy}, v_{iz}\}$ and $\{v_{cx}, v_{cy}, v_{cz}\}$ are the coordinates of the vertex and the mesh centroid on X-axis, Y-axis and Z-axis, respectively. Using Eq. (2), the furthest vertex with the position v_d to the mesh centroid can be found out and its corresponding distance D is denoted as

$$D = \sqrt{(v_{dx} - v_{cx})^2 + (v_{dy} - v_{cy})^2 + (v_{dz} - v_{cz})^2} .$$ (3)

We refer to the distance D as the largest dimension of the mesh model and the quantization step S is chosen as

$$S = D / N, \tag{4}$$

where N is a specified value. The distance from a given face to the mesh centroid is defined as the Euclidean distance from the centroid of the face to that of the mesh. Furthermore, the centroid position of a given face f_i with u edges can be obtained by

$$v_{ic} = \frac{1}{u} \sum_{j=1}^{u} v_{ij}, \tag{5}$$

where v_{ij}, $j \in \{1, 2, \cdots, u\}$ is the vertex position in the face f_i. The distance d_{fi} from the face f_i to the mesh centroid can be calculated by

$$d_{fi} = \sqrt{(v_{icx} - v_{cx})^2 + (v_{icy} - v_{cy})^2 + (v_{icz} - v_{cz})^2}. \tag{6}$$

Subsequently, we obtain the integer quotient Q_i and the remainder R_i by

$$Q_i = d_{fi} / S, \tag{7}$$

and

$$R_i = d_{fi} \% S. \tag{8}$$

To embed one bit value $w(i)$, we modify the position v_{ic} of the face centroid so that Q_i is an even value for the bit value 0, and an odd value for 1. In order to make $Q_i \% 2 = w(i)$ always hold meanwhile reducing the false-alarm probability, we modulate the distance d_{fi} according to the bit value in the following way:

$$d_{fi}' = \begin{cases} Q_i \times S + S/2 & if \quad Q_i \% S = w(i) \\ Q_i \times S - S/2 & if \quad Q_i \% S = \overline{w(i)} \ \& \ R_i < S/2 \ , \\ Q_i \times S + 3S/2 & if \quad Q_i \% S = \overline{w(i)} \ \& \ R_i \geq S/2 \end{cases} \tag{9}$$

where d_{fi}' is the modulated distance from the face f_i to the mesh centroid. Suppose the face f_i consists of u vertices with the centroid position v_{ic}, the position v_{is} of one selected vertex in f_i will be adjusted using d_{fi}' by

$$v_{is}' = (v_c + (v_{ic} - v_c) \times \frac{d_{fi}'}{d_{fi}}) \times u - \sum_{j=1, j \neq s}^{u} v_{ij}, \qquad (10)$$

where v_{ij} refers to the vertex position in f_i and v_{is}' is the adjusted position of the selected vertex.

The watermark information embedded in our method is inherently invariant to affine transformations that include any transformation preserving collinearity (i.e., all points lying on a line initially still lie on a line after transformation) and ratios of distances (e.g., the midpoint of a line segment remains the midpoint after transformation). So the ratio between the distance from each surface face to the mesh centroid and the quantization step, which is proportional to D, remains the same after the model is translated, rotated or uniformly scaled. Otherwise, if the mesh model is processed by other operations that change the ratios, the formula $Q_i\%2 = w(i)$ will not always hold and the embedded watermark will be changed. Since we need to detect a trivial modification on the mesh model, the integer value Q_i should be sensitive to the distance from the mesh centroid to the surface face. In practice, we assign N a large value to obtain a small quantization step S. Please note that the precision of the arithmetic operations must be regarded; otherwise, it may increase the false-alarm probability.

The face index of the mesh model is used to represent the connectivity of vertices. If there is any change in the mesh topology, such as mesh decimation or mesh resampling, the face index will be modified and the information hidden in the distances to the mesh centroid would be undermined, therefore the unauthorized modification on mesh topology can be detected.

2.2   The Encoding Process

In this subsection, we will elaborate on how to adjust the mesh surface faces and eventually move them to the desired positions. Please note that faces share edges and vertices with their neighbors, adjusting one face's position may also modify its neighbors' positions. To successfully retrieve the embedded information and preserve the mesh geometry, the centroid position and the largest dimension of the mesh model must remain the same during the encoding process.

Since the position of a given face depends on the coordinates of its vertices, we can lock the face position by locking the coordinates of its vertices. To move one face centroid to the desired position so that one bit of the watermark information is embedded, only one vertex position in the face need to be adjusted. To avoid the embedded information is changed by the following encoding operations, the vertices in the moved faces need to be marked. Once one watermark bit value is hidden in the distance from the face to the mesh centroid, all the vertices in the face will be marked and their positions can not be modified any more.

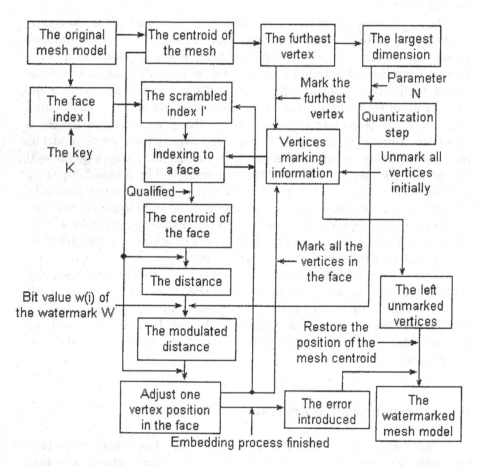

Figure 1. The flow chart of the encoding process

The overall encoding process is as shown in Fig. 1. At first, all vertices in the original mesh are unmarked, the position of mesh centroid is obtained by Eq. (1). Then the furthest vertex to the mesh centroid is found out using Eq.

(2) and its corresponding distance to the mesh centroid is calculated using Eq. (3). Since the value of D should not be changed, the furthest vertex is marked and its coordinate will not be modified. The quantization step S can be chosen by specifying the value of N. Using a key, we scramble the face index I to obtain the scrambled index I', which we will follow in the following encoding process. Before one bit information is embedded in the distance from a face to the mesh centroid, all vertices in the face need to be checked. If there is at least one unmarked vertices in the face f_i, it is qualified to carry one bit value. The distance from the face to the mesh centroid is calculated by Eq. (6) and modulated by Eq. (9) according to the bit value. Noting that the value of D must be maintained in the encoding process, if the modulated distance exceeds it, twice of the quantization step should be subtracted from it so that the embedded bit value is held. Then the coordinate of one unmarked vertex is adjusted using Eq. (10), whereby the face centroid is moved to the desired position. At the end of the embedding operation, all vertices in f_i will be marked. If there is no unmarked vertex in a face, which means the face is not qualified, the checking mechanism will skip to the next face until all watermark information is embedded.

The above embedding process inevitably introduces the distortion of the mesh geometry as some of the vertex coordinates are changed. However, the distortion can be limited to a predefined range, since the elongate or the reduction of the distance from a face to the mesh centroid is no more than twice of the quantization step in the proposed embedding algorithm. The distortion of the mesh geometry also changes the position of the mesh centroid, although adjusting the vertex coordinate may counteract each other. So in the encoding process, not all faces can be used to embed the watermark information. Otherwise, the centroid position of the mesh model will be lost. A small portion of the vertices are needed to restore it after the embedding process. We refer to this process as the centroid restoration process, which modifies the coordinates of the unmarked vertices in the last faces indexed by I' to compensate the error introduced by the embedding process.

The centroid restoration process begins with the calculation of the introduced error E using

$$E = \sum\nolimits_{j=1}^{m} v_j' - \sum\nolimits_{j=1}^{m} v_j, \tag{11}$$

where v_j is the original vertex position while v_j' is the adjusted vertex position after the embedding process. Since the value of D should be maintained in the encoding process, the distance from the mesh centroid to the adjusted vertex should not exceed it. So we adjust the unmarked vertices in the centroid restoration process by the following way:

Firstly, we calculate the admissible adjusting radius r_j of an unmarked vertex with the position v_j by

$$r_j = D - \sqrt{(v_{cx} - v_{jx})^2 + (v_{cy} - v_{jy})^2 + (v_{cz} - v_{jz})^2} \,. \qquad (12)$$

Then we use the value of r_j to weight the adjusting vector of each unmarked vertex to ensure that the vertex will not be moved outside its admissible range. Suppose the sum of the unmarked vertices used in the centroid restoration process is L, the individual adjusting weight e_j can be obtained by

$$e_j = E \cdot \frac{r_j}{\sum_{k=1}^{L} r_k}. \qquad (13)$$

Subsequently, we subtract the individual adjusting weight from vertex position v_j to restore the position of the mesh centroid by

$$r_j' = r_j - e_j, \qquad (14)$$

where v_j' represents the adjusted vertex position after the centroid restoration process and v_j the original one. The encoding process ends as the centroid position of the mesh model is restored.

2.3   The Authentication Process

In the authentication process, only the parameter N, the key K and the original watermark W are needed to authenticate the watermarked mesh geometry. The detailed procedure is shown in Fig. 2.

At first, similar to the encoding process, all the vertices of the original mesh are unmarked, the centroid position v_c' of the suspect mesh geometry is obtained by Eq. (1), which should equal to the centroid position v_c of the original mesh. Then the furthest vertex to the mesh centroid is found using Eq. (2) and its corresponding distance D' is calculated by Eq. (3), which should equal to the largest dimension D of the original mesh model. The quantization step is calculated by $S'=D'/N$ with the provided parameter N. Since the furthest vertex is marked before the embedding process, it should also be marked before the authentication process. Then the face index I of the mesh is scrambled using the key K to produce the scrambled index I'. Before retrieving one bit value from the distance from a given face to the mesh centroid, the vertex marks in the face need to be checked. If there is at least one unmarked vertex in a face f_i', the face will be qualified to extract the embedded bit information and its centroid position v_{ic}' will be calculated using Eq. (6). Then the distance D_{fi} from the face f_i' to the mesh centroid can be calculated by

$$D_{fi} = \sqrt{(v_{icx}'-v_{cx})^2 + (v_{icy}'-v_{cy})^2 + (v_{icz}'-v_{cz})^2}, \quad (15)$$

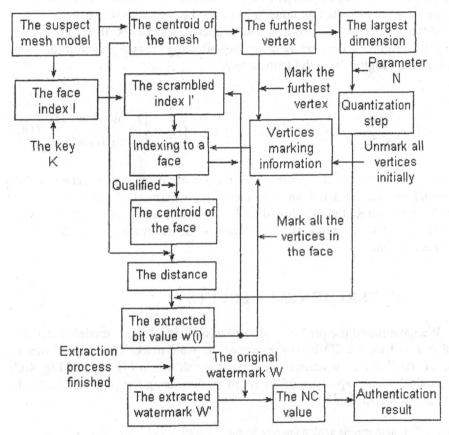

Figure 2. The flow chart of authentication process

and the integer quotient Q_i' can be obtained by

$$Q_i' = D_{fi} / S'. \quad (16)$$

The embedded bit information $w'(i)$ can be extracted by

$$w'(i) = Q_i'\%2.\tag{17}$$

At the end of the extracting operation, all the vertices in f_i' will be marked. If there is no unmarked vertex in a face, no information is extracted and the authentication mechanism will automatically skip to the next face indexed by I'. Since the original watermark W is known, the extraction process will cease once the extracted bit number matches the embedded bit number.

After the extraction process, the extracted watermark W' is compared with the original watermark W using the following cross-correlation function, given their lengths are both identical to K:

$$NC = \frac{1}{K}\sum\nolimits_{i=1}^{K} I(w'(i), w(i)), \quad I(w'(i), w(i)) = \begin{cases} 1 & w'(i) = w(i) \\ 0 & w'(i) \neq w(i) \end{cases},\tag{18}$$

where NC refers to the normalized cross-correlation value between the original and the extracted watermarks. If the watermarked mesh model has not been tampered, the NC should be 1; otherwise, it will be less than 1. We claim the mesh geometry as being tampered if the resulting NC from Eq. (18) is less than 1.

3.   EXPERIMENTAL RESULTS

We have tested the proposed algorithm on several mesh models listed in Table 1 and used a 2D binary image as the watermark. The original mesh model "dog" and its watermarked version are shown in Fig. 4a and Fig. 4b, respectively. The capacities of the mesh models using the proposed method are also shown in Table 1.

Table 1. The mesh models used in the experiments [*]

Models	vertices	faces	capacity(bits)
dog	7158	13176	4219
wolf	7232	13992	4450
raptor	8171	14568	5695
horse	9988	18363	5731
cat	10361	19098	6149
lion	16652	32096	10992

[*] About 1% vertices of each mesh model are used in the restoration process.

To evaluate the imperceptibility of the embedded watermark using the proposed algorithm, we used the Hausdorff distance between the original and the watermarked mesh models to measure the distortion introduced by the encoding process, upon the fact that the mesh topology is not changed during the watermarking process. Fig. 3 describes the amount of the distortion subject to the parameter of N, given that the percent of vertices used for the restoration operation is about 1%. The Hausdorff distance is normalized by the largest dimension D of the mesh geometry. From the experimental results, it can be seen that the introduced distortion on mesh model decreases as the parameter N is increased.

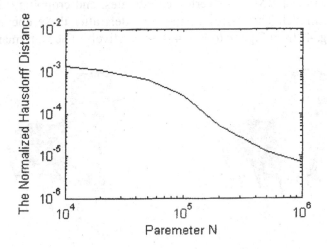

Figure 3. The normalized Hausdorff Distance subject to the parameter N

In our proposed approach, the global characteristics such as the centroid position and the largest dimension of the mesh model are used. If these global characteristics are slightly altered, the watermark information will be dramatically changed and the modifications on the watermarked mesh model can not be located. However, it is easy to locate the tampering if it has little impact on the global characteristics. Since we used a 2D binary image as the

Table 2. The NC values between the original and extracted watermarks [*]

Models	Affine transformation	changing one vertex position	reducing one face	adding 0.0001% noise	cropping 0.1% faces	extracting without the key
dog	1.0000	0.5375	0.0215	0.6721	0.0133	0.0200
wolf	1.0000	0.9276	0.0685	0.5934	0.0083	0.0029
raptor	1.0000	0.9996	0.0063	0.6972	0.0024	0.0225
horse	1.0000	0.8249	0.0737	0.4661	0.0039	0.0102
cat	1.0000	0.9993	0.0308	0.7072	0.0195	0.0103
lion	1.0000	0.9996	0.0905	0.6363	0.0059	0.0088

[*] About 1% vertices of each mesh model are used in the restoration process and N=1,000,000.

watermark, the impact of trivial modifications can be visualized in the extracted watermark image while severe modifications make it meaningless. With the extracted watermark, we can detect the unauthorized modifications and estimate the strength of tampering, if any.

In the experiments, the watermarked mesh models went through affine transformations (including translation, rotation and uniformly scaling), modifying one vertex coordinate with the vector $\{D/500, D/500, D/500\}$, reducing one face from the mesh, adding noise signal that is uniformly distributed within $[-S,S]$ to all vertex coordinates, and cropping 0.1% faces of the mesh model. The processed mesh models after these operations are shown in Fig. 4 (from Fig. 4c to Fig. 4g), respectively. The watermarks are

Figure 4a. The original mesh model "dog"   *Figure 4b.* The watermarked mesh model

Figure 4c. The watermarked mesh model after modifying one vertex position

Figure 4d. The watermarked mesh model after reducing one face

Figure 4e. The watermarked mesh model after adding noise

Figure 4f. The watermarked mesh model after cropping 0.1% faces

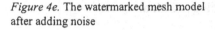

Figure 4g. The watermarked mesh model after affine transformations

extracted from the processed mesh models with and without the key and the *NC* values between the original and the extracted watermarks are calculated using Eq. (11). The results are listed in the Table 2.

4. CONCLUSION

In this paper, we have proposed a new fragile mesh watermarking method to authenticate the integrity of 3D mesh model. The watermarking process is conducted in spatial domain and applies to all the mesh models without any restriction. The experimental results have demonstrated that the proposed method is able to imperceptibly and adaptively embed a considerable amount of information into the mesh model, and the embedded watermark can be blindly extracted from the watermarked mesh model to authenticate the watermarked mesh model. In our method, the distortion introduced by the encoding process is quite small and can be controlled within a predefined range. Compared to the previous works, the embedded watermark using our method is invariant to integrity-reserved affine transformation, but sensitive to other processing that alters the mesh model. Therefore, unauthorized modifications of the mesh models can be successfully detected and estimated.

ACKNOWLEDGEMENT

For the use of the 3D models, we would like to thank the web sources of Department of Computer Science, the University of North Carolina at Chapel Hill, USA. Also, many thanks go to Dr. Zheming Lu for helpful discussions at Harbin Institute of Technology, China.

REFERENCE

1. B. Chen and G. W. Wornell, Quantization index modulation: A class of provably good methods for digital watermarking and information embedding, *IEEE Trans. Inform. Theory*, **47**, 1423-1443 (2001).
2. B. Chen and G. W. Wornell, Digital watermarking and information embedding using dither modulation, *IEEE Second Workshop on Multimedia Signal Processing*, 273-278 (1998).
3. Z. Q. Yu, H. H. S. Ip and L. F. Kwork, A robust watermarking scheme for 3D triangle mesh models, *Pattern Recognition*, **36**(12), 2603-2614 (2003).
4. M. M. Yeung and B. L. Yeo, Fragile watermarking of three dimensional objects, *Proc. 1998 Int'l Conf. Image Processing, ICIP98*, **2**, 442-446 (IEEE Computer Society, 1998).
5. B. L. Yeo and M. M. Yeung, Watermarking 3D objects for verification, *IEEE Comput. Graph. Applicat*, 36-45 (1999).
6. F. Cayre and B. Macq, Data hiding on 3D triangle meshes, *IEEE Trans. Signal. Processing*, **51**(4), 939-949 (2003).
7. HsuehYi Lin, Hongyuan Mark Liao, ChunShien Lu and JaChen Lin, Fragile Watermarking for Authenticating 3D Polygonal Meshes, *Proc. 16th IPPR Conf on CVGIP*, 298-304 (2003).
8. C. Fornaro and A. Sanna, Public Key Watermarking for Authentication of CSG Models, *Computer-Aided Design*, **32**, 727-735 (2000).
9. O. Benedens, Watermarking of 3D polygon based models with robustness against mesh simplification, *Proc. SPIE: Security Watermarking Multimedia Contents*, 329-340 (1999).
10. O. Benedens, Geometry based watermarking of 3D models, *IEEE Comput. Graph., Special Issue on Image Security*, 46-55, Jan./Feb. 1999.
11. O. Benedens, Two high capacity methods for embedding public watermarks into 3D polygonal models, *Proc. Multimedia Security Workshop ACM Multimedia*, 95-99 (1999).
12. O. Benedens and C. Busch, Toward blind detection of robust watermarks in polygonal models, *Proc. EUROGRAPHICS Comput. Graph. Forum*, **19**(C), 199-208 (2000).
13. M. G. Wagner, Robust watermarking of polygonal meshes, *Proc. Geometric Modeling Processing*, Hong Kong, 201-208 (2001).

14. R. Ohbuchi, H. Masuda and M. Aono, Watermarking Three Dimensional Polygonal Models, *Proc. ACM Multimedia*, Seattle, 261-272 (1997).
15. R. Ohbuchi, H. Masuda and M. Aono, Watermarking Three Dimensional Polygonal Models Through Geometric and Topological Modifications, *IEEE J. Select. Areas Commun.*, **16**, 551-560 (1998).
16. R. Ohbuchi, H. Masuda and M. Aono, Geometrical and Non-geometrical Targets for Data Embedding in Three Dimensional Polygonal Models, *Computer Communications*, Elsevier, **21**, 1344-1354 (1998).
17. R. Ohbuchi, H. Masuda and M. Aono, A shape preserving data embedding algorithm for NURBS curves and surfaces, *Proc. Comput. Graph. Int.*, June, 1999.
18. R. Ohbuchi, S. Takahashi, T. Miyasawa and A. Mukaiyama, Watermarking 3D polygonal meshes in the mesh spectral domain, *Proc.Graphics Interface*, Ottawa, 9-17 (2001).
19. H. Date, S. Kanai and T. Kishinami, Digital watermarking for 3D polygonal model based on wavelet transform, *Proc. ASME Des. Eng. Techn. Conf.*, Sept 1999.
20. S. Kanai, H. Date and T. Kishinami, Digital watermarking for 3D polygons using multi-resolution wavelet decomposition, *Proc. Sixth Int.Workshop Geometric Modeling: Fundamentals Applicat.*, Sept 1998.
21. E. Praun, H. Hoppe and A. Finkelstein, Robust mesh watermarking, *Proc. SIGGRAPH*, 69-76 (1999).
22. Kangkang Yin, Zhigeng Pan, Jiaoying Shi and David Zhang, Robust mesh watermarking based on multi-resolution processing, *Computers & Graphics*, **25**, 409-420 (2001).
23. F. Cayre, P. RondaoAlface, F. Schmitt, B. Macq and H. Maitre, Application of Spectral Decomposition to Compression and Watermarking of 3D Triangle Mesh Geometry, *Signal Processing: Image Communications*, **18**(4), 309-319 (2003).
24. Liangjun Zhang, Ruofeng Tong, Feiqi Su and Jinxiang Dong, A Mesh Watermarking Approach for Appearance Attributes, *Pacific Conference on Computer Graphics and Applications*, Beijing, 450-451 (2002).
25. H. Hoppe, T. DeRose, T. Duchamp, J. McDonald and W. Stuetzle, Mesh optimization, *Computer Graphics (SIGGRAPH '93 Proceedings)*, 19-26 (1993).
26. Z. Karni and C. Gotsman, Spectral compression of mesh geometry, *Proc. SIGGRAPH*, 279-286 (2000).
27. I. Guskov, W. Sweldens and P. Schroeder, Multi-resolution signal processing for meshes, *Proc. SIGGRAPH*, 325-334 (1999).
28. P. J. Burt and E. H. Adelson, Laplacian pyramid as a compact image code, *IEEE Transactions on Communications*, 532-540(1983).

NEW PARADIGM IN GRAPH-BASED VISUAL SECRET SHARING SCHEME BY ACCEPTING REVERSAL IN BLACK-WHITE IMAGES

Yuji SUGA

Canon Inc., PF Technology Development Center, 30-2, Shimomaruko 3-Chome, Ohta-ku, Tokyo 146-8501, Japan

Abstract: The visual secret sharing scheme (for short the VSS scheme) with access structure based on graph has been proposed as one of the (2,n)-threshold visual secret sharing schemes. Ateniese et al.[1] showed a decomposition method into star graphs from a given graph which edges are specified by qualified sets, that is, two different participants (two vertices in the graph) have a common edge if and only if they can decrypt the secret image by stacking each share images. In this paper, we expand the definition of black-white visual secret sharing scheme and propose new decomposition methods by splitting complete n-partite graphs. These methods improve contrast of the decoded secret image. Moreover, we obtain several optimal examples and evaluate on graph-based VSS schemes.

Key words: visual secret sharing scheme; n-partite graph; complete n-partite graph

1.   INTRODUCTION

The visual secret sharing scheme (abbreviated as VSS scheme) proposed by Naor and Shamir[11] is a method to distribute secret image S into n shadow images w_i ($1 <= i <= n$) called shares. Shares are printed to materials with permeability that can be stacked physically like OHP sheets and each participant receives one share in secret. Any qualified participants can reconstruct the secret image visually by stacking shares, but forbidden participants cannot obtain any information about secret image. In the (k, n)-

threshold VSS scheme, any k out of n participants can decrypt the secret image, but any k-1 or fewer participants cannot decode.

The VSS schemes with various access structures (which differs from threshold schemes) for reconstruction have proposed. Ateniese et al.[11] proposed graph-based access structure, that is, vertices on a given graph are identified as participants with the following property. Two vertices have a common edge if and only if participants can decrypt the secret image by stacking shares. Graph-based access structure scheme with a complete graph, which any different two vertices have a common edge, is as same as (2, n) - threshold VSS scheme, so this implies that graph-based VSS scheme can be considered as an extension of (2, n) -threshold VSS scheme.

Ateniese et al. proposed "star graph decomposition method", this method means that given graph is divided into a collection of star graphs. This method has advantage that one can construct graph-based VSS scheme systematically for arbitrary given graph, but also has disadvantage of inefficient pixel expansion (a measure for contrast of reconstructed image).

One of approach is improving graph decomposition of their method, but we introduce new paradigm of VSS schemes as follows; we can accept "reversal image" which every pixel color is opposite in decoded image. In ordinary case, we use black-color in the object and white-color in the background, so we decrypt by stacking shares and understand secret image by recognizing more black as object in pre-image. In our new paradigm, we can accept reversal image, that is, we can understand the secret image by recognizing white areas as the object in pre-image. To introduce the weaker definition than ordinary definition of VSS schemes leads to get efficient constructions.

The rest of the paper is organized as follows. Section 2 mentions previous construction and defines our new definition of VSS schemes. Section 3 gives various efficient constructions in our new paradigm. Section 4 gives evaluation of our schemes and implies efficiency. Section 5 concludes this paper.

2. THE VSS SCHEME FOR GRAPH ACCESS STRUCTURE

2.1 Preliminaries

In this section, now we recall some terminologies on graph theory. A graph is a pair $G = (V,E)$ consisting of a set V, referred to as the vertex set of G and a set E of 2-subsets of V, referred to as the edge set of G. Assume that our graph does not contain loops, undirected edges and multiple edges. For given G, we define the adjacency matrix (a_{ij}) (whose rows and columns are indexed by the elements of V) where the (i,j)-th entry $a_{ij} = 1$ if and only if (x_i, x_j) is a vertex in E.

We say that $G' = (V', E')$ is a subgraph of $G = (V,E)$ if V' is a subset of V and E' is a subset of E. Furthermore, subgraph $G' = (V', E')$ is called induced subgraph of G if it satisfies that E' consists of E that have both vertices in V'. Let Ind(G) be the collection of induced subgraphs of G, we define that Ind(G) include G, but Ind(G) does not contain an empty graph (a graph with no edge). A complete graph is a graph in which each pair of distinct vertices is joined by an edge, and the complete graph on n vertices is denoted by K_n. For any graphs G, a complete subgraph of G is called a clique of G. The number of vertices in a largest clique of G is denoted by c(G).

A graph G is called n-partite if the vertex set V can be partitioned into k nonempty sets $V_1, V_2, ... , V_n$ such that every edge of G joins vertices from different subsets. The n-partite graph G is called complete n-partite if, for each i, j (i does not equal j), every vertex of V_i is adjacent to every vertex of V_j, and the complete n-partite graph is denoted by $K_{a1,a2, ... ,an}$ where $|V_i|=a_i$ for each i. Especially, G is called complete bipartite graph if k=2. The n-partite graph for $G = (V(G),E(G))$ (denoted by $K_{a1,a2, ... ,an}(G)$) is a subgraph of $K_{a1,a2, ... ,an}$ if every vertex of V_i is adjacent to every vertex of V_j such that (v_i, v_j) in E(G) where a vertex set $V(G) = \{v_1, v_2, ... , v_n\}$ is correspond with a partitioned subsets $\{V_1, V_2, ... , V_n\}$.

2.2 The model

We assume that a secret image is a black-white image which is encoded to n images w_i ($1 <= i <= n$). Each pixel (in an original image) expands to m subpixels (in distributed images) and parameter m is called pixel expansion. In expression of images, we denote white pixel and black pixel by 0 and 1 respectively, this notation is used for both of a secret image and shares. By stacking two shares, we can decode a secret image visually because of the difference of the number of black pixel in the OR-operated subpixels.

We introduce basis matrices containing two matrices denoted by S0 and S1 written how to share shadow images.

2.2.1   Basis matrices

When we generate shares, we use basis matrices which row vectors are indexed by a set of shares $W = \{w_i \mid 1 <= i <= n\}$. These matrices are expressed in n by m binary matrices where m is pixel expansion. We denote graph-based VSS scheme with graph G and pixel expansion m by GVSSS-(G, m).

For any vector v, w(v) is the Hamming weight of v, that is, the number of "1" in v. For any binary matrix B which the i-th row vector of B denotes b_i, we define symmetric matrix R(B) which the (i,j)-th element equals $w(b_i) + w(b_j)$. For any matrix A, we define normalized matrix norm(A) such that $(norm(A))_{xy} := 0$ if $A_{xy} = 0$, $(norm(A))_{xy} := 1$ if A_{xy} does not equal to 0.

Definition 1 [Basis matrices of GVSSS-(G, m)]

|V(G)| by m basis matrices S0, S1 with respect to GVSSS-(G, m) satisfies that norm(R(S1) - R(S0)) = Adj(G) where Adj(G) is an adjacency matrix of G.

Example 2

$$Adj(G) = \begin{bmatrix} 0 & 1 & 1 & 1 & 0 & 0 \\ 1 & 0 & 1 & 0 & 1 & 0 \\ 1 & 1 & 0 & 0 & 0 & 1 \\ 1 & 0 & 0 & 0 & 1 & 1 \\ 0 & 1 & 0 & 1 & 0 & 1 \\ 0 & 0 & 1 & 1 & 1 & 0 \end{bmatrix}, \; S_0 = \begin{bmatrix} 1 & 1 & 0 \\ 1 & 1 & 0 \\ 1 & 1 & 0 \\ 1 & 0 & 0 \\ 1 & 0 & 0 \\ 1 & 0 & 0 \end{bmatrix}, \; S_1 = \begin{bmatrix} 1 & 1 & 0 \\ 1 & 0 & 1 \\ 0 & 1 & 1 \\ 0 & 0 & 1 \\ 0 & 1 & 0 \\ 1 & 0 & 0 \end{bmatrix}.$$

$$R(S_0) = \begin{bmatrix} 2 & 2 & 2 & 2 & 2 & 2 \\ 2 & 2 & 2 & 2 & 2 & 2 \\ 2 & 2 & 2 & 2 & 2 & 2 \\ 2 & 2 & 2 & 1 & 1 & 1 \\ 2 & 2 & 2 & 1 & 1 & 1 \\ 2 & 2 & 2 & 1 & 1 & 1 \end{bmatrix}, \; R(S_1) = \begin{bmatrix} 2 & 3 & 3 & 3 & 2 & 2 \\ 3 & 2 & 3 & 2 & 3 & 2 \\ 3 & 3 & 2 & 2 & 2 & 3 \\ 3 & 2 & 2 & 1 & 2 & 2 \\ 2 & 3 & 2 & 2 & 1 & 2 \\ 2 & 2 & 3 & 2 & 2 & 1 \end{bmatrix}.$$

Restriction of Definition 1 is "weaker" than original difinition[1] because of the following viewpoints; 1) We do NOT consider the case of three or more shares and 2) We can allow "reversal image" (which every pixel color is opposite) in decoded image. Due to above two restrictions, we can reduce an increase of pixel expansion and obtain higher contrast in the reconstructed image. Please keep in mind that theorem 7 mentioned latter does not match previous result (Th 5.2 in Ateniese paper[1]) because of the difference of definitions.

2.2.2 The minimum pixel expansion

For given graph G, let $m^*(G)$ be the minimum of m if GVSSS-(G, m) exists. We call GVSSS-(G, m) is optimal if $m=m^*(G)$. The following results are known on $m^*(G)$.

Theorem 3 [Th 7.3 in Ateniese paper[1]] $m^*(K_n) = \min \{ m \mid n <= {}_mC_{[m/2]}\}$.

Theorem 4 [Th 7.4 in Ateniese paper[1]] $m^*(G) >= m^*(K_{c(G)})$.

Theorem 4 gives a lower bound of $m^*(G)$, however problem to calculate the greatest clique for given graph G is known as a NP problem.

2.2.3 Independent graph-based VSS schemes

In some case of choosing basis matrices, we see the next boring example, which has same row vectors in basis matrices. This means that different participants have same shares.

Example 5

$$Adj(G) = \begin{bmatrix} 0 & 1 & 1 \\ 1 & 0 & 0 \\ 1 & 0 & 0 \end{bmatrix}, \; S_0 = \begin{bmatrix} 1 & 0 \\ 1 & 0 \\ 1 & 0 \end{bmatrix}, \; S_1 = \begin{bmatrix} 1 & 0 \\ 0 & 1 \\ 0 & 1 \end{bmatrix}.$$

Now we define new concept because we would like to exclude the above case.

Definition 6 [Independent]
The GVSSS-(G, m) is called independent if all shadow images are different each other, strictly speaking, i, j does not exist such that w_i-th vector in S0 equal w_j-th vector in S0 and w_i-th vector in S1 equal w_j-th vector in S1.

Note that example 2 is independent, but example 5 is not independent. We discuss independent graph-based VSS schemes and some results are obtained.

Theorem 7

If there exists an independent GVSSS-(G, 2), G is included in Ind(K_{22}).
[Proof]

(=>) In case of m=2, we enumerates all possible row vectors of S0,
S1 as follows:

$$
\bar{S}_0 = \begin{matrix} w_1) \\ w_2) \\ w_3) \\ w_4) \\ w_5) \\ w_6) \end{matrix} \begin{bmatrix} 0 & 0 \\ 1 & 0 \\ 1 & 0 \\ 0 & 1 \\ 0 & 1 \\ 1 & 1 \end{bmatrix}, \quad \bar{S}_1 = \begin{matrix} w_1) \\ w_2) \\ w_3) \\ w_4) \\ w_5) \\ w_6) \end{matrix} \begin{bmatrix} 0 & 0 \\ 1 & 0 \\ 0 & 1 \\ 1 & 0 \\ 0 & 1 \\ 1 & 1 \end{bmatrix}.
$$

So we can calculate R(~S1) - R(~S0) as follows;

$$
R(\bar{S}_1) - R(\bar{S}_0) = \begin{bmatrix} 0 & 0 & 0 & 0 & 0 & 0 \\ 0 & 0 & 1 & -1 & 0 & 0 \\ 0 & 1 & 0 & 0 & -1 & 0 \\ 0 & -1 & 0 & 0 & 1 & 0 \\ 0 & 0 & -1 & 1 & 0 & 0 \\ 0 & 0 & 0 & 0 & 0 & 0 \end{bmatrix}.
$$

We consider norm(R(~S1) - R(~S0)), and omit rows and columns
which each elements are all 0 (w_1) and w_6)). So we can obtain an adjacent
matrix of $K_{2,2}$.

(<=) It is enough that we show an example of GVSSS-($K_{2,2}$, 2) (See
example 8). Q.E.D.

Example 8

$$
Adj(G) = \begin{bmatrix} 0 & 1 & 1 & 0 \\ 1 & 0 & 0 & 1 \\ 1 & 0 & 0 & 1 \\ 0 & 1 & 1 & 0 \end{bmatrix}, \quad S_0 = \begin{bmatrix} 1 & 0 \\ 1 & 0 \\ 0 & 1 \\ 0 & 1 \end{bmatrix}, \quad S_1 = \begin{bmatrix} 1 & 0 \\ 0 & 1 \\ 1 & 0 \\ 0 & 1 \end{bmatrix}.
$$

Corollary 9

If there exists an independent GVSSS-(G, 3) such that G is not included
in Ind(K_{22}), GVSSS-(G, 3) is optimal.

We can see that example 8 is optimal. Moreover, we obtain the following
theorem and corollary straightforwardly.

Theorem 10

If there exists a GVSSS-(G, 2), G is included in Ind($K_{a1,a2, \ldots ,a4}(K_{2,2})$).

Corollary 11

If there exists a GVSSS-(G, 3) such that G is not included in Ind($K_{a1,a2, \ldots ,a4}(K_{2,2})$), GVSSS-(G, 3) is optimal.

3.   CONSTRUCTIONS BY GRAPH DECOMPOSITION

In this section, we treat some graph decomposition methods as constructions of basis matrices. First, we recall star graph decomposition method as a previous construction with no efficiency. Next, we propose new methods and show that our methods have usefulness at a point of view "whether to be optimal or not".

3.1   The star graph decomposition method

The star graph decomposition method is proposed in Ateniese paper[1], this method means that we divide given graph G into star graph $K_{1,a}$ which edges are joined to only one vertex (called the center vertex). We can construct GVSSS-($K_{1,a}$, 2) with basis matrices S0, S1 such that each row vector in S0 have {1,0}, a row vector with related to the center vertex in S1 has {1,0} and the others have {0,1}. Finally we concatenate basis (sub)matrices of all star graphs side by side, so we can construct GVSSS-(G, 2 b(G)) for any given graphs where b(G) is the number of decomposed star graphs.

Figure 1 [Difference of decomposition methods]

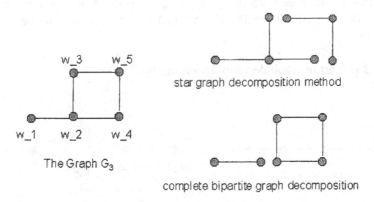

star graph decomposition method

The Graph G_3

complete bipartite graph decomposition

Example12 [The star graph decomposition method]
The star graph decomposition for graph G_3 in Blundo's paper[2] at figure 1(left) causes GVSSS-(G_3, 4) because of decomposition expressed at figure 1 (upper right) which basis matrices are as follows:

$$
S_0 = \begin{array}{c} w_1) \\ w_2) \\ w_3) \\ w_4) \\ w_5) \end{array}
\begin{bmatrix}
1 & 0 & 0 & 0 \\
1 & 0 & 0 & 0 \\
1 & 0 & 1 & 0 \\
1 & 0 & 1 & 0 \\
0 & 0 & 1 & 0
\end{bmatrix}, \;
S_1 = \begin{bmatrix}
0 & 1 & 0 & 0 \\
1 & 0 & 0 & 0 \\
0 & 1 & 0 & 1 \\
0 & 1 & 0 & 1 \\
0 & 0 & 1 & 0
\end{bmatrix}.
$$

3.2   The complete n-partite graph decomposition method

We propose new extended method that we decompose complete n-partite graphs instead of star graphs, that is, we treat GVSSS-($K_{a1,a2, ... ,an}$, $m^*(K_n)$). Actually, we use basis matrices derived from GVSSS-(K_n, $m^*(K_n)$) which row vectors are iterate a_i times for each i.

Example 13 [The complete n-partite graph decomposition method]
The decomposition for graph G_3 causes GVSSS-(G_3, 4) because of decomposition expressed at figure 1 (lower right) which basis matrices are as follows:

$$S_0 = \begin{matrix} w_1) \\ w_2) \\ w_3) \\ w_4) \\ w_5) \end{matrix} \begin{bmatrix} 1 & 0 & 0 & 0 \\ 1 & 0 & 1 & 0 \\ 0 & 0 & 1 & 0 \\ 0 & 0 & 1 & 0 \\ 0 & 0 & 1 & 0 \end{bmatrix}, \; S_1 = \begin{bmatrix} 1 & 0 & 0 & 0 \\ 0 & 1 & 1 & 0 \\ 0 & 0 & 0 & 1 \\ 0 & 0 & 0 & 1 \\ 0 & 0 & 1 & 0 \end{bmatrix}.$$

3.3   The n-partite graph decomposition method

We propose new extended method with decomposed graph which we consider the n-partite graph $K_{a1,a2, ... ,an}(G)$ (for given graph) instead of star graph, that is, we treat GVSSS-($K_{a1,a2, ... ,an}$, $m^*(G)$). Actually, we use basis matrices derived from GVSSS-(K_n, $m^*(G)$) which rows are iterate a_i times for each i.

Figure 2 [n-partite graph decomposition method]

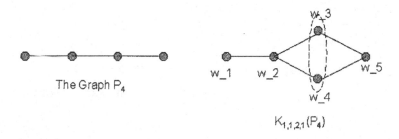

The Graph P_4

$K_{1,1,2,1}(P_4)$

Example 14 [The n-partite graph decomposition method]
Note that P_n is a path with n vertices. There is a GVSSS-(P_4, 3) which basis matrices are as follows:

$$S_0 = \begin{bmatrix} 1 & 0 & 0 \\ 1 & 1 & 0 \\ 1 & 1 & 0 \\ 0 & 1 & 0 \end{bmatrix}, \ S_1 = \begin{bmatrix} 1 & 0 & 0 \\ 0 & 1 & 1 \\ 1 & 1 & 0 \\ 0 & 0 & 1 \end{bmatrix}.$$

So, we can extend basis matrices by using our proposed method mentioned above, w_3 and w_4 have same share image each other.

$$S_0 = \begin{array}{c} w_1) \\ w_2) \\ w_3) \\ w_4) \\ w_5) \end{array} \left[\begin{array}{c|cc} 1 & 0 & 0 \\ 1 & 1 & 0 \\ 1 & 1 & 0 \\ 1 & 1 & 0 \\ 0 & 1 & 0 \end{array} \right], \ S_1 = \left[\begin{array}{c|cc} 1 & 0 & 0 \\ 0 & 1 & 1 \\ 1 & 1 & 0 \\ 1 & 1 & 0 \\ 0 & 0 & 1 \end{array} \right].$$

3.4   The edge-deletion method

This method means that we represent edge set (of given graph) as edge sets of $K_{a1,a2, \dots ,an}$ and $K_{a1,a2, \dots ,an}(G)$ with "difference for set", then we realize by exchange S0, S1 each other in "difference".

Figure 3 [The Graph $K_{2,3}$ - $K_{1,1}$]

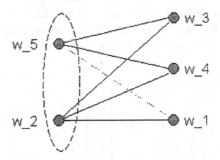

Example 15 [The edge-deletion method]

In figure 2, we can describe that $E(G_3)=E(K_{2,3})$ - (w_1, w_5) =$E(K_{2,3})$ - $E(K_{1,1})$. When we obtain basis matrices from concatenation of GVSSS-($K_{2,3}$, 2) and GVSSS-($K_{1,1}$, 2), we do the following process beforehand. The above process is the deletion of an edge (w_1, w_5), that is, exchange of S0, S1 in GVSSS-($K_{1,1}$, 2).

$$S_0 = \begin{array}{c} w_1) \\ w_2) \\ w_3) \\ w_4) \\ w_5) \end{array} \left[\begin{array}{c|cc} 1 & 1 & 0 \\ 1 & 0 & 0 \\ 1 & 0 & 0 \\ 1 & 0 & 0 \\ 1 & 0 & 1 \end{array} \right], \ S_1 = \left[\begin{array}{c|cc} 1 & 0 & 1 \\ 0 & 1 & 0 \\ 1 & 0 & 0 \\ 1 & 0 & 0 \\ 0 & 1 & 1 \end{array} \right].$$

Remark 16

Note that example 14,15 are optimal from corollary 11, because G_3 is not included in $\text{Ind}(K_{a1,a2,\ldots,a4}(K_{2,2}))$ for any a_i's. Furthermore we can see that $\text{Ind}(K_{2,2}) = \{P_2, P_3, C_4 (=K_{2,2})\}$ where graph C_n is a cycle with n vertices.

4.   EVALUATION AND VARIATION

We already see that some optimal examples exist, so it is natural to have been renewal of interest in clarification of optimal graph-based VSS schemes. Several studies have been made on classification of optimal case in ordinary (non visual) secret sharing schemes, Blundo et al.[2] restricted the number of participants and classified at small order. On the other hand, we choose the pixel-expansion-fix approach because this approach is more suitable than participant-fix approach.

4.1   Classification in the case of m* is at most 3

Theorem 7 means that classification of optimal graph-based VSS scheme GVSSS-(G, 2) have already finished, now we consider the case of $m^* = 3$ as same as theorem 7. So, we obtain C_6, P_5 as optimal cases and the basis matrices of GVSSS-$(C_6, 3)$ are the following example.

Example 17 [GVSSS-$(C_6, 3)$]

$$S_0 = \begin{bmatrix} 1 & 0 & 0 \\ 1 & 0 & 0 \\ 1 & 1 & 0 \\ 1 & 1 & 0 \\ 0 & 1 & 0 \\ 0 & 1 & 0 \end{bmatrix}, \; S_1 = \begin{bmatrix} 0 & 1 & 0 \\ 1 & 0 & 0 \\ 0 & 1 & 1 \\ 1 & 1 & 0 \\ 0 & 0 & 1 \\ 0 & 1 & 0 \end{bmatrix}.$$

Note that P_5 is subgraph of C_6, so the basis matrices of GVSSS-$(P_5, 3)$ is derived from the basis matrices of GVSSS-$(C_6, 3)$ by using only some 5 rows.

4.2   Evaluation on Graph-type VSS scheme

An extended scheme called by the Graph-type VSS scheme has been introduced[14], this scheme focuses on the distance of two vertices instead of the existence of edge. In the Hamming graph $L_2(3)$ with 9 vertices, we can reduce the parameter of pixel expansion from 21 to 6. Note that if we use star graph decomposition method, we need 30 as pixel expansion.

4.3   Reuse for the color/gray-scale images

We show the ability to reuse basis matrices of optimal graph-based VSS scheme for the black-white image and extend into the color/gray-scale images similarly. Assume that a secret image has different t colors ($\{c_1, c_2, \ldots, c_t\}$) and one of them (denoted by 1) is the strongest color which can cover other colors by stacking shares. For example, color sets are R(red), G(green), B(blue) and 1(black).

Our proposal method is as follows: 1) we change 0 and 1 in basis matrices of the black-white image into 1 and c_i (some color) for all non-black colors, then 2) we concatenate new basis matrices similarly mentioned above. In this case, the pixel expansion equals mt when we apply GVSSS-(G ,m).

5.   CONCLUSION

We proposed optimal constructions of graph-based VSS scheme over the new definition that brings higher constant of reconstructed secret image. In new definition, we obtain optimal GVSSS-(G ,m) such that m = 2,3 and suggested an extended construction for color images by re-using basis matrices of the black-white images.

REFERENCES

1. G.Ateniese, C.Blundo, A.D. Santis, D.R. Stinson, Visual Cryptography for General Access Structures, Information and Computation 129, 86-106, 1996.
2. C.Blundo, A.D.Santis, D.R.Stinson, U.Vaccaro, Graph decompositions and secret sharing schemes, EUROCRYPT'92, pp.1-24, 1992.
3. E. Bannai and T. Ito, Algebraic Combinatorics I :Association schemes, Benjamin / Cummings, Menlo Park, California, 1984.
4. C.Blundo, A.D.Santis, D.R.Stinson, On the Contrast in Visual Cryptography Schemes, Journal of Cryptology 12, 261-289, 1999.
5. J.Clark, D.A.Holton, A First Look at Graph Theory, World Scientific Publishing, 1991.
6. New results on visual cryptography, CRYPTO'96, pp.401-415, 1996
7. M.Iwamoto, H.Ymamamoto, A visual secret sharing scheme for plural images (in Japanese), SITA2001, pp.565-568, 2001.

8. T.Kato, H.Imai, An extended construction method of visual secret sharing scheme (in Japanese), IEICE Trans., vol. J79-A, no.8, pp.1344-1351, 1996.
9. H.Koga, H.Yamamoto, Proposal of a Lattice-Based VSSS for Color and Gray-scale Images, IEICE Trans. on Fundmentals, vol. E81-A, no.6, pp.1262-1269, 1998.
10. H.Kuwakado, H.Tanaka, Polynomial representation of visual secret sharing scheme for black-white images, 2001 Symposium on Cryptography and Information Security, pp.417-422, 2001.
11. M.Naor, A.Shamir, Visual Cryptography, EUROCRYPT'94, pp.1-12, 1994.
12. M.Naor, A.Shamir, Visual Cryptography 2, Lecture Notes in Computer Science 1189, pp.179-202, 1997.
13. A.Shamir, "How to Share a Secret", Commun.of ACM, Vol.22, No.11, pp.612-613, 1979.
14. Y.Suga, K.Iwamura, K.Sakurai, H.Imai, Extended Graph-type Visual Secret Sharing Schemes with Embedded Plural Secret Images (in Japanese), IPSJ JOURNAL Vol.42 No.08, pp.2106-2113, 2001.
15. E.R.Verheul, H.C.A.van Tilborg, Constructions and properties of k out of n visual secret sharing scheme, Designs, Codes, and Cryptography, vol.1, no.2, pp.179-196, 1997.

PART II   WORKSHOP PAPERS

OVERCOMING CHANNEL BANDWIDTH CONSTRAINTS IN SECURE SIM APPLICATIONS

John A. MacDonald[1], William Sirett[2] and Chris J. Mitchell[1]

[1]*Information Security Group, Royal Holloway, University of London, Egham, Surrey TW20 0EX, UK;* [2]*Smart Card Centre, Information Security Group, Royal Holloway, University of London, Egham, Surrey TW20 0EX, UK.*

Abstract: In this paper we present an architecture based on a Java (J2SE, J2EE, J2ME and Java Card) platform supporting a secure channel from a Mobile Operator to the SIM card. This channel offers the possibility of end to end security for delivery of large data files to a GSM SIM card. Such a secure channel could be used for delivery of high value content that requires a high bandwidth channel – perhaps either rendered for user infotainment, or processed in the client Mobile Station (device and SIM card) for remote device management. Our methodology overcomes the bandwidth constraints of the SIM Toolkit Security scheme described in GSM standard 03.48. To validate our proposal we have developed code to create DRM and Web Service test scenarios utilising readily available J2ME, Java Card, J2SE and J2EE platforms, Web Services tools from Apache, the KToolBar emulator from Sun, and a Gemplus Java Card.

Keywords: J2ME, Java Card, SAT Security, SIM card, Web Service Security, DRM.

1. INTRODUCTION

Since its inception in September 1994, the SIM Application Toolkit (SAT) (3GPP TS 31.111, 2004; GSM 11.14, 2001) and SIM Toolkit Security (3GPP TS 03.48, 2001) have been used extensively. They are primarily used to securely transfer device and network management information and simple user applications (such as device independent, Operator-specific, power-on menus) to the SIM card.

These two independent concepts – the SAT and GSM standard 03.48, have been a very successful marriage (Guthery and Cronin, 2002). SAT allows applications resident within the tamper proof SIM card to initiate actions, whilst GSM standard 03.48 provides security services for any SMS message. Together they have been a critical enabler of many network management and revenue generating services deployed by GSM operators worldwide.

However the availability of large capacity SIM cards and high performance 2.5G and 3G devices means that this once-successful combination is now proving to be a constraint for the following reasons:

- GSM standard 03.48 uses SMS as the transport mechanism. SMS stands for Short Message Service, and is a way of sending a maximum of 160 characters (140 bytes) to and from mobile devices. Despite the GSM standard 03.48 allowing the concatenation of up to 255 such SMS messages to increase the payload, it is reported (Guthery and Cronin, 2002) that most operators limit this to approximately 5, i.e. a maximum payload of only 700 bytes. This is due to uncertain and indeterminate device operation when receiving such a large concatenated SMS message. With 128kB Java Card devices now routinely deployed, this bandwidth limitation is equivalent to less than 1% of the capacity of current generation SIM cards.
- Although a significant innovation in 1994, the SAT instruction list comprises only 31 proactive commands. These commands provide only limited control over the user experience, e.g. PLAY TONE, DISPLAY TEXT, GET INKEY, more appropriate for the text-based devices of the mid 1990's. The devices typical of today's 2.5G and 3G market would benefit from greater application customisation capability between device and SIM card.

A secure channel capable of downloading high bandwidth, high-value data within an application framework that provides rich control over the host device could thus be advantageous. This paper proposes such a channel.

2.   THE JAVA FRAMEWORK

In recent years, Java enabled devices have become increasingly popular within the mobile market. Our proposal creates a high bandwidth secure channel for a Java platform, utilising Java Card (Chen, 2004) and J2ME (Topley, 2002) technologies.

In the GSM and UMTS system architectures, the Mobile Station (MS) may comprise two java components:

- the user device (often referred to as the handset). This typically comprises a Java runtime environment conforming to the J2ME Connected Limited Device Configuration (CLDC), complemented by additional classes from the Mobile Information Device Profile (MIDP). Java applications that run on MIDP compliant user devices are known as MIDlets.
- the SIM card provided by the network operator. The latest generation devices are typically UICC (3GPP TS 31.101, 2003) Java Cards where the SIM application (3GPP TS 31.102, 2003) is just one of the possible Java applications (ETSI TS 101 476, 2000) that the Java Card is capable of running. Java applications that run on Java Cards are known as Applets.

Recent work through the Java Community Process (JCP) has increased the utility of a mobile Java solution. The result of this technical innovation has been a rapid growth of complex, revenue generating, but largely fun-based J2ME applications within the gaming and entertainment sector. However, although some serious business applications exist (Itani and Kayssi, 2004), the Java environment has largely been ignored by the professional business and network management community because of concerns over security.

The fundamental problem is that the MIDlet runs within the Java implementation of the user device. The user device is unlikely to be trusted by the Operator to hold network level components and functions that protect valuable network assets. This distrust is likely to get worse as devices move from traditionally closed proprietary operating systems to more open operating systems capable of performing the file manipulation required by advanced 2.5G and 3G services. Securing a J2ME application currently requires the security keys, certificates and user identities to be stored within the user device. Many institutions within the Mobile Operator and Financial Service sectors are likely to consider this to be an unacceptable security risk.

In the GSM/3GPP mobile architecture, security and trust resides in two locations, the network HLR and the Operator issued tamper-resistant SIM card. The threat model is well researched and has resulted in the security services model at the heart of the GSM and 3GPP design (Hillebrand, 2002). What is needed is a methodology to extend this trust to the MIDlet environment.

3.   THE SAT SECURITY FRAMEWORK

Our proposal builds on the dual capabilities of SMS Security and SIM Application Toolkit (SAT). The former is defined in Security Mechanisms for SIM stage 2 (3GPP TS 03.48, 2001). It provides end to end security services for an SMS message going to or coming from the SIM card. The SAT API allows a SIM card application to be informed of events (referred to as *event download*) by the user device, and to issue commands (referred to as *proactive command*) to the user device.

We use the proactive command SET_UP_EVENT_LIST to register for the SMS_PP event. On occurrence of such an event, or when commanded by the *Protocol Identifier* of the SMS *Mandatory Header*, the received SMS is passed on to the SIM application as a compound TLV (Tag Length Value) in the data field of an ENVELOPE APDU command. The SMS's *Command Header* specifies how the payload data is secured. The SIM application's response to the ENVELOPE command is then returned to the sender in a *Response Packet*. By using this approach, and by concatenating five SMS messages, it is possible to securely deliver around 700 bytes of data from Server to SIM card, receiving a proof of delivery in acknowledgement. We use this capability to securely transfer the Operator domain certificate and long term symmetric keys necessary to establish and secure our high bandwidth channel to the SIM card.

4.   THE PROPOSED SECURE DATA TRANSFER TECHNIQUE

The MIDP 2.0 specification (JSR-118 JCP, 2002) introduces the concept of domains within a J2ME implementation (Block and Wagner, 2003). A Domain Protection Root Certificate controls application access to a domain. Any application within a domain enjoys a set of unique permissions and access to restricted and sensitive APIs provided by that domain. Before an application can be over the air (OTA) loaded into the Operator domain it must be digitally signed. The signature is checked against the SIM card resident root certificate and, if authorised, the application is loaded into the Operator domain of the untrusted device.

The Security and Trust Services API (JSR-177 JCP, 2004) provides an Operator domain J2ME application with the ability to access a connected trusted element (i.e. a SIM card within our scenario). Our proposal involves creating a J2ME and Java Card *Security Agent* application that is capable of implementing a secure high bandwidth channel between Server and SIM card endpoints. At no time does the J2ME application have access to any of the enabling cryptographic keys or functions. The bandwidth of the secure channel created by the *Security Agent* is only limited by the 2G/2.5G/3G network and the data rate resulting from the ENVELOPE APDU command.

The J2ME element of the *Security Agent* benefits from the processing power and I/O capabilities of the user device and has direct access to the results of secure SIM card computations executed by the Java Card Applet. Serious business applications such as DRM, e-commerce and securing web services can now be implemented by combining such J2ME and Java Card *Security Agent* applications. These business applications will additionally benefit from device vendor independence and potentially rapid rollout from OTA distribution and installation.

5.   PROTOCOL

Full details of our protocol are provided elsewhere (MacDonald et al., 2004). It uses both symmetric and asymmetric cryptographic techniques to provide the authentication, integrity and confidentiality services required to support a secure high data bandwidth channel from Server to SIM card. Our protocol has been designed on the assumption that the user device and SIM card are pre-issued and in the field. We assume that neither user device nor SIM card contain pre-installed application code to create the desired secure high bandwidth channel.

We choose to use symmetric rather than asymmetric cryptography for authentication and key agreement. Performance is critical in a mobile system and overhead must always be minimized wherever possible (Blanchard and Trask, 2002). The long term secret key K_{SC} shared by the Server and the SIM card, and used to support the secure channel to the SIM, is confidentially distributed from the Server to the SIM card endpoint with authentication and integrity services provided by GSM standard 03.48.

STEP 1 *Install MIDlet into Operator Domain, and Applet into SIM*

The first step is to prepare the SIM card so that the MIDlet can be installed within the Operator domain of the J2ME device. The MExE (3GPP TS 23.057, 2003) security framework, like other specialist services and applications that use the mobile network purely as a transport mechanism, relies on signature verification before the MIDlet can be installed within the target domain. We use GSM standard 03.48 to securely transfer the Operator Domain public key certificate $Cert_{OP\_DOM}$ to the SIM card. The MExE J2ME implementation on the user device receives the signed MIDlet. Successful verification of the signature using the public key in $Cert_{OP\_DOM}$ provides data origin authentication and integrity of the MIDlet JAD and JAR files. The

Security Agent MIDlet is installed in the Operator domain of the user device with full JSR 177 permissions, allowing APDU commands to be issued to SIM card resident Applets. To initiate installation of the Java Card *Security Agent* Applet, the MIDlet starts an http session with the server, and supplies it with the SIM card's unique identifier. The server responds with the SIM card Applet code, integrity protected with a MAC computed using the shared secret K_{SC}. The MIDlet *Security Agent* then transfers this data to the SIM card via the `Envelope` APDU command. An on-card installer application verifies the MAC and hence the origin authentication and data integrity of the Applet. If there is any discrepancy the installation process ceases; otherwise the *Security Agent* Applet is securely installed. This results in the creation of an applet instance and its registration with the Java Card runtime environment.

Note that neither MIDlet nor Applet carry any secret keys or other private data. Hence code encryption is not necessary. Integrity services to protect the MIDlet and Applet against virus insertion attack whilst in transit are required and are provided by the use of Digital Signatures and MACs respectively.

STEP 2 *Perform mutual entity authentication*

At some time later, i.e. after the http session of STEP 1 has closed and both Applet and MIDlet are installed, the Operator may choose to securely download bulk data from the Server to the SIM card. Before this begins, both endpoints verify each other's identity by means of a mutual entity authentication protocol. We use a three-pass mutual authentication protocol based on MACs and nonces, as specified in ISO/IEC 9798-4 (ISO/IEC9798-4, 1999).

STEP 3 *Set up session keys to protect bulk data transfer*

Following mutual entity authentication, both Server and SIM card derive session Integrity (IK) and Confidentiality (CK) keys to provide security services to protect the bulk data transferred between Server and SIM card. Both Server and SIM card Applet will contain identical functions $f1$ and $f2$ to calculate the session cipher and integrity keys using the nonces r_S and r_C exchanged as part of the authentication protocol in step 2, and the long term shared secret K_{SC}, as follows:

$$CK = f1_{K_{SC}}(r_S\|r_C) \quad \text{and} \quad IK = f2_{K_{SC}}(r_S\|r_C).$$

Once session keys have been established, the bulk data may be transferred between Server and SIM, encrypted for confidentiality with CK and concatenated with a MAC computed using IK for data origin authentication and integrity.

6. PROOF OF CONCEPT PROTOTYPE IMPLEMENTATION

To validate our proposal we have constructed the Proof of Concept model of Figure 1, based on readily available open source tools:

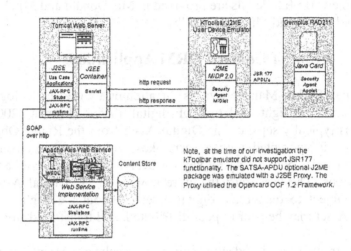

Figure 1. Proof of Concept Prototype Implementation

- A J2EE Servlet web application performs the Mobile Operator function and is packaged as a WAR file (Web Application Archive) for easy deployment on a Tomcat Apache Web Server.
- The J2ME Client is emulated by the Wireless KToolbar (Sun Microsystems, 2003) from Sun Microsystems, running our *Security Agent* MIDP 2.0 MIDlet on the reference J2ME implementation.
- The SIM card function is provided by a Gemplus GemXpresso RAD 211 Java Card with crypto package, connected to our demonstration environment via a USB card reader.
- A Web Service application communicates with the Mobile Operator function using SOAP over http. We used the jax-rpc API together with tools from Apache Axis to create the service WSDL and deploy the Web Service on a Tomcat Server.

The demonstration environment of our proof of concept model is implemented in J2SE. J2SE provides the necessary Java Swing classes for monitoring the various use case applications tested on our model. The model is designed so that each phase of a specific use case is initiated manually and monitored by visual feedback through the use of J2SE's GUI `LayoutManager` class and `ActionListener` interface.

A framework that provides a high bandwidth secure channel between Server and SIM card is a significant enabler for application deployment. For demonstration purposes we have deployed Digital Rights Management and Web Services Security platform applications onto this framework. We now review how these operations leverage the proposed high bandwidth channel and framework; full details are provided in MacDonald and Mitchell (2004a) and MacDonald and Mitchell (2004b).

6.1   Proof of Concept DRM Applications

Digital Rights Management is an attempt to use technology to limit piracy and copyright violation of digital media (Litman, 2001). DRM solutions typically separate the Digital Asset from the Rights Object. Often the Asset is encrypted with a secret key. A separately delivered Rights Object includes both the secret key for decryption of the Asset and the user permissions. The user must therefore have both the Digital Asset and the Rights Object to render the digital asset. Without the Rights Object, the Digital Asset may be peer to peer distributed, and transferred from device to device.

Our framework is ideally suited to such a content centric DRM application. Typically the user device would be notified of the receipt of such an encrypted Digital Asset by the asset's MIME type. This would invoke the *Security Agent* to store the encrypted Digital Asset in the, relatively plentiful, device memory, and then securely fetch the Rights Object from the Rights Fulfilment Server. The Rights Object would be securely transferred to the SIM card via our high bandwidth channel. The Digital Asset is recoverable only by the entity that holds the Rights Object.

At some time later, upon user request, the Digital Asset would be transferred (perhaps streamed) to the SIM card for authorisation, where it is decrypted and streamed back to the device for consumption and rendering. The Rights Object, comprising the root decryption key of the Digital Asset and the current user permissions, would always reside on the secure SIM card. Such an implementation greatly reduces the network resource cost incurred by the practice of streaming each rendering instance of the Digital Asset over the WAN 2.5G and 3G network.

6.2   Proof of Concept Web Services Security Application

Our framework can be extended to provide a mobile Operator endorsed authentication and payment platform for web services. For this vertical application the Server Servlet also provides the stub to the remote *Web Service* which is packaged as a WAR file and deployed on the Tomcat Server. Described by its WSDL we use the JAX-RPC API from Apache Axis to create the stubs to the service interface. Communication between Servlet and Web Service is according to the SOAP protocol using http as the transport mechanism.

In this application a high level user discovery process is provided by the J2ME *Security Agent*. User service selection initiates the mutual authentication step concluding with the creation of the high bandwidth secure channel between Server and SIM card. The Server may now issue an authentication token followed by an authenticated payment token when the user decides to consume the service. The authenticated payment token is exchanged for the web service, and the content associated with the service provided to the Server using SOAP over http. The service content may now be securely transferred to the SIM card via the high bandwidth J2ME and JavaCard *Security Agent* channel. This implementation provides:

- the user with a high level service discovery interface plus anonymity from Web Service providers;
- the Mobile Operator with a pivotal role and revenue generating opportunity in the provision of a web services security and payment platform;
- the Content Provider with a secure, scaleable distribution channel.

Note that, whilst it is possible to use the J2ME and Java Card *Security Agent* to create a secure high bandwidth channel, it may not be desirable to use these entities for service rendering and consumption. Extending connectivity to the personal area network of the J2ME device is particularly straightforward given the availability of the SAT OPEN CHANNEL proactive command.

7.   CONCLUSION

In this paper we have introduced a novel approach to securely transfer large data files from an application server to the mobile device SIM card. Our approach is based on a Java solution and overcomes a potential bandwidth restriction of the current GSM standard 03.48 and SAT Security process. We present a protocol and methodology that allows the secure channel to be created on capable, but unprepared, devices and SIM cards that are already issued. We have modelled our proposed solution and protocol using open source tools and indicate how it can be extended to apply to future application implementations such as DRM and Web Services.

ACKNOWLEDGEMENTS

The work of the first author was funded by Telefonica Móviles, España, S.A.U., Pza. de la Independencia, Madrid, Spain. The work of the second author was funded by the Smart Card Centre, Royal Holloway, University of London. This support is gratefully acknowledged.

REFERENCES

3GPP TS 03.48 (2001). *Technical Specification Group Terminals; Security Mechanisms for the SIM application toolkit; stage 2.* http://www.3gpp.org.
3GPP TS 23.057 (2003). *Technical Specification Group Terminals; Mobile Execution Environment (MExE); Functional description; Stage 2.* http://www.3gpp.org.
3GPP TS 31.101 (2003). *Technical Specification Group Terminals; UICC-terminal interface; Physical and logical characteristics.* http://www.3gpp.org.
3GPP TS 31.102 (2003). *Technical Specification Group Terminals; Characteristics of the USIM application.* http://www.3gpp.org.
3GPP TS 31.111 (2004). *Technical Specification Group Terminals; USIM Application Toolkit(USAT).* http://www.3gpp.org.
Blanchard, C. W. and Trask, N. (2002). Wireless security. In Temple, R. and Regnault, J., editors, *Internet and Wireless Security*, number 4 in BT Exact Communications Technology Series, chapter 9, pages 146-170. IEE, London.
Block, C. and Wagner, A. C. (2003). *MIDP 2.0 Style Guide.* Addison-Wesley, London.
Chen, Z. (2004). *Java Card Technology for Smart Cards.* Addison-Wesley, London.
ETSI TS 101 476 (2000). *Digital cellular telecommunication system (Phase 2+); Subscriber Identity Module Application Programming Interface (SIM API); SIM API for Java Card; Stage 2 (GSM 03.19).* ETSI, http://www.etsi.org.
GSM 11.14 (2001). *Digital cellular telecommunications system (Phase 2+); Specification of the SIM Application Toolkit for the Subscriber Identity Module-Mobile Equipment (SIM-ME) interface.* ETSI, http://www.etsi.org.
Guthery, S. B. and Cronin, M. J. (2002). *Mobile Application Development with SMS & the SIM Toolkit.* McGraw-Hill.
Hillebrand, F. (2002). *GSM and UMTS: The creation of global mobile communications.* John Wiley & Sons, Ltd.
ISO/IEC 9798-4 (1999). *Information technology – Security techniques – Entity authentication – Part 4: Mechanisms using a cryptographic check function.* International Organization for Standardization, http://www.iso.org, 2nd edition.
Itani, W. and Kayssi, A. (2004). J2ME application-layer end-to-end security for m-commerce. *Journal of Network & Computer Applications*, 27:13-32.
JSR-118 JCP (2002). *Mobile Information Device Profile, v2.0 (JSR-118).* Sun Microsystems, http://java.sun.com.
JSR-177 JCP (2004). *Security & Trust Services API (SATSA) (JSR-177).* Sun Microsystems, http://java.sun.com.
Litman, J. (2001). *Digital Copyright.* Prometheus Books, New York.
MacDonald, J. A. and Mitchell, C. J. (2004a). Content centric DRM for mobile vertical market. Information Security Group, Royal Holloway, University of London – Internal paper.

MacDonald, J. A. and Mitchell, C. J. (2004b). Web services security platform using mobile operator credentials. Information Security Group, Royal Holloway, University of London – Internal paper.

MacDonald, J. A., Sirett, W. G., and Mitchell, C. J. (2004). Establishing a security context between server & SIM: A 3 pass mutual AKE protocol with signature & MAC. Information Security Group, Royal Holloway, University of London – Internal paper.

Sun Microsystems (2003). *Wireless Toolkit, Version 2.1.* Sun Microsystems, http://java.sun.com/products/j2mewtoolkit.

Topley, K. (2002). *J2ME In a Nutshell.* O'Reilly.

ON THE PERFORMANCE OF CERTIFICATE REVOCATION PROTOCOLS BASED ON A JAVA CARD CERTIFICATE CLIENT IMPLEMENTATION

K. Papapanagiotou, K. Markantonakis, Q. Zhang, W.G. Sirett and K. Mayes
The Information Security Group Smart Card Centre (Founded by Vodafone, G&D and the ISG), Royal Holloway, University of London

Abstract: The use of certificates for secure transactions in smart cards requires the existence of a secure and efficient revocation protocol. There are a number of existing protocols for online certificate revocation and validation, among which OCSP and SCVP are the most widely used. However there are not any real applications testing the efficiency of these protocols when run in a smart card, even though the advantages of such an implementation are promising. In this paper we examine the details of the implementation of these protocols, emphasising on the issues arisen from the limitations of the smart cards. We also discuss the performance results from the implementation of OCSP and SCVP in a multi-application smart card environment. Results from two different Java Card platforms are presented and analyzed.

Key words: Certificate Revocation, OCSP, SCVP, Java Card

1. INTRODUCTION

In recent years, X.509 certificates have been used more and more in smart cards in order to validate users, establish secure channels or perform secure transactions. One of the most significant problems of Public Key Infrastructures (PKI) is certificate revocation and validation. Any PKI deployment, whether it includes smart cards or not, should provide efficient and secure mechanisms for certificate validation. Until now, protocols, such as the Online Certificate Status Protocol (OCSP) [15] and the Simple

Certificate Validation Protocol (SCVP) [12] have been designed and successfully implemented in many systems. Their weaknesses and general issues concerning their use have been discussed extensively in academic literature [2]. However, none of these protocols have been specifically designed or tested for smart cards.

In this paper we outline the design of a smart card certificate validation model and discuss the issues surrounding its implementation. We also implement OCSP and SCVP in two different Java Card platforms, which from now on will be referred to as Vendor A and Vendor B, using the Java Card API Ver. 2.1 [22]. Our goal is to create a transparent, lightweight, and independent application that will provide certificate validation function to any other application in a multi-application smart card environment. It will be completely trusted by all other applications on the card to perform certificate validation on their behalf. Thus, other applications should not need to be aware of the specifics of the underlying protocols. Our purpose is, firstly, to present some performance measurements concerning the implementation of the most widely used revocation protocols in a smart card. Secondly, we want to highlight the issues regarding the design and development of a validation protocol in such a limited processing and memory environment. Such protocols are an essential part of every PKI. It is important that we identify the existing limitations in their implementation, as input into future designs for optimised solutions.

We will briefly describe the certificate validation protocols and then analyze the issues surrounding certificate validation on a smart card. Subsequently, we will present the smart card certificate validation model and the entities involved. Furthermore, we provide the implementation details of the model along with providing performance results and timings. Finally, we provide some concluding remarks and discuss directions for further research.

2.   THE CONCEPT OF ON-LINE CERTIFICATE VALIDATION

Certificate revocation and validation using Certificate Revocation Lists [9] may in some cases be inefficient or indeed inadequate. Protocols for online certificate validation have been proposed to solve the issues surrounding certificate revocation. These protocols can in theory be used in a smart card architecture to validate certificates for card based applications.

2.1 Certificate Validation Protocols

Many methods and protocols have been proposed for online certificate validation [2]. Currently, the most widely used are OCSP [15] and SCVP [12], which are simple client-server protocols involving requests and responses that provide the current status of one or more certificates. A client can send a request to a server (usually called "responder"), asking for the status of one or more certificates. The server responds with a signed message containing the status of the certificate(s), the time when the responder last updated the status information and the time of the creation of the message. Responses should be signed by the CA, a trusted or an authorized responder.

In OCSP version 1 [15] (OCSPv1), the certificate is referenced using its serial number along with the issuer's name and public key. Thus, it is necessary for the client sending the request to construct and validate the full certificate path from the queried certificate to the root CA. Certificate path construction can be difficult to implement in environments with limited processing power, such as smart cards. As a result, other solutions were proposed, like OCSP version 2 (OCSPv2) [16] and SCVP [12]. OCSPv2 practically added two more possibilities to reference the certificate: the client can send the entire certificate or a hash of some specific fields from it. Consequently, OCSP clients do not have to do any certificate path construction. Nevertheless, the size of the request message is significantly increased. Even though the draft for OCSPv2 has expired since 2002 [16], we believe that it may be applicable to smart cards.

SCVP [12] is a protocol that can provide further information than just revocation status. An SCVP Server can perform certificate path construction and validation as well as revocation checking. The request message includes either the entire certificate or a hash of the certificate and can be signed if needed. The server can be instructed to provide certification path for the certificate, revocation status or both. SCVP Responses are always signed, unless an error message is given back. Practically, they contain the same information with OCSP responses, in terms of time and date values. SCVP messages are encapsulated in Cryptographic Message Syntax (CMS) [6] data structures, which results in larger and more complex messages.

2.2 Issues surrounding the Smart card Certificate validation

Nowadays, there are numerous proposals for smart card applications exploiting public key cryptography, which implies the handling of certificates by the smart card. Therefore, an efficient mechanism for certificate validation is considered essential. However, the implementation of such a mechanism has to address some significant issues, mainly caused by limitations of the smart card environment.

First of all, it is considered more secure and efficient for a smart card application to be able to determine the validity of a certificate without having to rely on an application that doesn't reside on the card. The tamper-resistant nature of the card makes resident application and data more trustworthy and secure. On the other hand, an off-card certificate validation application should be trusted by the card and would require a secure communication channel with the card.

Most difficulties regarding the implementation of certificate validation protocols on smart cards are the result of the limited processing power and memory of the card. The limited processing capability of a smart card makes the creation of a certification path on the card very time and memory consuming. Thus, the use of OCSPv1 is not recommended, as it requires a fully validated certification path. Apart from that, it is not very efficient to implement a fully functional ASN.1 [10] parser and DER [11] encoder in such a limited environment. X.509 Certificates, OCSP and SCVP protocols are all designed using ASN.1 notation. Thus, the creation and parsing of messages can be very time consuming. In addition OCSPv1 messages require extraction of fields from the queried certificates, which demands the presence of an ASN.1 parser or a package that handles X.509 certificates.

Smart cards have also a very limited memory. An X.509 certificate may occupy more than 1000 bytes [17] of memory. This is not such a problem for the card's capacity as it is for the communication channel between the card and the reader. An APDU data buffer [8] can hold up to 255 bytes of data, so a series of APDUs is needed for an X.509 certificate to be transmitted to the card. An OCSPv2 message also contains numerous other fields and a digital signature, which increase even more the total size of the messages. SCVP messages can contain only a hash of the certificate, so SCVP is expected to be more efficient than OCSP. Hashing the certificate is also possible in OCSPv2, however only a part of the certificate is hashed and thus, a package handling X.509 certificates would be needed. Nevertheless, a hash function requires more processing power and thus, is expected to need more time for execution. SCVP response messages also contain the certificate of the SCVP server, so they are expected to be as big as OCSP responses. Finally, both OCSP and SCVP protocols require some time checks to be done to determine the validity of the responses. Such checks cannot be done on-card as the card doesn't have a clock and, thus, knowledge of current time or date. Thus, in order to prevent replay attacks, nonces [12, 15] should be used.

The Open Mobile Alliance has already published a candidate version of an OCSP profile for mobile environments [18]. Its goal is to enable the use of OCSP in mobile devices with limited resources that use the Wireless

Application Protocol (WAP). This profile sets requirements and constraints on OCSP in order to have smaller, simpler and more easily processed messages. Nonetheless, it is not specifically designed for smart cards.

2.3 Motivation

As the need for the use of certificates in transactions with smart cards increases, the use of an online validation mechanism provided within the card becomes very attractive. An evaluation of the different online validation protocols is required in order to determine which one is more efficient for smart card use. The limitations of smart cards bring about the issues presented previously, which need to be met in a real world application.

Until now, there is little public information relating to the implementation of a validation protocol on the card. X.509 certificates can be relatively large in size, a fact that makes their management and manipulation in a smart card environment difficult. Nevertheless, X.509 is currently the most widely used certificate format and most validation protocols are to be used with such certificates [12, 15]. A smart card application that will implement these protocols can provide significant feedback concerning the practical and theoretical issues of certificate validation in smart cards. The implementation of a certificate validation protocol on a smart card can also facilitate the management of certificates within the card. The protocol can be implemented in a separate, stand-alone application which provides a shareable interface to all other applications in a multi-application smart card. As a result, any application can use the protocol, without being aware of its details. Many different validation servers can be registered to the card, which can decide where to send the validation request.

3. A SMART CARD CERTIFICATE VALIDATION MODEL

The design of a smart-card software solution can be easily split into three parts: the card side, the pc-client side and the server side. In our case we have two applications on the card side: a generic application and a validation client. The pc-client side acts as a gateway between the validation client on the card and the validation server. The entities that are involved and the technology that we used to implement them are described in this section.

3.1 The entities involved

The entities that take part in a certificate validation protocol on a smart card are the following:

- Smart card Application (SA): A third party application (applet) which has an X.509 certificate and wishes to use the card's validation functionality.
- Smart card Validation Client (SVC): It implements the validation protocol and provides its functionality to other applications through a shareable interface.
- PC-Client Terminal Application (PCAP): It communicates with the card and forwards Certificate Validation Requests from the card to the server, and Responses from the Server to the card. It also provides the APDU commands needed by the SVC to perform the validation.
- Server: This can be any server that supports OCSP, SCVP or other online certificate validation protocols.

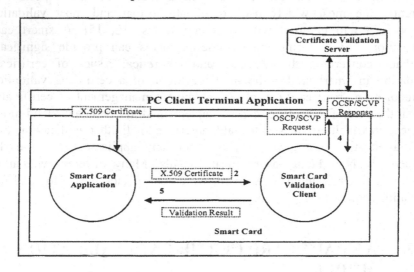

Figure 1. Message flow during protocol execution

Figure 1 illustrates the message flow during protocol execution. More specifically, in 1 the SA receives the certificate from the PCAP, and then in 2 invokes the method of SVC to send the certificate to the SVC using the Shareable Interface Object (SIO). Further on, the SVC formats an OCSP or SCVP Request and forwards it to the Server (3), which then issues an OCSP or SCVP Response (4). Finally, the SVC gets the response by the PCAP (4), verifies it and sends the result back to SA (5).

3.1.1 The Smart card Application

The SA can be any application using X.509 certificates. In our implementation it receives an X.509 certificate from the PCAP. The typical size of such a certificate is greater than 255 bytes, which is the size of the command APDU data buffer. Consequently, data will have to be sent in blocks and a series of APDUs can be used to send an X.509 Certificate. It should be noted that the Java Card environment restricts the maximum size of arrays. As a result, the maximum size of a certificate is not only limited by the small command APDU data buffer but also by the idiosyncrasies of the underlying Java Card platform.

Once receiving a command APDU containing a part of a certificate (*Send Certificate*), the SA calls a shareable interface method of the SVC which stores the certificate. Java Card SIOs [14] only allow passing of primitive types as parameters. Thus, for passing the certificate to the SVC, a global array, in our case the APDU buffer, had to be used [14]. The overall security of this approach is discussed and evaluated in [14]. A different command APDU (*Get Result*) returns the result of the validation protocol to the SA. The SA must be aware of the SVC's Shareable Interface methods.

In our implementation for the two smart card components, we used Java Card 2.1.1 [22], which is supported by our smart card application development tools. Java card is one of the most widely recognised and used smart card multi-application environments.

3.1.2 The Smart card Validation Client

This entity performs all the functions required by the validation protocol. First, it receives the certificate from the SA. The certificate is stored in a byte array throughout the execution of the protocol, as it is needed for the verification of the Server's response. A command APDU (*Create Request*) triggers the function that constructs a validation request, which is later sent to the PCAP (*Send Request*), broken into data blocks. The PCAP also forwards the validation response to the SVC (*Get Response*) where it is processed and verified (*Process Response*) so that the result can be returned to the SA. The SVC holds the Server's public key so that it can verify the digital signature in the response message. Finally, the SVC provides shareable interface methods to the SA, which are needed for passing the certificate and the result of its validation.

3.1.3 The PC-Client Terminal Application

The terminal application was implemented using PC/SC [19]. PC/SC is currently one of the most widely used and supported card terminal programming environments. PC/SC architecture is widely accepted and implemented by large and established companies such as Microsoft, Apple and Philips. Some of the supported programming languages include Visual Basic and C++. Most smart card manufacturers provide drivers for PC/SC.

The terminal application receives validation requests from the smart card in the form of multiple APDUs. Its role is to combine APDUs and send the validation request to the given validation server, as instructed by the card application. Then, it receives validation responses, breaks them into APDUs and sends them back to the card. The PC/SC terminal application is a completely passive component which simply facilitates the communication between the server and the card application, by selecting the applets of SA and SVC, and transmitting the appropriate command APDUs.

The programming language we used to implement the terminal application was Visual Basic 6. Microsoft's Visual Basic is currently very widely used and also directly supports PC/SC. For the purposes of this paper the terminal application was also configured to send an X.509 certificate to the card application. As Visual Basic does not support X.509 certificate handling and digital signatures, CAPICOM [13], a cryptographic library developed by Microsoft, was used to manipulate certificates.

3.1.4 The Server

A dummy OCSP and SCVP responder was implemented to handle OCSP and SCVP requests. The validation server checks the syntax of request messages and then issues digitally signed OCSP and SCVP responses as required. For the purposes of this paper the server was directly integrated with the PCAP. The language used for the implementation was Visual Basic v6 for compatibility and integration with the PCAP. For testing purposes the sever was configured to always send responses with either valid, invalid or unknown certificate status, regardless of the actual status of the certificate. CAPICOM was used to create digital signatures.

3.2 Implementation Details

The SVC was implemented to handle OCSP requests and responses. The implementation was based on OCSPv2. In order for no certificate manipulation to be required the entire certificate was included in the OCSP request. For maximum efficiency, a specific ASN.1 parser and DER encoder and decoder were implemented, which can only handle such messages. Moreover, only a few of the OCSP Response acceptance requirements specified in [16] were implemented: the certificate was compared with the original queried certificate and then its status was retrieved. The SVC was implemented to only accept messages from a specific trusted responder.

The verification of the digital signature in the validation response messages may be essential for the protocols, but is also a time-consuming function when performed on the card [20]. The purpose of this paper is to evaluate the performance of revocation protocols and not of the signature algorithms. As a result, and for testing and evaluation purposes, digital signatures were not verified on the card so as not to influence our results. However, smart cards running the SVC should support the most known algorithms for digital signatures and hash functions, as digital signature verification is required for correct execution of both OCSP and SCVP.

4.   RESULTS AND PERFORMANCE EVALUATION

In this section we present and evaluate the results and timings we took for each of the two protocols, using two different smart cards.

We have generated a set of results for OCSP and SCVP, using a specified, 573 byte, X.509 certificate. Two different high end Java cards were used provided by Vendor A and Vendor B. The smart card application development tools were also different respectively. Due to Java Card's interoperability, there are only minor changes to the implementation for each card, that don't affect the overall performance. Each protocol was executed 5 times for each card. Timings, expressed in milliseconds, are presented in table 1. They were taken using an APDU monitoring tool attached to a P4 Windows machine. Commands marked with * only involve data I/O. The functionality of each command is explained in section 3. We also include for reference timings required for sending the certificate to the card.

The timing results presented in Table 1 are coherent. Even though execution times for most commands differ between the two cards, we can reach into the same conclusions for the protocols we implemented. It should be noted, that the differences in each card's timings is attributed to the different nature of each card. The statistical analysis of the results that we presented leads us to the remarks that we analyze in this section.

First of all, we observe that for OCSP the *Create Request* command is the least time consuming of all. The OCSP protocol doesn't require any special functions for creating request messages, and thus, an OCSP request is created really fast. On the other hand, for SCVP it appears to be the most time consuming command, excluding commands that handle I/O, as a hash function has to be computed. The processing of an OCSP response (*Process Response*) is the most time-consuming command for the OCSP protocol, excluding the ones regarding I/O. The comparison of the certificate in the response with the original queried certificate is what makes this command more time-consuming than any other. SCVP responses in contrast, require the comparison of a much smaller byte array and thus demand less time, as a hash of a certificate is, of course, smaller than the certificate itself.

Table 1. Performance results

Command	Metrics (clk)	Vendor A		Vendor B	
		OCSP	SCVP	OCSP	SCVP
Send Certificate*	Average	13978,48		37614,04	
	Median	13994,71		37363,65	
	Std. Dev.	22,81		563,74	
Create Request	Average	8,56	1351,56	24,83	1932,78
	Median	8,53	1351,84	24,83	1934,21
	Std. Dev.	0,03	8,11	0,02	42,80
Send Request*	Average	28918,72	20,02	14064,73	25,56
	Median	28951,10	22,56	14035,67	25,56
	Std. Dev.	48,34	3,54	62,85	0,03
Get Response*	Average	20710,90	24776,68	57034,72	67655,90
	Median	20941,47	24829,20	56719,33	67417,51
	Std. Dev.	2423,76	113,20	718,87	702,24
Process Response	Average	68,17	48,32	421,00	322,18
	Median	66,62	55,94	458,39	312,54
	Std. Dev.	3,21	21,13	52,56	23,50
Get Result	Average	14,47		121,03	
	Median	14,50		114,88	
	Std. Dev.	0,05		12,72	
Total	Average	63699,30	40189,53	109280,35	107671,49
	Median	63976,93	40268,75	108716,75	107168,35

Furthermore, the most time-consuming functions of all are the ones that have to do with the input of the certificate (*Send certificate*), OCSP and SCVP responses (*Get Response*) to the card and the output of OCSP requests (*Send Request*). Concerning the transmission of the certificate to the card, it is obvious that even for a cut-off version of an X.509 certificate a significant amount of time is required. As a result, we have to consider the use of other certificate formats. The transmission of the response messages for both protocols requires the same amount of time, as their size is almost the same. On the other hand, a SCVP request is significantly smaller than an OCSP request. Thus, only a very small fraction of the time needed to send the OCSP request, is required to send a SCVP request message. Consequently, any decrease in the size of the messages will have a significant impact on the total time in which the protocol is executed. The *Get Result* command only involves changing the value of some variables, and thus, it doesn't interfere with the protocol run. Dring protocol runs it was observed that the cards quickly ran out of memory and the applets needed to be reinstalled. This was attributed to the fact that many large arrays are used. Thus, special care has to be taken for memory allocation and garbage collection.

Overall, SCVP runs faster than OCSP in both cards. This is mostly because a SCVP request is much smaller and thus can be sent much faster than an OCSP request. Even though the creation of a SCVP request is more time consuming, the fact that it can be transmitted in a single APDU makes SCVP more efficient. As we already mentioned, the commands that involve data I/O are the most time consuming and any alteration of the size of the messages has more impact on the time required for a complete protocol run than an improvement of any other command might have.

5. CONCLUSIONS

OCSP and SCVP protocols were implemented on two different Java Card platforms. Despite the issues that came up and the compromises regarding the certificate size, we have shown that it is feasible to implement and run known and widely used certificate validation protocols on a smart card. Even though the memory size of modern smart cards has significantly increased, an X.509 certificate is still quite large to be used in such a limited environment. The time that is required for a certificate to be loaded onto the card is clearly a major factor. However in future work there may be scope to reduce delays by exploiting faster card I/O options.

Furthermore OCSP is not very efficient for use in a smart card environment. OCSPv1 cannot be used at all, as certificate manipulation and path construction adds a significant overhead. We have shown that an implementation of OCSPv2 is feasible, even though the messages involved are quite large in size. Additionally, checks regarding time cannot be performed on a smart card, even though corresponding fields add up to the total size of the messages. SCVP, which is recommended for use in limited environments, uses a hash of the certificate, significantly reducing the overall size of all messages. This reduces the time required for I/O, but also increases the time required for a construction of a SCVP request. Nevertheless, in total, SCVP runs faster than OCSP.

Currently, we are experimenting with alterations to existing validation protocols as well as system architectures in order to have a more efficient protocol, specifically designed for smart cards. The suggestions of OMA

[18] are also considered, as so far there has not been any known implementation of their protocol. Moreover, recently a new IETF draft on OCSP was submitted [3], describing a lightweight implementation of OSCP, but it is not specifically designed for smart cards. The design of a smart card-specific certificate validation protocol is also examined, as well as the support of other certificate formats [1, 21, 23] that can facilitate certificate revocation and validation. In particular, we need to focus on formats that provide a more compact and efficient way of storing and managing certificates [17] and key pairs [4]. These will enable us to provide more accurate figures and comments on the performance of these protocols in a multi-application smart card environment.

REFERENCES

1. ANSI. X9.68 - 2001: Digital Certificates for Mobile/Wireless and High Transaction Volume Financial Systems: Part 2: Domain Certificate Syntax. 2001
2. A. Arnes. Public Key Certificate Revocation Schemes. PhD thesis. Norwegian University of Science and Technology, 2000
3. A. Deacon and R. Hurst. Lightweight OCSP Profile for High Volume Environments, IETF, 2004
4. N. Feyt and M. Joye. A Better Use of Smart Cards in PKIs. Gemplus Developer Conference, Singapore. Springer Verlag, 2002
5. P. Hoffman. RFC 2634 - Enhanced Security Services for S/MIME. IETF, 1999
6. R. Housley. RFC 2630 - Cryptographic Message Syntax. IETF, 1999.
7. R. Housley, W. Polk, W. Ford and D. Solo. RFC 3280 - Internet X.509 Public Key Infrastructure Certificate and Certificate Revocation List (CRL) Profile. IETF, 2002
8. ISO. ISO/IEC 7816-4, Information technology – Identification cards - Integrated Circuit(s) cards with contacts – Interindustry Commands for Interchange. ISO, 1995
9. ITU-T Recommendation X.509. Information Technology - Open Systems Interconnection – The Directory: Public-key and attribute certificate frameworks. 1997
10. ITU-T Recommendation X.681. Information technology - Abstract Syntax Notation One (ASN.1): Information object specification. 1997
11. ITU-T Recommendation X.690. Information Technology - ASN.1 Encoding Rules: Specification of Basic Encoding Rules (BER), Canonical Encoding Rules (CER) and Distinguished Encoding Rules (DER). 2002
12. A. Malpani, R. Housley and T. Freeman. Simple Certificate Validation Protocol (SCVP). IETF, 2003
13. Microsoft. CAPICOM Reference. http://msdn.microsoft.com/library/en-us/security/Security/capicom_reference.asp
14. M. Montgomery and K. Krishn. Secure Object Sharing in Java Card. USENIX, 1999
15. M. Myers, R. Ankney, A. Malpani, S. Galperin, C. Adams. RFC 2560 - X.509 Internet Public Key Infrastructure Online Certificate Status Protocol - OCSP. IETF, 1999
16. M.Myers, A. Malpani, D.Pinkas. X.509 Internet Public Key Infrastructure Online Certificate Status Protocol version 2. IETF, 2002
17. M. Nyström and J. Brainard. An X.509-Compatible Syntax for Compact Certificates. In Proc. Int. Exhibition and Congress on Secure Networking '99, Springer-Verlag, 1999

18. Open Mobile Alliance. OCSP Protocol Mobile Profile Candidate V1.0, 2004
19. PC/SC Workgroup. Interoperability Specification for ICCs and Personal Computer Systems. http://www.pcscworkgroup.com/, 1997
20. J-J. Quisquater and M. De Soete. Speeding up smart card RSA computations with insecure coprocessors. in Smart Card 2000. Amsterdam, 1991
21. RSA Labs. PKCS #15 v1.1: Cryptographic Token Information Syntax Standard, 2000
22. Sun Microsystems. Java Card 2.1.1 Application Programming Interface, Rev. 1.0. 2000
23. WAP Forum. WAP Certificate and CRL Profiles Specification, 2001

ON-THE-FLY FORMAL TESTING OF A SMART CARD APPLET

Arjen van Weelden, Martijn Oostdijk, Lars Frantzen, Pieter Koopman, Jan Tretmans

Institute for Computing and Information Sciences
Radboud University Nijmegen - The Netherlands

{arjenw, martijno, lf, pieter, tretmans}@cs.ru.nl

Abstract: This paper presents a case study on the use of formal methods in specification-based, black-box testing of a smart card applet. The system under test is a simple electronic purse application running on a Java Card platform. The specification of the applet is given as a Statechart model, and transformed into a functional form to serve as the input for the on-the-fly test generation, -execution, and -analysis tool GAST. We show that automated, formal, specification-based testing of smart card applets is of high value, and that errors can be detected using this model-based testing.

Key words: model-based testing; smart cards; Java Card; automatic test generation; executable specification.

1.   INTRODUCTION

Smart devices are often used in critical application domains, such as electronic banking and identity determination. This implies that their quality, such as their safety, security, and interoperability, is very important. Such devices commonly implement a Java Card virtual machine, which is able to execute Java Card applets. Each application is then implemented as a separate applet.

One way to increase the quality of applets is the use of formal methods. Such applets are sufficiently small to make a complete formal treatment with current day formal technology feasible. Systematic testing is another method, predominantly used to check the quality of smart devices in an experimental way.

In this paper we combine testing and formal methods: we test a Java Card applet, in a black-box setting, based on a formal specification of its required behavior. Compared with formal verification, testing has the advantage that it examines the real, complete system consisting of applet, platform and hardware together, whereas formal verification is usually restricted to a model of the applet only. Compared with traditional, manual testing, formal testing has as first advantage that the formality reduces the ambiguities and misinterpretations in the specification so that it is clearer what should be tested. Secondly, formal specifications allow to completely automate the testing process: test cases are algorithmically generated from the formal specification, and test results can automatically be analyzed. This makes it possible to generate and execute large quantities of large tests in a short time. It is mainly this second advantage that we will pursue in this paper.

Our investigation of formal testing is conducted using a case study. The applet that we test is a simple electronic purse application, implemented as a Java Card applet, with a limited set of methods like asking the value on the card, debiting, crediting, etc. The formal specification of the electronic purse applet is given as a Statechart[1]. The case and its formal specification are described in Section 2.

Automatic test generation, test execution, and test result analysis are performed in an *on-the-fly* fashion, using the test tool GAST[2]. The test tool GAST is described in Section 3. Section 4 describes how the Statechart specification is transformed into Clean[3], a functional programming language used as the input language for GAST. How GAST, the applet under test, and the platform are connected is described in the test architecture, which is given in Section 5.

We constructed one (assumed to be) correct applet implementation. From this implementation we derived 22 mutants by inserting subtle bugs to see whether such bugs would be detected by our automated, formal testing method. A summary of the performed tests is given in Section 6. Finally, Section 7 and 8 discuss related work, conclusions, and possible future extensions.

2.   CASE STUDY

We demonstrate our testing methodology by applying it to a simple electronic purse application as a case study. The basic events that the electronic purse can receive are:

- set an initial value n via `setValue(n)`

- query the actual value via `getValue()`
- pay an amount of n via `debit(n)`
- authenticate with a `pin` (personal identification number) via `authenticate(pin)` before charging the card
- charge the card with an amount of n via `credit(n)`
- reset the card using a `puk` (personal unlocking key) via `reset(puk)`

   All these events are input events for the card, because they are sent from the Card Accepting Device (CAD, also called *terminal*) to the card. To every input event, the card answers with a corresponding output event:

- acknowledge an operation via `ackOK` or `ack(n)`
- report an error via `error(n)`

   Figure 1 shows the specification of the purse, modeled as a Statechart.

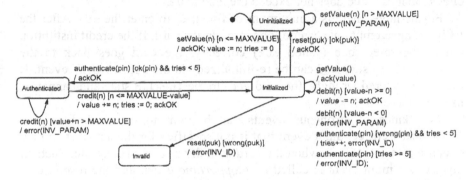

Figure 1. Statechart model of the purse applet.

The transition labels between two states s_1 and s_2 are of the form:

$$s_1 \xrightarrow{\; i[g]/\,act \;} s_2$$

with i being an input event, g being a guard, and *act* representing a sequence of actions. We exemplify the semantics with this transition:

$$Autenticated \xrightarrow{\; credit(n)[n \le MAXVALUE - value]/value += n;tries:=0;ackOK \;} Initialized$$

The input event (i) is `credit(n)`. The argument n represents the amount of money to be added to the card. The applet uses signed 16-bit integer shorts and it gives an `error(INV_PARAM)` on negative values. We abstract from that in the Statechart to keep it concise. The actual value of the card is saved in the variable `value`. A transition can only fire when the corresponding guard g holds. In this example, one can only increase the value of the card by n, when n does not exceed the `MAXVALUE-value`. If the transition is taken, the actions *act* are performed. In this case, the variable `value` is incremented by n, the `tries` variable is reset to zero, and the acknowledgment `ackOK` is sent to the terminal.

Intuitively, the purse works as follows. At first, the card is in the `Uninitialized` state. It is initialized by the credit institution, which issues the card to the customer by putting a certain amount n of money on it via the `setValue(n)` event.

In the `Initialized` state the customer can query the actual value via the `getValue()` event, or pay with the card via the `debit(n)` event. To increase the value, one must first authenticate at a terminal with a card-specific `pin`, leading to the `Authenticated` state. Being in that state, one can add money via the `credit(n)` event, leading back to the `Initialized` state. The card checks that its value does not exceed the `MAXVALUE`.

Furthermore, there is a maximum of five tries to enter the `pin`. After the fifth wrong attempt, one can no longer credit the card. If the credit institution enters the reset code (called `puk`) correctly, the card goes back to the `Uninitialized` state and can be re-initialized via the `setValue(n)` event. If the `puk` is entered wrongly, the card goes to the `Invalid` state and cannot be used anymore.

Two kinds of erroneous events can be sent to the card. Firstly, a syntactically correct input event that is not specified for the actual state may occur, e.g., a `credit(n)` when the card is in the `Initialized` state. Such an unspecified input event is called an *inopportune event*, and the response of the applet should be an error message `error(INV_CMD)`, whereas the applet remains in its actual state. Secondly, a syntactically incorrect event may occur, e.g., a command-APDU with a non-existing event-code. This is also implicitly assumed to lead to an error message, while the card stays in its current state.

3.   THE TEST TOOL GAST IN A NUTSHELL

The test tool GAST is designed to be open and extendable. For this reason it is implemented as a library rather than a standalone tool. The functional programming language Clean is chosen as host language due to its expressiveness.

GAST can handle two kinds of properties. It can test properties stated in logic about (combinations of) functions. GAST can also test the behavior of reactive systems based on Extended (Finite) State Machines, E(F)SM. Here we will only discuss the ability to test reactive systems.

An ESM as used by GAST comes quite close to the Statechart of Figure 1. It consists of states with labelled transitions between them. A transition is of the form $s \xrightarrow{i/o} t$, where s, t are states, i is an input which triggers the transition, and o is a, possibly empty, list of outputs. The domains of the inputs, outputs, and states can be given by arbitrarily complex recursive Algebraic Data Types (ADT). This constitutes the main difference with traditional testing with FSM's where the testing algorithms can only handle finite domains and deterministic systems[4].

A transition $s \xrightarrow{i/o} t$ is represented by the tuple (s, i, t, o). A relation based specification δ_r is a set of these tuples: $\delta_r \subseteq S \times I \times S \times O^*$. The transition function δ_f is defined by $\delta_f(s,i) = \{ (t, o) \mid (s, i, t, o) \in \delta_r \}$. Hence, $s \xrightarrow{i/o} t$ is equivalent to $(t, o) \in \delta_f(s, i)$. A specification is *partial* if for some state s and input i we have $\delta_f(s, i) = \varnothing$, otherwise it is *total*. A specification is *deterministic* if for all states and inputs the size of the set of targets contains at most one element: $\#\delta_f(s, i) \leq 1$. A *trace* σ is a sequence of inputs and associated outputs from a given state. Traces are defined inductively: the empty trace connects a state to itself: $s \xrightarrow{\varepsilon} s$. We can combine a trace $s \xrightarrow{\sigma} t$ and a transition $t \xrightarrow{i/o} u$ form the target state t, to trace $s \xrightarrow{\sigma;i/o} t$. We define $s \xrightarrow{i/o} \equiv \exists t.s \xrightarrow{i/o} t$ and $s \xrightarrow{\sigma} \equiv \exists t.s \xrightarrow{\sigma} t$. All traces from state s are: $traces(s) \equiv \{ \sigma \mid s \xrightarrow{\sigma} \}$. The inputs allowed in state s are given by $init(s) = \{ i \mid \exists o.s \xrightarrow{i/o} \}$. The states after trace σ in state s are given by s after $\sigma \equiv \{ t \mid s \xrightarrow{\sigma} t \}$. We overload *traces, init,* and *after* for sets of states instead of a single state by taking the union of the individual results. When the transition function, δ_f, to be used is not clear from the context, we will add it as subscript.

The basic assumption for our formal testing is that the Implementation Under Test, iut, is also a state machine. Since we do black box testing, the state of the iut is invisible. The iut is assumed to be *total*: any input can be applied in any state. Conformance of the iut to the specification spec is defined as (s_0 is the initial state of spec, and t_0 of iut):

$$\text{iut } conf \text{ spec} \equiv \forall \sigma \in traces_{\text{spec}}(s_0), \forall i \in init(s_0 \text{ after}_{\text{spec}} \sigma), \forall o \in O^* \bullet$$

$$(t_0 \text{ after}_{\text{iut}} \sigma) \xrightarrow{i/o} \Rightarrow (s_0 \text{ after}_{\text{spec}} \sigma) \xrightarrow{i/o}$$

Intuitively: if the specification allows input i after trace σ, the observed output of the iut should be allowed by the specification. If spec does not specify a transition for the current state and input, anything is allowed. This notion of conformance is very similar to the ioco relation[5] for Labeled Transition Systems (LTS). In an LTS each input and output is modeled by a separate transition. In our approach an input and all induced outputs up to *quiescence* are modeled by a single transition. Quiescence characterizes a state of the iut that will not produce any output before a new input is provided, i.e. a quiescent system waits for input and cannot do anything else.

In order to test conformance, a collection of input sequences is needed. At the beginning of each input sequence GAST resets the iut and the spec to their initial state. By applying the inputs of a sequence one by one, GAST investigates if it can be transformed to a trace of spec. The previous inputs and observed responses are remembered in trace σ. If $\delta_{spec}(s, i) \neq \emptyset$ for the current input i and some state s reachable from the いにちあ 1 state, S_0, by trace σ (i.e. $s_0 \xrightarrow{\sigma} s$), the input is applied to the iut, and the observed output is checked to be allowed by spec.

GAST has several algorithms for input generation, e.g.:
- Systematic generation of sequences based on the input type.
- Sequences that cover all transitions in a *finite* state machine.
- Pseudo random walk through the transitions of a specification.
- User defined sequences.

In this paper we will only use the third algorithm to generate input sequences. Testing is *on-the-fly*, which means that input generation, execution, and result analysis are performed in lockstep, so that only the inputs actually needed will be generated. The lazy evaluation of Clean used for the implementation of GAST makes this easy.

Within the test tool GAST, the mathematical state transition function, δ_f, specifying the desired behavior is represented by a function in the functional programming language Clean. Functional languages allow very concise representations of specifying functions and have well understood semantics. Using an existing language as notation for the specification prevents the need to design, implement and learn a new language. The rich data types and available libraries enable compact and elegant specifications. The advanced type system of functional languages enforces consistency constraints on the specification, and hence prevents inconsistencies in the specification.

Since the specification is a function in a functional programming language, it can be executed. This is convenient when one wants to validate the specification by observation of its behavior.

Any Clean type can be used to model the state, the input and the output of the function specifying δ_f, including user-defined data types. GAST uses generic programming techniques for generating, comparing, and printing of these types. This implies that default implementation of these operations can be derived without any effort for the test engineer. Whenever desired, these operations can be tailored using the full power of the functional programming language.

4.   THE PURSE SPECIFICATION FOR GAST

The specification given in the Statechart is transformed to the functional language Clean in order to let GAST execute and manipulate it. This section gives some details about the representation in Clean of the electronic purse from Figure 1. Due to space limitations we will only show snapshots of the (executable) specification. A parameterized enumeration type represents the state of the purse

```
:: PurseState = Uninitialized | Initialized Short Short
             | Authenticated Short | Invalid
```

We use one constructor for each state from the Statechart in Figure 1. The arguments of the constructor Initialized represent the tries counter and the value. The type Short represents signed 16-bit integers. This implies that there are actually $2^{16} \times 2^{16} = 2^{32}$ different initialized states, of which some are not reachable. There are similar types for input and output.

A transition function purse, similar to δ_f, in Section 3 models the transitions. The only difference with the mathematical specification is that the result is a list of tuples instead of a set of tuples. Some function alternatives specifying characteristic transitions are:

```
purse :: PurseState PurseInput -> [(PurseState, [PurseOutput])]
purse Uninitialized (SetValue n)
    = if (n >= 0 && n <= MAXVALUE)
        then [(Initialized 0 n, [AckOK])]
        else [(Uninitialized, [Error INV_PARAM])]
purse (Initialized tries value) Reset
    = [(Uninitialized, [AckOK])]
purse (Initialized tries value) GetValue
    = [(Initialized tries value, [Ack value])]
...
purse state any = [(state, [Error INV_CMD])]
```

The first alternative models both transitions for the input SetValue n from the state Uninitialized. The second and third alternative show two transitions form the state Initialized. The last alternative captures the informal requirement that inopportune events should cause no state transition and an error message as output. Since state and input any are variables, this alternative covers any combination not listed above. Since exactly one transition is defined for each combination of state and input, the specification is total and deterministic.

This specification is an ordinary definition in the functional programming language Clean. It is checked by the compiler before it is used by GAST. This guarantees well-defined identifiers and type correctness.

5.   TESTING JAVA CARDS WITH GAST

The tests, which will be described in Section 6, have been executed using the test architecture of Figure 2.

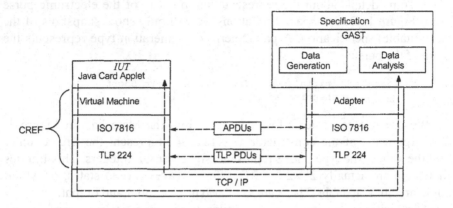

Figure 2. The general testing framework.

The IUT is the Java Card applet implementing our simple electronic purse. To make testing easier and more flexible, we used a simulation platform to execute the applet. The simulation environment is the *C-language Java Card Runtime Environment* (CREF), which comes with the *Java Card Development Kit*. CREF simulates a Java Card technology-compliant smart card in a card reader. It further consists of a Java Card Virtual Machine, and communication protocol entities to allow communication between the applet and the outside world.

To communicate with the applet under test, GAST was enhanced to be able to deal with these typical smart card communication protocols *ISO-7816-4* and *TLP-224* over TCP/IP. On top of these protocol entities an *adapter* (glue code), was implemented. The adapter transforms the high level inputs, generated by GAST, and represented as Clean data values, into the low-level APDUs, coded as appropriate byte codes, and then sent according to the ISO-7816-4 protocol. Vice versa, the adapter decodes the APDUs received from the applet under test to Clean data values, which are then analyzed and checked by GAST.

For data generation and analysis GAST uses the Clean EFSM specification, which was developed in Section 4. Except for the access to TCP/IP, the right-hand side of Figure 2 was entirely implemented in Clean.

The use of a simulation platform for testing is not a restriction with respect to testing of real smart cards. Since only standardized protocols are used, GAST cannot see the difference between testing on a simulator, and testing a real smart card. The test architecture could easily be adapted to test real cards by swapping CREF with a real card and its reader. The use of a simulation platform does facilitate easy switching between different applets, and saving and restoring applet state.

6.   RESULTS

The Statechart in Figure 1 and its implementation as an applet were developed in an incremental way. GAST appears to spot differences between the specified and actual behavior very rapidly. Once the specification and implementation were finished, the testing power of the GAST system was determined in a systematic way using *mutants*. Starting from the ideal (assumed to be correct) applet we injected typical programming errors into the applet, and analyzed how long it took GAST to find errors by generating, executing, and analyzing tests. The mutants are obtained by subtle changes like omitting checks or updates to the state of the applet. The test results for the 22 mutants used are listed in Table 1.

Table 1. Overview of test results.

nr.	paths	trans-itions	time (s)	inputs until error	comments
1	1	25	0.49	25	6 tries allowed in this mutant
2	2	66	0.09	31	incorrect overflow during `credit`
3	1	9	0.47	9	negative balance allowed in mutant
4	5	247	0.71	41	tries not reset after `authenticate`
5	8	406	0.38	51	tries not reset after `reset`
6	1	1	0.05	1	`credit` allowed without `authenticate`
7	1	1	0.52	1	`setvalue(0)` not allowed
8	1	4	0.06	4	`credit` with negative amount allowed
9	1	2	0.50	2	`debit` with negative amount allowed
10	11	542	0.48	23	no check for locked flag
11	7	327	0.80	26	not locked after 5 attempts
12	1	13	0.06	13	stays authenticated
13	21	1020	1.28	21	not locked after `reset`
14	1	16	0.09	16	MAXVALUE too low
15	1	24	0.07	24	`authenticate` does not authenticate
16	1	33	0.52	33	`reset` does not make it uninitialized
17	94	4757	3.82	29	`tries ≤ 5` instead of `tries < 5`
18	4	207	0.26	23	fresh card has nonzero balance
19	1	6	0.30	6	`setValue` allowed in initial state
20	3	145	0.18	44	`setValue` does not initialized/unlock
21	1	4	0.50	4	MAXVALUE too high
22	5	206	0.67	2	MAXVALUE balance not allowed
	7.8	366.4	0.56	19.5	averages
	100	5081	4.20	n/a	original applet, no counterexample

For instance, mutant 17 differs from the ideal applet by testing whether the number of remaining authentication tries is *less than or equal to* five rather than *less than* five before setting a flag indicating that the applet should no longer accept authentication attempts. This mutant was found after executing 94 paths, within 3.82 seconds, containing 4757 transitions in total. This mutant showed an invalid output after an input sequence of length 29 in path 94. To identify the error, the trace of inputs and associated responses are written to a file.

GAST was able to identify the 22 incorrect implementations without any help, using a minimum path length of 50 transitions and a maximum of 100 paths. It took an average of 0.56 seconds to generate and execute, on average, 366 transitions on a 1.4GHz Windows computer. Identifying incorrect behavior for all 22 mutants cost only 12 seconds in total. This shows that GAST is an efficient and effective test tool.

7.   RELATED WORK

Two approaches are closely related to ours due to the fact that both rely on tools that implement variants of the ioco testing relation[5]. Du Bousquet and Martin[6] use UML specifications, which are translated into Labeled Transition Systems to serve as input for the TGV tool[7]. Instead of an on-the-fly execution, TGV uses additional test purposes to generate test cases. The authors created a tool to automate the generation of test purposes based on common testing strategies. The generated test cases are finally translated into Java code, which communicates with the applet and executes the test. TGV does not treat data symbolically, which can easily lead to a state space explosion when dealing with large data domains. Because we generate test cases on-the-fly based on the (symbolic) EFSM, this problem does not occur.

To support symbolic treatment of data, Clarke et al.[8] use Input/Output Symbolic Transition Systems. The basic approach is similar to TGV, hence also here test purposes are needed. The test automation is done via a translation to C++ code, which is linked with the implementation. This restricts the IUT to be a C++ class with a compatible interface.

Rather than *testing* properties of the IUT, its implementation (i.e., the Java Card applet) can also be formally *verified*. Testing and verification are complementary techniques to check the correctness of systems, as explained in Section 1. A common technique used for verifying Java Card applets is to prove their correctness with respect to a specification in the Java Modeling Language (JML). State-based specifications similar to the one in Figure 1 can uniformly be translated to JML specifications, as shown by Hubbers,

Oostdijk, and Poll[9]. The resulting annotated Java Card applet can then be verified using one of the many JML tools[10], for instance, the ESCJava2 static analyzer[11]. Most Java Card applets are small enough to even attempt a formal correctness-proof using the Loop tool, as demonstrated by Jacobs, Oostdijk, and Warnier[12].

8. CONCLUSION AND FUTURE WORK

We have presented an approach to automate the testing of Java Card applets using the test tool GAST. The test case derivation is based on a Statechart specification of the applet under test. The specification can directly be translated into a corresponding GAST specification. Tests were completely automatically derived, executed, and analyzed. Discrepancies between the formal specification and its Java Card implementation were successfully detected, which shows the feasibility of this approach.

The direct translation from the Statechart model to the GAST specification, and the on-the-fly execution of the test cases enable the developer to start with automatic testing of the applet in the early stages of development. The co-development of the formal model and the implementation, and the facility to do automatic tests, has shown to be very useful. Both the code and the specification have evolved simultaneously, vastly improving the quality of the applet, and leading to a complete and reliable specification. Such a specification delivers insight on how to specify similar cases, and can serve as a pattern for these.

The tested mutants, representing typical programming errors, have increased our confidence in the error detecting power of the GAST algorithm. We are planning to compare this with other test tools, e.g., the ioco-based tool TorX[13], to test more complex applets, testing applets on real cards, and testing advanced aspects like the integration, interference, and feature interaction between different applets on one card.

Finally, we will compare the testing approach with the formal verification approach, e.g., using JML, to see how far we can get in unifying verification and testing techniques into one common framework, and to investigate the precise shape of their complementarities.

REFERENCES

1 UML resource page. http://www.uml.org.
2 P. Koopman and R. Plasmeijer. Testing reactive systems with GAST. In S. Gilmore, editor, *Trends in Functional Programming 4*, 111-129 (2004)

3 R. Plasmeijer and M. van Eekelen. *The Concurrent Clean Language Report*, version 2.0. http://www.cs.kun.nl/~clean.
4 D. Lee and M. Yannakakis. Principles and methods of testing finite state machines - a survey. *Proc. IEEE*, 84(8):1090--1126 (1996)
5 J. Tretmans. Test generation with inputs, outputs and repetitive quiescence. *Software-Concepts and Tools*, 17(3):103-120 (1996)
6 L. du Bousquet and H. Martin. Automatic test generation for Java-Card applets. In *4th Workshop on Tools for System Design and Verification*, (2000)
7 C. Jard and T. Jéron. TGV: theory, principles and algorithms. In *IDPT '02*, Pasadena, California, USA, Society for Design and Process Science (2002)
8 D. Clarke, T. Jéron, V. Rusu, and E. Zinovieva. Automated test and oracle generation for smart-card applications. In *Proceedings of the International Conference on Research in Smart Cards*, volume 2140 of *LNCS*, 58-70, Cannes, France (2001)
9 E. Hubbers, M. Oostdijk, and E. Poll. From finite state machines to provably correct java card applets. In D. Gritzalis, S. De Capitani di Vimercati, P. Samarati, and S.K. Katsikas, editors, *Proceedings of the 18th IFIP Information Security Conference*, Kluwer Academic Publishers, 465-470 (2003)
10 L. Burdy, Y. Cheon, D. Cok, M. Ernst, J.R. Kiniry, G.T. Leavens, K.R.M. Leino, and E. Poll. An overview of JML tools and applications. In Th. Arts and W. Fokkink, editors, *FMICS '03*, volume 80 of *ENTCS*, pages 73-89 (2003)
11 D. Cok and J. Kiniry. ESC/Java2: Uniting ESC/Java and JML: progress and issues in building and using ESC/Java2. Submitted for publication (2004)
12 B. Jacobs, M. Oostdijk, and M. Warnier. Source code verification of a secure payment applet. *JLAP*, 58:107--120 (2004)
13 J. Tretmans and E. Brinksma. TorX: Automated model based testing. In A. Hartman and K. Dussa-Zieger, editors, *First European Conference on Model-Driven Software Engineering*. Imbuss, Möhrendorf, Germany (2003)

A COMPUTATIONALLY FEASIBLE SPA ATTACK ON AES VIA OPTIMIZED SEARCH

Joel VanLaven[1], Mark Brehob[2], and Kevin J. Compton[3]
EECS Department
University of Michigan - Ann Arbor
Ann Arbor, MI 48109-2122, USA
[1]jvanlav@umich.edu, [2]brehob@umich.edu, [3]kjc@umich.edu

Abstract: We describe an SPA power attack on an 8-bit implementation of AES. Our attack uses an optimized search of the key space to improve upon previous work in terms of speed, flexibility, and handling of data error. We can find a 128-bit cipher key in 16ms on average, with similar results for 192- and 256-bit cipher keys. The attack almost always produces a unique cipher key and performs well even in the presence of substantial measurement error.

Keywords: AES, SPA, Rijndael, power attack.

1.   Introduction

In 2001 the National Institute of Standards and Technology selected the block cipher Rijndael as the Advanced Encryption Standard (AES), making it the standard for private key encryption. A cryptographic attack on AES, such as linear or differential cryptanalysis, appears intractable at this time. Therefore, some researchers have investigated side-band attacks, which use information about the physical manifestation of the hardware or software implementing the algorithm [1, 2, 6].

Side-band attacks assume access to the hardware performing the encryption. Timing attacks, proposed in 1996, assume only the ability to time the encryption (or perhaps sub-portions of the encryption) [4]. Such attacks are relatively easy to thwart by writing encryption software that uses a fixed sequence of operations. Power attacks, another style of side-band attack, are more difficult to thwart. The most common power attacks assume the ability to observe the power utilization of the processor or ASIC over time [5].

Power attacks provide both high-level information about the operations being performed on the chip and low-level information about the data being operated upon. The high-level information is similar to timing information and can be dealt with in a similar way. Low-level information about the data arises from sources such as asymmetry in the efficiency of the n and p transistors or the flipping of bits on a bus or in a register. Without great care in the chip design and addition of power inefficiencies, a CMOS chip will use a slightly different amount of power based on the data being calculated [8]. Microprocessors operate on a fixed number of bits at a time (usually words or bytes), so what is actually revealed is the sum of the bits, or the Hamming weight of the data.

The two main variants of power attacks are differential power analysis (DPA), which requires the plaintexts or ciphertexts in addition to the power traces for many encryptions with the same key; and simple power analysis (SPA), which exposes the secret key solely from power traces [5]. While in theory most SPA attacks could reveal the key from a single encryption, poor signal-to-noise ratio forces averaging of the error over many encryptions with the same key. DPA is applicable to most ciphers and implementing such an attack is relatively straightforward. SPA is greatly affected by the design of a cipher and the susceptibility of a cipher to this style of attack may not be obvious.

This paper details an SPA power attack on an 8-bit implementation of AES. We assume that the Hamming weights of the bytes of the expanded key can be measured, possibly with some error. Our approach exploits regularities in the AES key schedule, which could likely be utilized even if different information more specific to the implementation is exposed. It improves upon a previously published attack by Mangard [6] in terms of speed, flexibility, and handling of measurement error. Specifically our algorithm improves upon this work in four ways.

- It runs approximately 20000 times faster.
- It nearly always finds a unique solution rather than a handful of candidate solutions.
- It works on cipher-key sizes of 192 and 256 in addition to 128 bits.
- It performs well under a more realistic error model.

Table 1. Time and Discovery Rate of Cipher Keys

	128-bit key no error	256-bit key no error	128-bit key with error ($\sigma = 0.25$)
Average time	16ms	20ms	35s
% of attacks with a unique solution	100%	99.97%	96%

With regard to the second item, the previous work required a ciphertext/ plaintext pair to find the correct solution, thus negating a primary advantage of an SPA attack. While the algorithm in this paper cannot guarantee a unique solution, our results show that even with significant errors in the data, it almost always finds a unique solution. Table 1 provides a summary of our results.

2.   AES

In private-key cryptosystems such as AES, both the sender and receiver of a message require access to the same secret key. Public-key cryptosystems allow the sender and receiver to use different keys, only one of which needs to be secret, but require significantly more computational power as well as a significantly longer key. AES (like the DES standard that it replaced) is an iterative block cipher. This means that the data is manipulated in series of "mini-encryptions," called rounds, each of which uses its own key. In order to generate these *round keys* the AES algorithm expands the 128-, 192-, or 256-bit private key (also called the *cipher key*) into the needed number of 128-bit round keys using the key expansion algorithm described below.

Key Expansion in AES

Our attack exploits the relationships between the round keys resulting from patterns in the key expansion algorithm. As such, it is necessary to carefully describe the algorithm found in the AES specification [3]. The key expansion algorithm is slightly different depending upon the cipher key size. Though our attack works on all three different key sizes, for simplicity we will discuss only the 128-bit key expansion (which is the most commonly used). The 192-bit and 256-bit key expansions are similar, and the results of attacks on those key sizes are summarized in section 5.

The 128-bit cipher key is expanded into eleven 128-bit round keys, each of which can be thought of as 16 bytes arranged in a 4-by-4 block. Each successive round key is simply a transformation of the previous round key. Define $RK[N, R, C]$ for $N = 0,...,10$, $R = 0,...,3$ and $C = 0,...,3$, to be the byte found in the N-th round key at row R and column C. The first round key (i.e.

the round key for $N = 0$) is a copy of the 128-bit cipher key. When $N > 0$, the round key $RK[N, R, C]$ is equal to [24]

$$\begin{cases} RK[N-1,R,C] \oplus RK[N,R,C-1], \text{if } C > 0; \\ RK[N-1,R,C] \oplus SB[RK[N-1,R-1,3]], \text{if } R > 0 \text{ and } C = 0; \\ RK[N-1,R,C] \oplus SB[RK[N-1,3,3] \oplus RC[N], \text{if } R = 0 \text{ and } C = 0. \end{cases}$$

Here $\oplus$ is the XOR function; SB is an invertible function, called the *subbyte* function, which maps bytes to bytes; and $RC[N]$ is the N-th round constant, a fixed value independent of the cipher key. The AES standard gives the precise definitions of SB and $RC[N]$. Each byte other than those in the cipher key is computed from exactly two other bytes. For example, when $N > 0$ and $C > 0$,

$$RK[N,R,C] = RK[N-1,R,C] \oplus RK[N,R,C-1]$$

but then we also have

$$RK[N-1,R,C] = RK[N,R,C] \oplus RK[N,R,C-1]$$
$$RK[N,R,C-1] = RK[N,R,C] \oplus RK[N-1,R,C]$$

That is, the computational relationship between bytes is symmetric in the sense that each of the bytes is computable from the other two. This is also true in the cases where $N > 0$ but $C \neq 0$, the only difference being that we have slightly more complicated expressions involving the SB function and $RC[N]$ constants. We picture all of these computational relationships in the hypergraph of Figure 1.

---

[24] The notation here is not the same as the round key function $W[i,j]$ in [3]. The relationship between the two notations is $RK[N,R,C] = W[-R \bmod 4, 4N + C]$. Our notation was chosen to make our description of the key schedule structure clearer.

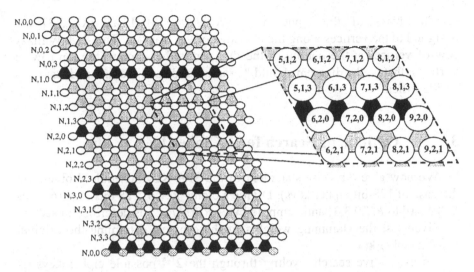

Figure 1. The Key Schedule Hypergraph for 128-bit Cipher Keys

A *hypergraph* is a pair (V, E) where V is the *vertex set* and $E \subseteq 2^V$ is the *hyperedge set*. If we require that all hyperedges contain exactly two vertices, we have the usual definition of a graph. In our hypergraph, the vertices are the bytes of the round keys and the hyperedges are the 3-element sets of computationally related bytes.

The ovals in the diagram represent the vertices of the hypergraph. For clarity, we have not labeled every vertex in the diagram. In the row labeled $N,0,0$, for example, the eleven vertices should be labeled

$$(0,0,0),(1,0,0),(2,0,0),...,(10,0,0)$$

and the bytes assigned to them are

$$RK[0,0,0],RK[1,0,0],RK[2,0,0],...,RK[10,0,0]$$

respectively. The figure shows a blowup of a small section of the hypergraph with the vertices labeled. The shaded triangles represent the hyperedges; the light shaded triangles represent the hyperedges where the subbyte relation is not used to compute the computational relationship. For example, since

$$RK[1,0,1] = RK[0,0,1] \oplus RK[1,0,0]$$

we have a light hyperedge $\{(1,0,1),(0,0,1),(1,0,0)\}$, which is the upper-leftmost shaded triangle. From

$$RK[1,1,0] = RK[0,1,0] \oplus SB[RK[0,0,3]]$$

we get a dark shaded hyperedge $\{(1,1,0),(0,1,0),(0,0,3)\}$.

The bytes of the cipher key $RK[0,0,0], RK[0,0,1], ..., RK[0,3,3]$ are assigned to the vertices along the left edge of the diagram. The slightly fuzzy row of vertices at the bottom of the diagram is the same as the top row of vertices; that is, the diagram should "wrap around" and the first and last rows be identified.

3.   Optimizing Search for a Cipher Key

We now give a precise statement of the SPA Key Schedule Problem (in the case of 128-bit cipher keys). Divide a cipher key of 128 bits into 16 bytes $RK[0,0,0]$ to $RK[0,3,3]$ and compute bytes $RK[N,R,C]$ as described in section 2. Given just the Hamming weights of these bytes, determine the original 128-bit cipher key.

An exhaustive search, cycling through the 2^{128} possible cipher keys, is clearly infeasible. Even if we cycle through only those keys where each byte of the cipher key has the correct Hamming weight, the number of possible keys could be as large as 2^{98}, still far too large to search. We need to utilize the Hamming weights of the entire expanded key to reduce the search space to a manageable number of keys.

A naive approach would be to search for the cipher key by sequentially assigning possible values for the bytes $RK[0,0,0]$ to $RK[0,3,3]$ (i.e., those bytes for which $N = 0$) and checking consistency with the Hamming weight information after each assignment. Inspection of Figure 1 shows that this is little better than an exhaustive search. After we have assigned values to $RK[0,0,0]$ and $RK[0,0,1]$, for example, we have no further information about Hamming weights of other bytes in the key schedule since they do not belong to a common hyperedge.

Suppose, instead, that we assign possible values for $RK[0,0,0]$ and $RK[1,0,0]$ corresponding to vertices in the bottom row of the hypergraph. We can then compute $RK[0,3,3]$ and check three values (rather than just two as before) for consistency with the Hamming weight information. This improves on exhaustive search because it eliminates many possible assignments. This is the main idea behind our search sequence optimization.

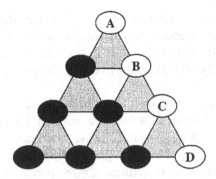

Figure 2. A Fragment of the Key Schedule Hypergraph

Systematic use of this idea results in a highly optimized search. Consider the small fragment of the hypergraph in Figure 2. Suppose we have assigned values consistent with the Hamming weight information for the six shaded vertices. If we then make an assignment to vertex A, we can compute values for vertices B, C, and D, then check that these values are consistent with the Hamming weights. Notice also that if we have values assigned to the six shaded vertices, assigning a value to any one of the vertices A, B, C or D allows us to compute values for the other three.

Thus, there are many ways to choose a sequence so that the maximum number of byte values can be computed after successive assignments to vertices in the sequence. However, it is not difficult to see that after each assignment (for at least the first 11 assignments), the pattern of computable values will be a triangular array of the type shown in Figure 2. That is, we can find a vertex sequence $S_0, S_1, ..., S_{15}$ so that after values have been assigned to $S_0, ..., S_i$, we can compute $(i+1)(i+2)/2$ byte values in a triangular array. When $i \geq 11$ a complete triangular array will not fit horizontally in the hypergraph shown in Figure 1 so the increase in the number of computable values is not as great. However, by this stage so many values are determined that maximizing the number of computable values is not so important (Figure 4 illustrates this). After assignments to $S_0, ..., S_{15}$, all 176 bytes can be computed because of the wrap-around in the hypergraph.

Besides maximizing the number of computable values after each assignment, speedups can be gained by taking advantage of information available from the subbyte operation used in the dark hyperedges. Maximizing the number of dark hyperedges contained within the triangular array of computable values results in additional pruning in the early stages of the search. This can give approximately an order of magnitude speedup. Because the subbyte is applied to the top vertex in any dark hyperedge, it is possible to easily extract this information without determining the two lower vertices of such a hyperedge. Maximizing the number of top vertices of dark hyperedges determined instead of whole dark hyperedges, allows a further speedup by a factor of about 2. The search sequence we use is: (9,0,3), (8,0,3), (7,0,3), (6,0,3), (5,0,3), (4,0,3), (3,0,3), (2,0,3), (1,0,3), (0,0,3), (0,1,0), (0,1,1), (0,1,2), (0,1,3), (0,2,0), (0,2,1). There are many other optimal sequences.

We can now give a precise description of the search algorithm. Let $S_0,S_1,...,S_{15}$ be a fixed optimal search sequence of 16 vertices as described above. Suppose that at some time during the search, values have been assigned to $S_0,...,S_i$ and are stored in a global array A. Let consistent(i) be a Boolean function that returns true precisely when the values computed from these $i+1$ values are consistent with the Hamming weight information and the information from dark hyperedges mentioned in the previous paragraph. Thus, when $i < 11$, consistent checks the consistency of $(i+1)(i+2)/2$ values.[25]

The search algorithm is a standard backtrack algorithm. Pseudocode for a recursive version of the algorithm is given in Figure 3. (Our implementation was iterative, to optimize performance, but the recursive version here is a little more transparent.) Function search(n) cycles through possible assignments to S_n, storing them in an array A at index n. For those bytes that are consistent with the Hamming weight information, the search goes on to search(n+1). For those that are not consistent, it goes on to the next possible byte. If all the bytes have been checked, it returns to the last calling search. Whenever n reaches 16, it writes out a possible solution stored in A. To run the algorithm we initially call search(0).

```
void search(n)
{
        if (n==16) write A;
        else
                foreach byte w
                {
                        A[n]=w;
                        if (consistent(n))
                                search(n+1);
                }
}
```

Figure 3. Pseudocode for the Recursive Search Algorithm

[25] In fact, it is really only necessary to check the consistency of the $i+1$ values added since the last consistency check.

4.   The Attack in the Presence of Error

While work by Mayer-Sommer has shown that it is possible to determine Hamming weights in "an unequivocal manner" [7], measurement and data collection will inevitably have an associated error rate. We model this error by adding Gaussian noise with a mean of zero for each of the measured Hamming weights. This model is reasonable, because even if the actual distribution of noise for a single run is not Gaussian, averaging a number of independent noise measurements together should yield an overall distribution that approaches Gaussian. A mean of zero should be obtainable as part of the method used to calibrate the measurements of the Hamming weights.

The measured Hamming weights with this error assumption will be real-valued rather than discrete. Further, it is not possible to search for the exact key that matches the given Hamming weights: all keys match, but some keys are more likely than others. Our attack provides all cipher keys (if any exist) whose round key expansions have Hamming weights differing from the measured Hamming weights by less than some bound (using a sum of the squares metric). Given our assumptions about the nature of the error, the sum of squares difference is a maximum likelihood estimator. This means that any key not reported is less likely than any key that is reported. The bound can be chosen to guarantee with some confidence (e.g. 95%), given an expected amount of error, that the true key will be returned.

Changes to the Algorithm

When dealing with error, the definition of the function `consistent` from Figure 3 needs to be modified. Specifically, `consistent` will return `false` if the assignment to S_n gives a sum-of-squares difference greater than a certain bound. Our implementation computes an optimistic estimate of this value. It is computed as the sum of two values:

- The sum of squares difference between the Hamming weights of those bytes determined by S_n and their measured values.
- The minimum sum of squares difference between the measured values of all those bytes which are not determined by S_n and the integer values closest to those measured values.

The bound is determined by adding a fixed value based on desired confidence to the minimum sum-of-squares difference between the measured values and the integer values closest to those values. This method of determining the bound helps to make the work required by the algorithm more uniform than using a fixed bound based on desired confidence.

5.   Results

We ran all of our simulations on a 500MHz Sun Blade 100 with 512MB of DRAM. A synopsis of our simulation results can be found in Table 2. As noted in section 4, the bound for the search is determined by the desired confidence given an expected amount of error. The last four entries in the table all assume a different amount of error. The expected amount of error is expressed as a standard deviation of the expected Gaussian noise. For all of the runs with error we targeted a confidence rate of 95%.

Table 2. Time and Discovery Rate of Cipher Keys

Attack type	Average time per attack	% of attacks with a unique solution
128-bit no error	16ms	100%
192-bit no error	60ms	100%
256-bit no error	20ms	99.97%
128-bit, $\sigma = 0.20$	4s	95%
128-bit, $\sigma = 0.25$	35s	96%
128-bit, $\sigma = 0.30$	38 min	**
128-bit, $\sigma = 0.35$	15 hours	**

For the entries labeled **, not enough data was collected to provide a meaningful value

Notice that the run time of our algorithm increases exponentially relative to the expected amount of noise. This is because a higher expected error rate forces us to have a looser bound, and that reduces the amount of pruning of the search space that can be accomplished. Figure 4 graphically shows how large an impact this has. An implication of Table 2 and Figure 4 is when a large amount of noise is present our algorithm's runtime will become untenable.

Another interesting result from Table 2 is that the 192-bit implementation takes longer than either the 128 or 256 bit implementations. This is because the 192-bit version of AES has fewer instances of *subbyte* per expanded key byte than the 128 or 256 bit versions.

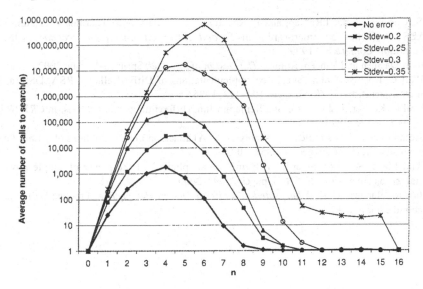

Figure 4. Average number of calls to search as a function of n

6.   Conclusions and Future Work

We have shown that AES is susceptible to very efficient attacks based solely upon Hamming weights of the bytes of the expanded key. These Hamming weights would likely be exposed in the case of an 8-bit implementation, even if a pre-expanded key were used. Further, this algorithm works well even in the presence of significant Gaussian noise.

This work can be extended in the following ways:

- Modifying the algorithm to work with 16 or 32 bit implementations without a pre-expanded key.
- Proving information theoretic bounds about the feasibility of this SPA attack on 16 and 32 bit implementations with pre-expanded keys.
- Significantly improving the algorithm's efficiency in the face of error to handle larger errors.
- Gathering the data and performing the attack on a real system.

REFERENCES

[1] E. Biham and A. Shamir. Power analysis of the key scheduling of the AES candidates. In *Second Advanced Encryption Standard (AES) Candidate Conference*, 1999.

[2] S. Chari, C. Jutla, J.R. Rao, and P. Rohatgi. A cautionary note regarding evaluation of AES candidates on smart-cards. In *Second Advanced Encryption Standard (AES) Candidate Conference*, 1999.

[3] Joan Daemen and Vincent Rijmen. *The Design of Rijndael*. Springer-Verlag, 2002.

[4] Paul C. Kocher. Timing attacks on implementations of Diffie-Hellman, RSA, DSS, and other systems. In *CRYPTO*, pages 104–113, 1996.

[5] Paul C. Kocher, Joshua Jaffe, and Benjamin Jun. Differential power analysis. In *CRYPTO*, pages 388–397, 1999.

[6] Stefan Mangard. A simple power-analysis (SPA) attack on implementations of the aes key expansion. In *ICISC*, pages 343–358, 2002.

[7] Rita Mayer-Sommer. Smartly analyzing the simplicity and the power of simple power analysis on smartcards. In *CHES*, pages 78–92, 2000.

[8] Jan M. Rabaey. *Digital Integrated Circuits: a Design Perspective*. Prentice-Hall, Inc., second edition, 2002.

THE PROOF BY 2M-1: A LOW-COST METHOD TO CHECK ARITHMETIC COMPUTATIONS

Sylvain Guilley and Philippe Hoogvorst
Ecole Nationale des Télécommunications, 46, rue Barrault, 75634 Paris CEDEX 13, France.
E-mail :sylvain.guilley@enst.fr ,philippe.hoogvorst@enst.fr

Abstract: Injecting faults into an arithmetic device is a way of attacking cryptographic devices.The *proof by 2^m-1* is a method to detect arithmetic errors induced by this attack without having to duplicate the computations. This method is simple and not too expensive, in terms of computation power when the arithmetic in software and in terms of both silicon surface and power consumption when the arithmetic operations are performed by a hard-wired operator. In that the *proof by 2^m-1* is well-suited for martcards, in which these resources are limited. The *proof by 2^m-1* is scalable, in that the designer can choose the parameter m, which determines the level of protection offered and the resources needed for the verification.

Key words: Fault injection, modular computations, Security, Modular Computations.

1. INTRODUCTION

The injection of errors during a cryptographic computation is an efficient attack provided the attacker has access to the target. A smartcard is its typical target. Its power is such that a single uccessful fault induction is enough to break RSA cite [Boneh et *al., 1997, Siefert, 2002].*

The *proof by 2^m-1* is a self-test method which can verify arithmetic computations either in Z or in Z/nZ, for some (big) integer n and thus detect such fault induction. It is specific to countering fault injection: it does not address any other class of attack, such as DPA [Kocher et *al.,* 1999], SPA, timing attack [Kocher et *al.,* 1996] or EMA [Quisquater and Samyde, 2001].

The *proof by 2^m-1* is scalable, in that the designer can choose the parameter m, which determines the level of protection offered and the resources needed for the verification.

All cryptographic applications based on arithmetic operations can take advantage of the *proof by 2^m-1*, such as, [Rivest, Shamir and Adleman, 1978], the Diffie & Hellman protocol [Diffie and Hellman, 1976].

Section 2 exposes the principles of using arithmetic residue codes to check arbitrary size arithmetic computations. Section 3 shows the case of decimal computation.Then we switch to a larger modulus. Section 4 exposes how to check modular multiplications. Section 5 concludes the article.

2.   PRINCIPLES OF THE *PROOF BY 2^M-1*

Let $(Z,+,\times)$ be the ring of integers, $M > 1$ an integer, $(Z/MZ,\hat{+},\hat{\times})$ the ring of residual classes modulo M, and $P_M(.)$ be the canonical ring homomorphism $(Z,+,\times) \rightarrow (Z/MZ,\hat{+},\hat{\times})$.

The *proof by 2^m-1* uses the properties of $P_M(.)$ to verify the computations performed by a cryptographic device subject to fault injection with a small computational overhead, which is nearly independent of the size of the operands.

Other works [Noufal and Nicolaidis, 1999] use similar concepts but focuses on the synthesis of hardware multipliers. The authors use dedicated structures for fixed-width data paths that cannot be scaled to an arbitrary operand size.

Our solution can be implemented using either regular software or off-shelf hardware structures, available in CAD vendors libraries.

3. VERIFYING COMPUTATIONS IN Z

3.1 Proof by 9

When no pocket calculator is available, a simple way exists to check multiplications, which is a lot simpler than doing them twice: the *proof by 9*, which consists in repeating the multiplication modulo 9. If the actual result of the multiplication and the one obtained with the reduced operands are not congruent modulo 9, one of the results is wrong.

Calculating in $Z/9Z$ is easy because any operation involves only 1-digit numbers. As for any positive integer i, 10^i is congruent to 1 modulo 9, reducing a number n modulo 9 comes down to adding its decimal digits and reiterating the process until the sum is strictly smaller than 10. At the end a result equal to 9 is replaced by a 0 if necessary.

This process is guaranteed to stop because, if n consists $p > 1$ digits in base 10, i.e. $n = \sum_0^{p-1} n_i 10^i$, then $n' = \sum_0^{p-1} n_i$ is strictly smaller than n . The successive sums form a strictly decreasing sequence of integers, which must fall below 10 after a finite number of steps.

The process of reduction is roughly linear in complexity with the size of the operands but the complexity of the multiplication in $Z/9Z$ is independent of this size as it always consist of always a single 1-digit by 1-digit multiplication, followed by the reduction of a 2-digit number.

The *proof by 9* **does not prove correctness**. To prove correctness the process should be repeated with different prime numbers until the chinese remainder theorem could be used. However $P_9(x)P_9(y) \neq P_9(xy)$ proves that an error occurred. Otherwise a random mistake is detected with probability 1/9. In addition, the *proof by 9* will detect any error which affects a single digit, except if a 9 was replaced by a 0 or vice versa.

3.2   Proof by M-1

A 1/9 probability of missing an error is not safe enough for security purposes. Besides, if the operation $A \times B$ was to be checked, computing $P_9(A)$ and $P_9(B)$ would involve a lot of computations, each of them possibly subject to other attacks.

However changing the modulus will at the same time yield a much better fault coverage and make it very easy to compute the proof. In particular, any error affecting a single digit in base M will be detected except if a $M-1$ is replaced by a zero or vice versa. As no single-bit fault can do that, any single-bit fault will always be detected. Again the cost of the computation modulo $M-1$ will be a lot lower than the one of the real computation. The *proof by* $M-1$ is thus a lot cheaper than the repetition of the computations.

3.3   Notations

Let $M>1$ an integer. Though M will usually be 2^m, with $m \in \{8,16,32,64\}$, we will not use this fact in the rest of the article more than deriving the name of the method from it.

An uppercase letter X represents a positive integer, whose base B decomposition is denoted $\left(..., X^{[2(B)]}, X^{[1(B)]}, X^{[0(B)]}\right)$, in which the $X^{[i(B)]}$ are an infinite sequence of integers in $[0, B-1]$ among which only a finite number of them is non-zero. Given a positive integer X, its size in base B, denoted $|X|_B$, is the smallest integer x such that $\forall j \geq x, X^{[j(B)]} = 0$.

To simplify the notations from now on we will use some shorthands. Unless otherwise specified:
- the canonical ring homomorphism $Z \rightarrow Z/(M-1)Z$ is denoted $P(.)$ instead of $P_{M-1}(.)$,
- $\hat{+}$ and $\hat{\times}$ operate in $Z/(M-1)Z$,
- the digits of X in base M are denoted X^i instead of $X^{(i[M])}$,
- the size of X in base M is denoted $|X|$ instead of $|X|_M$.

3.4   Mathematical basis

It is easy to prove that, for any positive integer n and any number M,
$$\forall n > 0, (M^n - 1) = (M-1)(M^{n-1} + M^{n-2} + ... + M + 1).$$
Computing modulo $M - 1$, this identity becomes:
$$\forall n > 0, M^n \equiv 1 \bmod (M-1) \tag{1}$$
From Eq. (1), given $A = \sum_{i=0}^{|A|} A^{[i]} M^i$, the evaluation of $A \bmod (M-1)$

consists in the addition of the $A^{[i]}$ with the following peculiarities:

1. a Carry$_{out}$ has weight M, thus is congruent to 1 modulo $(M-1)$, it can be re-injected as the Carry$_{in}$ of the next addition;
2. the reduction ends with the addition of a zero to ensure that the last Carry$_{out}$ is effectively added. This last addition can produce no Carry$_{out}$ because, at the preceding addition, each of the operands was at most $(M-1)$.Thus the value of the result was at most $2M - 2$, which can be rewritten as $M + (M-1) - 1$ and, given than $M \equiv 1 \bmod (M-1)$, the result of the last addition will be at most $(M-1)$ and no Carry$_{out}$ can be generated;
3. at the end if the result is exactly $(M-1)$, it is replaced by zero.

The number of additions needed to reduce a number A modulo $(M-1)$ is $|A| + 1$. It is proportional to $\log A$ and inversely proportional to $\log M$.

If this reduction has to be implemented in hardware, the data path consists in a conventional adder, in which the Carry$_{out}$ is connected to the Carry$_{in}$, together with the control circuitry which instructs the device to perform the final addition of a zero and, if necessary, the replacement of a $(M-1)$ by 0.

3.5   Checking additions and multiplications in Z

When computing with integers, the verification process is straightforward: given two operands A and B, an operation $*$ (which can be any of $+$, $-$ or $\times$) is performed and verified as follows:

1. evaluate $a = P(A)$ and $b = P(B)$,
2. evaluate $R = A * B$ in Z,
3. evaluate $r = a \hat{*} b$ in $Z/(M-1)Z$,
4. check that $P(R) = r$.

The operations involved are: the evaluations of $P(A)$ and $P(B)$, which cost $|A|+|B|+2$ additions, the evaluation of $P(R)$, which costs $|R|+1$ additions, the $\hat{*}$ operation between a and b, which is a single addition or multiplication, the reduction modulo $(M-1)$ of the result, which costs two additions at most.

Although they add overhead, the reductions of the operands must be performed only once, for the original operands. During the computation the reduced results of any operation will be kept together with the real results for use in the next operations. In that the overhead will be nearly independent of the computation performed.

3.6 Security of the test

Again the *proof by 2^m-1* can only prove non-correctness: that the results of the integer and of the modular operations are not congruent proves that an error occurred. Otherwise, the probability of an undetected error is exponentially decreasing with $|M|_2$ as shown on the table below.

1. $|M|_2$	1. Probability of an undetected error
2. 8	2. $2^{-8} = 3.92 \times 10^{-3}$
3. 16	3. $2^{-16} = 31.53 \times 10^{-5}$
4. 32	4. $2^{-32} = 2.33 \times 10^{-10}$
5. 64	5. $2^{64} = 5.42 \times 10^{-20}$

4.   CHECKING MODULAR COMPUTATIONS

Cryptographic operations seldom involve multiprecision computations in Z itself. Most of the time, the computations are done in Z/nZ, for some integer number n. As there exists no non-trivial ring homomorphism from Z/nZ into $Z/(M-1)Z$ (unless n is a multiple of $(M-1)$) the checking of the computations is a little more difficult.

4.1 Modular Addition

Given two operands in Z, A and B, each of them being in $[0, N-1]$, their addition modulo N is performed in three steps:

1. compute $R' = A + B$ in Z. As $A < N$ and $B < N$, $R' < 2N$;
2. if $R' < N$ set $R = R'$ else set $R = R'-N$.

None of these low-level operations is modular. It is thus possible to check the modular addition with two operations in $Z/(M-1)Z$:

1. compute $r' = P(A) \hat{+} P(B)$;
2. if $R' < N$ set $r = r'$ else set $r = r'-P(N)$.

Normally $P(N)$ will have been precomputed when N was set. Verifying a modular addition costs thus only one addition more than verifying it in Z.

4.2   Modular Multiplication

The multiplication modulo N of A and B, both in $[0, N-1]$ can be performed as:

1. let $T = A \times B$,
2. let Q and R be respectively the quotient and the remainder of the division of T by N,
3. the result is R.

As $A \times B = Q \times N + R$, the relation $P(A) \hat{\times} P(B) = P(Q) \hat{\times} P(N) \hat{+} P(R)$ must hold, which allows us to check the modular multiplication. However, a modular multiplication is never computed like that: the division is too expensive.

4.3   Binary modular multiplication

When no hard-wired multiplier is available, the multiplication in Z consists in a sequence of additions, together with left shifts for the multiplicand and bit tests for the multiplicator. Note that in all figures, the notation (X, x) stands for the couple $(X, P(X))$.

To transform the binary integer multiplication into a verifiable modular multiplication, we will first pretend to compute in Z but in base N. Each of B and P will be represented as digits in base N. As we are computing modulo N, we assume that $A < N$ and $B < N$, thus we have: $P = A \times B < N^2$ and $|P|_N \le 2$. As for B, its value will be under N^2 until the beginning of the last round of computation. The last doubling may yield a wrong result but we don't use this last B. Fig. 4.3.1 shows the multiplication with base N computations. For the sake of readability, we have replaced $P^{[1(N)]}$ by P_h, $P^{[0(N)]}$ by P_l, $B^{[1(M)]}$ by B_h and $B^{[0(M)]}$ by B_l.

1. mulmod $\left((A,a),(B,b),(N,n)\right)$	1.
1. {	2.
1. $P_h \leftarrow 0; P_l \leftarrow 0; B_h \leftarrow 0; B_l \leftarrow 0$;	3. // Initialize
1. **for**($i \leftarrow 0$; $i < \lvert A \rvert_2$; $B_{li} \leftarrow 0\ i{+}1$) {	4. // Loop on bits of A
1. **if**($A^{[i(2)]} \neq 0$) {	5. // Addition necessary?
1. $P_h \leftarrow P_h + B_h; P_l \leftarrow P_l + B_l$;	6.
1. **if**($P_l \geq N$){ $P_h \leftarrow P_h +1; P_l \leftarrow P_l - N$;}	7. // Correction modulo N
1. }	8.
1. $B_h \leftarrow B_h + B_h; B_l \leftarrow B_l + B_l$;	9. // Double B
1. **if**($B_l \geq N$){ $B_h \leftarrow B_h +1; B_l \leftarrow B_l - N$;}	10.// Correction modulo N
1. }	11.
1. Check that	12.// Final check
$a \hat{\times} b \equiv P(P_h) \hat{\times} P(N) \hat{+} P(P_l)$;	
1. **return** $\left(P_l, P(P_l)\right)$;	13.// Return both real and
2. }	14.// reduced results.

Fig. 4.3.1 : binary integer multiplication in base N.

As only the projection of P_h into $Z/(M-1)Z$ is used in the verification, we can compute this value directly in $Z/(M-1)Z$ by replacing from the beginning B_h and P_h by their projections, respectively denoted b and p, into $Z/(M-1)Z$, which are single-word integers. Fig. 4.3.2 shows the final binary multiplication algorithm with verification of the result using the proof by $2^m - 1$, in which we have yet more simplified the notations by replacing P_l by P and B_l by B.

1. $\text{mulmod}\big(\,(A,a),(B,b),(N,n)\,\big)$ 1.

1. {   2.

1. $p \leftarrow 0; P \leftarrow 0;$ 3. // Initialize

1. $a' \leftarrow 0\;;\; a'' \leftarrow 1\;;$ 4. // Init eval of $P(A)$

1.   **for**($i \leftarrow 0\;;\; i < |A|_2\;;\; i \leftarrow i+1$) {   5. // Loop on bits of A

1.   **if**($A^{[i(2)]} \neq 0$) {   6. // Addition necessary?

1. $p \leftarrow p+b; P \leftarrow P+B;$ 7. // Add B to P, then

2. $\text{if}(P \geq N)\{ p \leftarrow p+1; P \leftarrow P-N\;;\}$ 8. // correct modulo N,

3.   **if**($p \geq M-1$) $p \leftarrow p-(M-1)$;   9. // then modulo $M-1$.

1. $a' \leftarrow a'+a''$;   10.// Eval $P(A)$, then

2.   **if**($a' \geq M-1$) $a' \leftarrow a'-(M-1)$;   11.// correct modulo $M-1$

1.   }   12.

1. $a'' \leftarrow a''+a''$;   13.// Double a'', then correct

2.   **if**($a'' \geq M-1$) $a'' \leftarrow a''-(M-1)$;   14.// $\text{modulo }M-1$

1. $b \leftarrow b+b; B \leftarrow B+B;$ 15.// Double B, then correct

2.   **if**($B \geq N$) $\{ b \leftarrow b+1; B \leftarrow B-N\;;\}$ 16.// modulo N, then modulo

3.   **if**($b \geq M-1$) $b \leftarrow b-(M-1)$;   17.// $M-1$

1.   }   18.

1.   **Check that** $a \equiv a'$;   19.// Final check on tests

1.   **Check that** $a \,\hat{\times}\, b \equiv p \,\hat{\times}\, n \,\hat{+}\, P$;   20.// Final check

1.   **return** $(P_l, P(P_l))$;   21.// Return both real and

2. }   22.// reduced results.

Fig. 4.3.2 : binary modular multiplication with verification.

4.4   Computational overhead

The computational overhead consists in:
1. the precomputation of the projections of the operands and of the modulus, the result and each of the parameters consist of two fields: the number itself and its projection in $Z/(M-1)Z$;
2. one single-precision addition is added to each multiple-precision operation, with possibly a reduction modulo $M-1$;
3. the final checking test implies one multiplication in $Z/(M-1)Z$.

Out of these sources of overhead, a single one is significant: the added single-precision operands. If the operands are 1024-bit wide and the word size is 32 bits, each *big integer* consists of 32 words. Thus the multiple-precision addition consists of 32 additions and, consequently, as the overhead consists of a single operation, its relative value is $1/32$.

4.5   Evaluation of security

We already know that a random error is detected with probability $1 - \frac{1}{M-1}$. However can an induced fault generate a non-random error, i.e. an error which would preserve the result of the final test?

Obviously no arithmetic operation operates at the same time on the real values and on the reduces ones. Only the tests will have an effect on both.
A single error, which changes a value x into x' such that $x' \equiv x \bmod (M-1)$ will not be detected. However the only pair of numbers in $[0:M-1]$ which are congruent modulo $(M-1)$ are 0 and $(M-1)$. A single error on a single bit will always be detected because 2^i cannot be a multiple of $(M-1)$ if $M = 2^m$ for some m , which is always the case on a binary processor.

Artificially changing the result of the test on the bits of A will add (resp. subtract) $B \times 2^i$ for some i to (resp. from) the result. Thus, the error will be detected if $B \neq 0 \bmod (M-1)$, which happens with probability $1 - \frac{1}{M-1}$ if the attacker cannot inject a specific B . Otherwise, a specific check must be done on this test and it is why we added the statements to eval $P(A)$ and check it against the value passed as a parameter together with A .

A perturbation of a reduction modulo N will change p the reduced value 1 and the real value by N in the reverse direction. As a real N will never be a multiple of M, the error will be detected.

A perturbation of a reduction modulo $(M-1)$ will also be detected for the same reason.

4.6 Modular exponentiation

This operation is basically a sequence of modular multiplications. If each of the multiplications is properly protected, there is a single waek point in the exponentiation: the test on the bits of the exponent. However, the reduced value of the exponent can again be directly computed from its value and computed from the actual results of the tests of these bits as the bits of A were protected in Fig. 4.3.2.

As the reduced modulo $(M-1)$ value of the result is returned at the same time as the actual result of the modular multiplication, the additional penalty for using the proof by $2^m - 1$ in a modular exponentiation is just the initial reduction of the number to be exponentiated. In the case of the Diffie-Hellmann protocol\cite{DH76}, even this number is constant.

5. CONCLUSION

From a simple arithmetic trick, the *proof by 9*, and a property of rings with a unit, we have constructed coherent schemes, based on the *proof by* $2^m - 1$, to protect the binary implementation of the modular multiplication and the modular exponentiation from fault injection.

Even if the math behind them is relatively simple, these schemes will resist any single-bit fault and a random fault will have an exponentially low probability of not being detected.

Thus the *proof by* $2^m - 1$ is thus a cheap way to render the attack by fault injection very chancy if the builtin arithmetic is 8-bit or 16-bit and impractical if the builtin arithmetic is 32-bit or 64-bit.

Besides its cost is negligible in front of the cost of multiprecision computations in case of a software implementation and, in case of a hardware implementation, it requires very little additional hardware.

Further work, to be published soon, will extend this protection to the multiplication of Montgomery.

REFERENCES

[Boneh at *al.*, 1977] Boneh, Dan, DeMillo, Richar A. and Lipton, Richar J. (1977) On the importance of checking cryptographic protocols for faults. *LNCS*, 1233:37-51.

[Diffie and Hellmann, 1976] Diffie W. and Hellman M.E. (1976) New directions in cryptography. In *IEEE Transations on Information Theory*, volume 22, pages 644-654.

[Kocher et *al.*, 1999] Kocher, Paul C., Jaffe, Joshua and Jun, Benjamin (1999) Differential Power Analysis. *LNCS*, 1666:388-397.

[Kocher et *al.*, 1996] Kocher, Paul C., Jaffe, Joshua and Jun, Benjamin (1996) Tioming Attacks on Implementations of Diffie- Hellman, RSA, DSS and other systems. *LNCS*, 1109:104-113.

[Noufal and Nicolaidis, 1999] Noufal, I. Alzaher and Nicolaidis M. (1999). A CAD Framework for Generating Self-checking Multipliers Based on Residue Codes. In *Date'99*, pages R122-129.

[Quisquater and Samyde, 2001] Quisquater J.J. and Samyde D. (2001) Electromagnetic Analysis (EMA) measures and counter- measures for Smart Cards. *E-smartcard Programming and Security, I. Attali and T. Jensen, editors*, 2140:200-210.

[Rivest, Shamir and Adleman, 1978] R.L.Rivest, A.Shamir and L. Adleman (1978) A Method for Obtaining Digital Signatures and Public-Key Cryptosystems. *Communications of the ACM*, 21(2):120-126.

[Siefert, 2002] C. Siefert, C. Aumüller, P. Bier, W. Fischer, P. Hofreiter and J.P. (2002) Fault Attacks on RSA with CRT: Concrete Results and Practical Countermeasures. In *LNCS-CHES 2002*, vol. 2523, pp 260-275.

STREAMTO: STREAMING CONTENT USING A TAMPER-RESISTANT TOKEN

Jieyin Cheng[1], Cheun Ngen Chong[1], Jeroen M. Doumen[1], Sandro Etalle[1], Pieter H. Hartel[1] and Stefan Nikolaus[2]
[1]University of Twente, P.O.Box 2100, 7500 AE Enschede, The Netherlands, {chengj,chong,doumen,pieter}@cs.utwente.nl;

[2]WIBU-SYSTEMS AG, Rueppurrer Strasse 52-54, 76137 Karlsruhe Germany, Stefan.nikolaus@wibu.com

Abstract: StreamTo uses a tamper resistant hardware token to generate the key stream needed to decrypt encrypted streaming music. The combination of a hardware token and streaming media effectively brings tried and tested Pay-TV technology to the Internet. We present two prototype implementations with a performance assessment, showing that the system is both effective and efficient.

Key words: streaming; content protection; tamper-resistant hardware.

1. INTRODUCTION

To enforce usage rights and to prevent copyright violations, digital content needs to be protected. As shown in Fig. 1, content protection has three objectives (Judge and Ammar, 2003): (1) *protected distribution*, which protects content when it is accessed online by a content renderer, e.g. a streaming mechanism; (2) *protected storage*, which protects content while being stored locally, e.g. safe disc; and (3) *protected output*, which protects content after it is being rendered by a content renderer at the content output (say a sound card), e.g. Microsoft Secure Audio Path (SAP).

Content protection is difficult on a personal computer (PC) because most of the PC components (i.e., content renderer and content output) are open (i.e. programmable) and thus not trustworthy. When protected content is being used locally on a PC, an attacker might be able to retrieve the actual content by circumventing the protection mechanism (Greene, 2001).

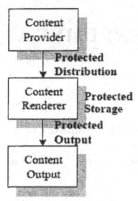

Figure 1. Three phases of content protection.

However, if content is stored on a server while being used via a streaming mechanism (SM), the security of the content can be guaranteed to a certain extent because the *entire* content is not sent to the user's PC directly but only piecemeal as a stream of packets (Holankar and Stamp, 2004). This stream of packets is interpreted and rendered at the user's PC as they arrive. Therefore, SM helps to achieve *protected distribution* of content, provided that the stream cannot be captured easily.

Compared to a PC, a consumer electronic (CE) device is relatively more trustworthy because its components can be manufactured compliant and non-programmable (Eskicioglu and Delp, 2001). Therefore, it is more difficult to circumvent the protection mechanisms applied to CE devices. A common example of such a content protection mechanism is the Conditional Access System (CAS) (Kravitz and Goldschlag, 1999). A Pay-TV system (Jain et al., 2002) applies CAS to control users' access to broadcast TV. Similar to SM, CAS is able to achieve *protected distribution* of the content.

In this paper, we propose StreamTo, which combines aspects of CAS and SM to design a content protection approach, supported by a tamper-resistant hardware token, e.g. a USB dongle. A tamper-resistant hardware token can also provide *protected storage* for the content.

In addition, StreamTo has the following benefits:

- It allows using content without an active Internet connection or when the user does not have sufficient bandwidth.
- It allows flexible sharing of content between users. The provider can control access to different parts of the content by different users. This is useful for business-to-business (B2B) and business-to-consumer (B2C), for instance, when paying users can enjoy full content access at near CD quality, while non-paying users can only listen to clips.

StreamTo is able to solve some of the security threats faced by CAS and SM. This will be discussed later in section 4. Like most streaming mechanisms, StreamTo is not easily scalable. Scalability could be achieved by using Broadcast Encryption techniques as pioneered by Fiat and Naor

(1994). However, this is beyond the scope of the present paper. Here, we show that StreamTo is applicable, practical and secure (within limits).

The remainder of the paper: Section 2 briefly explains CAS and SM, which inspired StreamTo. Section 3 describes StreamTo in detail. Section 4 implements a prototype on a CM-Stick and an iButton. Section 5 assesses the performance of the prototype. The last section concludes and presents future work.

2. CAS AND SM

A Conditional Access System (CAS) is a smart-card-based technology (Guillou, 1984), which is used in Pay-TV systems. The smart-card stores subscription information and a secret key. A set-top-box (STB) is required to interface with the smart-card and the television (TV).

The provider encrypts a TV program using a content key (which is the same for all users) and broadcasts the encrypted TV program, as shown in Fig. 2.

The key management system (KMS), which is responsible for billing, subscriber and key management, transmits the universal content key to the authorized subscribers. A content key is encrypted with the unique secret key stored on a smart-card (Macq and Quisquater, 1995). The smart-card decrypts and stores the content key received from the provider (via the STB). The STB decrypts the encrypted TV program with the content key, and displays the program on the TV.

The provider updates the content key used to encrypt the TV program/channel on a frequent basis (normally each 5 to 20 seconds). Once the key is updated, the KMS must retransmit the updated content key to the subscribers within seconds.

In a streaming mechanism (SM), as shown in Fig. 3, the provider encrypts the content with different content keys for different users. The content is encrypted and transmitted to the user.

A user has a renderer, which is a software application that establishes a secure channel with the provider. The content key is transmitted to the renderer when a secure streaming session is established. The renderer then decrypts the content packet by packet with the content key and renders it, as it is received, leaving behind no residual copy of the content at the renderer (assuming that the renderer is not hacked).

The characteristics of the content key of CAS, SM and StreamTo differ as shown in Table 1. We list the two most important characteristics of a content key: (1) uniqueness (whether the key is unique for different content and users), and (2) update (whether the key is updated on a regular basis).

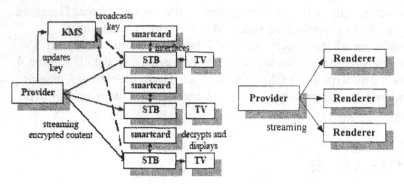

Figure 2. An abstract view of a conditional access system (CAS).

Figure 3. An abstract view of a streaming mechanism.

Table 1. Comparison of CAS, SM and StreamTo with respect to the characteristics of the content key.

	Uniqueness	Update
CAS	A content key is shared among all authorized users.	The content key is updated frequently.
SM	A unique content key is assigned to a user.	The content key is not updated in a streaming session.
StreamTo	A unique content key is assigned to a user.	The content key is updated frequently.

3.   STREAMTO

In this section, we discuss StreamTo as outlined in Fig. 4.

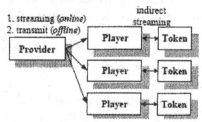

Figure 4. An abstract view of StreamTo.

We use a *token*, which has a cryptographic co-processor and tamper-resistant storage. The token is dispatched physically by the *provider* to a user in the same way as a Pay-TV smart-card. The provider also serves encrypted content. A user has a customized *player* (a software application) that interfaces with the token, and which can play encrypted content. The player depends on the token for providing the key stream necessary to decrypt the content stream.

StreamTo can handle two methods of rendering the content, as shown in Fig. 4: *online* and *offline*. For online rendering, the provider streams the content to the player; whereas for offline rendering, the provider transmits the entire encrypted content to the player. For both access methods, the player plays the content *piecemeal*, waiting for each subsequent block of the key stream from the token. We call this *indirect streaming*.

StreamTo has the characteristics of both the CAS and SM:

- The provider generates a unique content key for a user (SM provider).
- The player decrypts and plays the content piecemeal (SM renderer).
- The token stores a unique secret key (CAS smart-card).
- The content key is updated frequently (CAS provider).
- The token transmits the updated key for decryption to the player (CAS KMS and smart-card).

To explain the StreamTo protocols in more detail, we use the notation listed in Table 2.

Table 2. The notation of the StreamTo protocols.

Notation	Meaning
$SecK$	A secret key shared between the token and the provider.
K_i	The content key for the ith content frame.
S_i	The key stream for the ith content frame.
P_i	The ith frame of content (plaintext).
C_i	The corresponding i^{th} frame of encrypted content (ciphertext).

3.1 Keys

We use three different key types: a secret key, a content key and a key stream.

- A secret key ($SecK$) is a secret shared between the provider and the token. We assume that an attacker cannot read, modify or access this key stored on the token; it never leaves the token, and is preloaded on the token in a secure environment of the provider.
- A content key (K_i) is used for generating the key stream. The first content key K_0 is generated randomly by the provider and sent encrypted (with the secret key) to the user along with the encrypted content.
- A key stream (S_i) is used to en/decrypt the content. The key stream is derived from the content key and the content.

The size of the content key is short (e.g. 128 bits) so that a provider can send it to a player efficiently. The size of the key stream is equal to the size of the content so that stealing the key stream is inconvenient.

As a refinement, the provider could partition the content, using a different K_0 for each partition. This would allow for example free use of trailers but paid for use of the remaining content. In this paper, for simplicity, we only use one content key to explain StreamTo in the subsequent sections.

3.2   Encryption Process

Streaming content, e.g. an MPEG audio/video has a special structure: it is composed of multiple frames, each of which has a descriptive header. This header contains the particular information for the corresponding frame, e.g. bit-rate, sample-rate, etc. StreamTo exploits this special feature of streaming content as follows:

$$S_i = \textbf{generate}(K_i, SecK) \tag{1}$$

$$K_{i+1} = \textbf{transform}(K_i) \tag{2}$$

$$C_i = P_i \oplus S_i \tag{3}$$

The encryption process, as shown in Fig. 5 is performed by the provider. The provider generates a first content key K_0 randomly. The generate function (Eq. 1) takes the content key K_i and the secret key to produce a block of key stream for the current frame (P_i). The encrypted frame (C_i) is then XORed with the block of key stream, as shown in Eq. 3. Finally, the

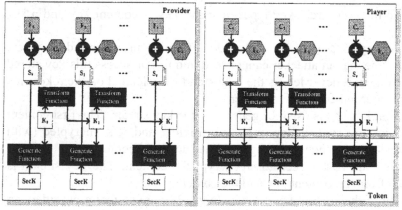

Figure5. Encryption of streaming content, frame by frame, at the provider with a key stream that is generated from an initial content key.

Figure 6. Decryption of encrypted streaming content, frame by frame, with the regenerated key stream using the regenerated content key.

next content key is calculated by the transform function. If the output of the generate function is shorter than the frame size, it is repeated to form the required length.

The encrypted frames $(C_0,...,C_n)$ are written to a new content file, preceded by a header. The header contains the first content key (K_0) (encrypted with the secret key *SecK* of the token), padding, and information about the **generate** and **transform** functions.

3.3 Decryption Process

The player receives an encrypted content file from the provider. When the player renders the encrypted content, the decryption process is executed as shown in Fig. 6.

The player interprets the header information of the encrypted content to retrieve the encrypted first content key K_0 and other information. The player then asks for a valid token. Authentication can be achieved between the player and the token with standard methods (Kelsey and Schneier, 1999); this falls outside the scope of this paper. The player then feeds the token with the encrypted first content key (K_0), so that the token can decrypt it.

The token uses the generate function (Eq. 1) to re-generate the key stream, and sends it to the player. The player retrieves a frame C_i from the encrypted content, and decrypts it (by XOR-ing) with key stream S_i generated by the token, as shown in Eq. (4).

$$P_i = C_i \oplus S_i \tag{4}$$

The player then updates the content key using the transform function (Eq. 2), sends it to the token to generate the next block of the key stream, and the next frame is decrypted with this key stream block. At the same time, the player plays the previously decrypted frame P_{i-1}. The decrypted frames will be overwritten by newly decrypted frames after they are played (again, assuming the player has not been hacked). Ideally, the token would perform the transform function itself, but our hardware (the CM-stick) is not capable of doing this.

4. PROTOTYPES

In this section, we discuss the implementation of our prototype. We use streaming audio (MP3) in our prototype because it is less demanding on resources than video. If StreamTo can be applied practically to protect streaming audio, we can investigate if StreamTo can support other streaming content as well.

The architectural overview of our prototype is given in Fig. 7. The Provider and the Player are the two applications we have created by using Windows Media Format SDK, iB-IDE, CM-Stick SDK (WIBU, 2003), and JavaZoom JLayer SDK.

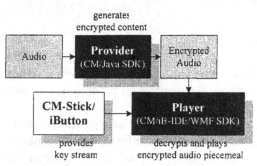

Figure 7. Architectural Overview of the prototype.

The Provider executes the encryption process discussed in section 3.2. It takes as input an MP3 audio file and produces an encrypted audio file as output. The Player performs the decryption process discussed in section 3.3. It asks the token (i.e., CM-Stick or iButton) continuously for blocks of key stream to decrypt the audio.

The hardware token we use in our prototype are a CodeMeter Stick (CM-Stick) and an iButton (as shown in Table 3).

Table 3. Comparison of the iButton and the CM-Stick.

	CM-Stick	iButton
Manufacturer	WiBu-Systems AG, Germany	Dallas Semiconductor,America
Processor Speed	24 MHz	10–20 MHz (Kingpin, 2002)
Non-volatile memory	128 kBytes	134 kBytes
Cryptographic algorithms	AES, Triple-DES (for communication), ECC, SHA-256	DES, Triple-DES, RSA and SHA-1
Interface	USB connection	Serial/Parallel and USB connection

We use the standard Counter-mode (CTR-mode) symmetric encryption (Lipmaa and Rogaway, 2000) to implement StreamTo by virtue of the simplicity, efficiency and proven security of CTR-mode encryption.

The content key (K_n) is the counter of CTR-mode encryption, which is initialized to a random n-bit string. The implementation of the **transform** function of the content key (Eq. 2) is simple:

$$K_{i+1} = K_i + 1$$

The generation of the key stream (S_n) (Eq. 1) is the encryption of the counter in CTR-mode encryption. We use AES encryption, which is the only symmetric encryption supported by the CM-Stick; and DES encryption on the iButton as the **generate** function (Eq. 1).

In our prototypes, each time a new frame is decrypted a click is audible. This allows us to point out during demonstrations when decryption happens.

5. PERFORMANCE ASSESSMENT

To justify the practicality of StreamTo, we assess the performance of our prototype. Our prototype is built on a platform with an Intel Pentium 4, 1.4 GHz, 512 MBytes RAM, 20 GBytes hard disk space, running Windows XP. We use a 1-minute 192 kbps MP3 audio as the sample for our performance assessment. The sample has 2300 frames, each of which contains 623 bytes.

In our prototype, we use a CM-Stick, which is attached with a USB interface; and an iButton, with two different interfaces to the platform, namely a serial port connection (with the adapter DS9097U) and USB connection (with the USB iButton holder DS9490B).

5.1 Content Key Size

The key stream is generated on the tokens by using the firmware symmetric encryption algorithm. Therefore, to determine if the content key size influences the performance of our prototype, we assess the performance of symmetric encryption on the iButton and the CM-Stick.

From our previous experience, we know that the cryptographic operations on the iButton are relatively slow (Chong et al., 2003). DES encryption of 128 bytes on the iButton takes roughly 200 ms (Chong et al., 2004).

We also need to measure the time required by the CM-Stick to perform AES encryption, which we use to generate the key stream. We use an LSQ-fit equation to summarize the result of 10 measurements as follows:

$$t = (0.5 \pm 0.002) \times d + (36 \pm 26) \text{ ms}$$

Here, t is the time required in milliseconds and d is the data size in bytes. Thus, it takes approximately 100 ± 25 ms to encrypt 128-byte of data on the CM-Stick, making the CM-Stick about twice as fast as the iButton. This is consistent with the cryptographic co-processor speed (Table 3).

If we use 128 bytes of content key, i.e., 1024 bits, the iButton requires approximately $2300 \pm 0.2 = 460$ seconds to generate the key stream, whereas the CM-Stick needs roughly 230 seconds. For a 1-minute MP3 this is too long, hence, we must sacrifice security for performance by (1) using a smaller content key size; and (2) en/decrypting every *n*-th frame of the audio sample only.

In our prototypes, we choose a content key of size 8 bytes (64 bits) for the iButton and 32 bytes (256 bits) for the CM-Stick. On the CM-Stick, it takes approximately 40 ± 26 ms to generate a block of key stream. However, for the iButton, it takes approximately 70 ms due to the slower co-processor. Therefore, we also use the second tradeoff on the iButton prototype, as will be discussed in next section.

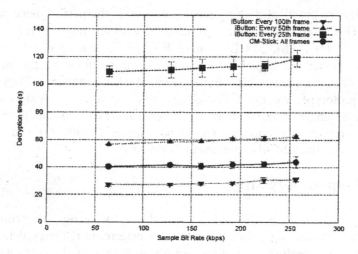

Figure 8. The time required to decrypt the encrypted audio sample frame by frame with the iButton and the CM-Stick.

5.2   Sample Bit Rate

The MP3 sample bit rate refers to the transfer bit rate for which an audio file is encoded. The sampling frequency refers to the number of samples of an audio taken per unit time, i.e., the rate at which audio signals are sampled into digital form.

The frame size depends on the sample bit rate and sampling frequency according to the MPEG-3 standard. We use 6 different sample bit rates (with the same sampling frequency of 44.1 KHz) of our experiment, which include 64 kbps, 128 kbps, 160 kbps, 192 kbps, 224 kbps and 256 kbps. Each frame has standard constant time length of 26 ms. A 1-minute MP3 has approximately 2300 frames.

It takes roughly 80 ms to generate a block of key stream. Therefore, for decrypting the audio sample 192 kbps of 2300 frames (1 minute of play time) by using a content key of 64 bits, theoretically the iButton needs approximately $2300 \times (0.08 + 0.2 \times 2) = 1104$ seconds in total to generate and transmit the key stream to the player. We have rerun the test using the USB iButton holder. However, there is no obvious improvement of the speed due to the slow cryptographic operations on the iButton.

To overcome this problem, we choose appropriate values of n, en/decrypting every n-th frame of the audio sample. Fig. 8 shows the measurement for $n = 25$, 50 and 100. We report the average of 10 measurements. The decryption time measured includes the time required to upload the updated content key; to generate and transmit a block of the key stream; and XOR-ing of the encrypted frame. The actual play time of the audio sample is 60 seconds (i.e., $y = 60$).

As can be seen in Fig. 8, the graphs of the iButton are slightly slant, indicating that the time required to decrypt the audio frames increases with faster sample bit rate. This is caused by the preprocessing of the encrypted audio file, i.e. reading the audio frames from an encrypted audio file.

When $n = 100$, the iButton is able to handle the key stream generation and audio frames decryption comfortably in real time. When $n = 50$, the decryption time is marginally parallel with the play time of the audio sample. This means that real time playback is possible but only when every n-th frame is encrypted and $n \geq 50$.

On the other hand, the CM-Stick, due to its faster cryptographic co-processor, has better performance than the iButton, as shown in Fig. 8. In conclusions, the CM-Stick is able to provide real time playback at $n = 1$.

6. CONCLUSIONS AND FUTURE WORK

We propose a streaming content protection approach, namely StreamTo, which combines the technology of the Internet streaming mechanism (SM), Pay-TV Conditional Access System (CAS) and a tamper-resistant hardware token.

We implement StreamTo on two commercial tokens, namely the iButton and the CM-Stick, by using the CTR-mode of symmetric encryption. Thus, we show the applicability of StreamTo. We also evaluate the performance of the implementation to justify the practicality of StreamTo. The CM-Stick has a better performance than the iButton due to its faster cryptographic coprocessor.

REFERENCE

Buchheit, M. and Kgler, R. (2004). Secure music content standard – content protection with codemeter. In *4th Open Workshop of Interactive Music Network Multimedia MUSICNETWORK*, page Paper 10.

Chong, C. N., Peng, Z., and Hartel, P. H. (2003). Secure audit logging with tamper-resistant hardware. In Gritzalis, D., di Vimercati, S. D. C., Samarati, P., and Katsikas, S. K., editors, *18th IFIP International Information Security Conference (IFIPSEC)*, volume 250 of *IFIP Conference Proceedings*, pages 73–84. Kluwer Academic Publishers.

Chong, C. N., Ren, B., Doumen, J., Etalle, S., Hartel, P. H., and Corin, R. (2004). License protection with a tamper-resistant token. In Lim, C. H. and Yung, M., editors, *5th Workshop on Information Security Applications (WISA 2004)*, volume 3325 of *LNCS*, pages 224–238. Springer-Verlag.

Eskicioglu, A. M. and Delp, E. J. (2001). An overview of multimedia content protection in consumer electronics devices. *Signal Processing: Image Communication*, 16:681–699.

Fiat, A. and Naor, M. (1994). Broadcast encryption. In *Advances in Cryptology (CRYPTO'03 Proceedings*, volume 773 of *LNCS*, pages 480–491. Springer-Verlag.

Greene, T. C. (2001). MS digital rights management scheme cracked. *TheRegister.co.uk.*

Guillou, L. C. (1984). Smart cards and conditional access. In *Advances in Cryptology EUROCRYPT 84)*, volume 209 of *LNCS*, pages 480–485. Springer-Verlag.

Holankar, D. and Stamp, M. (2004). Secure streaming media and digital rights management. In *Proceedings of the 2004 Hawaii International Conference on Computer Science*, pages 85–96. ACM Press.

Jain, P. C., Joshi, S., and Mitra, V. (2002). Conditional access in digital television. In *The 8th National Conference Communications (NCC) 2002*, Technical Session paper 30.

Judge, P. and Ammar, M. (2003). The benefits and challenges of providing content protection in peer-to-peer systems. In *Int. Workshop for Technology, Economy, Social and Legal spects of Virtual Goods*, paper 12, Ilmenau. Germany.

Kelsey, J. and Schneier, B. (1999). Authenticating secure tokens using slow memory access (extended abstract). In *USENIXWorkshop on Smart Card Technology*, pages 101–106. USENIX Press.

Kingpin (2002). A practical introduction to the dallas semiconductor ibutton. Technical report,@Stake, Inc.

Kravitz, D. W. and Goldschlag, D. M. (1999). Conditional access concepts and principles. In *Proceedings of the 3rd International Conference on Financial Cryptography*, volume 1648 of *LNCS*, pages 158–172. Springer-Verlag.

Lipmaa, H. and Rogaway, P. (2000). Comments to NIST concerning AES-modes of operations: CTR-mode encryption. In *Symmetric Key Block Cipher Modes of Operation Workshop*, Electronic Proceedings.

Macq, B. M. and Quisquater, J.-J (1995). Cryptology for digital tv broadcasting. *Proceedings of IEEE*, 83(6):944–957.

WIBU (2003). *CodeMeter Developer's Guide*. WIBU-SYSTMES AG, Rueppurrer Str.53-54 76137 Karlsruhe, Germany, 1.0 edition.